PHOTODETECTORS

Edited by **Sanka Gateva**

Photodetectors
Edited by Sanka Gateva

Published by InTech
Janeza Trdine 9, 51000 Rijeka, Croatia

Publishing Process Manager Jana Sertic

Technical Editor Teodora Smiljanic

Cover Designer InTech Design Team

Additional hard copies can be obtained from orders@intechopen.com

Photodetectors, Edited by Sanka Gateva
p. cm.
ISBN 978-953-51-0358-5

For EU product safety concerns:
IN TECH d.o.o., Prolaz Marije Krucifikse Kozulić 3, 51000 Rijeka, Croatia,
info@intechopen.com or visit our website at intechopen.com.

Meet the editor

Dr Sanka Gateva graduated from Sofia University with a Master's degree in atomic physics, prepared her doctoral thesis at the Institute of Electronics, Bulgarian Academy of Sciences and specialized in development of diode lasers for spectroscopy of trapped ions at NPL, England. Her main areas of interest and her main publications are in the fields of development of different types of lasers (gas, dimer lasers and diode lasers), mode selection and frequency stabilisation in lasers, high resolution spectroscopy, coherent population trapping, and optical magnetometry. She is an Associate Professor, Head of the Laser Systems Laboratory and she was recently assigned as Director of the Institute of Electronics, Bulgarian Academy of Sciences.

Contents

Preface XI

Part 1 Photodetectors 1

Chapter 1 **ZnO Metal-Semiconductor-Metal UV Photodetectors on PPC Plastic with Various Metal Contacts 3**
N.N. Jandow, H. Abu Hassan, F.K. Yam and K. Ibrahim

Chapter 2 **UV-Vis Photodetector with Silicon Nanoparticles 33**
J.A. Luna-López, M. Aceves-Mijares, J. Carrillo-López, A. Morales-Sánchez, F. Flores-Gracia and D.E. Vázquez Valerdi

Chapter 3 **Silicon Photodetectors Based on Internal Photoemission Effect: The Challenge of Detecting Near-Infrared Light 51**
Maurizio Casalino, Luigi Sirleto, Mario Iodice and Giuseppe Coppola

Chapter 4 **Silicon Photomultipliers: Characterization and Applications 77**
Marco Ramilli, Alessia Allevi, Luca Nardo, Maria Bondani and Massimo Caccia

Chapter 5 **High Sensitivity Uncooled InAsSb Photoconductors with Long Wavelength 101**
Yu Zhu Gao

Chapter 6 **Terahertz Emitters, Detectors and Sensors: Current Status and Future Prospects 115**
M. Ghanashyam Krishna, Sachin D. Kshirsagar and Surya P. Tewari

Chapter 7 **Far-Infrared Single-Photon Detectors Fabricated in Double-Quantum-Well Structures 145**
Takeji Ueda and Susumu Komiyama

Chapter 8 **Broadband Photodetectors Based on *c*-Axis Tilted Layered Cobalt Oxide Thin Films 165**
Shufang Wang and Guangsheng Fu

Chapter 9 **Geiger-Mode Avalanche Photodiodes in Standard CMOS Technologies 175**
Anna Vilà, Anna Arbat, Eva Vilella and Angel Dieguez

Part 2 Theoretical Modelling and Simulations 205

Chapter 10 **Avalanche Process in Semiconductor Photo Detectors in the Context of the Feedback Theory 207**
Vasily Kushpil

Chapter 11 **Multiscale Computer Aided Design of Microwave-Band P-I-N Photodetectors 231**
Mikhail E. Belkin

Chapter 12 **Mathematical Modeling of Multi-Element Infrared Focal Plane Arrays Based on the System 'Photodiode – Direct-Injection Readout Circuit' 251**
I.I. Lee and V.G. Polovinkin

Part 3 Photodetection Systems 279

Chapter 13 **Ultrafast Imaging in Standard (Bi)CMOS Technology 281**
Wilfried Uhring and Martin Zlatanski

Chapter 14 **Photoconductors for Measuring Speckle Motion to Monitor Ultrasonic Vibrations 307**
Jonathan Bessette and Elsa Garmire

Chapter 15 **Remote Optical Diagnostics of Nonstationary Aerosol Media in a Wide Range of Particle Sizes 341**
Olga Kudryashova, Anatoly Pavlenko, Boris Vorozhtsov, Sergey Titov, Vladimir Arkhipov, Sergey Bondarchuk,Eugeny Maksimenko, Igor Akhmadeev and Eugeny Muravlev

Chapter 16 **Atmospheric Clock Transfer Based on Femtosecond Frequency Combs 365**
Lingze Duan and Ravi P. Gollapalli

Chapter 17 **Measurement of the Polarization State of a Weak Signal Field by Homodyne Detection 389**
Sun-Hyun Youn

Chapter 18 **Spin Photodetector: Conversion of Light Polarization Information into Electric Voltage Using Inverse Spin Hall Effect 405**
Kazuya Ando and Eiji Saitoh

Chapter 19 **Shape of the Coherent Population Trapping Resonances Registered in Fluorescence 417**
Sanka Gateva and Georgi Todorov

Chapter 20 **Photodetectors in Calorimeters for the Linear Collider 441**
Jaroslav Cvach and CALICE Collaboration

Preface

Photodetectors are devices which sense light or electromagnetic energy and convert it into a response, usually electrical. As part of photonics systems, they are developing very fast with photonics applications, e.g. photosensors, telecommunications, information processing, metrology, spectroscopy, medicine, military technology, laser material processing, biophotonics, agriculture, robotics.

There are different types of photodetectors: optical and chemical detectors, photoresistors, photodiodes, thermocouples, phototubes, solar cells, etc. The choice of a particular detector depends on the requirements of a specific application – e.g. the wavelength of the light, the sensitivity, the speed of response. The variety of photodetectors is so wide, that it is impossible for a single book to cover the state-of–art, the latest developments and the new trends in the field including the theoretical aspects, design and simulation, experimental results, and applications of all photodetectors.

In this book some recent advances in photodetectors and photodetection systems for specific applications are examined. The book is divided into three sections.

Section 1 includes nine chapters where recent developments on different photodetectors and their characteristics are described. In *Section 2* a theoretical model and some simulations are presented. In *Section 3* eight examples of photodetection systems for different applications are given: imaging, particle size analysis, transfers of time, and measurement of vibrations, magnetic field, polarization of light, and particle energy.

The book is addressed to students, engineers, and researchers working in the field of photonics and advanced technologies.

Dr. Sanka Gateva
Associate Professor, Institute of Electronics,
Bulgarian Academy of Sciences,
Bulgaria

Part 1

Photodetectors

1

ZnO Metal-Semiconductor-Metal UV Photodetectors on PPC Plastic with Various Metal Contacts

N.N. Jandow*, H. Abu Hassan, F.K. Yam and K. Ibrahim
Nano-Optoelectronics Research and Technology Laboratory School of Physics Universiti Sains Malaysia, Minden, Penang Malaysia

1. Introduction

The unique and attractive properties of the II-VI compound semiconductors have triggered an enormous incentive among the scientists to explore the possibilities of using them in industrial applications. Zinc Oxide (ZnO) is one of the compound semiconductors of the II-VI family with a direct band gap of 3.37 eV at room temperature, and a large excitation binding energy (60 meV). ZnO is low cost and easy to grow. It is also sensitive to the UV region because of its ultra violet absorbance and has high photoconductivity (Young et al., 2006). These properties make ZnO a promising photonic material for several applications, such as transparent conducting electrodes, surface acoustic wave filters, gas sensors, light emitting diodes, laser diodes and ultraviolet detectors (Kumar et al., 2009; Lim et al., 2006; Janotti et al., 2009; Bang et al.,2003). ZnO is of value and importance due to wide chemistry, piezoelectricity and luminescence at high temperatures. In industry, it can be used in paints, cosmetics, plastics and rubber manufacturing, electronics, and pharmaceuticals [http://www.navbharat.co.in/Clients.htm].

Films of ZnO, indium tin oxide (ITO), and cadmium oxide (CdO) have recently been investigated as transparent conducting oxide (TCO) due to their good electrical and optical properties, their abundance in nature, their optical transmittance (>80%) in the visible region, and they are non-toxic (Angelats, 2006).

ZnO crystallizes in two different crystals structural. The first is the hexagonal wurtzite lattice with lattice constants of a = 3.249 Å and c = 5.207 Å which is mainly used in thin film industry as a transparent conducting oxide (TCO) or as a catalyst in methanol synthesis (Jin, 2003). The second structure, which is well known to geologists, is in the form of rock salt structure which is used in understanding the earth's lower mantle (Hussain, 2008).

ZnO possesses very similar properties to the nitride-based semiconductors such as GaN; yet it has many advantages over GaN (Koch et al., 1985; Paul et al., 2007). For examples: firstly, the growth of high quality single crystal of ZnO is a well understood technology; whereas

* Corresponding Author

for GaN it is difficult. Secondly, the ZnO has large exciton binding energy which means that it has a lower threshold power for lasing. Thirdly, it is one of the few oxides that show quantum confinement in an experimentally accessible range of the size of the particles.

2. Deposition of ZnO on various organic substrates

ZnO thin films traditionally have been deposited on mostly inorganic substrates such as quartz, silicon, glass, sapphire, GaAs, fluorite, mica, GaN, Al_2O_3, diamond, NaCl, and InP (Hickernell, 1976; Ianno et al., 1992; Craciun et al., 1994; Jin et al., 2001; Shan et al., 2004; Sans et al., 2004; Ghosh et al., 2004; Zhang et al., 2004; Tsai et al., 2007; Kiriakidis et al., 2007; Wang et al., 2008) by using different techniques (Mahmood et al., 1995; Auret et al., 2007; Rao et al., 2010; Chakraborty et al., 2008; Hwang et al., 2007; Nunes et al 2010; Sofiani et al., 2006).

The preparation of ZnO thin film on flexible substrates such as plastic has so far received much interest due to its wide variety of applications as in flexible sensors and curved detector arrays. Plastic substrates provide lighter, more resistant to damage, flexible, and durable devices (Nandy et al., 2010); these attributes make them suitable for portable devices such as smart cards, personal digital assistants, digital cameras, cell phones, remote control, and circuits camcorders (Brabec et al., 2001). The substrate materials, that are commonly used for the above applications, include polyethylene terephthalate (PET), polyolefin, polytetrafluoroethylene (Teflon), Polycarbonate (PC), polyarylate (PAR), polyestersulfone (PES), polyimide (PI), poly(ethylene naphthalate) (PEN), thermoplastic polymethyl methacrylate (Perpex, Plexiglas), polyethylene naphthalate, and cellulose triacetate (Ma et al., 2008).

Most of ZnO studies have focused on fabrication issues with little attention given to the structural and optical properties of these films on flexible substrates. Interfacial phenomena in ZnO/polymer heterostructures are expected to be very different from those observed for inorganic substrates, there is a need for more comprehensive investigations to assess the potential of ZnO as a material that can be used in flexible electronic applications. Deposition of ZnO on polymers may open a new opportunities for the creation of novel multifunctional polymer/semiconductor heterostructures.

ZnO thin films deposited on plastics had a number of advantages compared with those on inorganic substrates. From the literature, various types of organic substrates have been explored by researchers to deposited ZnO.

Many researchers (Ott & Chang, 1999; Banerjee et al., 2006; Tsai et al., 2006; Lu et al., 2007) have grown ZnO on PET plastic substrates using different techniques for exemple: Ott and Chang (Ott & Chang, 1999) reported that their samples showed highly transparent (T > 80%) and conductive ($\rho \sim 10^{-3}\ \Omega$ cm) and the best film grown on PET had a resistivity of 1.4×10^{-3} Ωcm. Banerjee *et al* (Banerjee et al., 2006) deposited ZnO thin films with two different thicknesses (~260 and ~470 nm) for 4 and 5 hours using DC sputtering technique. They found that X-Ray diffraction (XRD) pattern for 4 h deposited ZnO films showed a weak intensity as well as broad peak at (002). But the film deposited for 5 h confirmed the formation of crystalline ZnO and three other peaks originated from (100), (002) and (101) reflections of hexagonal ZnO. Their films showed almost 80% to more than 98% visible transmittance and the bandgap (E_g) values were 3.53 eV and 3.31 eV for the films with (~260 and ~470 nm) thicknesses. Room temperature conductivities of the films were found ranging around 0.05 to 0.25 S cm^{-1} and the maximum carrier concentrations around 2.8×10^{16} and 3.1×10^{20} cm^{-3} with a variation in the deposition time of 4 and 5 h respectively.

On the other hand, Tsai *et al* (Tsai et al., 2006) reported that their ZnO/PET films showed a strong (002) peak from XRD measurements and the scanning electron microscopy (SEM) morphology of the films showed that no crack or bend appeared on the films. The average transmittance in the visible spectrum was above 80% for both substrates and the value of E_g was found equal to 3.3 eV. The lowest resistivity obtained was 4.0×10^{-4} Ω cm. Lu *et al* (Lu et al., 2007) found that XRD results of their deposited thin films showed a strong peak at (002) orientation and the root mean square (rms) were in the range 2.63~11.1 nm determined by atomic force microscopy (AFM). The average transmittance of the film was obtained over 80% in the visible spectrum. The lowest resistivity obtained was 4.0×10^{-3} Ωcm.

Whereas, Liu *et al* (Liu et al., 2007) deposited ZnO thin film at the first time on Teflon substrate by the rf magnetron sputtering. They reported that the XRD pattern of the film showed a strong peak at $2\theta = 34.283°$ which corresponding to the (002) peak with full-width at half-maximum (FWHM) of 0.724° and the film also showed other peaks such as (100), (101), (102), (103) and the crystallite size was equal to 10 nm. The SEM image of ZnO thin film showed that the ZnO film structure consists of some columnar structured grains.

On the other hand, Kim et al (Kim et al., 2009) deposited ZnO films on PC and PES substrates by using rf sputtering system. They studied the effect of sputtering power ranged from 100 to 200 W on the characteristics of the films. XRD patterns of their deposited films showed strong 2θ peaks at 34.4° and they found that the intensities of the ZnO (002) peak and the grain size increased with increasing the sputtering power. The transmittance of the films on both substrates was 80-90%.

Also Liu et al (Liu et al., 2009) prepared Transparent conducting aluminum-doped zinc oxide (ZnO:Al) films on PC substrates by pulsed laser deposition technique at low substrate temperature (room-100 °C. they reported that their experiments were performed at various oxygen pressures (3 pa, 5 pa, and 7 Pa). they studied the influence of the process parameters on the deposited (ZnO:Al) films. X-ray diffraction for their prepared films showed polycrystalline ZnO:Al films having a preferred orientation with the c-axis perpendicular to the substrate were deposited with a strong single violet emission centering about 377–379 nm without any accompanying deep level emission. The average transmittances exceed 85% in the visible spectrum.

Other organic substrates such as polyimide (Paul et al., 2007 and Craciun et al., 1994), polyarylate (Ianno et al., 1992), PEN (Koch et al., 1985) have also been tried by other researchers to deposit ZnO thin film. These ZnO thin films have been used to fabricate thin film transistor (TFT). The fabricated ZnO TFT on various organic substrates showed very encouraging results.

In summary, ZnO thin film deposited on organic substrates by different techniques are in the polycrystalline form with (002) orientation as the dominant peak. The thin films grown on these substrates usually demonstrated good optical transmittance characteristic, i.e. above 80% and the bandgap could be varied from 3.31-3.53 eV. In addition, the carrier concentration is dependent on the growth condition. Overall the ZnO thin films quality on organic substrates is comparable to those in organic substrates.

3. An overview of photodetectors

Photodetectors are basically semiconductor devices that convert the incident optical signal into an electrical signal which is usually revealed as photocurrent. The photodetectors can detect the optical signals over a range of the electromagnetic spectrum that is usually predominantly defined based on the material properties. A detector is selected based on the requirements of a particular application. The general requirements include wavelength of light to be detected, sensitivity level needed, and the response speed. In general, photodetectors respond uniformly within a specific range of the electromagnetic spectrum. Consequently, the wavelength of light detected determines the selection of the photodetector material and the target application and defines the structure (Sze, 2002). This will be explained further in the following sections.

When the incident photons with energy higher than the bandgap energy of a semiconductor; some of them will be absorbed within the semiconductor layer. Such a successful absorption process results in the generation of a free electron-hole pair. The energy gained by the electron which is called the work function must be sufficient to make the electron cross the barrier height between the metal contact and the semiconductor with kinetic energy (E_e). The kinetic energy can be given as (Sze, 2002).

$$E_e = \frac{hc}{\lambda} - \phi_m \tag{1}$$

where c is the light velocity, ϕ_m is the metal work function and λ is the incident wavelength.

Since the photoelectric effect is based on the photon energy $h\nu$, the wavelength of interest is related to energy transition ΔE in the device operation, with the following relationship:

$$\lambda = \frac{hc}{\Delta E} = \frac{1.24}{\Delta E(eV)} \quad (\mu m) \tag{2}$$

where ΔE is the transition of energy levels.

Because the photon energy $h\nu > \Delta E$ can also cause excitation, Eq. 2 is often the minimum wavelength limit for detection. The transition energy ΔE, in most cases, is the energy gap of the semiconductor. It depends on the type of photodetector, and it can be the barrier height as in a metal-semiconductor (MS) photodiode. Alternatively, the transition energy can be between an impurity level and the band edge as in an extrinsic photoconductor. The type of photodetector and the semiconductor material are normally chosen and optimized for the wavelength of interest. The absorption of light in a semiconductor is indicated by the absorption coefficient. The transition energy does not only determine whether light can be absorbed for photoexcitation, but it also indicates where light is absorbed. A high value of absorption coefficient indicates light is absorbed near the surface where light enters. Whilst, a low value means the absorption is low that light can penetrate deeper into the semiconductor (Sze & Kwok 2006).

3.1 Semiconductors for UV photodetection

UV detection has usually been applied to narrow bandgap semiconductor photodiodes, thermal detectors, photomultiplier tubes (PMT), or charge-coupled devices (CCD) because they exhibit high gain and low noise and they can be rather visible-blind. However, PMT is a fragile and bulky device which requires high power supplies. On the other hand, CCD detectors are slow and their response does not depend on the wavelength. Semiconductor photodetectors require only a mild bias, and can be benefited for their small size and light weight, and being insensitive to magnetic fields. Their low cost, good linearity and sensibility, and capability for high-speed operation make them excellent devices for UV detection (Monroy et al., 2003).

The main disadvantage of these narrow-bandgap semiconductor detectors is device aging due to the exposure to radiation that has much higher energy than the semiconductor bandgap. Moreover, passivation layers, typically SiO_2, reduce the quantum efficiency in the deep-UV range, and are also degraded by UV illumination (Caria et al., 2001)[43]. Another disadvantage of these devices is their sensitivity to low energy radiation, eventually, filters are required to block out visible and infrared photons, resulting in a significant loss of the effective area of the device. Finally, for high-sensitivity applications, the detector active area must be cooled to reduce the dark current. The cooled detector behaves as a cold trap for contaminants which leads to a lower detectivity (Monroy et al., 2003). Such problems make the wide-bandgap semiconductors, such as gallium nitride and ZnO, attractive alternatives. Therefore, ZnO has been used in optoelectronic, high power and high frequency devices.

3.1.1 Wide-bandgap semiconductors for UV photodetection

Detection of ultraviolet (UV) radiation is increasingly becoming important in a number of areas, such as flame detection, water purification, furnace control, UV astronomy, UV radiation dosimetry (Lakhotia et al., 2010).

Even though the responsivity of Si-based optical photodetectors in the UV region is low, they are still being used for light detection (Lakhotia et al., 2010). This has promoted some researchers to use wide direct band gap materials to fabricate optoelectronic devices that are sensitive in the UV region. Hence, GaN-based UV photodetectors have already become commercially available (Hiramatsu et al., 2007). ZnSe-based UV photodetectors, which is another wide direct band gap material, have also been manufactured (Hanzaz et al., 2007). Fabrication and characterization of low-intensity ultraviolet metal-semiconductor-metal (MSM) photodetectors based on AlGaN have also been reported (Gökkavas et al., 2007). ZnO is another semiconductor of wide direct bandgap that is also sensitive in the UV region and is of low cost and easy to manufacture. Therefore, ZnO will be focused in this chapter.

The importance of semiconductor UV PD has expanded the semiconductor industry and emphasized the development of low-light-level imaging systems for military and civilian surveillance applications. These detectors should:

1. not be sensitive to light at visible wavelengths (commonly referred to as being solar blind),
2. have a large response at the wavelength to be detected and have high quantum efficiency,
3. have a small value for the additional noise introduced by the detector.

There are many different types of semiconductor ultraviolet photodetectors such as: photoconductive or photoconductors detectors, p-n junctions photodiode, and MSM (metal-semiconductor-metal) photodetector (Liu et al., 2010). In the field of optical devices, several trends are pushing research to use new materials. For example, the UV PD is successfully fabricated based on wide bandgap semiconductors (E_g > 3.0 eV). Photon detectors may be further subdivided according to their physical effects that make the detector responsive.

3.1.2 Metal-Semiconductor-Metal (MSM) photodiodes

Metal-semiconductor-metal (MSM) is a type of PD used for UV detection. MSM PDs consist of two interdigitated Schottky contacts which are called fingers deposited on top of an active layer as shown in Figure 1. These devices have a fast response and simple structure compared to other photodetectors of the same active area because of the interdigitated structure which reduces carrier transit time through close spacing of the electrodes, while maintaining a large active area.

The MSM PD operates when the incident light is directed on the semiconductor material between the fingers, electrons will be generated in the conduction band, and thus creating holes in the valance band of the undoped region of the semiconductor. This results in creating a photocurrent by means of one of two processes with the operative process determined by the magnitude of the incident photon energy (hv) relative to the energy bandgap (E_g) of the semiconductor and the metal work function (ϕ_{Bn}). If E_g is greater than hv and hv is greater than ϕ_{Bn}, then photoelectric emission of electrons from the metal to the semiconductor occurs. Alternatively, for the second process, if hv is greater than E_g, then photoconductive electron-hole pairs are produced in the semiconductor. The generated electrons and holes are separated by an electric field intrinsically formed between the fingers.

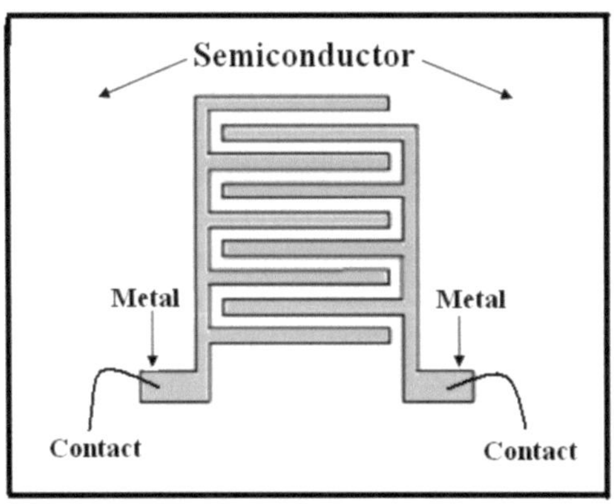

Fig. 1. Top view of an MSM PD planar interdigitated structure

4. ZnO as a UV detector

As mentioned before, ZnO is one of the most prominent semiconductors in the metal-oxide family because of its excellent properties which attract many researcher groups to use this material in UV detection applications. Below, a briefly discussion of ZnO as a UV detector will present and some earlier studies about the fabrication of ZnO as UV detector will also be provided.

ZnO was one of the first semiconductors to be prepared in a rather pure form after silicon and germanium. It was extensively characterized as early as in the 1950's and 1960's due to its promising properties (Plaza et al., 2008). Wide band gap semiconductors have so far gained much attention for the last decade because they can be used in optoelectronic devices in the short wavelength and also in the UV region of the electromagnetic spectrum. These

Wide band gap semiconductor	**Crystal structure**	**Lattice parameter (Å)**		**E_g (eV) at RT**	**Melting temp. (K)**	**Excitation binding energy (meV)**
		a	*c*			
ZnO	Wurtzite	3.25	5.206	3.37	2248	60
GaN	Wurtzite	3.189	5.185	3.4	1973	21
ZnSe	Zinc-blende	5.667	-	2.7	1790	20
ZnS	Wurtzite	3.824	6.261	3.7	2103	36
4H-SiC	Wurtzite	3.073	10.053	3.26	2070	35

Table 1. Comparison of different semiconductors (Tüzemen & Gür, 2007; Nause & Nemeth, 2005) Note: Where RT is room temperature, meV is millielectron volt.

Sab.	**D.M.**	**PD T**	**R. C.**	**O. C.**	**B.V. (V)**	**I_d (A)**	**I_{ph} (A)**	**λ (nm)**	**R (A/W)**	**Ref**
Sapp.	MOCVD	MSM UV	-	Al	5	1×10^{-6}	-	-	1.5	(Liang et al., 2001)
GaAs	rf	p-n HJ	-	-	~3.0	-	$\sim 2 \times 10^{-3}$	325	-	(Moon et al., 2005)
p-Si (100)	rf		Au-Al	In	30	-	-	310	0.5	(Jeong et al., 2004)
Sapp.	Sol-gel	MSM UV	Au	-	-	-	-	350	0	(Basak et al., 2003)
Sapp.	MOCVD	MSM	Al	Al	5	450×10^{-6}	-	-	400	(Liu et al., 2000)
Quar.	rf	MSM UV	Au	Au	3	250×10^{-6}	-	360	30	(Liu et al., 2007)
Sapp.	PA MBE	UV		Al-Ti	20		-	374	1.7	(Mandalapu et al., 2007)
SiO_2	rf	MSM UV	Au	Au	3	1×10^{-3}	-	-	0.3	(Jiang et al., 2008)
Sapp.	rf	MSM UV	Ir	Ir	-	-	-	370	0.2	(Young[a] et al 2007)
Sapp.	rf	MSM UV	Pd	Pd	1	-	-	370	0.1	(Young[b] et al., 2007)
Sapp.	MBE	MSM UV	Ru	Ru	-	8×10^{-8}	1.8×10^{-5}	-	-	(Lin et al., 2005)

Sab = Substrate; D.M. = Deposition method; PD T. = PD type; B.V. = Bias voltage R C. = Rectifying contacts; O. C. = Ohmic contact; I_d = Dark current; I_{ph}= Photocurrent; λ = Wavelength; Sapp. = Sapphire; Quar. = Quartz; ED = Electrochemical deposition; PAMBE = Plasma-assisted molecular-beam epitaxy; p-nHJ = p-n homojunction

Table 2. Summarizes the characteristics of ZnO UV PDS reported by other researchers fabricated by different methods

semiconductors which include ZnO, GaN, ZnSe, ZnS, and 4H-SiC have shown similar properties with their crystal structures and band gaps (Tüzemen & Gür, 2007).

Table 1 below shows a summary of some of the important properties of these wide band gap semiconductors. Initially, ZnSe and GaN based technologies made significant progress in the blue and UV light emitting diode and injection laser. No doubt, GaN is considered to be the best candidate for the fabrication of optoelectronic devices. However, ZnO has great advantages for light emitting diodes (LEDs) and laser diodes (LDs) over the currently used semiconductors. Recently, it has been suggested that ZnO is promising for various technological applications, especially for optoelectronic short wavelength light emitting devices due to its wide and direct band (Nause & Nemeth, 2005).

From the literature ZnO UV PDs have been widely investigated by many researchers on different substrates through different methods. Table 2 compiles and summarizes the structural of ZnO UV PDS reported by other researchers.

5. Schottky barrier height calculation of MSM PD

A MSM PD is a unipolar device with two back-to-back Schottky junctions formed on the same semiconductor surface. Under the application of bias voltage, one of the diodes becomes reverse biased, forming a depletion region that tends to sweep out photocarriers. The other diode becomes forward biased, allowing the collected photocurrent to flow out just as an ohmic contact. Under sufficiently high bias voltage, the depletion region extends and touches the small space-charge region under the forward biased electrode.

Figure 2 shows the energy-band diagram of the MSM PD in the biased state. The vertical displacement of the electrode metals indicates the bias voltage applied to the device resulting in the forward biased condition of the left-hand metal-semiconductor interface and the reverse bias of the right-hand interface. Upon biasing, the semiconductor between the electrodes becomes fully depleted of free carriers. The reversed biased interface prevents the current from flowing through the device when there is no optical signal. The depleted regions between the electrodes are the photodetector active regions. As mentioned before, a photon with energy greater than the band gap of the semiconductor will be absorbed by an electron; this electron will get excited to the conduction band as shown by the process labeled as (1) in Figure 2. The photogenerated electron and hole are swept by the high applied fields to the positive and negative electrodes resulting in an electronic output signal (Haas, 1997).

One can assume that other current transport processes also contribute to the movement of electrons within the depletion region and across the barrier in Schottky contacts to determine the barrier height. The equations usually used to determine the barrier height in a Schottky diode. Assuming pure thermionic emission and V>3KT, the general I–V equations usually used to determine the barrier height in a Schottky diode are represented by (Daraee et al., 2008)

$$I = I_0 \exp[qV / (nKT)] \tag{3}$$

$$I_0 = A^* A T^2 \exp[-q\phi_B / (KT)] \tag{4}$$

where I_0 is the saturation current, n is the ideal factor, K is the Boltzmann's constant, T is the absolute temperature, ϕ_B is the barrier height, A is the area of the Schottky and A^* is the effective Richardson coefficient. Eq. 4 indicates the dependence of the barrier height on the saturation current (I_o). The theoretical value of A^* can be calculated using Eq. (5) below.

$$A^* = 4\pi m^* q K^2 / h^3 \tag{5}$$

where h is Planck's constant and $m^* \sim 0.27m_o$ is the effective electron mass for n-type ZnO so that $A^* \sim 32\ A/cm^2K^2$ (Liang et al 2001).

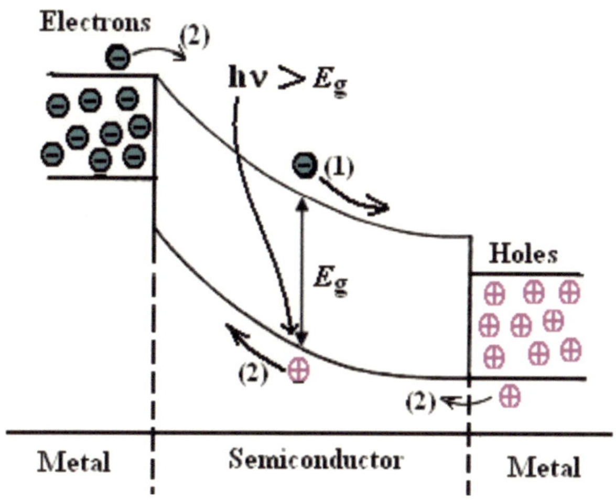

Fig. 2. Energy-band diagram of MSM detector indicating; (1) the photogeneration of signal charges and (2) the thermally-generated carriers overcoming barrier heights adding to device dark current

MSM-PD performance is critically dependent on the quality of the Schottky contacts. Therefore, it is necessary to measure the Schottky barrier height of the actual contact which is a constituent part of the MSM PD under investigation. As mentioned before, a MSM-PD essentially consists of two Schottky contacts connected back-to-back. When a bias is applied within the MSM; this will put one Schottky barrier in forward direction (anode) and the other is reverse direction (cathode).

Following the analyses of Sze (Sze et al., 1971), one can get the approximate dark current formula of the MSM PD. The current transported over the Schottky barrier height as a function of the applied voltage by considering both electron and hole current components here has the general expression as (Sze & Kwok 2006)

$$I_{da} = A_1 A_n^* T^2 \exp\left(\frac{-q\phi_{Bn}}{KT}\right) + A_2 A_p^* T^2 \exp\left(\frac{-q\phi_{Bp}}{KT}\right) \tag{6}$$

where A_1 and A_2 are the anode and cathode contact areas respectively; A_n^* and A_p^* are the effective Richardson constants; and ϕ_{Bn} and ϕ_{Bp} are the barrier heights for electrons and holes respectively. For a semiconductor with wide bandwidth, the Schottky barrier of the hole is high, and the hole is a minor carrier. Hence the hole current can be neglected, assuming that the dark current can mostly consist of the electron current (Jun at al., 2003). The dark current of the MSM PD is approximately as shown below (Yam & Hassan, 2008)

$$I = I_0 \exp[qV/(nKT)][1-\exp(-qV/KT)] \tag{7}$$

Thus, Equation (6) can be re-written follows:

$$\frac{I\exp[qV/(KT)]}{\exp[qV/(kT)]-1} = I_0 \exp[qV/(nKT)] \tag{8}$$

Based on Eq. (8), the plot of $\ln\{I\exp(qV/\{KT\})/[\exp(qV/\{KT\})-1]\}$ vs. V results in a straight line; I_0 is derived from the interception with y-axis as shown in Figure 3. Schottky barrier height ϕ_B at the MS interface can be obtained by substituting I_0 value in Eq. (4).

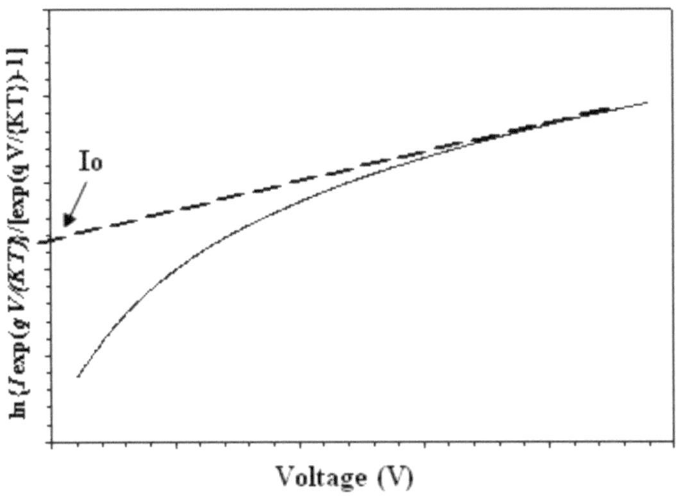

Fig. 3. $\ln\{I\exp(qV/\{KT\})/[\exp(qV/\{KT\})-1]\}$ vs.V of a MSM PD

6. Photodetectors characteristics

There are many characteristics that describe the performance of a PD. These performance characteristics indicate how a detector responds. The response of the detector should be great at the wavelength to be detected. Whilst, the additional noise created by the detector should be small. The response speed should be high so that the variations in the input optical signal can also be detected. The performance characteristics of the photodetectors are summarized in the following subsections.

6.1 Responsivity (R)

Responsivity is defined as the ratio of the photocurrent output (in amperes) to the incident optical power (in watts). It is expressed as the absolute responsivity in amps per watt (A/W). Therefore, responsivity is to measure the effectiveness of the detector for converting the electromagnetic radiation to the electrical current. Responsivity depends on wavelength, bias voltage, and temperature. Reflection and absorption characteristics of the detector's material change with wavelength and hence the responsivity also changes with wavelength (Sze & Kwok 2006).

The responsivity in terms of the photo current can be written as:

$$R = \frac{I_{ph}}{P_{inc}} = \frac{\eta q}{h\nu} \tag{9}$$

where I_{ph} is the photo current (A), P_{inc} is the incident optical power (W), η is the external quantum efficiency of the PD.

The responsivity in terms of wavelength can also be written as:

$$R = \frac{\eta \lambda q}{hc} = \frac{\eta \lambda(\mu m)}{1.24} \text{ (A/W)} \tag{10}$$

where λ is the incident wavelength.

Responsivity is an important parameter that is usually specified by the manufacturer. Through responsivity, the manufacturer can determine how much detector's output is required for a specific application.

6.2 Quantum efficiency

Quantum efficiency of a photodetector (QE) (η %) is defined as the ratio of the electron generation rate to the photon incidence rate. QE is related to the photodetector responsivity by the following equation (Sze & Kwok 2006):

$$\eta = \frac{I_{ph}/q}{P_{inc}/h\nu} = \frac{I_{ph}}{q} \cdot \frac{h\nu}{P_{inc}} \tag{11}$$

7. Characteristics of ZnO thin films deposited on PPC plastic

From the literature, the deposition of ZnO thin film on organic substrate such as PPC has not been reported by other groups. Therefore, PPC shows great potential as a substrate for ZnO thin film. The PPC has been chosen for its excellent properties such as low cost, high dielectric strength and high surface resistivity; and it is considered as one of the important polymers for biomedical and engineering application (Zhang et al., 2007). It also exhibits high transparency and superior mechanical strength (Song et al., 2009). These attractive

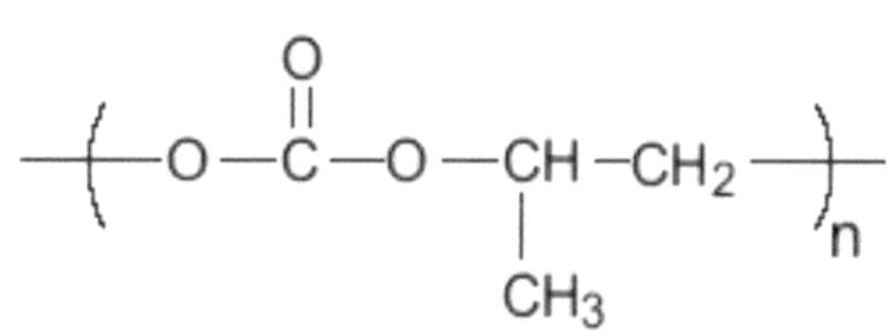

Fig. 4. The chemical structure of Poly (propylene carbonate) (PPC)

properties make the PPC a good substrate for a variety of microelectronic applications. Figure 4 shows the chemical structure of PPC plastic (Lu et al., 2005). This section briefly presents the structural, optical and electrical properties of ZnO deposited on PPC substrate.

7.1 The structural properties

ZnO thin film had been deposited on PPC plastic substrate with thickness of the 1μm. In order to study the structure of the film, XRD diffraction pattern of the deposited ZnO thin film on PPC plastic substrate was measured as shown in Figure 5. The XRD pattern shows that the film is ZnO as compared to the International Center for Diffraction Data (ICDD) library, and it is oriented in c-axis, which is the preferred orientation axis for such a material.

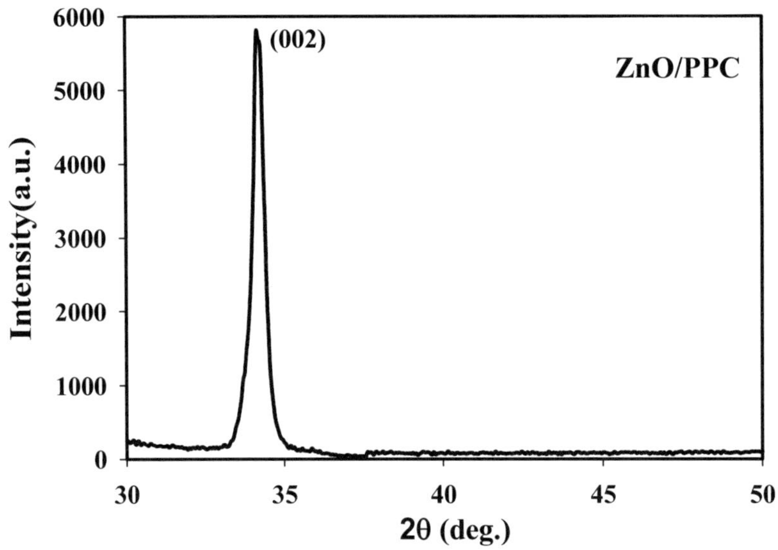

Fig. 5. XRD diffraction for ZnO thin film on PPC plastic substrate

The film shows a strong peak at 2θ = 34.27° which is correlated to the characteristic peak of the hexagonal ZnO (002) with full width at half maximum (FWHM) of 0.31°. The small FWHM of the ZnO (002) XRD peak again indicates good crystal quality of the sample (Jandow et al., 2010a).

The grain size also calculated from the FWHM of the XRD spectrum based on Scherrer formula as in Eq. 12 (Tan et al., 2005) was found to be about 26.8 nm.

$$D = \frac{K\lambda}{B\cos\theta} \tag{12}$$

where B is the full width at half maximum[(FWHM) in radians] intensity of XRD, λ is the X-ray wavelength (Cu $K\alpha$ =0.154 nm), θ is the Bragg diffraction angle, and K is a correction factor which is taken as 0.9 (Tan et al., 2005). Thus we can conclude that the film deposited on PPC plastic is nanostructured crystal. This finding is in agreement with the result reported by Myoung *et al* (Myoung et al., 2002).

The surface morphology of the film was studied using SEM as shown in Figure 6. The figure shows that the ZnO has smooth surface morphology and relatively smaller particles, which are well connected to each other; it strongly adheres to the substrate and has tightly bounded particles. Inset in Figure 6 shows the SEM image reported by Ergin *et al* (Ergin et al., 2009). A close visual inspection reveals that the prepared sample in this work has similar surface morphology as reported by Ergin. These good surface properties have strong effect on the optical properties such as transmittance and absorbance of the UV light when this material is used as UV detector (Jandow et al., 2010b).

Fig. 6. SEM image of ZnO film on PPC substrate

The elemental analysis of the sample was investigated by EDX as shown in Figure 7.The EDX result shows that Zn, O and C elements were present in the sample. Zn and O elements came from ZnO film, on the other hand, C element was not anticipated in the film; obviously, the presence of C was due to the PPC substrate. Similarly, Ergin *et al* (Ergin et al., 2009) who studied the properties of ZnO deposited on glass substrate pointed out that Si and Ca elements were not expected to be in ZnO film and these two elements could have come from the glass substrates. The inset in Figure 7 shows the EDX image reported by them.

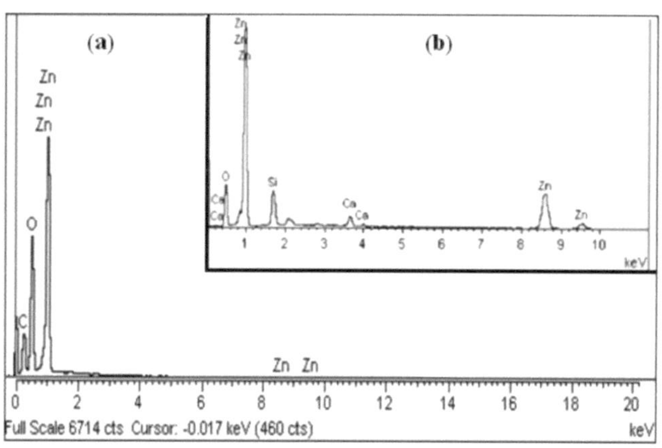

Fig. 7. (a) EDX image of ZnO film on PPC substrate. (b) Inset shows the EDX image from Ref. (Ergin et al., 2009)

The film surface was examined by AFM as shown in Figure 8. A typical AFM image shows that ZnO thin film consists of some columnar structure grains with root mean square (rms) equal to 10 nm; this nanostructure of the film surface may be useful in the absorption of the UV light when this material is used as a UV detector (Jandow et al 2010c).

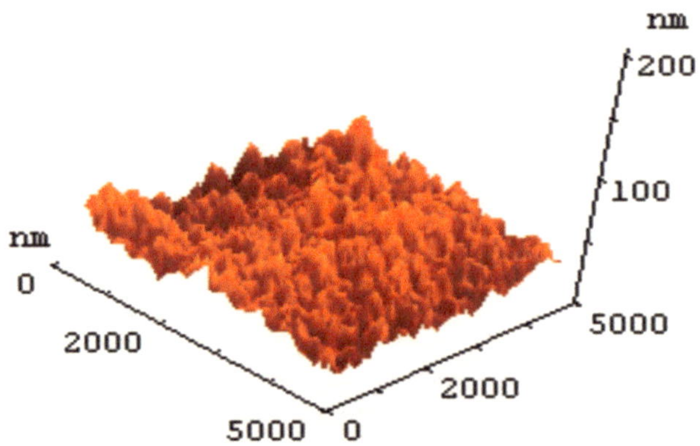

Fig. 8. AFM image of ZnO on PPC substrate (rms =10 nm)

7.2 The optical properties

The PL of the prepared ZnO thin film on PPC plastic was measured at room temperature. The result is given in Figure 9. Two luminescence peaks can be found in the figure. The first luminescence peak is the UV emission of ZnO thin films at 379.5 nm and as mentioned before it corresponds to the near band edge emission (NBE) due to the electronic transition from the near conduction band to the valence band as reported by Young *et al* and Gao and Li (Young et al., 2006; Gao &Li 2004).

The other luminescence peak is the blue-green emission ranged from 452.0 to 510.0 nm as shown in the inset in Figure 9, which is due to the defect related to deep level emission (Tneh et al., 2010). This result is in agreement with Wei *et al.* and Wu *et al* (Wei et al., 2007; Wu et al., 2007), which is attributed to the transition of electron from defect level of Zn interstitial atoms to top level of the valence band. In addition, the high UV to the visible emission ratio indicates a good crystal quality of the film which means a low density of surface defects (Jandow et al., 2010b).

The transmission and absorption spectra of ZnO film are shown in Figure 10. It can be seen that the transmission values of the film are low at short wavelengths (≤ 380nm) and high at long wavelengths. Therefore, the film behaved as an opaque material because of its high absorbing properties at short wavelengths as shown in the same figure and as a transparent material at long wavelengths.

This situation is related to the energy of the incident light; when energies of photons are smaller than the bandgap of ZnO film, they are insufficient to excite electrons from the valence band to the conduction band. However, ZnO has oxygen vacancies and interstitial Zn atoms, which act as donor impurities (Jandow et al., 2010d).

These impurities may be ionized by these low energies, so the film has low absorbance and high transmittance values at long wavelengths. The transmittance values increased higher than 80% in the visible region remarkably as the wavelength increased and this range refers to the fundamental absorption region (Shan et al., 2005). This indicates that the film could be used as transparent windows for UV light or as electrodes in a metal-semiconductor-metal MSM PDs (Jandow et al., 2010e).

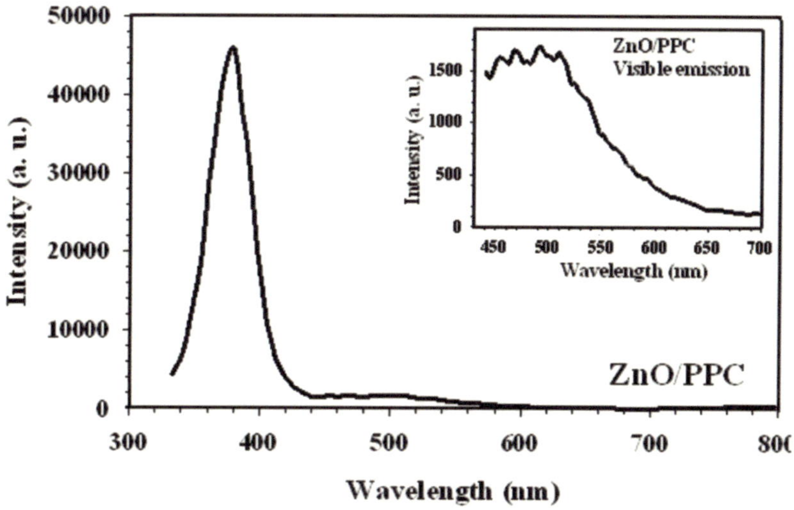

Fig. 9. Room temperature PL spectrum of DC- sputtered ZnO on PPC substrate. Inset shows the expansion of visible emission of ZnO thin film

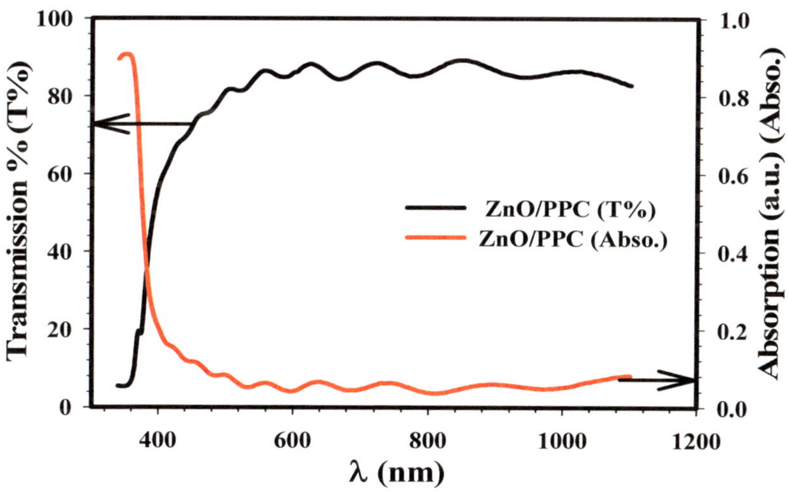

Fig. 10. Transmission and absorption spectra of ZnO thin film on PPC substrate

As mentioned earlier; ZnO is a wurtzite structure semiconductor with a direct band gap of 3.37 eV at room temperature. The absorption coefficient of the direct band gap material is given by the Eq. (Shan et al., 2005)

$$\alpha(h\nu) \propto (h\nu - E_g)^{1/2} \tag{13}$$

The dependence of $(\alpha h\nu)^2$ against the photon energy (hν) is plotted as in Figure 11. By extrapolating the linear part of the plot to $(\alpha h\nu)^2 = 0$ the value of the energy gap was found to be about 3.33 eV.

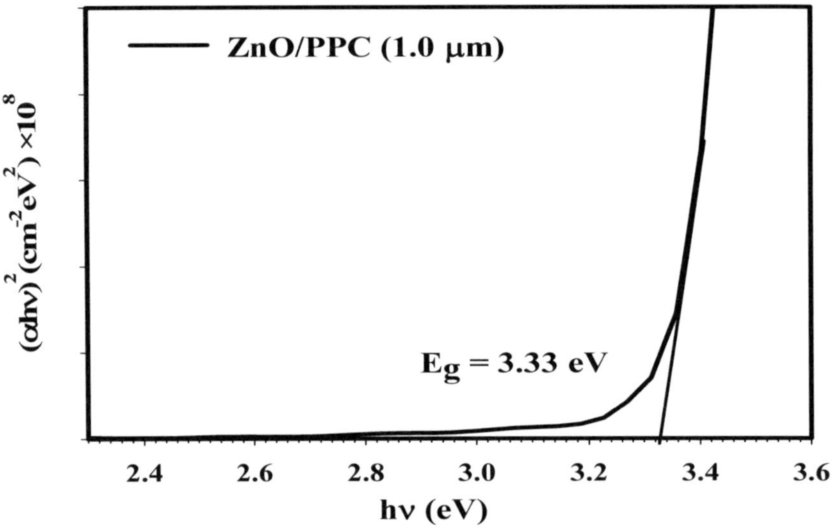

Fig. 11. The bandgap derivation for ZnO thin film on PPC substrate from the dependences of $(ah\nu)^2$ on $h\nu$

Similar results were reported by Banerjee *et al* (Banerjee et al., 2006) who deposited ZnO thin film on PET plastic substrate and also by many other authors who prepared their ZnO thin films on glass and silicon (Gümüş et al., 2006; Khoury et al., 2010; Kang et al., 2007; Lai et al., 2008).

7.3 The electrical properties

The Hall measurements show that the film is n-type with resistivity, ρ, of about 1.39×10^{-1} Ω-cm and mobility, μ, of 26 cm^2/V-s. The carrier concentration, n, was measured and it was found to be 1.72×10^{18} cm^{-3}.

Substrate	ZnO Electrical properties Resistivity, ρ (Ω-cm)	Mobility, μ(cm^2/V-s)	Carrier concentration, n (cm^{-3})	Ref.
PPC	1.39×10^{-1}	26.00	1.72×10^{18}	In this work
PET	1.00×10^{-3}	N.A.	N.A.	(Ott et al., 1999)
PET	N.A.	19.82	2.80×10^{16}	(Banerjee et al., 2006)
PET	4.0×10^{-3}	N.A.	N.A.	(Tsai et al., 2006)

Table 3. Some of the reported ZnO electrical properties on organic substrates

From the literature, the electrical properties of ZnO thin film deposited on organic substrates have rarely been reported. Table 3 summarizes some of the reported electrical properties. From the table, it was found that the resistivity value for our sample is about two orders of magnitudes higher than typical reported values (Ott et al., 1999; Tsai et al., 2006) and the mobility, μ, is 1.3 times higher than the results in (Banerjee et al., 2006). On the other hand, the carrier concentration, n, of our sample is about two orders of magnitudes higher as compared to the reported value in Ref (Banerjee et al., 2006).

8. The characteristics of ZnO UV PDs prepared on PPC

ZnO UV PDs with different metal contacts i.e. Pd, Ni and Pt have been fabricated on PPC plastic substrates. Figure 12 shows the fabricated ZnO UV PD with Pd contacts. The UV detector fabricated on PPC is very flexible and low cost.

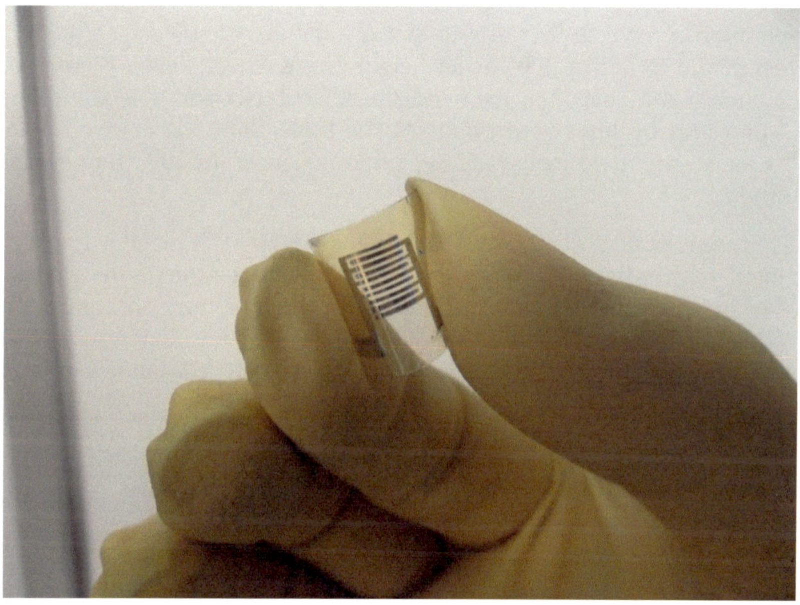

Fig. 12. The ZnO UV PD fabricated on PPC plastic with Pd contacts

8.1 I–V characteristics

Figure 13 shows the I-V characteristics of the fabricated ZnO MSM PDs (Pd/ZnO, Ni/ZnO and Pt/ZnO) on the PPC plastic measured without and with UV illumination (385 nm with power of 58.4 µW). Under dark environment, the current (I_d) at 0.8 volt was equal to 0.44, 1.72 and 1.90 □A and the ideality factor (n) was derived and found to be 1.37, 1.76 and 1.78 for PD with Pd, Ni and Pt contacts respectively..

On the other hand, when the sample was illuminated with UV wavelength of 385 nm with power of 58.4 µW; the photocurrent (I_{ph}) (at same voltage) biased at 0.8V was 5.27, 7.44 and 8.80 µA, and n was found to be 1.21, 1.46 and 1.50, respectively for the fabricated ZnO PD with Pd, Ni and Pt contact electrodes. Table 4 summarized the I-V characteristics with different values of I_d, I_{ph} and n for the PDs with Pd, Ni and Pt metal contacts.

Metal contact	Dark Environment			Illuminated Environment			Change of SBH (meV)
	Dark current, I_d (μA)	Ideality factor, n	SBH, ϕ_B (eV)	Photo current, I_{ph} (μA)	Ideality factor, n	SBH, ϕ_B (eV)	
Pd	0.44	1.37	0.738	5.27	1.21	0.700	38
Ni	1.72	1.76	0.705	7.44	1.46	0.672	33
Pt	1.90	1.78	0.700	8.80	1.50	0.668	32

Table 4. A summary of the I_d, I_{ph} and n for the PDs with Pd, Ni and Pt metal contacts

The results obtained for this study can be explained as follows: when light impinges onto the MSM UV detector, high-energy photons will be absorbed by the ZnO thin film, and with an appropriate bias, photon-generated carriers will drift toward the contact electrodes and a photocurrent will be observed (Young et al., 2006).

Further application of reverse bias acts to increase the electric field magnitude within the depletion region and to reduce the barrier height (Sze & Kwok 2006). When the energy of the incident photon is higher than the bandgap of ZnO, electron-hole pairs are generated inside ZnO thin film by light absorption. At the same time, the electron-hole pairs are separated by the electric field inside the depletion region of the ZnO thin film to generate the photocurrent.

In this study, the application of a bias on each of the Pd, Ni or Pt metallic fingers will create an electric field within the underlying ZnO thin film that sweeps the photo generated carriers out of the depletion region. The speed and the collection efficiency of the device vary, depending upon the magnitude of the applied bias, the finger separation and the average depth at which the photo generated carriers are produced (Jandow et al., 2010b).

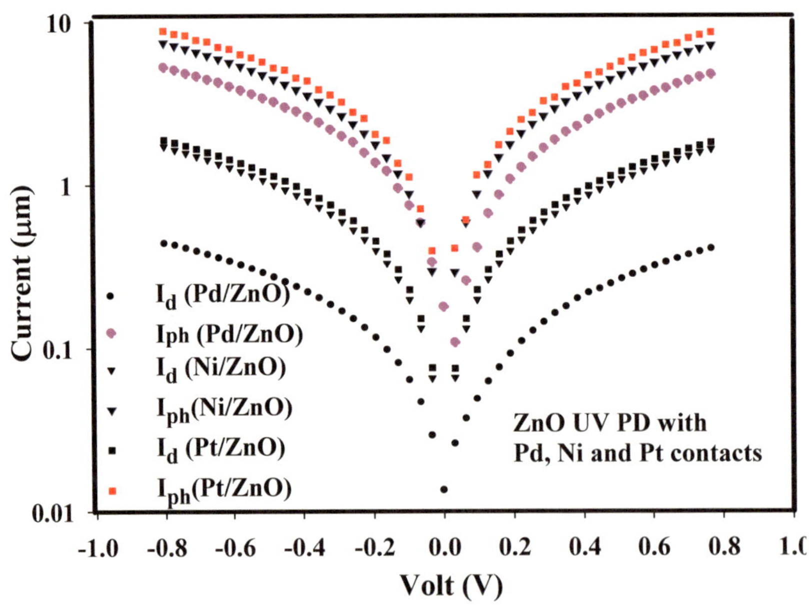

Fig. 13. Dark current (I_d) and photocurrent (I_{ph}) as a function of bias voltage (V) characteristics of the fabricated ZnO MSM PD with different electrodes type

Comparing among the three PDs, PD with Pd has the lowest dark current, this follows by Ni and Pt. the dark current for PT and Ni contacts are about 4.3 and 3.9 times of Pd. The difference of the dark current for PT and Ni contacts is relatively insignificant. This could be anticipated from the SBHs, as shown in Table 4. The difference of SBH value for these contacts is very small i.e. 5 meV.

8.2 Schottky barrier height calculation

As mentioned earlier, one of the most interesting properties of a metal-semiconductor (MS) interface is its Schottky barrier height (SBH), which is a measure of the mismatch of the energy levels for majority carriers across the MS interface. The SBH controls the electronic transport across MS interfaces and is, of vital importance to the successful operation of any semiconductor device (Tung et al., 2001). SBH plays an important role to modulate the dark and photo currents.

In particular, the forward-bias portion of the I-V characteristics has often been used to deduce the magnitude of SBH (Cho et al., 2000). The transport of carriers across a MS interface is very sensitively dependent on the magnitude of the energy barrier, SBH.

SBH for the three PDs could be determined by using Eqs. (3-8). Based on Eq. 8, the plot of $\ln\{I\exp(qV/\{KT\})/[\exp(qV/\{KT\})-1]\}$ vs. V for Pd contact is shown in Figure 14. I_0 was derived from the intercept with y-axis and by substituting this I_0 value in Eq. 4, ϕ_B was found to be 0.738 eV. Similarly for Ni/ZnO and Pd/ZnO, the ϕ_B determined from the plot under dark and illuminated conditions are summarized in Table 4.

Many metals such as Ag, Au, Pd and Pt have been used as Schottky contacts for ZnO, and resulted in SBH of between 0.6-0.8 eV (Gür et al., 2007).

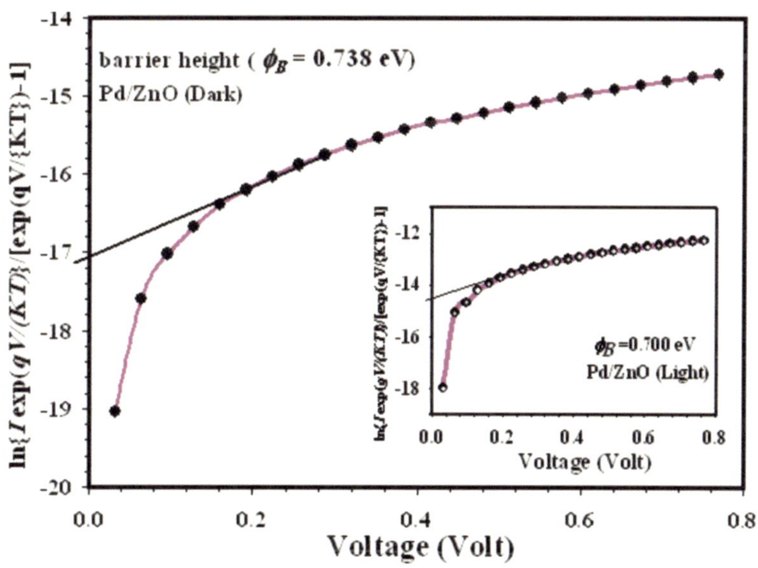

Fig. 14. $\ln\{I\exp(qV/\{kT\})/[\exp(qV/\{kT\})-1]\}$ vs V (under dark condition) of the fabricated ZnO MSM PD with Pd electrodes on PPC substrate. The inset shows ϕ_B calculation under illumination

From Figure 13 it can be found that the ZnO MSM UV PD with Pt electrode has the highest light current. That is because it has the lowest Schottky barrier height at the Pt/ZnO interface. While still the light current for the PD with Pd contact is the lowest for the same reason.

One can find that the difference among the electrical characteristics of the three PDs may mean that their barriers are different. Since the barrier height of the PD with Pd contact is higher than the barrier height for the PDs with Ni and Pt contacts, the integrated number of carriers above the barrier height for the PD with Pd contact will be less than the others, which caused the total current over the barrier to be lower. Eventually the lowest dark current was obtained with the Pd contact compares with the PDs with Ni and Pt contacts.

From Table 4, it was found that the calculated SBH value for the PD with Pd and Ni contacts is higher than that with Pt contact although the work function of Pt is the highest. This indicated that the barrier height is independent of the work function and many research groups (Kim et al., 2010; Wright et al., 2007; Ip et al., 2005; Liu et al 2004; Dong and Brillson, 2008; Kahng, 1063; Vanlaar and Scheer, 1965; Andrews and Phillips, 1975; Mead and Spitzer, 1963; Yıldırım et al., 2010; Roccaforte et al., 2010; Brillson et al., 2008; Rabadanov et al., 1982; Mead, 1966; Coppa et al., 2005) have reported that the barrier heights did not correlate with the metal work functions, suggesting many possible influences such as surface states, surface morphology, and surface contamination play important roles in the electrical properties of the contacts.

The increase of current when PD is illuminated by an UV source can be explained by work function of metal and semiconductor in the energy band diagram (Brillson et al., 2008; Rabadanov et al., 1982; Kim et al., 2010). In this case the metal and semiconductor are taken as Pd and ZnO.

ZnO which has a work function of ϕ_{ZnO}= 4.1 eV is also known to be a natural n-type semiconductor due to the oxygen vacancy which acts as trap center. The Pd work function (ϕ_{Pd} = 5.6 eV) is higher than that of ZnO. When a contact is formed electrons flow from ZnO to Pd metal until the Fermi levels align resulting in band bending as shown in Figure 15, where E_0, E_C, E_{FS}, E_V and □ are the vacuum level, the conduction band, the Fermi level, the valence band and the work function of the semiconductor (S) which in our case is ZnO, respectively, and E_{FM} is the Fermi level of the metal contact (M) which is Pd as shown in the figure, respectively. In ZnO, the trapping mechanism administrates the photoconduction. The oxygen molecules in the ZnO thin film surface capture electrons from the semiconductor. Under UV illumination, the photon energy releases the trapped electrons and also generates photo-induced electrons, causing an enhancement of the current.

The results obtained in this work are in agreement with the results reported by Young *et al* (Young et al., 2006) who fabricated ZnO MSM PD with Ag, Pd and Ni contact electrodes, and the barrier height for Ag/ZnO, Pd/ZnO and Ni/ZnO interfaces were 0.736, 0.701 and 0.613 eV, respectively although the work function for Ag =4.74, Pd =5.60 and Ni =5.42 eV. This result showed that although the work function of Ag is the lowest but the barrier height for Ag/ZnO was the highest, which indicated that the barrier height independent on the work function. Furthermore, Polyakov *et al* (Polyakov et al., 2003) results also showed that Schottky barrier heights were equal to 0.65-0.70 eV from capacitance-voltage measurements by using Au and Ag Schottky contacts on the bulk n-ZnO crystals; Au/ZnO SBH was lower than Ag/ZnO even though the work function of Au 5.47 is higher than the Ag. As well as, Neville and Mead (Neville & Mead 1970) results showed the barrier energy for Au is 0.66 eV and for Pd is 0.60 eV dependent on the forward current-voltage characteristics and as mentioned earlier the work function of Pd is higher than Au.

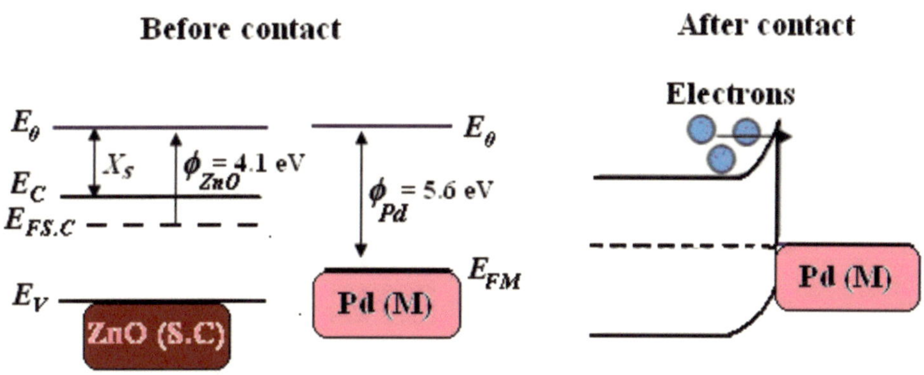

Fig. 15. The energy band diagram of ZnO and Pd.

The independence of SBH on the metal work function has been explained by many researchers (Brillson et al., 2008; Rabadanov et al., 1981; Mead, 1966; Coppa et al., 2005), they pointed out SBH could be affected by the interface structure and the associated interface states

8.3 Responsivity and quantum efficiency

Figure 16 shows the responsivity as a function of the incident wavelength for the three ZnO MSM UV PDs with Pd, Ni and Pt contact electrodes. At 0.8 V and by applying Eq. 9, it was found that the maximum responsivity for the three PDs were found to be 0.08, 0.09 and 0.11 A/W, which corresponds to quantum efficiency (η %) of 27.1 % 31.7 % and 37.7% respectively which are shown in Figure 17. It is observed that the PD with Pt contacts has higher responsivity and quantum efficiency comparing to the PDs with Pd and Ni contacts due to the lowest barrier height and highest photocurrent. Table 5 summarizes the responsivity and quantum efficiency for Pd, Ni and Pt contacts, respectively.

As shown in the figure, the PD responsivity was nearly constant in the UV region ~ 320–360 nm; it started to increase from 365 nm, of which the responsivity peaked at 385 nm and began to decrease whenever became close to the visible region. This can be explained as follows; when ZnO detector is irradiated by UV light with energy higher than the bandgap (3.37 eV for ZnO), electron-hole pairs will be generated, as a result these excess charge carriers contribute to photo current and result in the response to the UV light. The cut-off at wavelength of 385 nm is nearly close to the ZnO energy bandgap of 3.37 eV; the responsivity decreased at the shorter wavelength range due to the decrease of the penetrating depth of the light, resulting in an increase of the surface recombination (Young et al 2007; Jandow et al., 2010a; Yan et al., 2004).

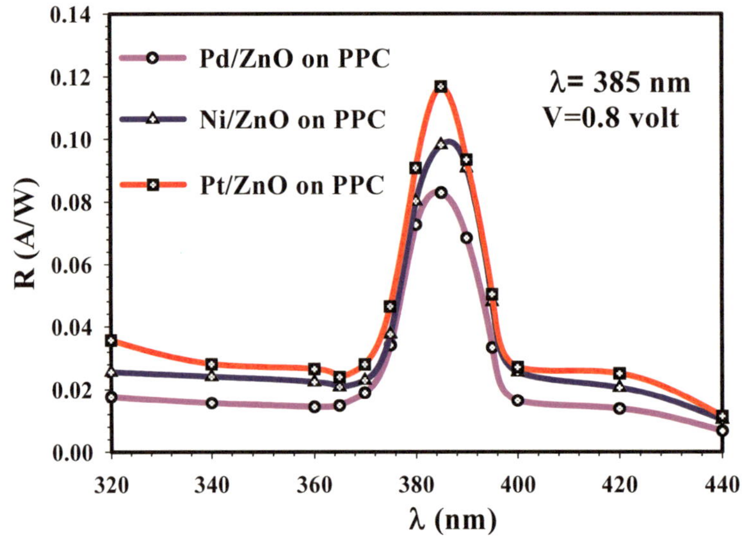

Fig. 16. Measured spectral responsivity of the three fabricated ZnO MSM PDs with different electrodes on PPC substrate

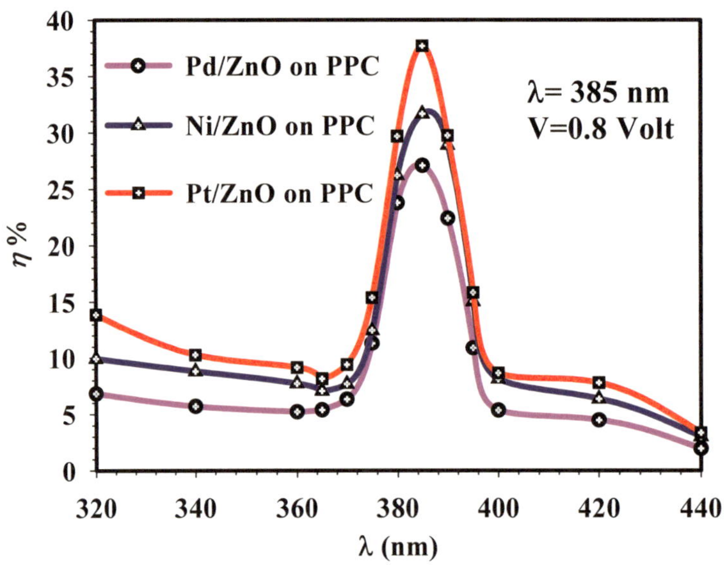

Fig. 17. Quantum efficiency of the three fabricated ZnO MSM PDs with different electrodes on PPC substrate

Metal contacts	Responsivity (R) (A/W)	Quantum efficiency (η %)
Pd	0.082	27.1
Ni	0.098	31.7
Pt	0.116	37.7

Table 5. A summary of the responsivity (R) and quantum efficiency (η %) values for the PDs with Pd, Ni and Pt metal contacts at 0.8 volt

9. Conclusion

In summary, an overview of the characteristics of the deposited ZnO thin film on various organic substrates such as polyethylene terephthalate (PET), polyolefin, polytetrafluoroethylene (Teflon) and Polycarbonate (PC) and their potential applications in various areas has been presented. A review of semiconductor PDs, types of PDs as well as their characteristics has been demonstrated. Apart from that, the properties of the ZnO thin films deposited on PPC plastic substrate have been expressed. ZnO UV detectors prepared on PPC substrate with different electrodes i.e. Pd, Ni and Pt have been fabricated and investigated. The results showed that the deposited ZnO thin film had good structural and optical properties. In addition ZnO UV PDs fabricated on PPC with different metal contacts showed that Pt/ZnO MSM UV PD has the highest quantum efficiency.

10. Acknowledgments

This work was conducted under the grant no. (100/PFIZIK/8/4010) support from Universiti Sains Malaysia is gratefully acknowledged.

11. References

Andrews. J. & Phillips, J. (1975). Chemical bonding and structure of meta semiconductor interfaces. *Phys. Rev. Lett.*, Vol.35, No.1, pp. 56-59

Angelats, L. (2006). "Silva Study of structural, electrical, optical and magnetic properties of ZnO based films produced by magnetron sputtering" PhD thesis, university of Puerto Rico UPR

Auret, F., Meyer, W., Rensburg, P., Hayes, M., Nel, J., Wenckstern, H., Schmidt, H., Biehne, G., Hochmuth, H., Lorenz, M., Grundmann, M. (2007). Electronic properties of defects in pulsed-laser deposition grown ZnO with levels at 300 and 370meV below the conduction band. *Physica B*, Vol.401-402, pp. 378-381

Banerjee, A., Ghosh, C., Chattopadhyay, K., Minoura, H., Sarkar, A., Akiba, A., Kamiya, A., & Endo, T. (2006). Low-temperature deposition of ZnO thin films on PET and glass substrates by DC-sputtering technique. *Thin Solid Films*, Vol.496, pp. 112-116

Banerjee, A., Ghosh, C., Chattopadhyay, K., Minoura, H., Sarkar, A., Akiba, A., Kamiya, A., & Endo, T. (2006). Low-temperature deposition of ZnO thin films on PET and glass substrates by DC-sputtering technique. *Thin Solid Films*, Vol.496, pp. 112-116

Bang, K.-H., Hwang, D.-K., & Myoung, J.-M. (2003). Effects of ZnO buffer layer thickness on properties of ZnO thin films deposited by radio-frequency magnetron sputtering. *Appl. Surf. Sci.*, Vol.207, pp. 359-364

Basak, D., Amin, G., Mallik, B., Paul, G., & Sen, S. (2003). Photoconductive UV detectors on sol-gel-synthesized ZnO films. *J. Cryst. Growth*, Vol.256, pp. 73-77

Brabec, C., Sariciftci, N., & Hummelen, J. (2001). Plastic Solar Cells. *Adv. Funct. Mater.*, Vol.11, No.1, pp. 15-26

Brillson, L., Mosbacker, H., Hetzer, M., Strzhemechny, Y., Look, D., Cantwell, G., Zhang, J., & Song, J. (2008). Surface and near-surface passivation, chemical reaction, and Schottky barrier formation at ZnO surfaces and interfaces. *Appl. Surf. Sci.*, Vol.254, pp. 8000-8004

Brillson, L., Mosbacker, H., Hetzer, M., Strzhemechny, Y., Look, D., Cantwell, G., Zhang, J., & Song, J. (2008). Surface and near-surface passivation, chemical reaction, and Schottky barrier formation at ZnO surfaces and interfaces. *Appl. Surf. Sci.*, Vol.254, pp. 8000–8004

Caria, M., Barberini, L., Cadeddu, S., Giannattasio, A., Lai, A., Rusani, A., & Sesselego, A. (2001). Far UV responsivity of commercial silicon photodetectors. *Nucl. Instrum. Meth. A*, Vol.466, No.1, pp. 115-118

Chakraborty, A., Mondal, T., Bera, S., Sen, S., Ghosh, R., & Paul, G. (2008). Effects of Aluminium and Indium incorporation on the structural and optical properties of ZnO thin films synthesized by spray pyrolysis. *Mater. Chem. Phys.*, Vol.112, pp. 162-166

Chang, S., Chang, S., Chiou, Y., Lu, C., Lin, T., Lin, Y., Kuo, C., & Chang, H. (2007). ZnO photoconductive sensors epitaxially grown on sapphire substrates. *Sensor. Actuator., A*, Vol.140, pp. 60-64

Cho, H., Leerungnawarat, P., Hays, D., Pearton, S., Chu, S., Strong, R., Zetterling, C.-M., Östling, M., & Ren, F. (2000). low-damage dry etching of SiC. *Appl. Phys. Lett.*, Vol.76, No.6, pp. 739-741

Coppa, B., Fulton, C., Kiesel, S., Davis, R., Pandarinath, C., Burnette, J., Nemanich, R., & Smith, D. (2005). Structural, microstructural, and electrical properties of gold films and Schottky contacts on remote plasma-cleaned, n-type ZnO{0001} surfaces. *J. Appl. Phys.*, Vol.97. pp. 103517-1-13

Coppa, B., Fulton, C., Kiesel, S., Davis, R., Pandarinath, C., Burnette, J., Nemanich, R., & Smith, D. J. (2005). Structural, microstructural, and electrical properties of gold films and Schottky contacts on remote plasma-cleaned, n-type ZnO{0001} surfaces. *J. Appl. Phys.* Vol.97, pp. 103517-1-13

Craciun, V., Elders J., Gardeniers, J., & Boyd, I. (1994). Characteristics of high quality ZnO thin films deposited by pulsed laser deposition. *Appl. Phys. Lett.*, Vol. 65, No.23, pp. 2963-2965

Daraee, M., Hajian, M., Rastgoo, M., & Lavasanpour, L. (2008). Study of electrical characteristic of surface barrier detector with high series resistance. *Adv. Studies Theor. Phys.*, Vol.2, No.20, pp. 957-964.

Dong, Y. & Brillson, L. (2008). First-principles studies of metal (111)/ZnO {0001} interfaces. *J. Eectron. Mater.*, Vol.37, No.5, pp. 743-748

Ergin, B., Ketenci, E., & Atay, F. (2009). Characterization of ZnO films obtained by ultrasonic spray pyrolysis technique. *Int. J. Hydrogen Energ.*, Vol.34, pp. 5249-5254

Gao, W., & Li, Z. (2004). ZnO thin films produced by magnetron sputtering. *Ceram. Int.*, Vol.30, pp. 1155-1159

Ghosh, R., Basak, D., & Fujihara, S. (2004). Effect of substrate-induced strain on the structural, electrical, and optical properties of polycrystalline ZnO thin films. *J. Appl. Phys.*, Vol.96, No.5, pp. 2689-2692

Gökkavas, M., Butun, S., Tut, T., Biyikli, N., & Ozbay, E. (2007). AlGaN-based high-performance metal–semiconductor–metal photodetectors. *PNFA.*, Vol.5, pp. 53-62

Gümüş, C., Ozkendir, O., Kavak, H., & Ufuktepe, Y. (2006). Structural and optical properties of zinc oxide thin films prepared by spray pyrolysis method. *J. Optoelectron. Adv. M.*, Vol.8, No.1, pp. 299-303

Gür, E., Tüzemen, S, Kılıç, B., & Coşkun, C. (2007). High-temperature Schottky diode characteristics of bulk ZnO. *J. Phys.: Condens. Matter.*, Vol.19, pp. 196206-196214

Haas, F. (1997). Principles of Semiconductor Devices. RL-TR-96-217, In-House Report

Hanzaz, M., Bouhdada, A., Vigué, F., & Faurie, J. (2007). ZnSe-and GaN-based Schottky barrier photodetectors for blue and ultraviolet detection. *J. Act. Pass. Electron. Dev.*, Vol.2, pp. 165-169

Hickernell, F. (1976). Zinc-Oxide Thin-film surface-wave transducers. *Proceedings of the IEEE*, Vol. 64, No.5, pp. 631-635

Hiramatsu, K., & Motogaito, A. (2003). GaN-based Schottky barrier photodetectors from near ultraviolet to vacuum ultraviolet (360-50 nm). *phys. stat. sol. (a)*, Vol.195, No.3, pp. 496-501

Hussain, S. (2008). "Investigation of Structural and Optional Properties of Nanpcrystalline ZnO" PhD thesis, Department of Physics, Chemistry and Biology Linköping University

Hwang, K-S. , Kang, B-A., Jeong, J-H. , Jeon Y-S. , & Kim B-H. (2007). Spin coating-pyrolysis derived highly c-axis-oriented ZnO layers pre-fired at various temperatures. *Curr. Appl. Phys.*, Vol.7, pp. 421-425

Ianno, N., McConville, L., Shaikh, N., Pittal, S., & Snyder, P. (1992). Characterization of pulsed laser deposited Zinc oxide. *Thin Solid Films*, Vol.220, pp. 92- 99

Ip, K., Khanna, R., Norton, D., Pearton, S., Ren, F., Kravchenko, I., Kao, C., & Chi, G. (2005). Thermal stability of W2B and W2B5 contacts on ZnO. *Appl. Surf. Sci.*, Vol.252, pp. 1846-1853

Jandow N. N., F. K. Yam, S. M. Thahab, H. Abu Hassan, & Ibrahim, K. (2010a). Characteristics of ZnO MSM UV photodetector with Ni contact electrodes on Poly Propylene Carbonate (PPC) plastic substrate. *Curr. Appl. Phys.*, Vol.10, pp. 1452-1455

Jandow, N., Ibrahim, K., Abu Hassan, H. (2010b). I-V Characteristic for ZnO MSM Photodetector with Pd Contact Electrodes on PPC Plastic, *AIP Conf. Proc.*, Vol.1250, pp. 424-427

Jandow, N., Ibrahim, K., Abu Hassan, H., Thahab, S., & Hamad, O. (2010c). The electrical properties of ZnO MSM Photodetector with Pt Contact Electrodes on PPC Plastic. *JED*, Vol.7, pp. 225-229

Jandow, N., Ibrahima, K., Yam, F., Abu Hassan, H., Thahab, S., & Hamad, O. (2010d). The study of ZnO MSM UV photodetector with Pd contact electrodes on (PPC) plastic. *J.Optoelectron. Adv. Mat.-Rabid*, Vol.4, No.5, pp. 726-730

Jandow, N., Yam, F., Thahab, S., Ibrahim, K., Abu Hassan, H. (2010). The characteristics of ZnO deposited on PPC plastic substrate. *Mater. Lett.*, Vol.64, pp. 2366-2368

Janotti, A., & Van de Walle, C. (2009e). Fundamentals of zinc oxide as a semiconductor. *Rep. Prog.* Phys., Vol. 72, pp. 126501-126530

Jeong, I.-S., Kim, J., Park, H.-Ho, & Im, S. (2004). n-ZnO/p-Si UV photodetectors employing AlO_x films for antireflection. *Thin Solid Films*, Vol.447-448, pp. 111-114

Jiang, D., Zhang, J., Lu, Y., Liu, K., Zhao, D., Zhang, Z., Shen, D., & Fan, X. (2008) Ultraviolet Schottky detector based on epitaxial ZnO thin film. *Solid State Electron.*, Vol.52, pp. 679-682

Jin, B., Woo, H., Im, S., Bae, S., & Lee, S. (2001). Relationship between photoluminescence and electrical properties of ZnO thin films grown by pulsed laser deposition. *Appl. Surf. Sci.*, Vol.169-170, pp. 521-524

Jin, C. (2003). "Growth and characterization of ZnO and ZnO-based Alloys-MgxZn1-xO and MnxZn1-xO". PhD thesis, Department of materials science and Engineering, North Carolina state university, Raleigh

Jun, W., Degang, Z., Zongshun, L., Gan F., Jianjun, Z., Xiaomin, S., Baoshun, Z., & Hui, Y. (2003). Metal-semiconductor-metal ultraviolet photodetector based on GaN. *Sci. China Ser. G*, Vol.46, No.2, pp. 198-203

Kahng, D. (1963). Conduction properties of the Au-n-type-Si Schottky barrier. *Solid State Electron.*, Vol. 6, pp. 281-295

Kang, S., Joung, Y., Chang, D., & Kim, K. (2007). Piezoelectric and optical properties of ZnO thin films deposited using various $O_2/(Ar+O_2)$ gas ratios. *J. Mater. Sci: Mater Electron.*, Vol.18, pp. 647-653

Khoury, A., al Asmar, R., Abdallah, M., El Hajj Moussa, G., & Foucaran, A. (2010). Comparative study between zinc oxide elaborated by spray pyrolysis, electron beam evaporation and rf magnetron techniques. *Phys. Status Solidi A*, Vol.207, No.8, pp. 1900-1904

Kim, H., Kim, H., & Kim, D.-W. (2010). Silver Schottky contacts to a-plane bulk ZnO. *J. Appl. Phy.*, Vol.108, pp. 074514-1-5

Kim, J., Lee J., Lee, J., Lee, D., Jang, B., Kim, H., Lee W., Cho, C., & Kim, J. (2009). Characteristics of ZnO films deposited on plastic substrates at various rf sputtering powers. *J. Korean Phys. Soc.*, Vol.55, No.5, pp. 1910-1914

Kim, J., Yun, J., Kim, C., Park, Y., Woo, J., Park, J., Lee, J., Yi, J., & Han, C. (2010). ZnO nanowire-embedded Schottky diode for effective UV detection by the barrier reduction effect. *Nanotechnology*, Vol.21, pp. 115205-115210

Kiriakidis, G., Suchea, M., Christoulakis, S., Horvath, P., Kitsopoulos, T., & Stoemenos, J. (2007). Structural characterization of ZnO thin films deposited by dc magnetron sputtering. *Thin Solid Films*, Vol.515, pp. 8577-8581

Koch, U., Fojtik, A., Weller, H., & Henglein, A. (1985). Photochemistry of semiconductor colloids, preparation of extremely small ZnO particles, fluorescence phenomena and size quantization effects, *Chem.Phys.Lett.*, Vol.122, pp. 507-510

Kumar, S., Kim, H.., Sreenivas, K., & Tandon, P. (2009). ZnO based surface acoustic wave ultraviolet photo sensor. *J. Electroceram*, Vol.22, pp. 198-202

Lai, L.-W., & Lee, C.-T. (2008). Investigation of optical and electrical properties of ZnO thin films, *Mater. Chem. Phys.*, Vol.110, pp. 393-396

Lakhotia, G., Umarji, G., Jagtap, S., Rane, S., Mulik, U., Amalnerkar, D., & Gosavi, S. (2010). An investigation on TiO_2-ZnO based thick film 'solar blind', photo-conductor for 'green' electronics. *Mater. Sci. Eng. B*, Vol.168, pp. 66-70

Liang, S., Sheng, H., Liu, Y., Huo, Z., Lu, Y., & Shen, H. (2001). ZnO Schottky ultraviolet photodetectors. *J. Cryst. Growth*, Vol.225, pp. 110-113

Lim, J., Kang/ Ch., Kim, K., Park, I., Hwang, D., & Park, S. (2006). UV Electroluminescence Emission from ZnO light-emitting diodes grown by high-temperature radiofrequency sputtering. *Adv. Mater.*, Vol.18, pp. 2720-2724

Lin, T., Chang, S., Su, Y., Huang, B., Fujita, M., & Horikoshi, Y. (2005). ZnO MSM photodetectors with Ru contact electrodes. *J. Cryst. Growth*, Vol.281, pp. 513-517

Liu, K., Ma, J., Zhang, J., Lu, Y., Jiang, D., Li, B., Zhao, D., Zhang, Z., Yao, B., & Shen, D. (2007). Ultraviolet photoconductive detector with high visible rejection and fast photoresponse based on ZnO thin film. *Solid State Electron.*, Vol.51, pp. 757-761

Liu, K., Sakurai, M., & Aono, M. (2010). ZnO-based ultraviolet photodetectors. *Sensors*, Vol.10, pp. 8605-8630

Liu, Y., Egawa, T., Jiang, H., Zhang, B., Ishikawa, H., & Hao, M. (2004). Near-ideal Schottky contact on quaternary AlInGaN epilayer lattice-matched with GaN, *Appl. Phys. Lett.*, Vol.85, No.24, pp. 6030-6032

Liu, Y., Gopla, C., Liang, S., Emanetoglu, N., Lu, Y., Shen, H., & Wraback, M. (2000). Ultraviolet detectors based on epitaxial ZnO films grown by MOCVD. *J. Electron. Mater*, Vol.29, No.1, pp. 69-74

Liu, Y., Li, Q., & Shao, H. (2009). Properties of ZnO-Al films deposited on polycarbonate substrate. *Vacuum*, Vol.83, pp. 1435–1437

Liu, Y.-Y., Yuan, Y. -Z., Gao, X.-T., Yan, S.-S., Cao, X.-Z., & Wei, G.-X. (2007). Deposition of ZnO thin film on polytetrafluoroethylene substrate by the magnetron sputtering method. *Mater. Lett.*, Vol.61, pp. 4463-4465

Lu, X., Zhu, Q., & Meng, Y. (2005). Kinetic analysis of thermal decomposition of Poly(Propylene Carbonate), *Polym. Degrad. Stab.*, Vol.89, pp. 282-288

Lu, Y.-M., Tsai, S.-Y., Lu, J.-J., & Hon, M.-H. (2007). The structural and optical properties of zinc oxide thin films deposited on PET substrate by rf. magnetron sputtering. *Solid State Phenom.*, Vol.121-123, pp. 971-974

Ma, C., Taya, M., & Xu, C. (2008). Flexible electrochromic device based on poly (3, 4-(2, 2-dimethylpropylenedioxy) thiophene). *Electrochim. Acta*, Vol.54, pp. 598-605

Mahmood, F., Gould, R., Hassan, A., & Salih, H. (1995). DC. properties of ZnO thin films prepared by rf. magnetron sputtering. *Thin Solid Films*, Vol.770, pp. 376-379

Mandalapu, L., Xiu, F., Yang, Z., & Liu, J. (2007). Ultraviolet photoconductive detectors based on Ga-doped ZnO films grown by molecular-beam epitaxy. *Solid State Electron.*, Vol.51, pp. 1014-1017

Mead, C. (1966). Metal-semiconductor surface barriers. *Solid State Electron.*, Vol.9, pp. 1023-1033

Mead, C., & Spitzer, W. (1963). Fermi level position at semiconductor surfaces. *Phys. Rev. Lett.*, Vol.10, No.11, pp. 471-472

Monroy, E., Omnés, F., & Calle, F. (2003). Wide-bandgap semiconductor ultraviolet photodetectors. *Semicond. Sci. Technol.*, Vol.18, pp. R33-R51

Moon, T.-H., Jeong, M.-C., Lee, W., & Myoung, J.-M. (2005). The fabrication and characterization of ZnO UV detector. *Appl. Surf. Sci.*, Vol.240, pp. 280-285

Myoung, J.-M., Yoon, W.-H., Lee, D.-H., Yun, I., Bae, S.-H., & Lee, S.-Y. (2002). Effects of thickness variation on properties of ZnO thin films grown by pulsed laser deposition, *Jpn. J. Appl. Phys.*, Vol.41, pp. 28-31

Nandy, S., Goswami, S., & Chattopadhyay, K. (2010). Ultra smooth NiO thin films on flexible plastic (PET) substrate at room temperature by rf magnetron sputtering and effect of oxygen partial pressure on their properties, *Appl. Surf. Sci.*, Vol.256, pp. 3142-3147

Nause, J., & Nemeth, B. (2005). Pressurized melt growth of ZnO boules. *Semicond. Sci. Technol.*, Vol.20, pp. S45-S48

Nav Bharat Metallic Oxide Industries Pvt. Limited. n.d. *Some of the popular applications of Zinc Oxide*. Available from: <http://www.navbharat.co.in/Clients.htm>

Neville, R., & Mead, C. (1970). "Surface barriers on zinc oxide". *J. Appl. Phys.,* Vol.41, No.9, pp. 3795- 3800

Nunes, P., Fortunato, E., & Martins, R. (2001). Influence of the annealing conditions on the properties of ZnO thin films. *Int. J. Inorg. Mater.*, Vol.3, pp. 1125-1128

Ott, A., & Chang, R. (1999). Atomic layer-controlled growth of transparent conducting ZnO on plastic substrates. *Mater. Chem. Phys.*, Vol.58, pp. 132-138

Paul, G., Bhaumik, A., Patra, A., & Bera, S. (2007). Enhanced photo-electric response of ZnO/polyaniline layer-by-layer self-assembled films, *Mater. Chem. Phys.*, Vol.106, pp. 360-363

Plaza, J., Martínez, O., Carcelén, V., Olvera, J., Sanz, L., & Diéguez, E. (2008). Formation of ZnO and Zn1-xCdxO films on CdTe/CdZnTe single crystals. *Appl. Surf. Sci.*, Vol.254, pp. 5403-5407

Polyakov, A., Smirnov, N., Kozhukhova, E., Vdovin, V., Ip, K., Heo, Y., Norton, D., & Pearton, S. (2003). "Electrical characteristics of Au and Ag Schottky contacts on n-ZnO". *Appl. Phys. Lett.*, Vol.83, No.8, pp. 1575-1577

Rabadanov, R., Guseikhanov, M., Aliev, I., & Semiletov, S. (1981). Properties of metal-zinc oxide contacts. *Russ. Phys. J.*, Vol.24, No.6, pp. 548-551

Rabadanov, R., Guseikhanov, M., Aliev, I., & Semiletov, S. (1982). Properties of metal-zinc oxide contacts. *Russ Phys J.*, Vol.24, No.6, pp. 548-551

Rao, T., Kumar, M., Safarulla, A., Ganesan, V., Barman, S., & Sanjeeviraja, C. (2010). Physical properties of ZnO thin films deposited at various substrate temperatures using spray pyrolysis. *Physica B*, Vol.405, pp. 2226-2231

Roccaforte, F., Giannazzo, F., Iucolano, F., Eriksson, J., Weng, M., & Raineri, V. (2010). Surface and interface issues in wide band gap semiconductor electronics. *Appl. Surf. Sci.*, Vol.256, pp. 5727-5735

Sans, J., Segura, A., Mollar, M., & Marí, B. (2004). Optical properties of thin films of ZnO prepared by pulsed laser deposition. *Thin Solid Films*, Vol.453-454, pp. 251-255

Shan, F., Liu, G., Lee, W., Lee, G., Kim, I., Shin, B., & Kim, Y. (2005). Transparent conductive ZnO thin films on glass substrates deposited by pulsed laser deposition. *J. Cryst. Growth*, Vol.277, pp. 284-292

Shan, F., Shin, B., Jang, S., & Yu, Y. (2004). Substrate effects of ZnO thin films prepared by PLD technique. *J. Eur. Ceram. Soc.*, Vol.24, pp. 1015-1018

Sofiani, Z., Derkowska, B., Dalasińsk, P., Wojdyła, M., Dabos-Seignon, S., Alaoui Lamrani, M., Dghoughi, L., Bała, W., Addou, M., Sahraoui, B. (2006). Optical properties of ZnO and ZnO:Ce layers grown by spray pyrolysis. *Opt. Commun.*, Vol.267, pp. 433-439

Song, P.-f., Wang, S.-j., Xiao, M., Du, F.-g., Gan, L.-q., Liu, G.-q., & Meng, Y.-z. (2009). Cross-linkable and thermally stable aliphatic polycarbonates derived from Co_2, propylene oxide and maleic anhydride. *J Polym Res*, Vol.16, pp. 91-97

Sze, S. (2002). Semiconductor devices, physics and technology, 2nd ed., John Wiley & Sons New York , USA

Sze, S., & Kwok K. (2006). Physics of semiconductor devices, 3rd ed., United States of America

Sze, S., Coleman, D., & Loya, J., (1971). Current transport in metal-semiconductor-metal (MSM) structures. *Solid State Electron.*, Vol.14, pp. 1209-1218

Tan, S., Chen, B., Sun, X., Fan, W., Kwok, H., Zhang, X., & Chua, S. (2005). Blueshift of optical band gap in ZnO thin films grown by metal-organic chemical-vapor deposition. *J. Appl. Phys.*, Vol.98, pp. 013505-013505-5

Tneh, S., Hassan, Z., Saw, K., Yam, F., Abu Hassan, H. (2010). The structural and optical characterizations of ZnO synthesized using the "bottom-up" growth method. *Physica B*, Vol.405, pp. 2045-2048

Tsai, H-Y. (2007). Characteristics of ZnO thin film deposited by ion beam sputter. *J. Materi. Process. Technol.*, Vol.192-193, pp. 55-59

Tsai, S.-Y., Lu, Y.-M., Lu, J.-J., & Hon, M.-H. (2006). Comparison with electrical and optical properties of zinc oxide films deposited on the glass and PET substrates. *Surf. Coat. Technol.*, Vol.200, pp. 3241-3244

Tung, R. (2001). Recent advances in Schottky barrier concepts. *Mater. Sci. Eng., R*, Vol.35, pp. 1-138

Tüzemen, S., & Gür, E. (2007). Principal issues in producing new ultraviolet light emitters based on transparent semiconductor zinc oxide. *Opt. Mater.*, Vol.30, pp. 292-310

Vanlaar, J. & Scheer, J. (1965). Fermi level stabilization at semiconductor surfaces. *Surf. Sci.* Vol.3, pp. 189-201

Wang, Q., Pflügl, C., Andress, W., Ham, D., & Capasso, F. (2008). Gigahertz surface acoustic wave generation on ZnO thin films deposited by radio frequency magnetron sputtering on III-V semiconductor substrates. *J. Vac. Sci. Technol. B*, Vol.26, No.6, pp. 1848-1851

Wei, X., Zhang, Z., Liu, M., Chen, C., Sun, G., Xue, C., Zhuang, H., & Man, B. (2007) Annealing effect on the microstructure and photoluminescence of ZnO thin films. *Mater. Chem. Phys.*, Vol.101, pp. 285-290

Wright, J., Stafford, L., Gila, B., Norton, D., Pearton, S., Wang, H.-T., & Ren, F. (2007). Effect of cryogenic temperature deposition of various metal contacts on bulk single-crystal n-type ZnO, *J. Electron. Mater.*, Vol.36, No.4, pp. 488-493

Wu, L., Tok, A., Boey, F., Zeng, X., & Zhang, X. (2007). Chemical Synthesis of ZnO Nanocrystals. *IEEE. Trans. Nanotechnol*, Vol.6, No.5, pp. 497-503

Yam, F., & Hassan, Z. (2008). The investigation of dark current reduction in MSM photodetector based on porous GaN. *J. Optoelectron Adv. M.*, Vol.10, pp. 545-548

Yan, F., Xin, X., Aslam, S., Zhao, Y., Franz, D., Zhao, J., & Weiner, M. (2004). 4H-SiC UV Photo detectors with large area and very high specific detectivity. *IEEE J. Quantum Elect.*, Vol.40, No.9, pp. 1315-1320

Yıldırım, N., Ejderha, K., & Turut, A. (2010). On temperature-dependent experimental I-V and C-V data of Ni/n-GaN Schottky contacts. *J. Appl. Phys.*, Vol.108, pp. 114506-1-8

Young, S., Ji, L., Chang, S., & Su, Y. (2006). ZnO metal-semiconductor-metal ultraviolet sensors with various contact electrodes. *J. Cryst. Growth*, Vol.293, pp. 43-47.

Young, S., Ji, L., Chang, S., Chen, Y., Lam, K., Liang, S., Du, X., Xue, Q., & Sun, Y. (2007). ZnO metal-semiconductor-metal ultraviolet photodetectors with Iridium contact electrodes. *IET Optoelectron.*, Vol.1, No.3, pp. 135-139

Young, S., Ji, L., Fang, T., Chang, S., Su, Y., & Du X. (2007). ZnO ultraviolet photodiodes with Pd contact electrodes. *Acta Mater.*, Vol.55, pp. 329-333

Zhang, B., Wakatsuki, K., Binh, N., Usami, N., & Segawa, Y. (2004). Effects of growth temperature on the characteristics of ZnO epitaxial films deposited by metalorganic chemical vapor deposition. *Thin Solid Films*, Vol.449, pp. 12-19

Zhang, J., Kang, J., Hu, P., & Meng, Q. (2007). Surface modification of poly(propylene carbonate) by oxygen ion implantation. *Appl. Surf. Sci.*, Vol.253, pp. 5436-5441

2

UV-Vis Photodetector with Silicon Nanoparticles

J.A. Luna-López[1], M. Aceves-Mijares[2], J. Carrillo-López[1],
A. Morales-Sánchez[3], F. Flores-Gracia[1] and D.E. Vázquez Valerdi[1]
[1]*Science Institute-Research Center for Semiconductor Devices-Autonomous Benemérita University of Puebla*
[2]*Department of Electronics, National Institute of Astrophysics, Optics and Electronics INAOE*
[3]*Centro de Investigación en Materiales Avanzados S. C., Unidad Monterrey-PIIT, Apodaca, Nuevo León, México*

1. Introduction

Nowadays, photodetector devices are important components for optoelectronic integration. In the past, various photodetector structures have been developed from pn junction, pin diode, bipolar transistor, avalanche photodiodes (APD), and metal-semiconductor-metal (MSM) structures (Ashkan et al., 2008; Hwang & Lin, 2005; DiMaria et al., 1984a, 1984b; Sabnis et al., 2005; Foster et al., 2006). In these structures, different semiconductors such as Si, III-V and II-VI compounds have been used, depending on the wavelength range to be detected. Nevertheless, silicon is the most common and important semiconductor in the integrated circuit technology, but it has an indirect band gap inhibiting optical functions. Silicon sensors are usually used in the visible-to near infrared (VIS-NIR) range. However, some commercial Si sensors have been enhanced to detect in the ultraviolet (UV) range, but they have one or more of the following weaknesses: very expensive, reduced responsivity in the VIS-NIR range, lack of compatibility with IC processes and a complex technology. So, most of the available materials for UV detection are not silicon but compound semiconductors (Hwang & Lin, 2005). Then, many works have been done to study Si-based optoelectronics materials to overcome the drawback of silicon. Silicon rich oxide (SRO) is one of such materials. SRO is a variation of silicon oxide, in which the content of silicon is changed. The main characteristic to enhance the UV detection in this material is the formation of silicon nanostructures. Among the different techniques to synthesize Si-nps in SRO films, the low pressure chemical vapour deposition (LPCVD) technique offers the films with the best luminescence properties, compared with those obtained by plasma enhanced CVD and silicon implantation into SiO_2 films (Morales et al., 2007).

The optical and electrical properties of SRO-LPCVD films have been extensively studied as a function of the silicon excess, temperature and time of thermal annealing (Morales et al., 2007, Zhenrui et al., 2008). Silicon nanoparticles (Si-nps) with different sizes in SRO-LPCVD films have been observed by transmission electron microscope (TEM) technique (Luna et al., 2009). Depending on the excess of Si content, SRO possesses some special properties such as charge trapping, carrier conduction and luminescence. Some novel devices have been proposed using these properties (Zhenrui et al., 2008, Luna et al., 2006, Berman et al., 2008). In Luna et al., 2006, a photodetector using SRO and an induced PN junction were presented, and it was shown that even when the active area is covered by opaque aluminium the photocurrent is considerable. However, it is not clear the reason of such a high response. Also, Aceves et al. proposed (Aceves et al., 1999) that an induced pn junction could be used as a photodetector and that if high resistivity silicon is used a high photoresponse could be expected. However, this idea has not been corroborated.

In this chapter, optical properties as absorption coefficient and optical band gap of SRO with Si-nps and different silicon excess were also studied in order to be able to apply them in optical photodetectors. A photodetector made of a simple Al/SRO/Si MOS-like grid structure is reported. The devices fabricated on high resistivity substrate possess high responsivity in photodetection from UV to VIS-NIR radiation. This structure requires few fabrication process steps and is completely compatible with the complementary metal-oxide-semiconductor (CMOS) technology. The possible mechanisms involved in the light detection and the extended photosensitivity of this structure are analysed and discussed.

2. Experimental procedure

The experimental procedure is divided in two parts. The first part is the following: SRO films were deposited on silicon substrates in a horizontal LPCVD hot-wall reactor using nitrous oxide (N_2O) and silane (SiH_4) as the reactant gases at 700°C. The partial pressure ratios Ro = $[N_2O]/[SiH_4]$, were Ro = 10, 20, and 30. The thicknesses of the SRO films were about 500 nm. Silicon Excess in these films are about 12.7, 8.0 and 5.5 at.% for Ro = 10, 20 and 30, respectively (Luna et al., 2007). After deposition, the films were densified at 1000 °C for 30 minutes in nitrogen atmosphere, and then some of them were thermally annealed at 1100°C for 180 minutes in nitrogen atmosphere.

The XRD measurements were done using a Bruker AXS D8 Discover diffractometer with Cu Kα radiation. The surface morphology of SRO films was studied using a nanosurf easy Scan AFM system version 2.3, operated in contact mode. A 4 × 4 μm^2 scanned area was used for each topographic image, and a 450-µm-long single-crystal Si cantilever operated at 12 kHz (type vista probes CL contact mode AFM probes) was used. Five different scans were done for each sample, showing good reproducibility. AFM images were analysed using scanning probe image processor (SPIP) software (Jogensen, 2002).

For TEM measurements, an energy-filtered TEM (EFTEM) was used. The images were obtained using an electronic microscope (JEOL JEM 2010F); all the EFTEM images were measured using a Si plasmon of 17 eV and cross-section views. EFTEM enables us to detect amorphous Si nanoclusters embedded in an oxide matrix (Spinella et al, 2005). Also, an FEI Tecnai F30 high-resolution TEM (HRTEM) with acceleration voltage of 300 kV and line resolution of 0.2 nm was used to study the microstructure of the Si nanocrystals.

Room-temperature photoluminescence (PL) of the SRO films was measured using a Perkin-Elmer spectrometer LS-50B model with a xenon source and a monochromator. The samples were excited using 250-nm radiation, and the emission signal was collected from 400 to 900 nm with a resolution of 2.5 nm. A cut-off filter above 430 nm was used to block the light scattered from the source.

Optical transmittance measurements were made in the UV range to near infrared at room temperature, using a spectrophotometer Perkin-Elmer LMBD 3B UV/VIS.

In the second part of the experimental procedure SRO films were deposited on n-type Silicon (100) substrates with resistivity of 2000-5000 Ω-cm and with a N+ implanted region on the back side. SRO layers were deposited in a horizontal LPCVD hot wall reactor using SiH_4 (silane) and N_2O (nitrous oxide) as reactive gases at 700 °C. The gas flow ratio Ro was used to control the amount of silicon excess in the SRO films. Ro = 10, 20, and 30, corresponding to a silicon excess from 12 to 5 %, were used for this experiment. After deposition, the samples were thermally annealed at 1000° C in N_2 atmosphere for 30 minutes. Aluminium grids with area of A = 0.073 cm^2 were patterned on the SRO surface by evaporation and standard photo-lithography. Figures 1(a) and 1(b) show a cross-section view scheme, with an approximation at only 3 grid electrodes and an image of the surface of the fabricated devices. The surface Al grid electrodes have several fingers with 40 μm width and distanced at 205 μm, as observed in figure 1(b). It is important to remark that the zones between fingers are covered by the SRO film.

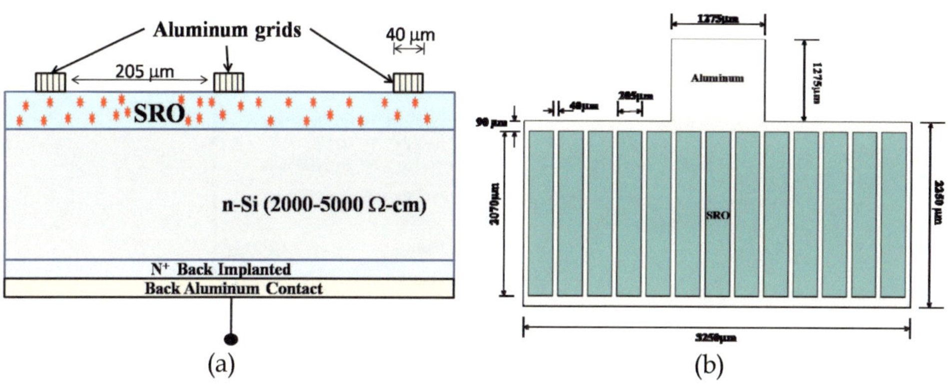

Fig. 1. Scheme of the Al/SRO/Si MOS-like device, (a) a cross-section view scheme of the fabricated devices with an approximation at only 3 grid electrodes and (b) dimensions of a device and of the grids electrodes.

Ellipsometric measurements were made with a Gaertner L117 ellipsometer to obtain the thickness and refractive index of the SRO films before annealing, whose values are shown in Table 1.

Ro	Refractive index	Thickness (Å)
10	1.78 ± 0.01	720 ± 28
20	1.55 ± 0.03	755 ± 25
30	1.46 ± 0.01	591 ± 3

Table 1. Refractive index and thickness of SRO films.

Current versus voltage (I-V) measurements were performed at room temperature, illuminated with UV or white light and under dark conditions, using a computer controlled Keithley 6517A electrometer. The voltage sweep was done with a rate of 0.1 V/s. Illumination was performed with an UV lamp (UVG-54, 5 to 6 eV approximated range) and a white light lamp (1.7 to 4 eV approximated range) with an output power of 4.3 mW/cm^2 and 3.2 mW/cm^2, respectively. The power of lamps was measured using a radiometer (International Light, USA. IL1 400A). On the other hand, other current versus voltage (I-V) measurements were performed at room temperature using a computer controlled Keithley 236 source-meter, under illuminations at specific wavelengths of 400, 650 and 1000 nm conditions, with a Sciencetech 9040 monochromator. The measured power were 83.6 $\mu W/cm^2$, 238.9 $\mu W/cm^2$ and 11.5 $\mu W/cm^2$ for 400, 650, 1000 nm wavelengths, respectively.

3. Results

3.1 Structural, optics and photoelectric results

Figure 2 shows the XRD patterns of SRO films with Ro = 10 and 20, annealed at 1100 °C for 180 minutes on silicon substrate. Peaks at 21, 28.5° and 47.4° are observed for SRO films with Ro = 10 and they are ascribed to SiO_2, Si (111) and (220), respectively. However, those peaks at 28.5° and 47.4° disappeared for Ro = 20. As Ro becomes higher (Ro = 30), the diffraction peaks were not observed (not shown here).

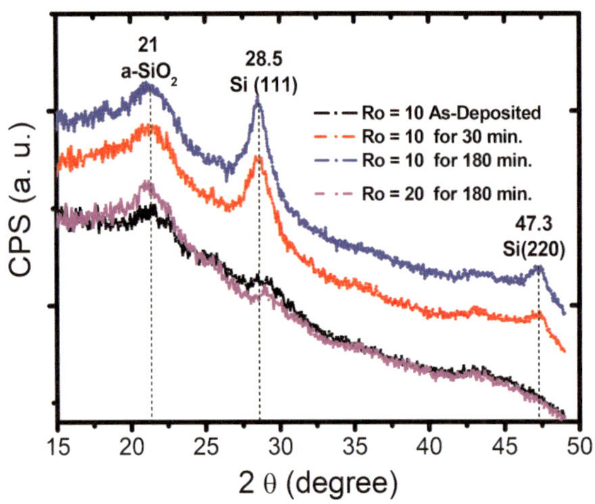

Fig. 2. XRD patterns of SRO films with Ro = 10 and 20 annealed at 1100 °C for 180 minutes on silicon substrate.

The average crystal size (D), dislocation density (δ) and microtension (ε) of the silicon nanocrystals (Si-ncs) for Ro = 10 were estimated from the Si (111) peak using the Scherrer formula (Karunagaran, B., et al., 2002, Comedi,D., et al., 2006,) and are listed in Table 2. δ and ε decreased when the annealing time increases for SRO_{10} films, which is related to a reduction on the defect concentration and imperfections of the nanocrystal cell (Yu et al, 2006).

		Samples of SRO on silicon			
SRO films	2θ (°)	[hkl]	(D) (nm)	δ (10^{12} cm^{-2})	ε (10^{-3})
10TT30	28.65	Si [111]	4.6	4.6	8.0
10TT180	28.55	Si [111]	5.4	3.4	7.0
20TT180	29.1	Si [111]	2.8	12.2	13.0

Table 2. Structural parameters of the SRO Films obtained from the XRD measurements for Si (111) peak.

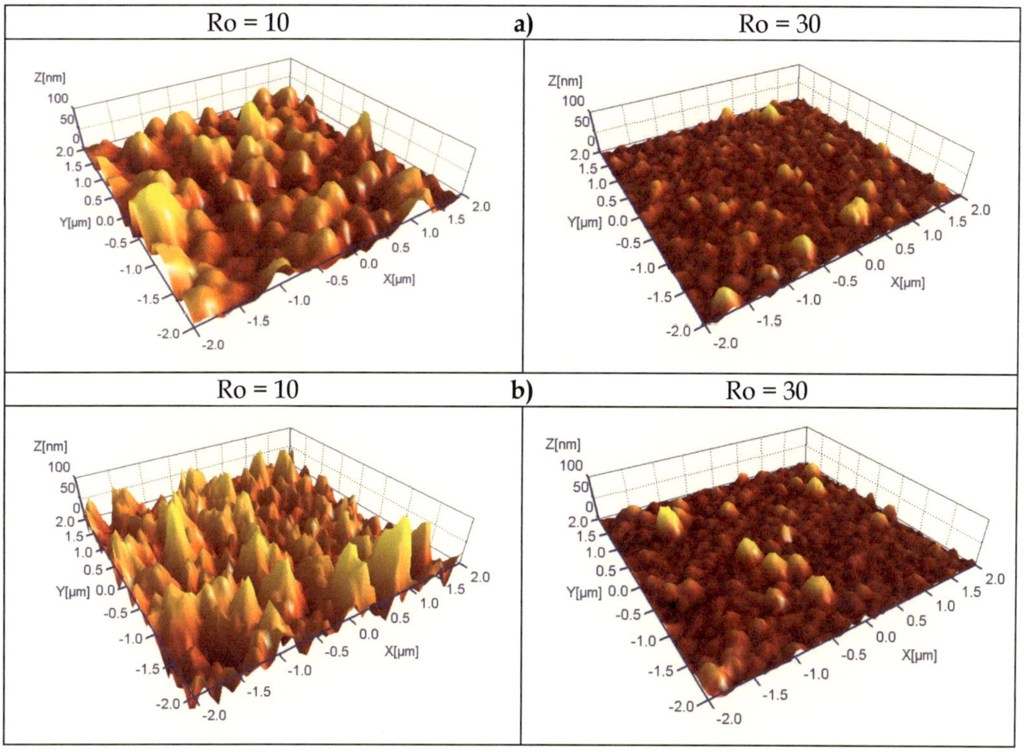

Fig. 3. 3D AFM images of a) as-deposited and b) thermally annealed at 1100 °C SRO_{10} and SRO_{30} films on silicon. Scanned area: 4 × 4 µm².

Figure 4 shows a comparison of the average roughness <Sa> for the SRO films with different Si excess for as-deposited, densified and annealed films.

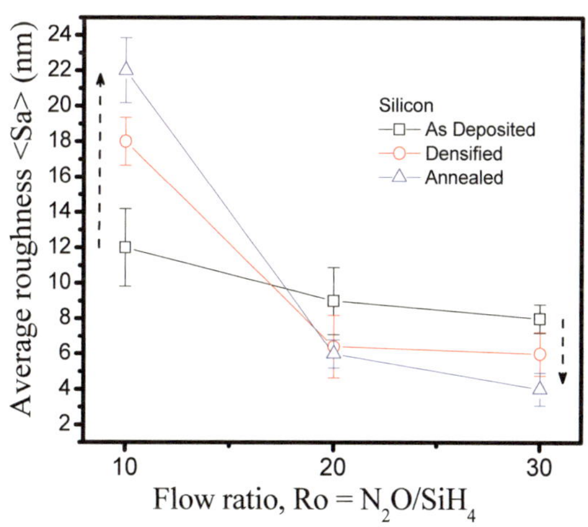

Fig. 4. Average roughness <Sa> as a function of the flow ratio (Ro) for as-deposited, densified, and thermally annealed SRO films deposited on silicon (lines are plotted as an eye-guide). Scanned area: 4 × 4 µm².

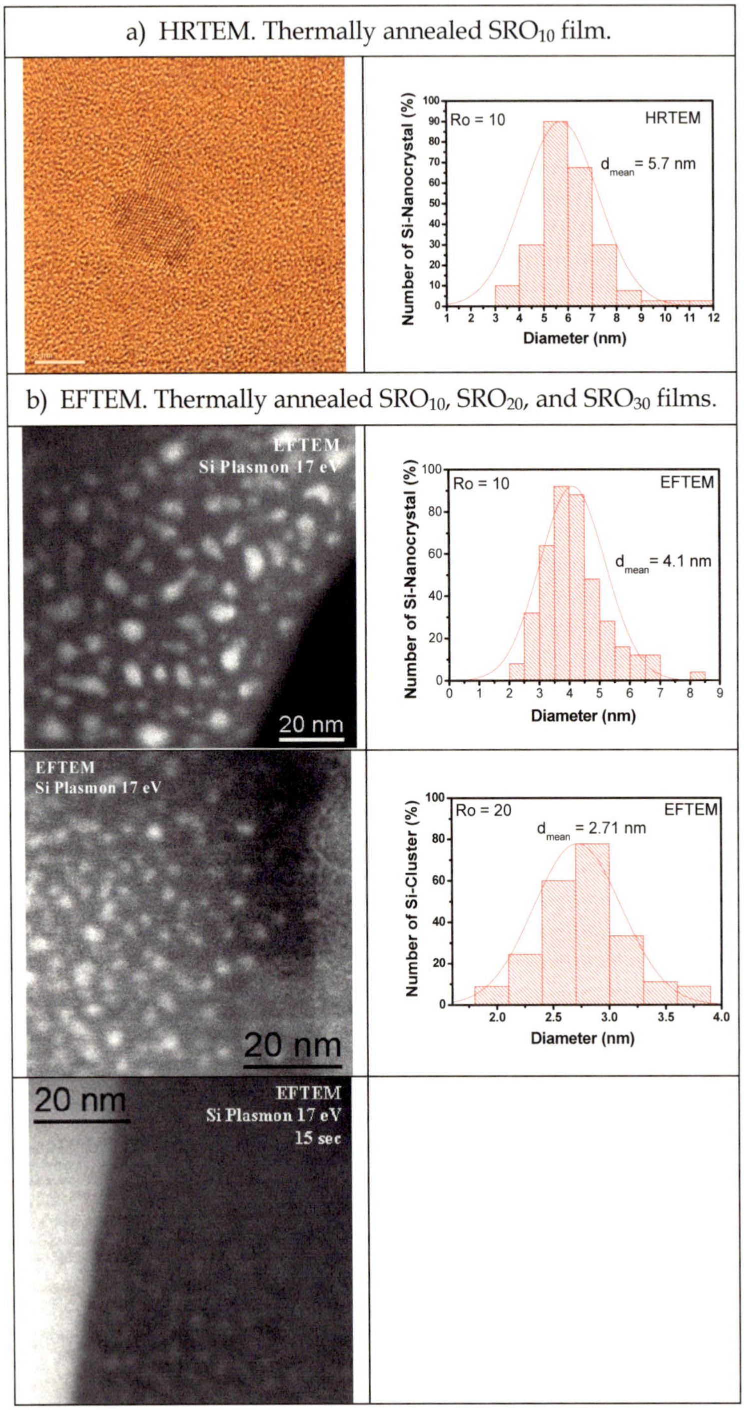

Fig. 5. Silicon nanocrystals in SRO_{10} films and silicon clusters in SRO_{10}, SRO_{20}, and SRO_{30} films were observed by a) HRTEM and b) EFTEM, respectively, after thermal annealing at 1100 °C.

Figure 3 displays three-dimensional (3D) AFM images of as-deposited and thermally annealed SRO films with different excess silicon. All images exhibit a rough surface. For as-deposited SRO films, the roughness decreases with increasing Ro, as shown in Fig. 3 (a). After thermal annealing, roughness increased for SRO films with Ro = 10, as depicted in Fig. 3 (b); for Ro = 20 and 30, no significant influence is visible from 3D images.

Figure 5 shows the HRTEM and EFTEM images of the annealed SRO films. For Ro=10, HRTEM and EFTEM images clearly show Si nanocrystals (Si-ncs) with a mean size of ~ 5.7 nm, as seen in Figures 5 (a) and 5 (b). Si-ncs were not observed for Ro = 20 and 30. However, bright zones associated with amorphous silicon clusters were observed by EFTEM on SRO films with Ro = 20 and 30, as shown in Fig. 5 (b). Si clusters with a mean size of ~ 2.7 nm were observed for Ro = 20, whereas for Ro = 30, because of low contrast and sharpness, the shape was not very clear, but Si clusters with size ≤ 1 nm could be identified.

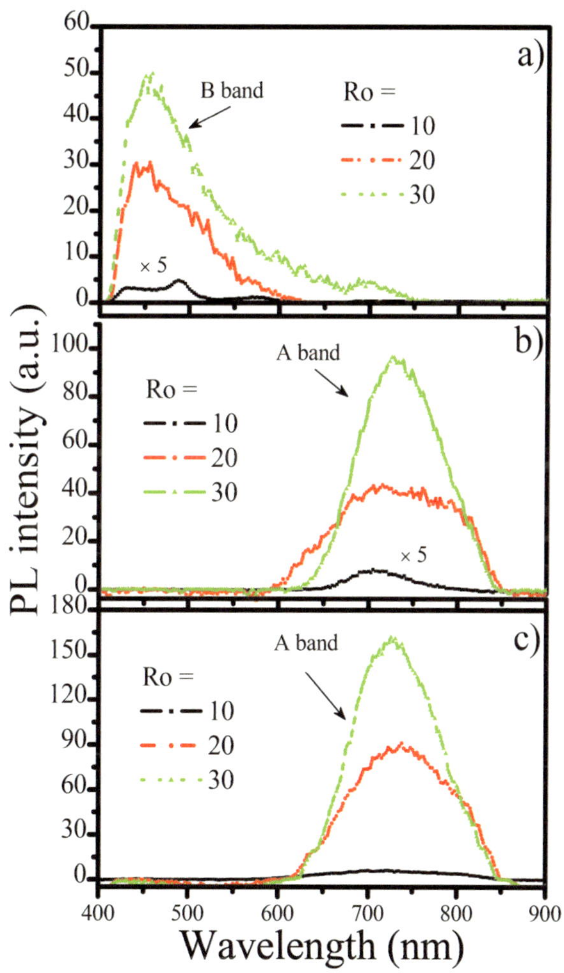

Fig. 6. PL spectra of SRO films with Ro = 10, 20 and 30 and a) as- deposited, b) densified at 1000 °C for 30 min, and c) thermally annealed at 1100° C for 180 minutes deposited on silicon substrate.

Figures 6 (a)-6 (c) show the PL spectra of the as-deposited, densified, and annealed SRO films, respectively. Two PL bands can be observed: B band (400-600 nm) and A band (600-850 nm). For as-deposited SRO films, the intensity of the B band increases with Ro, as shown in Fig. 6 (a). After densification, the B band disappears and the A band becomes visible [see Fig. 6 (b)]; prolonged annealing at higher temperature of 1100 °C for 180 min enhanced its intensity, as shown in Fig. 6 (c). SRO films with large Ro (low excess Si) always show higher PL intensity. On the other hand, the PL peak position did not show any significant shift with Ro.

Transmittance spectra were measured to obtain the optical band gap of SRO films. Transmittance spectra for Ro = 10, 20 and 30 films deposited on sapphire are shown in Figure 7. As can be observed, the transmittance of all these films is relatively high (> 80 %) between 600 and 900 nm, and reduces to zero for wavelengths below 600 nm. The annealing time does not produce a clear variation; however, the Ro causes a clear shift of the curves towards lower wavelengths. These samples have different behaviours, in agreement with the Ro value and annealing time.

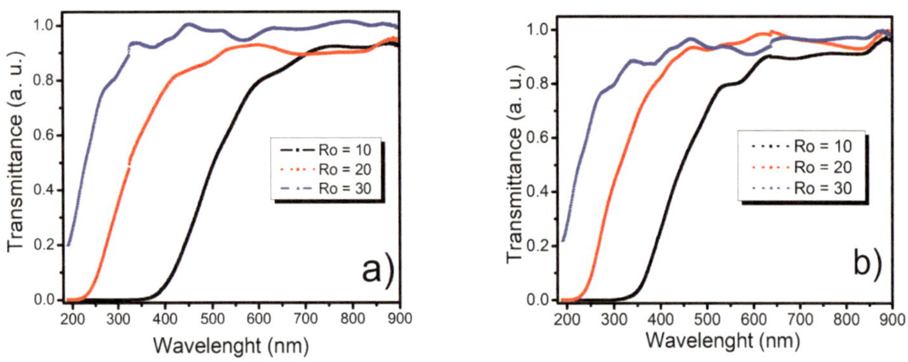

Fig. 7. UV/Vis transmittance spectra for Ro = 10, 20, and 30 on sapphires substrates. a) densified at 1000 °C and b) annealed at 1100 °C.

The absorption coefficients were determined from transmission spectra of Figure 7, and they are shown in Figure 8. The absorption coefficients show different behaviour with the Ro values.

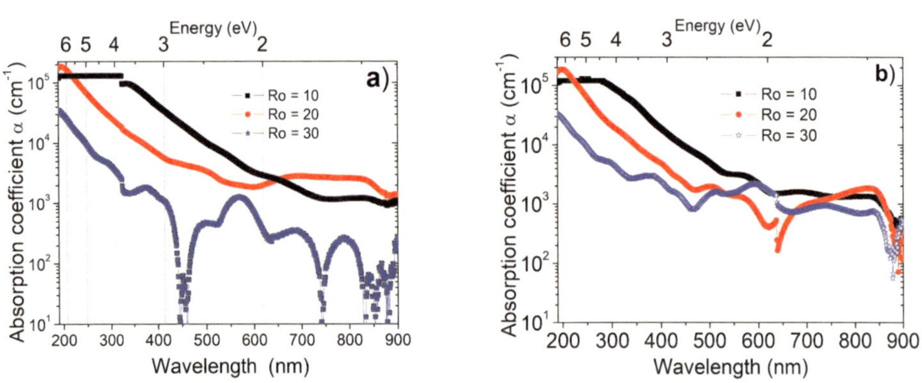

Fig. 8. Absorption coefficients in the range of the U-V to NIR for a) densified and b) thermally annealed SRO films.

The fundamental absorption edge in most semiconductors follows an exponential law (Pankove, 1975). The absorption coefficient α is correlated with the transmittance T and the reflectance R of a sample with thickness d through the relation:

$$T \approx (1-R)^2 e^{-\alpha d} \tag{1}$$

The absorption coefficient α shown in Figure 8 was estimated as:

$$\alpha(\hbar\nu) = \frac{-\ln(T(\hbar\nu))}{d} \tag{2}$$

From the Tauc law:

$$(\alpha\hbar\nu)^{1/2} \propto (\hbar\nu - E_g) \tag{3}$$

Graphs of $(\alpha\hbar\nu)^{1/2}$ versus photon energy ($\hbar\nu$) and the corresponding optical band gap are shown in Figure 9.

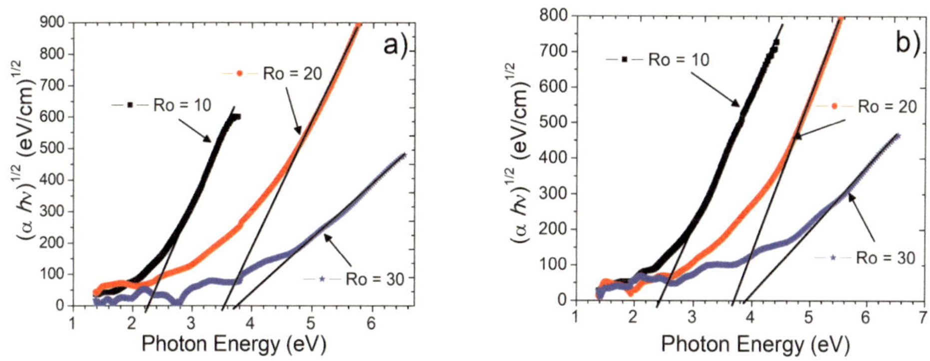

Fig. 9. $(\alpha\hbar\nu)^{1/2}$ vs. photon energy a) densified and b) thermally annealed SRO films with different Ro. Straight lines are used to estimate the optical band gap.

It is observed that the position of the absorption edge moves towards higher energy when Ro increases. The values obtained for the optical band gap are listed in Table 3.

Ro	Band gap (Eg)			
	Densified (1000°C)	Annealed at 1100°C		
		30 min	60 min	180 min
10	2.28±0.04	2.4±0.04	2.4±0.02	2.43±0.04
20	3.52±0.03	3.57±0.03	3.6 ±0.06	3.69 ±0.03
30	3.76 ±0.07	3.73 ±0.06	3.86±0.04	3.89±0.04

Table 3. Energy of the optical bang gap obtained from the Tauc's method for SRO films with different Ro values and times of thermal treatment.

The optical band gap increases with Ro. In this case, as the silicon excess reduces the absorption coefficient also reduces, as shown in Figure 8. As Ro moves towards low values, the band gap moves towards that of the bulk silicon; consequently, it is expected that the properties of an indirect semiconductor be kept.

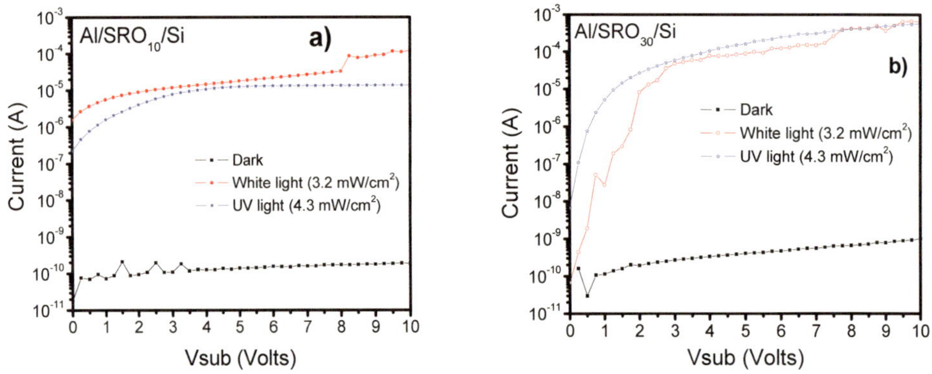

Fig. 10. I-V curves of the Al/SRO/Si MOS-like grid structure for a) SRO_{10} and b) SRO_{30} under dark and illumination conditions.

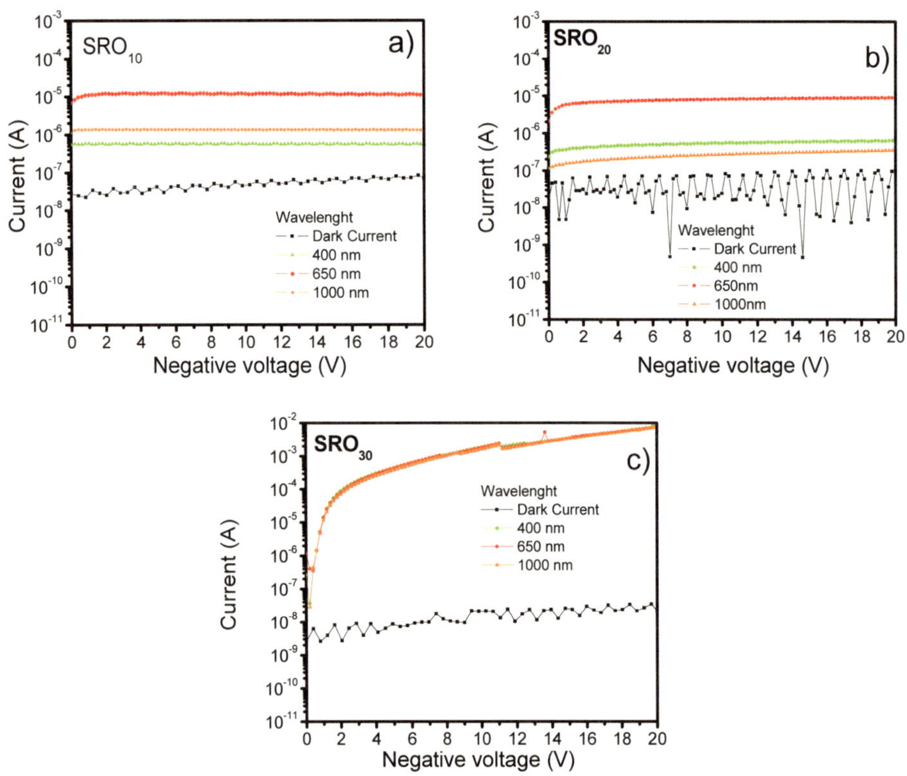

Fig. 11. Current-voltage characteristics of the Al/SRO/n-Si MOS-like structures with a) SRO_{10}, b) SRO_{20}, and c) SRO_{30} films under dark and illuminated with 250, 400, 650, 1000 nm wavelengths.

MOS-like structures were fabricated and measured. Figures 10(a) and 10(b) show, respectively, the typical I-V characteristics of devices with Ro = 10 and 30 at the surface inversion condition (negative voltage at the gate respect to the substrate), in dark and under UV and white light illumination. The dark current is on the order of 10^{-10} A, indicating a low leakage current in these structures. However, a large photocurrent was obtained when the structures were illuminated, this fact indicative of a high optical sensitivity in this simple structure. Similar photo-response was obtained with structures made with SRO of Ro = 20.

I-V characteristics of devices with Ro = 10, 20 and 30 at the surface inversion condition and under illumination at specific wavelengths are shown in Figures 11(a), 11(b) and 11(c), respectively. The order of the dark current is 10^{-8} A indicating a low leakage current in these structures. This current value changed from that obtained in Figure 10 because of the fact that different devices were measured and these ones have a bigger dark current. Nevertheless, at both cases a large photocurrent was obtained when the structures were illuminated. Therefore, these simple structures show high optical sensitivity. Different photocurrents were obtained for the different Ro's in each wavelength, being the SRO structures with Ro = 30 which have the highest photo-response.

4. Analysis and discussion

The results show that SRO films properties depend on the silicon excess and the annealing conditions. XRD results showed that in SRO films with thermal annealing, diffusion of the silicon excess took place creating silicon nanoparticles indicated mainly by the peaks, and their size depends upon the silicon excess in the SRO films. Diffractograms on Figure 2 show crystalline orientations (111), (220) and (311) of the nc-Si. It is clear that, when the silicon excess is higher, the agglomerated Si is present and with thermal annealing the generation of nc-Si is obtained, the diffusion coefficient of Si in the SRO at 1100 °C was calculated as $\sim 1 \times 10^{-16}$ cm^2s^{-1}. Diffraction peaks indicate the formation of nc-Si and it should be interpreted as a phase separation between nc-Si and SiO_2 induced by thermal annealing.

Si-ncs sizes obtained by XRD are similar to those observed by HRTEM and EFTEM. For Ro = 10, HRTEM images clearly show Si-ncs with a mean size of 5.7 nm but Si-ncs were not observed for Ro = 20 and 30. However, bright zones associated with amorphous silicon clusters were observed by EFTEM on SRO films with Ro = 20 and 30. Si clusters with a mean size of 2.7 nm were observed for Ro = 20, whereas for Ro = 30 the shape was not very clear, because of low contrast and sharpness, but Si clusters with 1 nm size could be identified.

AFM images of the SRO films on silicon substrates show that the shape and size of the roughness change with Ro. For SRO films with Ro = 10, the roughness increased after thermal annealing. However, for SRO with Ro = 20 and 30, the roughness reduced, as shown in Figure 4. For Ro = 10 the roughness can be as high as a maximum of 24 nm. However, for Ro = 30 the surface looks smooth and the maximum height is about 5 nm. Then, for Ro = 10, the roughness observed through AFM can be associated to Si-ncs. Nevertheless, for SRO films with Ro = 20 and 30 the roughness observed can be ascribed to silicon compounds, or silicon clusters, rather than to Si-ncs. Therefore, elemental Si, SiOx, and SiO_2 phases separate after annealing, and depending on the silicon excess, one of those phases can be dominant.

Si clusters in the SRO films are produced by diffusion of the Si excess at high temperature. That is, when SRO is annealed, the silicon particles diffuse creating silicon agglomerates around a nucleation site. If the Si excess is high enough, the Si clusters will be crystallized forming Si-ncs. However, when the Si excess reduces, Si clusters are amorphous due to the large lattice mismatch between SiO_2 and Si. If the Si excess is low enough, there are no pure Si clusters created, instead, Si-O compounds are formed in SRO films. Then a more homogeneous material is obtained. Of course, these two mechanisms are not exclusive each other, and they can exist simultaneously. However, one of them will dominate depending on the silicon excess.

On the other hand, there are different factors that have an important impact on the optical properties of the SRO films. Therefore, it is necessary to understand these factors for designing materials to be used in photodetectors. In our case, factors such as substrate type, silicon excess, annealing temperature and composition are very important for the optical properties of SRO films. PL also depends on Ro and thermal-annealing conditions. In Figure 6, all the as-deposited SRO films show a weak PL band at 400–600 nm (B band). After thermal annealing, the B band disappears and another PL band (A band) dominates. The intensity of the A band increases with the time and annealing temperature. The PL property of SRO has been extensively studied in literature. Two major mechanisms for PL in this kind of materials are generally accepted: quantum confinement effects in the Si-ncs and defect-related effects. In our samples, intense PL (A band) in SRO_{30} was observed; however, SRO_{30} has silicon compounds rather than Si-ncs. Therefore, the A band cannot be associated with quantum confinement effects; instead, it should be associated with Si-related defects. As reported previously (Lee et al., 1979, Lin et al., 2005), the B band is also associated with F and E defects and different types of oxygen vacancies. Although there are many reports about the origin of PL, most of them relates PL with the bulk microstructures of SRO. A very few studies report on the correlation between the PL and the surface morphology of the SRO films. Other authors (Torchynska et al, 2002) have reported that the diameter and area of the grains on the porous silicon film surface have an important influence on the PL spectra. In our case, SRO is not porous Si but it is considered to be Si nanoclusters embedded in an oxide matrix with very stable optical and electrical properties—quite different than porous silicon. However, from AFM images, the surface is completely irregular, with bumps of different heights and shapes. So, these bumps can be treated as grains on the surface. The grains vary in size depending on the Ro value and annealing time. They can then be related to the emission, as it was done for porous silicon. The height of the roughness, equivalently to the height of the grains, was already related to the silicon cluster size (Luna et al, 2007).

It is observed that the main influence on the structural and optical properties is due to the Ro value of the SRO films, their thickness, annealing time and substrate type. In XRD and AFM measurements important changes in the structure were observed, which indicates that the substrate has influence on the structural properties of the deposited SRO films. In the PL spectra the unusual high intensity of the A band is remarkable. The explanation for this observation is that different oxidation mechanisms lead to different termination of the tetrahedron SiO_4 and Si_4 of the silicon oxide and silicon, respectively. Si can be terminated with different combinations of $Si\text{-}Si_{4-n}O_n$, compounds (with n = 0 to 4), it can be terminated by double bond to an oxygen atom (Si=O), oxygen vacancies or by a bond to a hydroxyl group (Si-O-H).

Taking these results into account, it seems that the types of radiative centres producing the luminescence in SRO films deposited on silicon substrate are similar in the A band, but different in the B band. Therefore, depending on the substrate type, specific structural defects are formed in the SRO films. The Si content in the SRO system to form the phase (crystalline or amorphous) and the structure (crystal size and morphology) of the Si-nps, and especially of their surrounding $Si\text{-}SiO_2$ interface, has significant influence on the optical properties. Also, the optical band gap tends towards the optical band gap of the Bulk silicon by increasing the silicon excess, and with a lower silicon excess the optical band gap will tend towards that of the SiO_2.

The large variety of factors that can be used to influence the structural and optical properties makes these materials very interesting for optoelectronics applications. A shift of the optical band gap is possible by varying the silicon excess. Also, the PL emission can be tuned by thermal annealing, silicon excess and substrate type.

Therefore, MOS-like structures were designed to obtain high photocurrent. The results show that Al/SRO/Si grid structures have high photocurrent at different wavelengths. The large photocurrent is due to the photocarriers generation in the SRO films and also in the Si substrate. The mechanism of the carrier transport in the SRO layers can be via tunneling (Yu et al, 2008). High density of Si-nps and defects in the SRO films create conduction paths and then the photogenerated carriers can move through them allowing a large photocurrent.

This simple structure presents high optical sensitivity for UV and the visible to near infrared range (from 250 nm to 1000 nm), as depicted in Figures 10 and 11. The UV response of the SRO layer has been demonstrated in different reports, apparently the silicon nanoparticles (embedded or not) are in some way sensible to the UV. In (Nayfeh et al, 2004) the authors use silicon nanocrystals (not embedded) to develop an UV sensor in the 250 nm to 350 nm range. Also, our group has developed an UV-Vis-NIR sensor using SRO (with embedded Si nps) (Berman et al., 2008). Moreover, it is well known that red emission is obtained when SRO is irradiated with UV. So, UV response of the SRO could cooperate in two ways to the observed response of the devices reported in this work. First, the red photoluminescence produced by the SRO impinges into the silicon, which has a high sensitivity to these wavelengths. Second, the SRO itself has demonstrated some conduction properties under illumination, so this current component could be added to the photocurrent (Luna et al, 2010).

Now, the high photoresponse can only be explained if the whole device is considered photosensible. That includes the components due to the SRO (already discussed), and the components in the silicon. As Figure 12 shows, there are two zones in the silicon substrate where the light could produce electron-hole pairs (EHP): (1) under the Al fingers and (2) in the exposed regions between the fingers. Due to the bias of the device, under the Al fingers the silicon surface is in inversion (layers of holes formed), forming an induced-PN junction. However, owing to the fact that this PN junction is covered by an opaque Al layer, very few photons will produce photocurrent; then the current component due to this layer should be small. So, the PN junctions could act as a collector of the minority carrier produced somewhere else, and then contribute to the photocurrent through the SRO that allows the electrons to move from the substrate to the Al electrode.

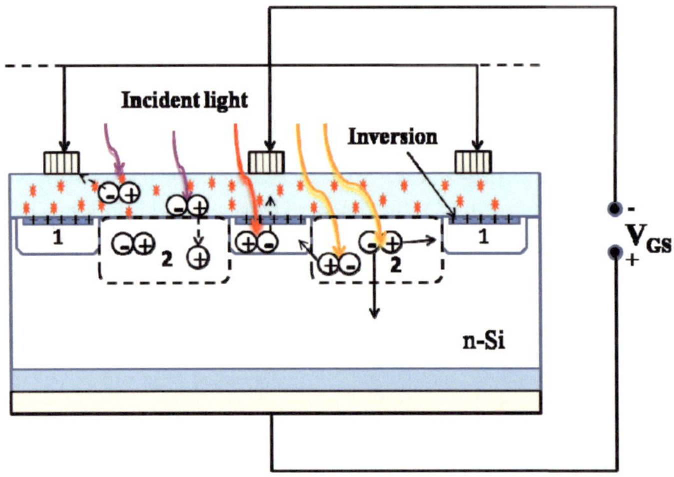

Fig. 12. A cross-section scheme of the zones in the SRO films and silicon high resistivity substrate where the light could produce electron hole-pairs (EHP). (1) Under the Al fingers the silicon surface is in inversion due to the bias of the device, in this case only a few photons arrive to the edge of the region, and (2) in the SRO films in the Al uncover between the fingers. The dash lines marks the maximum length where a photon can reach.

On the other hand, the SRO traps charge both positive and negative, so that the surface in this zone could be in accumulation if the SRO is positively charged, in inversion if the SRO is negatively charged, or neutral if there is no charge in the SRO. The resistivity of the silicon substrates used for this experiment is 2000-5000 ohms-cm, so small quantities of trapped charge could bias the surface. Moreover, the charge could get to be trapped during the fabrication process, especially during the high temperature process.

The inversion or accumulation of the surface produces channels where the carriers could move toward the fingered zone where they contribute to the photocurrent. However, in our devices the dark current is very low ($< 10^{-8}$ A) meaning that the channels do not exist.

Then, the only possibility to increase so much the photocurrent is that the uncovered neutral region under illumination produces electron-hole pairs able to arrive at the Al covered area. As known, only the EHP produced within a diffusion length, $L=\sqrt{D\tau}$ (D is the diffusivity and τ is the lifetime), will arrive at the PN junctions. In this case, only the holes which are the minority carriers are significant. The diffusivity can be estimated from the mobility using the Einstein's relationship. Also, the resistivity of the silicon substrate used here is very high and the mobility (510 cm^2/Vs) does not vary much for resistivities greater than 100 Ω-cm (Cronemeyer, 1957). Therefore, a conservative value of the diffusivity is 13.184 cm^2/s (Grove, 1967). Härkönen et al. (Härkönen et al., 2006) studied the lifetime of high resistivity silicon under low and high injection levels of carriers. They found that for high resistivity silicon the recombination lifetime is longer than 1 ms. Just for reference in our group, generation lifetime of high resistivity silicon was estimated to be between 1 and 2 ms (Luna et al., 2005). Then, a value of diffusion of 0.114 cm, or 1140 μm is calculated.

Now, for UV radiation (200 - 300 nm) the absorption coefficient is approximately, $1x10^6$ cm^{-1}, and the absorption depth is 0.01 μm (Sze, 2006). For other wavelengths, for example around 600 nm, the absorption depth is about 5 μm. In spite of fact that the absorption is quite superficial for UV, most of the generated holes reach the induced PN junctions due to their long diffusion length, and in general this is true for all the detection range: UV-Vis-NIR. So, the high sensitivity observed in this device is due to a large silicon volume where all the generated carriers contribute to the photocurrent and, as a complement, some photo-effects in the SRO also contribute to the photocurrent.

The structure with Ro = 30 produces the highest current. However, it is well known that the lower Ro values increase the SRO conductivity, so if the current goes through the SRO, the more conductive structure will be that with Ro = 10. On the other hand, it is also known that silicon excess of 5% (for Ro = 30) produces the highest PL emission (Morales et al., 2005, Luna et al., 2009), and then the more sensible to radiation, where the PL (red emission) generates e-h pares in the induced PN junction. Then, these two mechanisms have to be further studied to determine which one is more adequate to improve the photoresponse of our sensor.

5. Conclusion

The influence of different factors on the structural (Si-nc, a-Si, Si-nanocluster, roughness) and optical properties (PL, absorption coefficients, optical band gap) of the SRO films has been discussed from experimental data. PL increased with the annealing time and decreased with a greater amount of silicon excess. In our case, it has been discussed that PL is not due to Si-ncs. Si-clusters are the main nanoparticles that produce maximum PL due to the different type of defects. The formation and growth of Si-ncs and the formation of a-Si in samples with different silicon excess after different annealing times were investigated with XRD and HRTEM. Interestingly, it was found that the optical band gap strongly correlates with the silicon excess and it is shifted over a large energy range via the reactants ratio.

These interesting properties of the SRO films were used to fabricate MOS-like structures, and an Al/SRO/Si MOS-like grid structure was fabricated and tested as photodetector. Silicon-rich oxide films with different silicon excesses were used as dielectric layers and deposited on silicon substrates of very high resistivity. This structure shows high optical sensitivity for the whole UV-Vis-NIR range. The high photocurrent results from three different components. One of them is due to photo-carriers generated in the SRO films; another one comes from the red light that is photo generated in the SRO, as mentioned above. This red light impinges into the high resistivity silicon producing electron-hole pairs. The third component comes from the electron-hole pairs generated in the silicon by the light absorbed in the silicon volume. In spite of the fact that the UV radiation produces electron-hole pairs near the silicon surface, they contribute to the photocurrent because there is not a death zone. Due to the high resistivity of the silicon a long diffusion length is obtained, and the photo generated minority carriers are able to reach the induced PN junctions under the Al grids.

6. Acknowledgment

This work has been partially supported by CONACyT-154725, PROMEP-231 and VIEP-BUAP-2011. The authors acknowledge technicians of INAOE laboratory for their help in the samples preparation and measurements.

7. References

Aceves, M., O. Malik, V. Grimalsky, Application of silicon Rich Oxide Films in New Optoelectronic Devices, Physics and Chemistry of Solid State 2004, 5(2), pp. 234-240.

Aceves, M., W. Calleja , C. Falcony, J. A. Reynoso–Hernández, The Al/Silicon Rich Oxide/Si P-N Induced Junction As A Photodetector, Revista Química Analítica, 1999, 18, 102.

Ashkan, B.; Jason, J.; Yongho, C.; Leila, N.; Gunham, M.; Wu, Z.; Andrew, G.; Kapur, P. ; Kriehna, C. & Ant, U. (2008). Metal-Semiconductor-metal photodetectrors based on single-walled carbon nanotube film-GaAs Schottky contacts, J. Appl. Phys., Vol.13, (June 2008), pp. 114315, ISSN 0021-8979

Berman D. M., M. Aceves-Mijares, L. R. Berriel-Valdos, J. Pedraza, A. Vera-Marquina, Fabrication, characterization, and optimization of an ultraviolet silicon sensor, Optical Engineering, 2008, 47(10), 104001.

Chen Y., T. Cheng, . Cheng, C. Wang, C. Chen, C. Wie, Y. Chen, Highly sensitive MOS photodetector with wide band responsivity assisted by nanoporous anodic aluminum oxide membrane, Optics Express 2010, 18(1), pp. 56-62.

Comedi, D., O.H.Y. Zalloum, E. A. Irving, J. Wojcik, T. Roschuk, M. J. Flynn, and P. Mascher, X-Ray-Diffraction study of crystalline Si nanocluster formation in annealed silicon-rich silicon oxides, J. Appl. Phys. 99, 023512 (2006).

Comedi, D., Zalloum, O.H.Y., Irving, E. A., Wojcik, J., Roschuk, T., Flynn, M. J., and Mascher, P., (2006), X-Ray-Diffraction study of crystalline Si nanocluster formation in annealed silicon-rich silicon oxides, J. Appl. Phys. 99, 023512 (2006), ISSN 0021-8979

Di Maria, D. J., Kirtley, J. R., Pakulis J., Dong, D. W., Kuan, D., Pesavento, F. L., Theis, T. N., Cutro, N., Brorson, S. D.; (1984). Electroluminescence studies in silicon dioxide films containing tiny silicon islands, J. Appl. Phys.,Vol. 56(2) (1984), pp. 401-416, ISSN 0021-8979.

DiMaria, D. J., Dong, , Pesavento, D. W., F. L.; Enhanced conduction and minimized charge trapping in electrically alterable read-only memories using off-stoichiometric silicon dioxide films, J. Appl. Phys., 1984, 55(8), pp. 3000-3019, ISSN 0021-8979.

Donald C., Cronemeyer, Hall and Drift Mobility in High-Resistivity Single-Crystal Silicon, Phys. Rev., 1957, 105(2), 522.

Foster J., J. K. Doylend, P. Mascher, A. P. Knights, and P. G. Coleman, Optical attenuation in defect-engineered silicon rib waveguides, J. Appl. Phys., 2006, 99, pp. 073101, ISSN 0021-8979.

Grove, A. S., Physic and technology of semiconductor devices, John Wiley and sons, USA 1967.

Harkonen, J., E. Tuovinen, Z. Li, P. Luukka, E. Verbitskaya, V. Eremin, Recombination lifetime characterization and mapping of silicon wafers and detectors using the microwave photoconductivity decay (mPCD) technique, Materials Science in Semiconductor Processing, 2006, 9, 261.

Hwang, J. & Lin, C. (2005). Gallium nitride photoconductive ultraviolet sensor with a sputtered transparent indium–tin–oxide ohmic contact, Thin Solid Films, Vol.491 (May 2010), pp. 276-279, ISBN 842-6508-233.

Jøgensen, J. F., The Scanning Probe Image Processor (SPIP), Denmark, 2002, www.imagemet.com.

Karunagaran, B., Rajendra, R. T., Mangalaraj, D., Narayandass, K., Mohan, G., (2002), Influence of thermal annealing on the composition and structural parameters of DC magnetron sputtered titanium dioxide thin films, Cryst. Res. Technol., Vol. 37 (12), 2002, pp. 1285-1292.

Luna López, J. A., A. Morales Sanchez, M. Aceves-Mijares, Z. Yu, C. Dominguez, Analysis of surface roughness and its relationship with photoluminescence properties of silicon-rich oxide films, J. Vac. Sci. Technol. A, 2009, 27(1), 57.

Luna-López, A., M. Aceves-Mijares, O. Malik, R. Glaenzer, Low-and high-resistivity silicon substrate characterization using the Al/silicon-rich oxide/Si structure with comparison to the metal oxide semiconductor technique, J. Vac. Sci. Technol. A, 2005, 23 (3), 534-538.

Luna-López, J. A., Aceves M., O. Malik, Z. Yu, A. Morales, C. Dominguez and J. Rickards, Revista Mexicana de Física, S 53(7), 293 (2007).

Luna-López, J. A., M. Aceves-Mijares and O. Malik, Optical and electrical properties of silicon rich oxide films for optical sensor, Sensors and actuators A, 2006, 132, 278.

Luna-López, J. A., M. Aceves-Mijares, A. Morales-Sánchez, J. Carrillo-López, Photoconduction in silicon rich oxide films obtained by LPCVD, J. Vac. Sci. Technol. A 28(2), Mar/Apr 2010.

Morales A., J. Barreto, C. Domínguez, M. Riera, M. Aceves and J. Carrillo, (2005), Comparative study between silicon-rich oxide films obtained by LPCVD and PECVD, Physica E: Low-dimensional Systems and Nanostructures, 2007, 38(1), 54.

Nayfeh, O. M., S. Rao, A. Smith, J. Therrien, and M. H. Nayfeh, Thin Film Silicon Nanoparticle UV Photodetector, IEEE Photonics Technology Letters, 2004, 16(8), 1927.

Pankove, J. I., Optical Processes in Semiconductors, (Dover New York, 1975), Chap. 4-6.

Sabnis, V. A., Demir, H. V., Fidaner, O, Zheng, J. F., J S Harris, D A B Miller, N Li, T C Wu, H T Chen, Y M Houng (2005), Intimate Monolithic Integration of Chip-Scale Photonics Circuits, IEEE J. selected Topics in Quantum Electronics 2005, 11(6), pp. 1255.

Spinella C., C. Bongiorno, G. Nicotra, and E. Rimini, Appl. Phys. Lett. 87, 044102 2005.

Sze, S. M., Physics of semiconductor devices, John Wiley and sons, USA, (2006).

Torchynska, T. V., J. Appl. Phys. 92, 4019 2002.

Yu Z., M. Aceves-Mijares, J. A. Luna Lopez, J. Deng, Nanocrystalline Si-based metal-oxide-semiconductor, Proc. SPIE 7381 2009, pp. 7381 1H.

Yu Z., Mariano Aceves-Mijares, A. Luna-López, Enrique Quiroga and R. López Estopier, "Photoluminescence and Single Electron Effect Of Nanosized Silicon Materials". In:

Focus on Nanomaterials Research, Editor: B. M. Carota, ISBN 1-59454-897-8 Nova Science Publishers, Inc., 2008, 233.

Yu, Z., M. Aceves-Mijares, K. Monfil, R. Kiebach, R. Lopez-Estopier, J. Carrillo, Room temperature current oscillations in naturally grown silicon nanocrystallites embedded in oxide films, J. Appl. Phys., 2008, 103, 063706.

3

Silicon Photodetectors Based on Internal Photoemission Effect: The Challenge of Detecting Near-Infrared Light

Maurizio Casalino, Luigi Sirleto, Mario Iodice and Giuseppe Coppola
Istituto per la Microelettronica e Microsistemi, Consiglio Nazionale delle Ricerche, Naples Italy

1. Introduction

Silicon Photonics has emerged as an interesting field due to its potential for low-cost optical components integrated with electronic functionality. In the past two decades, there has been growing interest in photonic devices based on Si-compatible materials (Kimerling et al., 2004; Jalali & Fathpour, 2006) in the field both of the optical telecommunications and of the optical interconnects. In this contest, tremendous progresses in the technological processes based on the use silicon-on insulator (SOI) substrates have allowed to obtain reliable and effectiveness full complementary metal-oxide semiconductor (CMOS) compatible optical components such as, low loss waveguides, high-Q resonators, high speed modulators, couplers, and optically pumped lasers (Rowe et al., 2007 ; Vivien et al., 2006 ; Xu et al., 2007 ; Michael et al., 2007; Liu et al., 2007 ; Liu et al., 2006). All these devices have been developed to operate in the wavelength range from C optical band (1528 - 1561 nm) to L optical band (1561 - 1620 nm). However one of the crucial steps toward the integration of photonics with electronics resides in the development of efficient chip-scale photodetectors (PD) integrated on Si. Bulk photodetectors are perhaps the oldest and best understood silicon optoelectronic devices. Commercial products in Si operate at wavelengths below 1100 nm, where band-to-band absorption occurs. For the realization of photodiodes integrated in photonics circuits operating at wavelengths beyond 1100 nm silicon is not the right material because its transparency. In the last years, in order to take advantage of low-cost standard Si-CMOS processing technology, a number of photodetectors have been proposed based on different physical effects, such as: defect-state absorption (Bradley et al., 2005), two photon absorption (TPA) (Liang et al., 2002) and internal photoemission absorption (Zhu et al., 2008a). Physical effects, working principles, main structures reported in literature and the most significant results obtained in recent years were reviewed and discussed in our previous paper (Casalino et al., 2010a). In this paragraph, we go into more depth on photodetectors based on the internal photoemission effect (IPE). Silicon infrared photodiodes based on IPE are not novel, in fact PtSi/p-Si, Ir/p-Si and Pd_2Si/p-Si junctions are usually used in the infrared imaging systems (Kosonocky et al., 1985). The main advantages of these devices resides in their extremely high switching speed and in their simple fabrication process, but, due to high background current density these devices can only work at cryogenic temperature.

However, in recent years IPE has emerged as a new option for detecting also near infrared (NIR) wavelengths at room temperature. Unfortunately, the photoemission quantum efficiency is low compared to that of detectors based on inter-band absorption and this limits the application both in power monitoring and in the telecommunication field. Low quantum efficiency is a direct result of conservation of momentum during carrier emission over the potential barrier, in fact, the majority of excited carriers which do not have enough momentum normal to the barrier are reflected and not emitted. Moreover as incoming photons can excite carriers lying in states far below the Fermi energy, which can not overcome over the metal-semiconductor potential barrier, the quantum efficiency of these devices is further decreased. In recent years new approaches and structures are proposed in order to circumvent these limitations.

In this paragraph an overview of the state of the art of NIR all-Si photodetectors based on IPE is presented. First, the physical effects of IPE and the main figures of merit of devices based on IPE are elucidated. Then, the main structures reported in literature, starting from historical devices for imaging application at infrared wavelengths, up to new NIR devices which could be adapted for telecommunications and power monitoring, are described in detail. Finally the most significant results obtained in the last years are reviewed and discussed comparing the performances of devices based on different approaches.

2. IPE-based device performances

We hereafter present a theoretical background useful to clarify the physics behind the working principle of devices and we analyze crucial points affecting device performances.

2.1 Device efficiency and IPE theory

IPE is the optical excitation of electrons in the metal to energy above the Schottky barrier and then transport of these electrons to the conduction band of the semiconductor. The standard IPE theory is due to Fowler (Fowler, 1931). However, the Fowler's theory was obtained without taking into account the thickness of the Schottky metal layer. In 1973, Archer and Cohen (Archer & Cohen, 1973) proposed thinning the electrode to increase the emission probability. The enhancement of IPE in thin metal film was theoretically investigated by Vicker who introduce a moltiplicative factor to the Fowler's formula (Vickers, 1971). However the resulting electrode was so thin that it was semitransparent producing low metal absorbance. More recently Casalino et al. proposed to improve the absorbance in NIR photodetector based on IPE by using a microcavity Fabry-Perot (Casalino et al., 2008a, 2010b). Finally a further enhancement of IPE can be obtained due to image force effect which modify the Schottky barrier under a reverse bias favouring IPE. This effect can be taken into account by adding a barrier collection efficiency term. The resulting quantum efficiency of a photodetector based on IPE, can be written by the formula (Casalino et al., 2008b):

$$\eta = A_T F_e P_E \eta_c \tag{1}$$

where A_T is the total optical absorbance of the metal, F_e is the Fowler' factor, P_E is the Vicker's factor and η_c is the bias dependent barrier collection efficiency. Very often, device efficiency is described in term of responsivity, a very important property of a detector indicating the current produced (I_{ph}) by a certain optical power (P_{OPT}). Responsivity is strictly linked to a device's quantum efficiency by the formula:

$$\Re = \frac{I_{ph}}{P_{OPT}} = \frac{\lambda[nm]}{1242}\eta \tag{2}$$

Reasonable responsivities are necessary for an acceptable signal-to-noise ratio and to ease the design and realization of the amplifier circuitry that follows.

2.1.1 Standard fowler's theory of IPE

IPE is the optical excitation of electrons into the metal to an energy above the Schottky barrier and then transport of these electrons to the conduction band of the semiconductor (Fig. 1).

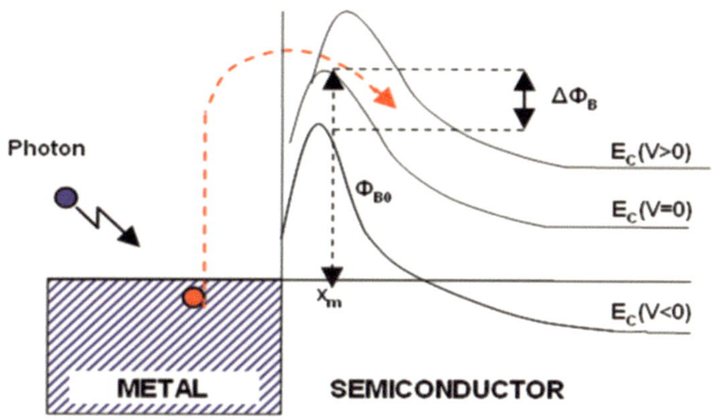

Fig. 1. Energy band diagram for a metal/n-semiconductor junction. "Reprinted with permission from M. Casalino *et al.*, "A silicon compatible resonant cavity enhanced photodetector working at 1.55 µm," Semicond. Sci. Technol., 23, 075001, 2008 (doi: 10.1088/0268-1242/23/7/075001). IOP Publishing is acknowledged."

The standard theory of photoemission from a metal into the vacuum is due to Fowler (Fowler, 1931). In a gas of electrons obeying the Fermi-Dirac statistic, if photon energy is close to potential barrier ($h\nu \approx \Phi_B$), the fraction (F_e) of the absorbed photons, which produce photoelectrons with the appropriate energy and momenta before scattering to contribute to the photocurrent, is given by:

$$F_e = \frac{\left[(h\nu - (\phi_{B0} - \Delta\phi_B))^2 + \frac{(kT\pi)^2}{3} - 2(kT)^2 e^{-\frac{h\nu-(\phi_{B0}-\Delta\phi_B)}{kT}} \right]}{8kTE_F \log\left[1 + e^{\frac{h\nu-(\phi_{B0}-\Delta\phi_B)}{kT}}\right]} \tag{3}$$

where hν is photons energy, Φ_{B0} is the potential barrier at zero bias, $\Delta\Phi_B$ is the lowering due to image force effect (as we will see later), E_F is the metal Fermi level, k is the Boltzmann's constant and T is the absolute temperature. As it is possible to see in Eq. 3, F_e is strongly depending from the potential barrier height of the metal-semiconductor interface.

2.1.2 IPE enhancement in thin metal films

In order to study the quantum efficiency for thin metal films, the theory must be further extended, taking into account multiple reflections of the excited electrons from the surfaces of the metals film, in addition to collisions with phonons, imperfections and cold electrons. Assuming a thin metal film, a phenomenological, semiclassical, ballistic transport model for the effects of the scattering mechanisms resulting in a multiplicative factor for quantum efficiency was developed by Vickers (Vickers, 1971) and recently reviewed by Scales *et al.* (Scales & Berini, 2010). According to this model the accumulated probability P_E that the electrons will have sufficient normal kinetic energy to overcome potential barrier is given by:

$$P_E \cong \frac{L_e}{d}\left[1 - e^{-\frac{d}{L_e}}\right]^{\frac{1}{2}} \tag{4}$$

where d is the metal thickness and L_e the metal mean free path. For example, as it can be seen by plotting Eq. 4 in Fig. 2 (assuming a copper mean free path of 45 nm (Chan et al., 1980a), the lower metal thickness, the higher P_E.

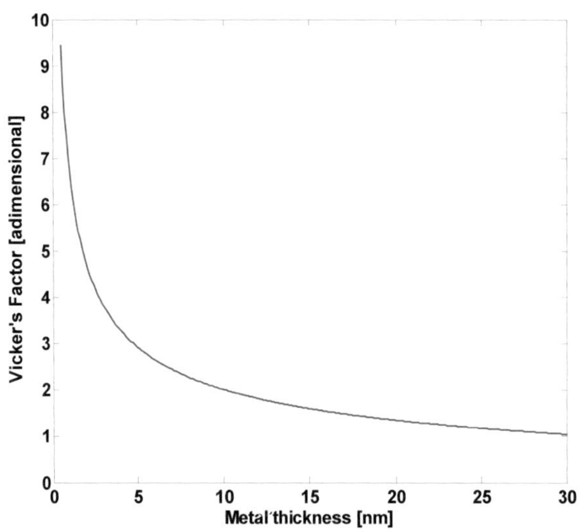

Fig. 2. Vicker's factor P_E versus metal thickness for copper metal.

It is worth noting that in a recent work (Scales & Berini, 2010), Scales and Berini show that a further enhancement of this probability emission can be obtained in structures realized with thin metal film buried in a semiconductor and forming two Schottky barriers.

2.1.3 IPE enhancement by optical cavity

The enhancement of IPE in thin metal film lead to a resulting electrode so thin to be semitransparent, producing low metal absorbance. In order to increase the absorbance in thin metal film, the use of a microcavity Fabry-Perot has been proposed. The sketch of proposed photodetector is shown in Fig. 3. The resonant cavity is a Fabry–Perot vertical to-the-surface structure. It is formed by a distributed Bragg reflector (DBR), a metallic top mirror and in between a Si cavity. A dielectric layer on top of the metal is generally considered for thin metal protection purpose.

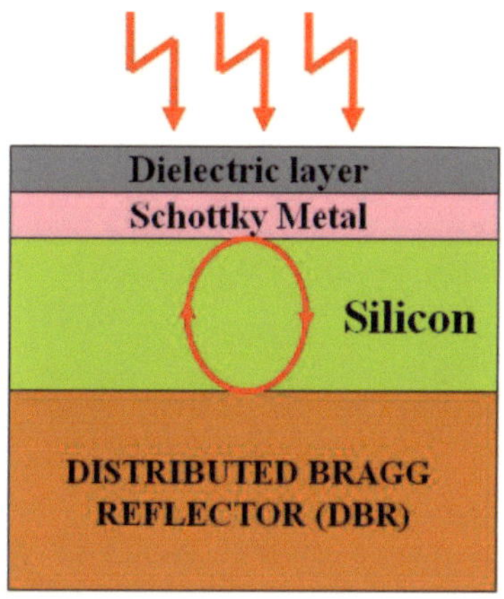

Fig. 3. Sketch of IPE photodetector based on optical cavity.

The proposed structure of Fig. 4 can be modelled by the multilayer shown in Fig. 3 and absorbance calculation can be carried out by the Transfer Matrix Method (TMM) (Muriel & Carballar, 1997).

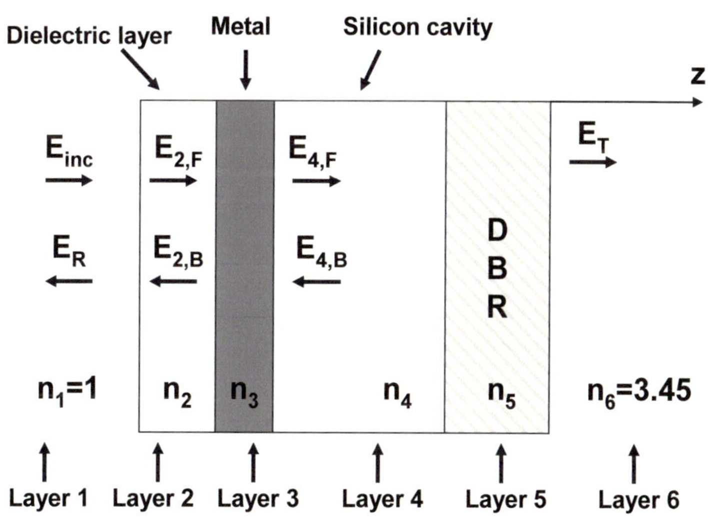

Fig. 4. Multilayer schematizing device in fig. 3 for absorbance calculation by TMM.

Normal incidence condition and the unidimensional variations of refractive index (n) along the propagation direction z, are taken into account. Let E_{inc} and E_R be the electric field complex amplitude of the incident and reflected waves at the interface between the air and the dielectric layer, and E_T the waves transmitted by final plane of structure, by TMM we obtain:

$$\begin{bmatrix} E_{inc} \\ E_R \end{bmatrix} = M_{TOT} \begin{bmatrix} E_T \\ 0 \end{bmatrix}$$
$$E_T = \frac{1}{M_{TOT_{11}}} E_{inc} \qquad E_R = \frac{M_{TOT_{21}}}{M_{TOT_{11}}} E_{inc} \tag{5}$$

where $M_{TOT_{ij}}$ are the element of the matrix M_{TOT} representing the whole multilayer:

$$M_{TOT} = M_{1/2} \cdot M_2 \cdot M_{2/3} \cdot M_3 \cdot M_{3/4} \cdot M_4 \cdot M_{4/5} \cdot M_5 \cdot M_{5/6} \tag{6}$$

where $M_{i/j}$ and M_i are the interface matrix between i and j layer and the i-layer matrix, respectively, which can calculated by knowing refractive index (n_i) and thickness (d_i) of every ith layer:

$$M_{i/j} = \frac{1}{2n_i} \begin{pmatrix} n_i + n_j & n_i - n_j \\ n_i - n_j & n_i + n_j \end{pmatrix} \qquad M_i = \begin{pmatrix} e^{jk_0 n_i d_i} & 0 \\ 0 & e^{-jk_0 n_i d_i} \end{pmatrix} \tag{7}$$

Let $E_{2,F}$ ($E_{4,F}$) and $E_{2,B}$ ($E_{4,B}$) be the frequency domain electric field complex amplitudes of the forward and backward travelling plane waves in the layer 2 (4), by TMM we obtain:

$$\begin{bmatrix} E_{2,F} \\ E_{2,B} \end{bmatrix} = M_A \begin{bmatrix} E_T \\ 0 \end{bmatrix}$$
$$E_{2,F} = \frac{M_{A_{11}}}{M_{TOT_{11}}} E_{inc} \qquad E_{2,B} = \frac{M_{A_{21}}}{M_{TOT_{11}}} E_{inc} \tag{8}$$

and

$$\begin{bmatrix} E_{4,F} \\ E_{4,B} \end{bmatrix} = M_B \begin{bmatrix} E_T \\ 0 \end{bmatrix}$$
$$E_{4,F} = \frac{M_{B_{11}}}{M_{TOT_{11}}} E_{inc} \qquad E_{4,B} = \frac{M_{B_{21}}}{M_{TOT_{11}}} E_{inc} \tag{9}$$

where M_A is the matrix calculated from interface between protection coating and metal to the final plane, while M_B is the matrix calculated from interface between metal and silicon cavity to the final plane,

$$\begin{aligned} M_A &= M_{2/3} \cdot M_3 \cdot M_{3/4} \cdot M_4 \cdot M_{4/5} \cdot M_5 \cdot M_{5/6} \\ M_B &= M_4 \cdot M_{4/5} \cdot M_5 \cdot M_{5/6} \end{aligned} \tag{10}$$

The optical power is linked at the electrical field by:

$$P = \frac{n}{2\eta_0} |E|^2 \tag{11}$$

where η_0 is the vacuum characteristic impedance of electromagnetic waves. The total power going in (P_{input}) and going out (P_{output}) from the metal is given by:

$$P_{input} = \frac{n_2}{2\eta_0}\left|E_{2,F}\right|^2 + \frac{n_4}{2\eta_0}\left|E_{4,B}\right|^2 = \left(n_2\left|\frac{M_{A_{11}}}{M_{TOT_{11}}}\right|^2 + n_4\left|\frac{M_{B_{21}}}{M_{TOT_{11}}}\right|^2\right)\frac{\left|E_{inc}\right|^2}{2\eta_0}$$
$$P_{output} = \frac{n_2}{2\eta_0}\left|E_{2,B}\right|^2 + \frac{n_4}{2\eta_0}\left|E_{4,F}\right|^2 = \left(n_2\left|\frac{M_{A_{21}}}{M_{TOT_{11}}}\right|^2 + n_4\left|\frac{M_{B_{11}}}{M_{TOT_{11}}}\right|^2\right)\frac{\left|E_{inc}\right|^2}{2\eta_0} \quad (12)$$

and metal absorbance can be written as (Casalino, 2006a, 2006b):

$$A_T = \frac{P_{input} - P_{output}}{P_{inc}} = \left\{\left(n_2\left|\frac{M_{A_{11}}}{M_{TOT_{11}}}\right|^2 + n_4\left|\frac{M_{B_{21}}}{M_{TOT_{11}}}\right|^2\right) - \left(n_2\left|\frac{M_{A_{21}}}{M_{TOT_{11}}}\right|^2 + n_4\left|\frac{M_{B_{11}}}{M_{TOT_{11}}}\right|^2\right)\right\} \quad (13)$$

where n_2 and n_4 are the refractive indices of the 2th and 4th layer, respectively.

In a work of 2006, Casalino *et al.* (Casalino et al., 2006a, 2006b) proposed a methodology in order to design an optimum device, in fact with a right choice of the multilayer thicknesses a maximum in absorbance can be obtained. In the device proposed by the authors DBR is considered formed by alternate layers of Si and SiO_2 having refractive index 3.45 and 1.45, and thickness of 340 nm and 270 nm, respectively, while a metallic layer of gold, having refractive index n_{Au}=0.174+j9.96 (Chan & Card, 1980b), mean free path L_e=55 nm (Chan et al., 1980a), Fermi level E_F=5.53 eV (Yeh, 1988) and thickness d=32 nm, was chosen as Schottky contact. We point out that in the simulations, semi-infinite first and last layers of air (n_1=1) and silicon (n_6=3.45), respectively, are considered. In Fig. 5 absorbance plotted against wavelengths is reported for the proposed device without and with DBR, showing a significant enhancement in absorbance, due to the optical cavity, of almost two order of magnitude at 1550 nm.

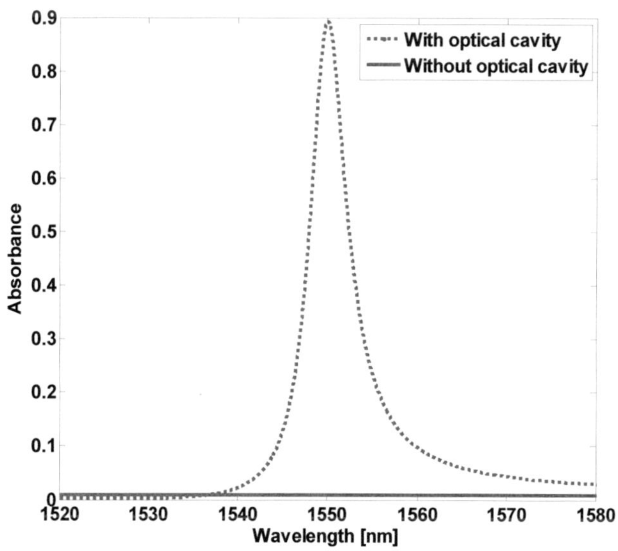

Fig. 5. Absorbance versus wavelengths for the device proposed by Casalino et al. (Casalino et al., 2006a, 2006b).

2.1.4 IPE enhancement by reverse voltage applied

In order to complete the transaction of the IPE theory, the image force between an electron and the metal surface must be taken into account. Due to the image force effect a lowering ($\Delta\Phi_B$) and displacement (x_m) of the metal-semiconductor interface potential barrier, is provided. These barrier lowering and displacement are given by (Sze, 1981):

$$\Delta\phi_B = \sqrt{\frac{q}{4\pi\varepsilon_{Si}}\frac{|V_{Bias}|}{W}} \qquad x_m = \sqrt{\frac{q}{16\pi\varepsilon_{Si}}\frac{W}{|V_{Bias}|}} \tag{14}$$

where ε_{Si} is the permittivity of silicon (10^{-12} C/cmV), W is the depletion width and V_{Bias} the applied bias voltage.

It is worth noting that while the potential barrier lowering can be taken into account by Fowler's theory (Eq. 3), the potential barrier displacement influences the probability that an electron undergoes scattering phenomena in Si travelling from the metal-semiconductor interface to the Schottky barrier maximum. This probability can be taken into account by the barrier collection efficiency (η_c):

$$\eta_c = e^{-\frac{x_m}{L_s}} \tag{15}$$

where L_s is the electron scattering length in the silicon. It is worth noting that increasing the bias voltage, a shift of Schottky barrier closer to metal/semiconductor interface is obtained. Therefore, the barrier collection efficiency increases. The improvement of the device efficiency by increasing reverse voltage has been theoretically investigated by Casalino et al. (Casalino et al., 2008b).

2.2 Bandwidth of IPE-based devices

The electrical properties of diodes based on Schottky junctions are determined by majority carrier phenomena, while for p-n diodes they are primarily determined by minority carriers. Therefore, the Schottky diodes can be switched faster because there are no minority carrier storage effects. The response time of Schottky barrier photodiodes can be determined by three parameters: the diffusion time in the quasi neutral region; the electrical frequency response or RC time required to discharge the junction capacitance through the resistance and the transit time across the depletion region. By designing the diode in such a way that depleted region length (W) equals to the device length (L), under suitable revere bias applied, the diffusion time can be neglected and the total frequency (f_{tot}) can be written as:

$$\frac{1}{f_{tot}} = \frac{1}{f_{tr}} + \frac{1}{f_{RC}} \tag{16}$$

where f_{tr} and f_{RC} are the transit time and time constant limited 3-dB bandwidth, respectively.

According to the model for small signal shown in Fig. 6 (Casalino et al., 2010b), the photodetector can be schematized as current generator (I_{ph}), resistance (R_j), capacitance (C_j) associated to the junction, series resistance (R_s) and load resistance (R_{load})

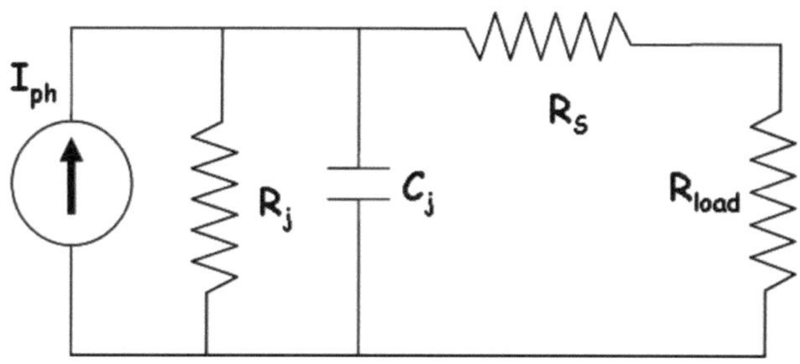

Fig. 6. Small-signal circuit associated to IPE-based device.

The RC limited bandwidth is given by:

$$f_{RC} = \frac{1}{2\pi\left[(R_s + R_{load}) // R_j\right]C_j} = \frac{1}{2\pi R_{Tot} C_j} \tag{17}$$

where R_j can be evaluated from the inverse derivative of the reverse current-voltage (I-V) electrical characteristic (tipical values are in the MΩ range) and R_s can be extracted from the forward I-V characteristic.

For high speed applications, the load resistance R_L, tipically 50 Ω, is much lower than series resistance (R_s) and junction resistance (R_j). Therefore, the device 3-dB frequency, becomes (Donati, 1999):

$$f_{3dB} = \frac{1}{2\pi R_s C_j} \tag{18}$$

Where the junction capacitor is linked to the photodetector area (A_{Ph}) from the following formula:

$$C_j = \frac{\varepsilon_{Si} A_{Ph}}{W} \tag{19}$$

with ε_{Si} is the silicon dielectric constant (10^{-12} C/cmV). It is worth noting that when the detector area is made sufficiently small, the influence due to the capacitance is reduced, the effect of the transit time dominates and high speed operation can be reached.

The intrinsic carrier-transit time limited 3-dB bandwidth for the device is given by (Donati, 1999):

$$f_{tr} = \frac{0.44}{t_d} = 0.44\frac{v_t}{L} \tag{20}$$

where v_t is the effective carrier saturation velocity (10^7cm/s in Si) and L is the carrier transit distance. In Fig. 7, the frequency due to RC time constant and transit time are reported as a function of the detector area for R_L= 50 Ω an W=1 μm. It is worth noting that when the detector area is made sufficiently small, i.e. smaller than A_{Ph}<65μm^2, the influence due to the capacitance is reduced of one order to magnitude with respect to the transit time, enabling device to work at GHz range.

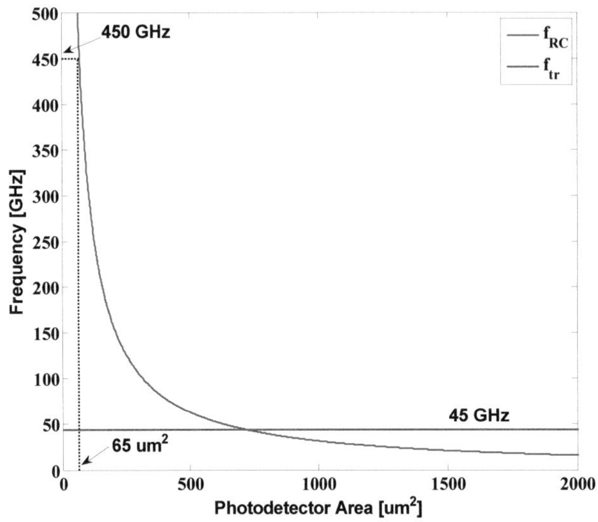

Fig. 7. Frequency due to the RC time and the transit time plotted against area detector for RL=50 Ω and W=1μm.

2.3 Noise and sensitivity of IPE-based devices

It is well known that the photodetector output is affected by a noise contribution which hinders the device sensitivity. The r.m.s. noise is due to two contributions: Johnson noise and shot noise. Johnson is the thermal noise associated to the resistance R_{load} and its r.m.s value can be defined as (Donati, 1999):

$$i_J = \sqrt{\frac{4kBT}{R_{load}}} \tag{21}$$

where R_{load} is the load resistance, k is the Boltzmann constant and T is the absolute temperature. On the other hand, the shot noise is associated to the discrete nature of the total current, i.e., the sum of the signal current and dark current (Donati, 1999):

$$i_S = \sqrt{2q(I_{ph}+I_d)B} \tag{22}$$

where q is the electron charge.

Because the two contributions are statistically independent their m.s. values can be added in order to get the total noise:

$$i_n = \sqrt{i_S^2 + i_J^2} = \sqrt{2q(I_{ph}+I_d)B + \frac{4kBT}{R}} \tag{23}$$

From Eq. 23, it should be clear that shot noise can always dominate the Johnson with a right choice of R_{load}. Dark current of infrared photodetectors based on the IPE effect is composed by the inverse saturation current (I_s) and by a background current (I_{bg}). The background current is due to the fact that at first approximation an infrared photodetector at a temperature T, can be view as a blackbody emitting a radiance r(λ) according to the Planck law (Donati, 1999):

$$r(\lambda) = \frac{2hv^2}{\lambda^3}\left[\frac{1}{e^{\frac{hv}{kT}} - 1}\right] \tag{24}$$

where hv and λ are the photon energy and wavelength, respectively. It could be shown that this radiance is proportional to a dc background current (I_{bg}) limiting infrared photodetector sensitivity. In order to get higher sensitivity, is necessary to reduce the background current contribute by reducing the emission radiance, i.e., by lowering the photodetector temperature (see Plank law). For this reason infrared detector needs to work at cryogenic temperature. On the other hand, photodetectors working at visible or NIR wavelengths are not able to detect blackbody emission and the Schottky junction saturation current limits the detector sensitivity. It is given by the Richardson-Dushamann equation (Yuan & Perera, 1995):

$$I_s = A_{ph}A^{**}T^2 e^{-\frac{\phi_B(V)}{kT}} \tag{25}$$

where $A^{**}=f_pA^*$, A^* is the Richardson constant (30 A/cm^2K^2 for p-type Si and 110 A/cm^2K^2 for n-type Si) (Sze, 1981), f_p is the barrier escape probability which as a first approximation is given by $f_p(V)=\exp(-x_m(V)/L_s)$, L_s is the electron scattering length, T is the absolute temperature and $\Phi_B(V)= \Phi_{B0}-\Delta\Phi_B(V)$ is the potential barrier which is the potential barrier at zero voltage minus the lowering due to the reverse bias. It is worth noting that both potential barrier and barrier escape probability are reverse voltage dependent due to force image effect. The plot of the saturation current density against reverse voltage applied for three different metal: gold (Au), silver (Ag) and copper (Cu), is reported in Fig. 8 (Casalino et al., 2008b). Schottky barrier used in the simulations for Au/p-Si, Ag/p-Si and Cu/p-Si interfaces are 0.78 eV, 0.78 eV and 0.58 eV, respectively. Due to the lowest potential barrier, copper shows the highest dark current density.

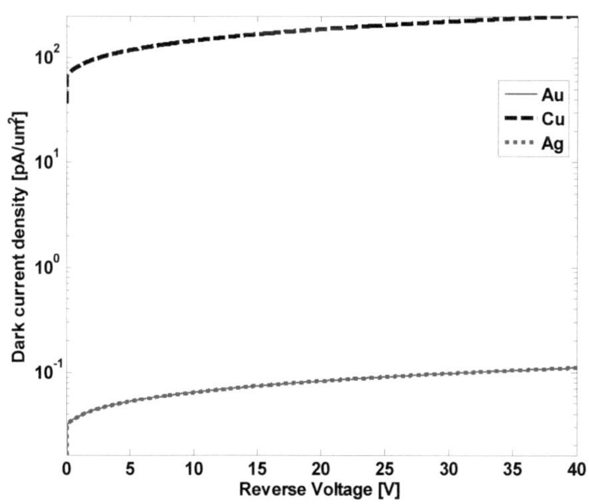

Fig. 8. Dark current density plotted against reverse voltage (semi-log scale) for three different metals: gold, silver and copper. Reprinted with permission from M. Casalino *et al.*, "A silicon compatible resonant cavity enhanced photodetector working at 1.55 μm," Semicond. Sci. Technol., 23, 075001, 2008 (doi: 10.1088/0268-1242/23/7/075001). IOP Publishing is acknowledged.

The photodetector sensitivity is directly linked to the r.m.s. noise current and can be represented in term of a well known parameter called NEP (Noise Equivalent Power) which is defined as the ratio of r.m.s. noise to responsivity (Donati, 1999):

$$\text{NEP} = \frac{i_n}{\Re} \quad [W] \tag{26}$$

NEP represents the lowest input power giving a unit signal/noise ratio. Because higher device performances correspond to smaller NEP, it is more convenient to define its inverse: the detectivity. It can be expected that the square of the r.m.s. noise (m.s. noise) is proportional to the electrical bandwidth (B) and detector area (A_{Ph}), for this reason in order to get a figure of merit not depending on these parameters, the detectivity is normalized to the square root of detector area and bandwidth (Donati, 1999):

$$\text{D} = \sqrt{A_{Ph} \cdot B}\frac{\Re}{i_n} \quad \left[\frac{cm\sqrt{Hz}}{W}\right] \tag{27}$$

3. Silicon photodetectors based on IPE

This section reviews the history and the progresses of the main structures reported in literature, starting from historical devices for imaging application at infrared wavelengths, up to new NIR devices adapted for telecommunications and power monitoring applications.

3.1 Infrared devices for imaging application

Schottky-barrier focal plane arrays (FPAs) are infrared imagers, fabricated by well-established silicon very-large-scale-integration (VLSI) process, representing one of the most effective technology for large-area high-density focal plane arrays for many near infrared (NIR, 1 to 3 μm) and medium infrared (MIR, 3 to 5 μm) applications. PtSi Schottky Barrier detectors (SBDs) represent the most established SBD technology but they must operate at 77 °K in order to reduce the dark current density in the range of a few nA/cm^2. Pd_2Si SBDs were developed for operation with passive cooling at 120 °K in the NIR band, while IrSi SBDs have also been investigated to extend the application of Schottky-barrier focal plane arrays into the far infrared (FIR, 8 to 10 μm) spectral range.

In 1973 Shepherd and Yang (Shepherd & Yang, 1973) proposed the concept of silicide Schottky-barrier detectors as much more reproducible alternative to HgCdTe FPAs for infrared thermal imaging. After being dormant for about ten years, extrinsic Si was reconsidered as a material for infrared imaging, especially after the invention of charge-coupled devices (CCDs) by Boyle and Smith (Boyle & Smith, 1970). For the first time it became possible to have much more sophisticated readout schemes and both detection and readout could be implemented on one common silicon chip. Since then, the development of the Schottky-barrier technology progressed continuously and currently offers large IR image sensor formats. These trends in IR FPA development show that the IR community today prefers more producible technologies with higher uniformity to the technology based on the narrow gap semiconductors, which still have serious material problems. Such attributes as: monolithic construction, uniformity in responsivity and signal to noise (the performance of an IR system ultimately depends on the ability to compensate the non uniformity of an FPA using external electronics and a variety of temperature references), and absence of discernible 1/f noise; make Schottky-barrier devices a formidable contender to the mainstream infrared systems and applications. (Shepherd, 1984; Shepherd, 1988; Kosonocky, 1991; Kosonocky, 1992; Shepherd, 1998). In the early years, the development of SBD FPA technology progressed from the demonstration of the initial concepts in the 1970's (Kohn et al., 1975; Capone et al., 1978; Kosonocky et al., 1978; Shepherd et al., 1979) to the development of high resolution scanning and staring devices in the 1980's and in the 1990's, that were at the basis of many applications for infrared imaging in the NIR, MIR and FIR bands. The first Schottky-barrier FPAs were made with thick Pd_2Si or PtSi detectors using about 600 Å of deposited palladium or platinum. These FPAs exhibited relatively small photoresponse. More than an order of magnitude improvement in photoresponse was demonstrated in 1980 with 50×50-element FPAs constructed with thin PtSi SBDs at the David Sarnoff Research Center (Taylor et al, 1980; Kosonocky et al., 1980). The SBDs in this FPA had an optical cavity in the form of a thin (20 - 40 Å) PtSi layer separated from an aluminium reflector by a layer of deposited SiO_2. The general concept of thin SBD with optical cavity was first described in 1973 by Archer and Cohen for SBDs in the form of Au on p-type Si (Archer & Cohen, 1973). The improved PtSi SBD structure was further developed and from 1980 to 1985 the fabrication process for PtSi SBDs was optimized at Sarnoff with the development of 32×63, 64×128 and 160×244 IR-CCD FPAs (Elabd et al., 1982; Kosonocky, 1985). Similar PtSi SBDs characteristics were also reported in the same years by Mitsubishi Corporation, Fujitsu, NEC, EG&G Reticon, Hughes, Loral Fairchild and Kodak.

As already discussed, the most popular Schottky-barrier detector is the PtSi detector, which can be used for the detection in the 3–5 μm spectral range (Kimata & Tsubouchi, 1995; Kimata, 2000). Radiation is transmitted through the p-type silicon and is absorbed in the metal PtSi (not in the semiconductor), producing hot holes, which are then emitted over the potential barrier into the silicon, leaving the silicide charged negatively. Negative charge of silicide is transferred to a CCD by the direct charge injection method (see Fig. 9).

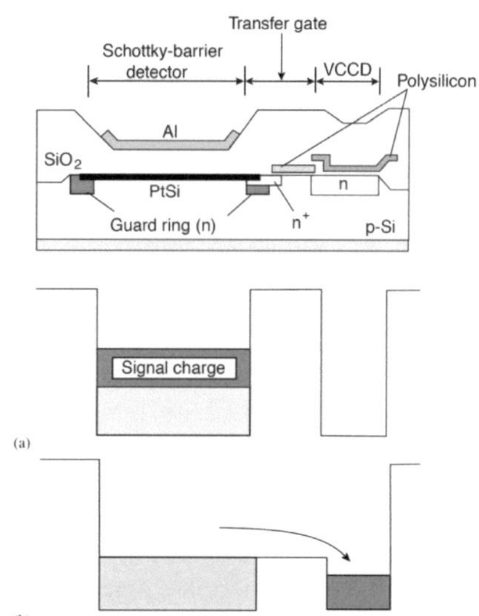

Fig. 9. Typical construction and operation of PtSi Schottky-barrier IR FPA designed with interline transfer CCD readout architecture. (a) and (b) show the potential diagrams in the integration and readout operations, respectively (Kimata, 1995, 1998). Reprinted with permission from M. Kimata, "PtSi Schottky-barrier infrared focal plane arrays," Opto-Electronics Review, 6, 1, 1998.

The effective quantum efficiency in the 3–5 μm window is very low, of the order of 1%, but useful sensitivity is obtained by means of near full frame integration in area arrays. The quantum efficiency has been improved by thinning PtSi film and implementation of an optical cavity. Due to very low quantum efficiency, the operating temperature of Schottky-barrier photoemissive detectors is lower than another types of IR photon detectors (Rogalsky, 1999).

Schottky photoemission is independent of such factors as semiconductor doping, minority carrier lifetime, and alloy composition, and, as a result of this, has spatial uniformity characteristics that are far superior to those of other detector technologies. Uniformity is only limited by the geometric definition of the detectors. The fundamental source of dark current in the devices is thermionic emission of holes over the potential barrier and its magnitude is given by Richardson's equation (Sze, 1982). The cooling requirements of photoemissive detectors are comparable to the extrinsic devices, and while an extension of the technology to the long wavelength band is possible using IrSi (see Fig. 10) this will require cooling below 77 °K (Shepherd, 1988).

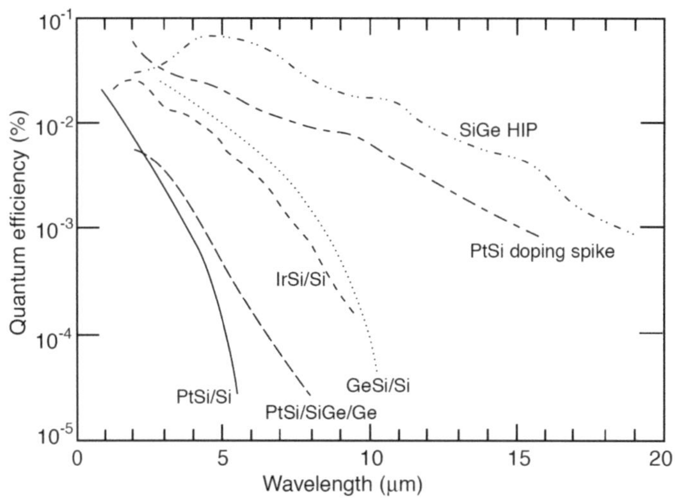

Fig. 10. Comparison of various infrared detectors based on IPE for PtSi, IrSi, PtSi/SiGe, PtSi doping spike, and SiGe/Si HIP (Kimata, 2000). Reprinted with permission from M. Kimata, Metal silicide Schottky infrared detector arrays. In: Infrared Detectors and Emitters: Materials and Devices, Kluwer Academic Publishers, Boston, USA, 2000.

The Schottky-barrier detector is typically operated in backside illumination mode. The quantum efficiency has been improved by thinning PtSi film. The thinning is effective down to the PtSi thickness of 2 nm. Another means of improving responsivity is implementation of an optical cavity. The optical cavity structure consists of the metal reflector and the dielectric film between the reflector and metal electrode of the Schottky-barrier diode. According to fundamental optical theory, the effect of the optical cavity depends on the thickness and refractive index of the dielectric films, and the wavelength (Kurianski et al., 1989). The main advantage of the Schottky-barrier detectors is that they can be fabricated as monolithic arrays in a standard silicon VLSI process. Typically, the silicon array is completed up to the Al metallization al step. A Schottky-contact mask is used to open SiO_2

surface to p-type <100> silicon (with resisitivity 30–50 Ω/cm) at the Schottky-barrier detector location. In the case of PtSi detectors, a very thin layer of Pt (1–2 nm) is deposited and sintered (annealed at a temperature in the range of 300–600°C) to form PtSi and the unreacted Pt on the SiO_2 surfaces is removed by dip etching in hot aqua regia. The Schottky-barrier structure is then completed by a deposition of a suitable dielectric (usually SiO_2) for forming the resonant cavity, removing this dielectric outside the Schottky-barrier regions, and depositing and defining Al for the detector reflector and the interconnects of the Si readout multiplexer. In the case of the 10 μm IrSi Schottky-barrier detectors, the IrSi was formed by in situ vacuum annealing and the unreacted Ir was removed by reactive ion etching. The progress of the Schottky-barrier FPA technology has been constant (Kimata et al., 1998). At the present time Schottky-barrier FPAs represent the most advanced FPAs technology for medium wavelength applications (see Table 1).

Array size	Readout	Pixel size (μm^2)	Fill factor (%)	Saturation (e^-)	NEDT/(f/#) (K)	Year	Company
512 × 512	CSD	26 × 20	39	1.3×10^6	0.07(1.2)	1987	Mitsubishi
512 × 488	IL-CCD	31.5 × 25	36	5.5×10^5	0.07(1.8)	1989	Fairchild
512 × 512	LACA	30 × 30	54	4.0×10^5	0.10(1.8)	1989	RADC
640 × 486	IL-CCD	25 × 25	54	5.5×10^5	0.10(2.8)	1990	Kodak
640 × 480	MOS	24 × 24	38	1.5×10^6	0.06(1.0)	1990	Sarnoff
640 × 488	IL-CCD	21 × 21	40	5.0×10^5	0.10(1.0)	1991	NEC
640 × 480	HB/MOS	20 × 20	80	7.5×10^5	0.10(2.0)	1991	Hughes
1040 × 1040	CSD	17 × 17	53	1.6×10^6	0.10(1.2)	1991	Mitsubishi
512 × 512	CSD	26 × 20	71	2.9×10^6	0.03(1.2)	1992	Mitsubishi
656 × 492	IL-CCD	26.5 × 26.5	46	8.0×10^5	0.06(1.8)	1993	Fairchild
811 × 508	IL-CCD	18 × 21	38	7.5×10^5	0.06(1.2)	1996	Nikon
801 × 512	CSD	17 × 20	61	2.1×10^6	0.04(1.2)	1997	Mitsubishi
1968 × 1968	IL-CCD	30 × 30	—	—	—	1998	Fairchild

Table 1. Specifications and performances of typical PtSi Schottky-barrier FPAs (Kimata, 2000). Reprinted with permission from M. Kimata, Metal silicide Schottky infrared detector arrays. In: Infrared Detectors and Emitters: Materials and Devices, Kluwer Academic Publishers, Boston, USA, 2000.

The details of the geometry, and the method of charge transfer differ for different manufacturers. The design of a staring Schottky-barrier FPAs for given pixel size and design rules, involves a trade-off between the charge handling capacity and the fill factor. Most of the reported Schottky-barrier FPAs have the interline transfer CCD architecture. The typical cross section view of the pixel and its operation in interline transfer CCD architecture is shown in Fig. 9. The pixel consists of a Schottky-barrier detector with an optical cavity, a transfer gate, and a stage of vertical CCD. The n-type guard ring on the periphery of the Schottky-barrier diode reduces the edge electric field and suppresses dark current. The effective detector area is determined by the inner edge of the guard ring. The transfer gate is an enhancement MOS transistor. The connection between detector and the transfer gate is made by an n+ diffusion. A buried-channel CCD is used for the vertical transfer. During the optical integration time the surface-channel transfer gate is biased into accumulation. The Schottky-barrier detector is isolated from the CCD register in this condition. The IR radiation generates hot holes in the PtSi film and some of the excited hot holes are emitted into the silicon substrate leaving excess electrons in the PtSi electrode. This lowers the electrical potential of the PtSi electrode. At the end of the integration time, the transfer gate is pulsed-on to read out the signal electrons from the detector to the CCD register. At the same time, the electrical potential of the PtSi electrode is reset to the channel level of the transfer gate. A unique feature of the Schottky-barrier IR FPAs is the built-in blooming control (blooming is a form of crosstalk in which a well saturates and the electrons spill over into neighbouring pixels). A strong illumination forward biases the detector and no further electrons are accumulated at the detector. The small negative voltage developed at the detector is not sufficient to forward bias the guard ring to the extent that electrons are injected to the CCD register through the silicon region under the transfer gate. Therefore, unless the vertical CCD has an insufficient charge handling capacity, blooming is suppressed perfectly in the Schottky-barrier IR FPA. The responsivity of the FPAs is proportional to their fill factor, and improvement in the fill factor has been one of the most important issues in the development of imagers. For improving the fill factor a readout architecture called the charge sweep device (CSD) developed by Mitsubishi Corporation is also used. Kimata and co-workers have developed a series of IR image sensors with the CSD readout architecture with array sizes from 256×256 to 1040×1040 elements. Specifications and performance of these devices are summarised in Table 2. The effectiveness of this readout architecture is enhanced as the design rule becomes finer. Using a 1.2 μm CSD technology, a large fill factor of 71% was achieved with a 26×20 μm^2 pixel in the 512×512 monolithic structure (Yagi et al., 1994). The noise equivalent temperature difference (NETD) was estimated as 0.033 °K at 300 °K. The 1040×1040 element CSD FPA has the smallest pixel size (17×17 μm^2) among two-dimensional IR FPAs. The pixel was constructed with 1.5 μm

Array size	256 × 256	512 × 512	512 × 512	512 × 512	801 × 512	1040 × 1040
Pixel size (μm^2)	26 × 26	26 × 20	26 × 20	26 × 20	17 × 20	17 × 17
Fill factor (%)	58	39	58	71	61	53
Chip size (mm^2)	9.9 × 8.3	16 × 12	16 × 12	16 × 12	16 × 12	20.6 × 19.4
Pixel capacitor	Normal	Normal	High-C	High-C	High-C	High-C
CSD	4-phase	4-phase	4-phase	4-phase	4-phase	4-phase
HCCD	4-phase	4-phase	4-phase	4-phase	4-phase	4-phase
Number of outputs	1	1	1	1	1	4
Interface	Non integration	Field integration	Frame/Field integration	Frame/Field integration	Flexible	Field integration
Number of I/O pins	30	30	30	30	25	40
Process technology	NMOS/CCD 2 poly/2 Al	NMOS/CCD 2 poly/2 Al	NMOS/CCD 2 poly/2 Al	NMOS/CCD 2 poly/2 Al	CMOS/CCD 2 poly/2 Al	NMOS/CCD 2 poly/2 Al
Design rule (μm)	1.5	2	1.5	1.2	1.2	1.5
Thermal response (ke/K)	—	13	—	32	22	9.6
Saturation (e)	0.7×10^6	1.2×10^6	—	2.9×10^6	2.1×10^6	1.6×10^6
NETD (K)	—	0.07	—	0.033	0.037	0.1

Table 2. Specifications and performance of 2-D PtSi Schottky-barrier FPAs with CSD readout (Kimata, 1998). Reprinted with permission from M. Kimata, "PtSi Schottky-barrier infrared focal plane arrays," Opto-Electronics Review, 6, 1, 1998.

design rules and has 53% fill factor (Kimata et al., 1992). If the signal charges of 1040×1040 pixels are readout from one output port at the TV compatible frame rate, an unrealistic pixel rate of about 40 MHz is required. Therefore, a 4-output chip design was adopted (Shiraishi et al., 1996). The array of 1040×1040 pixels is divided into four blocks of 520×520 pixels. Each block has a horizontal CCD and a floating diffusion amplifier. One-million pixel data at a 30 Hz frame rate can be readout by operating each horizontal CCD at a 10 MHz clock frequency. The NEDT of 1040×1040 element FPA at 300 °K with a 30 Hz frame is 0.1 °K. More recently, a high-performance 801×512-element PtSi Schottky-barrier infrared image sensor has been developed with an enhanced CSD readout architecture (Inoue et al., 1997; Kimata et al., 1997). The developed image sensor has a large fill factor of 61% in spite of a small pixel size of 17×20 μm^2. The NEDT was 0.037 °K at 300 °K. The total power consumption of the device was less than 50 mW. Current PtSi Schottky barrier FPAs are mainly manufactured in 150 mm wafer process lines with around 1 μm lithography technologies; the most advanced Si technology offers 200 mm wafers process with 0.25 μm design rules. Furthermore, 300 mm Si wafer processes with 0.15 μm fine patterns will soon available. However, the performance of monolithic PtSi Schottky-barrier FPAs has reached a plateau, and a slow progress from now on is expected.

3.2 IPE-based NIR devices

In this section, the main Near-Infrared silicon photodetectors structures based on the afore-mentioned IPE effect will be reviewed. The simplest detectors consist of a metal layer on a semiconductor forming a Schottky contact at the material interface, with the Schottky barrier energy Φ_B determined by the materials (Sze, 1981). If an infrared radiation reaches metal/semiconductor interface, the conduction electrons inside the metal can absorb photons gaining sufficient energy. These excited (hot) electrons are able to cross over the Schottky barrier, sweep out the depletion region of the semiconductor, and be collected as a photocurrent under reverse bias operation from an ohmic semiconductor-metal interface. In practice, p-type silicon is often used as the semiconductor because Schottky barriers are lower thereupon than on n-type silicon, allowing detection at longer wavelengths. The IPE-effect, as described in the section 2.1, is very fast allowing to reach high data rates. On the other hand, it is inherently weak so several photodetector designs have been proposed for achieving a suitable quantum efficiency. Elabd *et al.* proposed to use a very thin metal film (about 2 nm) to increase the escape probability of hot carriers due to their multiple reflections inside the metal film (Elabd et al., 1982). In particular, the proposed Schottky photodetector was characterized by a responsivity of about 250 mA/W for a wavelength λ=1500 nm. However, since the volume in which the photons interact with electrons in the metal is very small, only a small fraction of the incident photons actually causes photoemission. Several solutions have been proposed to enhance the efficiency of the IPE process. For example, in the 2001 Wang's group (Lee et al., 2001) investigated the spectral responsivity of Al-porous silicon Schottky barrier photodetectors in the wavelength range 0.4-1.7 μm. The structure of the PS photodetector was Al (finger type)/PS/Si/Al (ohmic), and the active area was 18 mm^2. The photodetectors show strong photoresponsivity in both the visible and the infrared bands, especially at 1.55 μm. The photocurrent can reach 1.8 mA at a reverse bias of 6 V under illumination by a 1.55-μm, 10-mW laser diode. The corresponding quantum efficiency is 14.4%; this high value comes from a very high surface-area-to-volume ratio, of the order of 200-800 m^2/cm^3 of porous Si. The dark current is ~5 μA at -10 V. Recently, Casalino *et al.* (Casalino et al., 2008a, 2010b) propose to enhance the IPE absorption by a resonant cavity effect. The conceptual scheme of the proposed device in shown in Fig. 11.

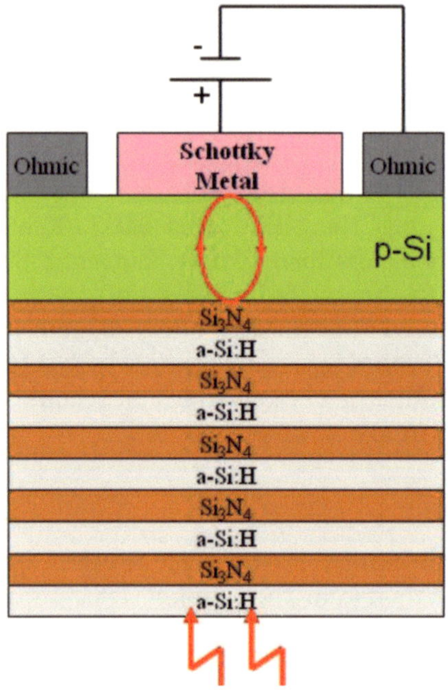

Fig. 11. Schematic cross section of the photodetector proposed by Casalino *et al.* (Casalino et al., 2010b) in 2010.

The resonant cavity is a vertical-to-the-surface Fabry-Perot structure. It is formed by a buried reflector, a top mirror interface and, in the middle, a silicon cavity. The buried reflector is a Bragg mirror, realized by alternating layers of amorphous hydrogenated silicon (a–Si:H) and silicon nitride (Si_3N_4). A Schottky metal layer (Cu), working both as active (absorbing) layer and as cavity mirror, is deposited above the silicon layer. The room temperature responsivity measurements on the device return a peak value of about 2.3 μA/W and 4.3 μA/W, respectively for 0 V and -10 mV of reverse bias applied (Casalino et al., 2008a). In (Casalino et al., 2010b) the authors propose both to scaling down and optimize the same device described in (Casalino et al., 2008a). In particular, the cavity finesse was increased allowing to reach a responsivity of 8 μA/W for a reverse bias of 100mV.

A substantial enhancement of the IPE process efficiency has been achieved increasing the interaction of light with the metal in the vicinity of the interface by the confinement of the infrared radiation into a silicon waveguide. This solution has effectively proven that Schottky diode photodetectors are good candidate for the highly integrated photonics circuits. An example of this approach has recently demonstrated by Coppola's group (Casalino et al., 2010c). The proposed device is schematically illustrated in Fig. 12. A rib waveguide was terminated on a deep trench that reaches down the buried oxide layer of the SOI wafer. A Cu/p-Si Schottky contact was fabricated on the vertical surface of the deep trench.

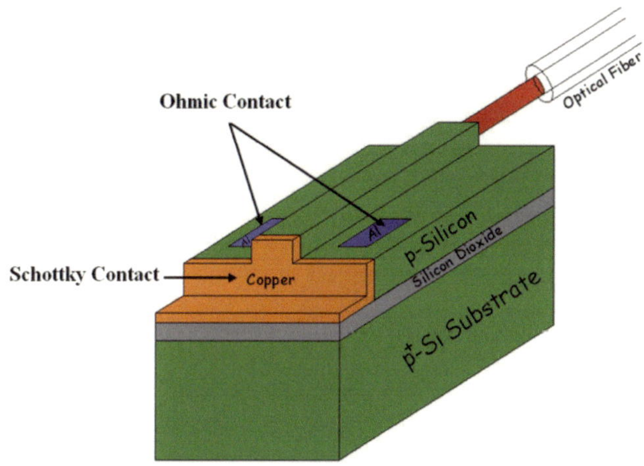

Fig. 12. Schematic view of the Cu/p-Si Schottky barrier-based integrated photodetector proposed by Casalino et al. (Casalino et al., 2010c). Reprinted with permission from M. Casalino *et al.*, "Cu/p-Si Schottky barrier-based near infrared photodetector integrated with a silicon-on-insulator waveguide," Appl. Phys. Lett., 96, 241112. Copyright 2010 American Institute of Physics.

By means of this technological solution a very narrow semiconductor/metal barrier transverse to the optical field coming out from the waveguide has been achieved. The integrated photodetector was characterized by a responsivity of 0.08 mA/W at a wavelength of 1550 nm with an reverse bias of -1V. Measured dark current at -1 V was about 10 nA. Moreover, the authors assert that the thinness of Cu/p-Si Schottky barrier could enable a speed operation in the gigahertz range. An indirect evaluation of the bandwidth of the detector was reported to confirm the operation speed potentialities. A bandwidth of about 3 GHz was measured by Zhu *et al.* in (Zhu et al., 2008a) on a Schottky barrier based integrated photodetector, where the junction was achieved by a nickel silicide layer ($NiSi_2$) on silicon (Zhu et al., 2008b, 2008c)(Fig. 13).

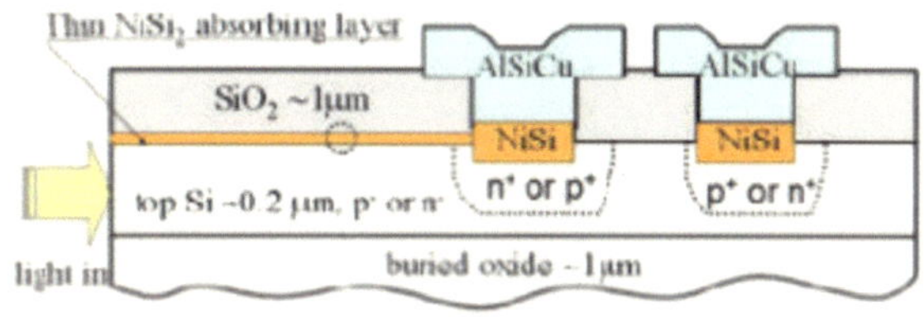

Fig. 13. Schematic structure of waveguide-based silicide Schottky-barrier photodetector proposed by Zhu et al. (Zhu et al., 2008b, 2008a). Reprinted with permission from S. Zhu *et al.*, "Near-infrared waveguide-based nickel silicide Schottky-barrier photodetector for optical communications," Appl. Phys. Lett., 92, 081103. Copyright 2008 American Institute of Physics.

In particular, the authors proposed the lengthening of a thin silicide layer on the surface of a SOI waveguide to achieve both a suitable optical absorption and an efficient photoexcitation of metal electrons or holes across the silicide/Si interface. A detailed analysis was reported on the influence of the silicide layer dimension on the performances of both $NiSi_2$/p-Si and $NiSi_2$/n-Si diodes. The overall performances of the $NiSi_2$/p-Si structure have been resulted better than that relative to the $NiSi_2$/n-Si interface, due to the lower Schottky barrier height of the latter structure. In particular, a responsivity of about 4.6 mA/W at a wavelength of 1550nm and a reverse bias of -1V was estimated for the $NiSi_2$/p-Si diode against the value of 2.3 mA/W of the $NiSi_2$/n-Si junction. Moreover, also the 3 nA of the measured dark current can be considered acceptable.

In order to increase further the capability of IPE-based silicon photodetector to detect long-wavelength (infrared) photons, the possibility to use the concept of surface plasmon polaritons (SPPs) has been explored. SPP are TM-polarized electromagnetic waves trapped at or guided along a metal-dielectric interface. This effect was discovered several decades ago but attention has been renewed by the phenomena of enhanced optical transmission through metallic films with nanostructure (Reather, 1988; Ebbesen et al., 1998). In fact, SPPs are shorter in wavelength than the incident light providing a significant increase in spatial confinement and local field intensity. In other words, while optical systems are basically diffraction limited, surface plasmon polaritons allow a tight localization of optical field to strongly subwavelength dimensions at the metal-dielectric interface (Barnes et al., 2003; Maier, 2006). Such property means that the infrared light is directly guided toward the active area of the IPE detector, confining the optical power at the boundary between the materials forming the Schottky contact, thereby increasing the interaction of light with the metal in the vicinity of the interface where the photoemission process takes place. Berini's group at University of Ottawa has recently applied these concepts confirming that the high light confinement of the SPPs structures can significantly improve the detection capability of the IPE-based photodetector integrated on Si wafers. In particular, Scales *et al.* in (Scales et al., 2004, 2009) described a Schottky diode photodetector obtained embedding a metal stripe of finite width in a homogeneous dielectric cladding (symmetric structure). The proposed device (see Fig. 14) was characterized by a responsivity of about 0.1 A/W, a dark currents of 21 nA, and minimum detectable powers of -22 dBm at a wavelength λ=1550 nm.

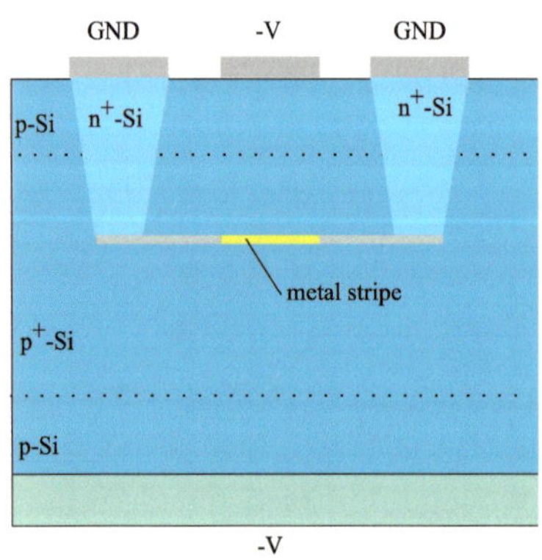

Fig. 14. SPPs Schottky detector based on a metal stripe surrounded by silicon.

An accurate investigation of the performance of the same device for different wavelengths and for several metals forming Schottky contacts were carried out in (Scales et al., 2011). It is worth noting that in these symmetric structures the realization of a thin metal film buried in a semiconductor become the fabrication process more complicated, however, despite of this drawback, the IPE is enhanced due to emission carriers occurring through two Schottky barriers (Scales & Berini, 2010). In order to simplify the fabrication process and to obtain a very short device, Berini's group has proposed an asymmetric SPPs-based photodetector (Akbari et al., 2009, 2010). The proposed structure supports highly confined and highly attenuated SPP modes; this latter feature allows to fabricate shorter devices compared to the symmetric photodetectors. The device, sketched in Fig. 15, consists of a metal stripe cladded at the bottom by a layer of Silicon and covered by air and exhibits for a wavelength λ=1280nm a maximum value of responsivity of about 1 mA/W with a dark current of 6 μA (Akbari et al., 2010).

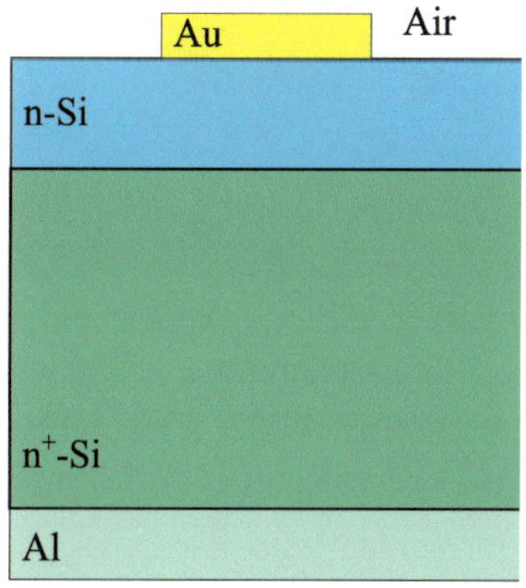

Fig. 15. Cross-section of the aymmetric SPPs Schottky detector proposed in (Akbari et al., 2010).

A detailed simulation analysis performed on asymmetric Schottky detector shows that a significant enhancement in the responsivity can be achieved for a thin metal stripe (about 5 nm) due to multiple internal reflections of excited carriers (Akbari et al., 2009). In particular, authors report that the enhancement is more noticeable for thin metal stripe on p-Si compared to device on n-Si. In this case, the energy range over which the hot electrons can experience multiple reflections is very small because of the larger Schottky barrier height. Finally, the same group has recently demonstrated that a considerable increase in responsivity can be reached with strong applied reverse bias (Olivieri et al., 2010). In particular, authors driving a no optimized detector like that shown in Fig. 15 into breakdown (V~-210 V), such that internal electronic gain is obtained by carrier multiplication, obtaining a responsivity of 2.35 mA/W.

As mentioned above the possibility to achieve a strong light confinement implies both an improvement of responsivity and an important advancement in device miniaturization enabling the realization of on-chip photodetectors on the nanoscale. In this contest, Goykhman et al. in (Goykhmanet al., 2011) have been characterized a nanoscale silicon surface-plasmon Schottky detector shown in Fig. 16.

The detector was fabricated employing a self-aligned approach of local-oxidation of silicon (LOCOS) on silicon on insulator substrate. This approach has been proved useful for fabricating in the same process both a low-loss bus photonic waveguide and the detector. Actually, the oxide spacers effectively define the nanometric area of metal-silicon interface and thus allow avoiding lateral misalignment between the silicon surface and the metal layer to form a Schottky contact. The so realized photodetector was characterized by a responsivity 0.25, 1.4, and 13.3 mA/W for incident optical wavelengths of 1.55, 1.47, and 1.31 μm, respectively.

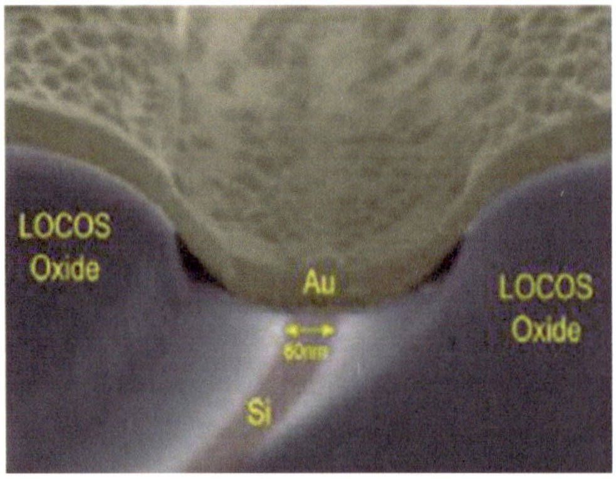

Fig. 16. SEM micrograph of the nanoscale Schottky contact proposed in (Goykhman et al., 2011). Reprinted with permission from I. Goykhmanet *al.*, "Locally Oxidized Silicon Surface-Plasmon Schottky Detector for Telecom Regime," Nano Lett. 11, 2219–2224. Copyright 2011 american Chemical Society.

4. Conclusion

In this chapter an overview on the NIR all-Si photodetectors based on the IPE has been presented. First, we have attempted to elucidate the IPE effects allowing Si absorption at sub band-gap wavelengths and the main figures of merit of IPE-based devices. Then, a quantitative comparison of the photodetectors proposed in the scientific literature, including both bulk and integrated devices, have been reviewed. Unfortunately, the efficiency of devices is low compared to that of detectors based on inter-band absorption. This property is a direct result of many factors: 1) the low absorption due to high reflectivity of the metal layer at NIR wavelengths, 2) the conservation of momentum during carrier emission over the potential barrier which lowers the carriers emission probability into semiconductor, 3) the excitation of carriers lying in states far below the Fermi energy, which get very low probability to overcome the Schottky barrier. While the first and second points can be partially improved by using an optical cavity and a thin metal film, respectively, the third point remains the main limiting factor of IPE. In 1971 Sheperd, Vickers and Yang (Sheperd et al., 1971) recommended replacing the metal electrodes IPE-based photodiodes with degenerate semiconductors. They reasoned that reducing the Fermi energy of electrode by substitution of a degenerate semiconductor could improve emission efficiency as much as 20-fold. Even if NIR IPE-based devices efficiency is still only adapted for power monitoring application, in our opinion that new structures based on the aforementioned insights could play a key role in telecommunication field and could open new frontiers in the field of low-cost silicon photonic.

5. References

Akbari A. & Berini, P. (2009). Schottky contact surface-plasmon detector integrated with an asymmetric metal stripe waveguide. *Appl. Phys. Lett.*, Vol.95, No.2, pp. 021104.

Akbari, A.; Tait, R. N. & Berini, P. (2010). Surface plasmon waveguide Schottky detector. *Opt. Express*, Vol.18, No.8, pp. 8505- 8514.

Archer, R.J. & Cohen, J. (1973). Schottky-Barrier Monolithic Detector Having Ultrathin Metal Layer. *U.S. Patent 3,757,123*

Barnes, W. L.; Dereux, A. & Ebbesen, T.W. (2003) Surface plasmon subwavelength optics. *Nature*, Vol.424, No.6950, pp. 824–830

Boyle, W.S. & Smith, G.E. (1970). Charge-coupled semiconductor devices. *Bell Syst. Tech. J.*, Vol.49, pp. 587–593

Bradley, J.D.B.; Jessop, P.E. & Knights, A.P. (2005). Silicon waveguide-integrated optical power monitor with enhanced sensitivity at 1550 nm. *Appl. Phys. Lett.*, Vol. 86, No.24, pp. 241103.

Capone, B.R.; Skolnik, L.H.; Taylor, R.W.; Shepherd, F.D.; Roosild, S.A.; Ewing, W.; Kosonocky, W.F. & Kohn, E.S. (1978). Evaluation of Schottky IRCCD Staring Mosaic Focal Plane. *22nd Int. Tech. Symp. Society of Photo-Optical Instrumentation Engineers (SPIE)*, San Diego, Aug. 28-29

Casalino, M. ; Sirleto, L. ; Moretti, L. ; Della Corte, F. & Rendina, I. (2006a). Design of a silicon RCE schottky photodetector working at 1.55 micron. *Journal of luminescence*, Vol.121, pp. 399-402.

Casalino, M. ; Sirleto, L. ; Moretti, L. ; Della Corte, F. & Rendina, I. (2006b). Design of a silicon resonant cavity enhanced photodetector based on the internal photoemission effect at 1.55 µm. *Journal of Optics A: Pure and applied optics*, Vol.8, pp. 909-913.

Casalino, M.; Sirleto, L.; Moretti, L.; Gioffrè, M.; Coppola, G. & Rendina, I. (2008a). Silicon resonant cavity enhanced photodetector based on the internal photoemission effect at 1.55 µm: Fabrication and characterization. *Appl. Phys. Lett.*, Vol.92, No.25, pp. 251104

Casalino, M.; Sirleto, L.; Moretti, L. & Rendina, I. (2008b). A silicon compatible resonant cavity enhanced photodetector working at 1.55 µm. *Semicond. Sci. Technol.*, Vol.23, No.7, pp. 075001

Casalino, M. ; Coppola, G. ; Iodice, M. ; Rendina, I. & Sirleto, L. (2010a). Near Infrared All-Silicon Photodetectors: State of the Art and Perspectives. *Sensors*, Vol.10, No.12, pp. 10571-10600

Casalino, M.; Coppola, G.; Gioffrè, M.; Iodice, M.; Moretti, L.; Rendina, I. & Sirleto, L. (2010b). Cavity enhanced internal photoemission effect in silicon photodiode for sub-bandgap detection. *J. Lightw. Technol.*, Vol. 28, No.22, pp. 3266-3272

Casalino, M.; Sirleto, L.; Iodice, M.; Saffioti, N.; Gioffrè, M.; Rendina, I & Coppola, G. (2010c). Cu/p-Si Schottky barrier-based near infrared photodetector integrated with a silicon-on-insulator waveguide. *Appl. Phys. Lett.*, Vol.96, No.24, pp. 241112.

Chan, E.Y; Card, H.C. & Teich, M.C. (1980a). Internal Photoemission Mechanism at interfaces between Germanium and Thin Metal Films. *IEEE Journal of Quantum Electronics*, Vol.QE-16, No.3, pp. 373-381

Chan, E.Y. & Card, H.C. (1980b). Near IR interband transitions and optical parameters of metal-germanium contacts. *Applied Optics*, Vol.19, No.8, pp. 1309

Donati, S. (1999). *Photodetectors: Devices, circuits, and applications*. Prentice Hall PTR, New Jersey, USA

Ebbesen, W. ; Lezec, H.J. ; Ghaemi, H.F. ; Thio, T. & Wolff, P.A. (1997). Extraordinary optical transmission through sub-wavelength hole arrays. *Nature, Vol.*391, pp 667-669

Elabd, H. & Kosonocky, W.F. (1982). Theory and measurements of photoresponse for thin film Pd_2Si and PtSi infrared Schottky-barrier detectors with optical cavity. *RCA Review*, Vol. 43, pp. 569–589

Elabd, H.; Villani, T. & Kosonocky, W.F. (1982). Palladium-Silicide Schottky-Barrier IR-CCD for SWIR Applications at Intermediate Temperatures. *IEEE Trans. Electron Devices Lett.*, Vol. EDL-3, pp. 89-90

Fowler, R.H. (1931). The Analysis of Photoelectric Sensitivity Curves for Clean Metals at Various Temperatures. *Physical Review*, Vol. 38, pp. 45-56

Goykhman, I.; Desiatov, B.; Khurgin, J.; Shappir, J. & Levy U. (2011). Locally Oxidized Silicon Surface-Plasmon Schottky Detector for Telecom Regime. Nano Lett. Vol. 11, pp. 2219–2224.

Inoue, M.; Seto, T.; Takahashi, S.; Itoh, S.; Yagi, H.; Siraishi, T.; Endo, K. & Kimata, M. (1997). Portable high performance camera with 801x512 PtSi-SB IRCSD, *SPIE Proc.*, Vol.3061, pp. 150–158

Jalali, B & Fathpour, S. (2006). Silicon Photonics. *J. Lightwave Technol.*, Vol.24, No.12, pp. 4600-4615

Kimata, M.; Yutani, N.; Tsubouchi, N. & Seto, T. (1992). High performance 1040x1040 element PtSi Schottky-barrier image sensor. *SPIE Proc.*, Vol.1762, pp. 350–360

Kimata, M. & Tsubouchi, N. (1995). Schottky barrier photoemissive detectors. In: *Infrared Photon Detectors*, A. Rogalski (Ed.), pp. 299-349, SPIE Optical Engineering Press, Bellingham

Kimata, M.; Ozeki, T.; Nunoshita, M. & Ito, S. (1997). PtSi Schottky-barrier infrared FPAs with CSD readout. *SPIE Proc.*, Vol.3179, pp. 212–223

Kimata, M.; Ueno, M.; Yagi, H.; Shiraishi, T.; Kawai, M.; Endo, K.; Kosasayama, Y.; Sone, T.; Ozeki, T. & Tsubouchi, N. (1998). PtSi Schottky-barrier infrared focal plane arrays, *Opto-Electronics Review*, Vol.6, pp. 1–10

Kimata, M. (2000). Metal silicide Schottky infrared detector arrays. In: *Infrared Detectors and Emitters: Materials and Devices*, Kluwer Academic Publishers, Boston, USA

Kimerling, L.C.; Dal Negro, L.; Saini, S.; Yi, Y.; Ahn, D.; Akiyama, S.; Cannon, D.; Liu, J.; Sandland, J.G.; Sparacin, D.; Michel, J.; Wada, K. & Watts, M.R. (2004). *Silicon Photonics: Topics in Applied Physics,* Springer, ISBN 3-642-05909-0, Berlin, Germania

Kohn, E.S.; Roosild, S.A.; Shepherd, F.D. & Yang, A.C. (1975). Infrared Imaging with Monolithic CDD-Addressed Schottky-Barrier Detector Arrays, Theoretical and Experimental Results. *Int. Conf. on Application on CCD's*, San Diego, Oct. 29-31

Kosonocky, W.F.; Kohn, E.S.; Shalleross, F.V.; Sauer, D.J.; Shepherd, F.D.; Skolnik, L.H.; Taylor, R.W.; Capone, B.R. & Roosild, S.A. (1978). Platinum Silicide Schottky-Barrier IR-CCD Image Sensors. *Int. Conf. on Application on CCD's*, San Diego, Oct. 25-27

Kosonocky, W.F.; Erhardt, H.G.; Meray, G.M.; Shallcross, F.V.; Elabd, H.A.; Cantella, M.; Klein, J.; Skolnik, L.H.; Capone, B.R.; Taylor, R.W.; Ewing, W.; Shepherd, F.D. & Roosild, S.A. (1980). Advances in Platinum-Silicide Schottky-Barrier IR-CCD Image Sensors. *SPIE Proc.*, Vol. 225, pp. 69-71

Kosonocky, W.F. ; Shallcross, F.V. & Villani, T.S. (1985). 160x244 Element PtSi Schottky-Barrier IR-CCD Image Sensor. *IEEE Trans. Electron Dev.*, vol. ED-32, No.8, pp. 1564

Kosonocky, W.F. (1991). Review of infrared image sensors with Schottky-barrier detectors. *Optoelectronics Devices and Technologies*, Vol.6, pp. 173–203

Kosonocky, W.F. (1992). State-of-the-art in Schottky-barrier IR image sensors. *SPIE Proc.*, Vol.1682, pp. 2–19

Kurianski, J.M.; Shanahan, S.T.; Theden, U.; Green, M.A. & Storey, J.W.V. (1989). Optimization of the cavity for silicide Schottky infrared detectors. *Solid-State Electronics*, Vol.32, pp. 97–101

Lee, M. K.; Chu, C. H.; Wang & Y. H. (2001) 1.55-μm and infrared-band photoresponsivity of a Schottky barrier porous silicon photodetector. *Opt. Lett.*, Vol.26, No.3, pp. 160-162

Liang, T.K.; Tsang, H.K.; Day, I.E.; Drake, J.; Knights, A.P. & Asghari, M. (2002). Silicon waveguide two-photon absorption detector at 1.5 μm wavelength for autocorrelation measurements. *Appl. Phys. Lett.*, Vol.81, No.7, pp. 1323

Liu, A.; Jones, R.; Cohen, O.; Hak, D. & Paniccia M. (2006). Optical amplification and lasing by stimulated Raman scattering in silicon waveguides. *J. Lightw. Technol.*, Vol.24, No.3, pp. 1440-1445

Liu, A.; Liao, L.; Rubin, D.; Nguyen, H.; Ciftcioglu, B.; Chetrit, Y.; Izhaky, N. & Paniccia, M. (2007). High-speed optical modulation based on carrier depletion in a silicon waveguide. *Opt. Express*, Vol.15, No. 2, pp. 660-668

Maier A. (2006). *Plasmonics: Fundamentals and Applications.* Springer, New York, USA

Michael, C.P.; Borselli, M.; Johnson, T.J.; Chrystal, C. & Painter, O. (2007). An optical fiber-taper probe for wafer-scale microphotonic device characterization. *Opt. Express*, Vol.15, No. 8, pp. 4745-4752

Muriel, M.A . & Carballar, A. (1997). Internal field distributions in fiber Bragg gratings. *IEEE Photonics technology letters*, Vol. 9, No.7, pp. 955, 1997

Olivieri, A.; Akbari A. & Berini, P. (2010) Surface plasmon waveguide Schottky detectors operating near breakdown. *Phys. Status Solidi RRL*. Vol 4, No.10, pp. 283 - 285

Reather, H. (1988). *Surface Plasmons on Smooth and Rough Surfaces and on Gratings.* Springer, Berlin, Germany

Rogalski, A. (1999). Assessment of HgCdTe photodiodes and quantum well infrared photoconductors for long wavelength focal plane arrays. *Infrared Phys. Technol.*, Vol.40, pp. 279–294

Rowe, L.K.; Elsey, M. ; Tarr, N.G. ; Knights, A.P. & Post, E. (2007). CMOS-compatible optical rib waveguides defined by local oxidation of silicon. *Electron. Lett.*, Vol.43, No.6, pp. 392-393

Scales, C. & Berini, P. (2004). Schottky Barrier Photodetectors, *U.S. Patent No. 7,026,701.*

Scales, C.; Breukelaar, I. & Berini, P. (2009). Surface-plasmon Schottky contact detector based on a symmetric metal stripe in silicon. *Opt. Lett.*, Vol. 35, No.4, pp. 529–531

Scales, C. & Berini, P. (2010). Thin-film Schottky barrier Photodetector Models. *IEEE Journal of Quantum Electronics*, Vol.46, No.5, pp. 633-643

Scales, C.; Breukelaar, I.; Charbonneau, R. & Berini, P. (2011). Infrared Performance of Symmetric Surface-Plasmon Waveguide Schottky Detectors in Si. *IEEE J. Lightw. Tech.*, Vol 29, No. 12, pp. 1852-1860

Sheperd, F.D.; Vickers, V.E. & Yang, A.C. (1971). Schottky Barrier Photodiode with a Degenerate Semiconductor Active Region. U.S. Patent No. 3.603.847

Shepherd, F.D. & Yang, A.C. (1973). Silicon Schottky Retinas for Infrared Imaging. *IEDM Tech. Dig.*, pp. 310-313

Shepherd, F.D.; Taylor, R.W.; Skolnik, L.H.; Capone, B.R.; Roosild, S.A.; Kosonocky, W.F. & Kohn, E.S. (1979). Schottky IRCCD Thermal Imaging. *Adv. Electron. Electron Phys., 7th Symp. Photo-Electronic Image Devices*, Vol. 22, pp. 495-512

Shepherd, F.D. (1984). Schottky diode based infrared sensors. *SPIE Proc.*, Vol. 443, pp. 42–49

Shepherd, F.D. (1988). Silicide infrared staring sensors. *SPIE Proc.*, Vol.930, pp. 2–10

Shepherd, F.D. (1998). *Platinum silicide internal emission infrared imaging arrays.* Academic Press, New York, USA

Shiraishi, T.; Yagi, H.; Endo, K.; Kimata, M.; Ozeki, T.; Kama, K. & Seto, T. (1996). PtSi FPA with improved CSD operation. *SPIE Proc.*, Vol.2744, pp. 33–43

Sze, S.M. (1981). *Physics of Semiconductor Devices*. John Wiley & Sons, New York, USA

Taylor, R.W.; Skolnik, L.H.; Capone, B.R.; Ewing, W.; Shepherd, F.D.; Roosild, S.A.; Cochrum, B.; Cantella, M.; Klein, J.; Kosonocky, W.F. (1980). Improved Platinum Silicide IRCCD Focal Plane. *SPIE Proc.*, Vol. 217, pp. 103-110

Vickers, V.E. (1971). Model of Schottky Barrier Hot-electron-Mode Photodetection. *Applied Optics, Vol.* 10, No.9, pp. 219

Vivien, L.; Pascal, D.; Lardenois, S.; Marris-Morini, D.; Cassan, E.; Grillot, F.; Laval, S.; Fedeli, J.M. & El Melhaoui, L. (2006). Light injection in SOI microwaveguides using high-efficiency grating couplers. *J. Lightw. Technol.*, Vol.24, No.10, pp. 3810-3815

Xu, Q.; Manipatruni, S.; Schmidt, B.; Shakya, J. & Lipson, M. (2007). 12.5 Gbit/s carrier-injection-based silicon micro-ring silicon modulators. *Opt. Express,* Vol.15, No. 2,pp. 430-436

Yagi, H.; Yutani, N.; Nakanishi, J.; Kimata, M. & Nunoshita, M. (1994). A monolithic Schottky-barrier infrared image sensor with 71% fill factor. *Optical Engineering,* Vol.33, pp. 1454–1460

Yeh, P. (1988). *Optical Waves in Layerer Media.* Wiley Interscience Publication, New York, USA

Yuan, H.X. & Perera, G.U. (1995). Dark current analysis of Si homojunction interfacial work function internal photoemission far-infrared detectors. *Appl. Phys. Lett.*, Vol.66, No.17, pp. 2262-2264

Zhu, S.; Yu, M.B.; Lo, G.Q. & Kwong, D.L. (2008a). Near-infrared waveguide-based nickel silicide Schottky-barrier photodetector for optical communications. *Appl. Phys. Lett.*, Vol.92, No.8, pp. 081103

Zhu, S.; Lo, G.Q. & Kwong, L. (2008b). Low-cost and high-gain silicide Schottky-barrier collector phototransistor integrated on Si waveguide for infrared detection. *Appl. Phys. Lett.*, Vol. 93, No.7, pp. 071108

Zhu, S.; Lo, G.Q. & Kwong, D.L. (2008c) Low-Cost and High-Speed SOI Waveguide-Based Silicide Schottky-Barrier MSM Photodetectors for Broadband Optical Communications. *IEEE Phot. Tech. Lett.*, Vol. 20, No.16, pp. 1396-1398

4

Silicon Photomultipliers: Characterization and Applications

Marco Ramilli[1], Alessia Allevi[1], Luca Nardo[1], Maria Bondani[2] and Massimo Caccia[1]
[1]*Dipartimento di Fisica e Matematica - Università degli Studi dell'Insubria*
[2]*Istituto di Fotonica e Nanotecnologie - Consiglio Nazionale delle Ricerche*
Italy

1. Introduction

Silicon Photo-Multipliers (SiPMs henceover) are photo-detectors based on a technology originally invented in Russia (Akindinov et al., 1997). They essentially consist of an *array of p-n junctions* operated beyond the breakdown voltage (McKay, 1954), in a Geiger-Müller (G-M) regime (Oldham et al., 1972), with typical gain of the order of 10^6 and on-cell integrated quenching mechanisms. Silicon photo-detectors with internal multiplication are in use since more than a decade (Lutz, 1995). Avalance Photo-Diodes (APDs) (Akindinov et al., 2005) are operated in a proportional regime, with typical gains of 10^4, Single-Photon Avalanche Diodes (SPADs) (Cova et al., 1996) are endowed with single-photon sensitivity and are tailored for high frequency counting with time resolutions down to 30 ps. However, being made of a single cell operated in binary mode, they do not carry any information about the intensity of the incoming light field.

SiPMs complement the family of existing sensors: with a cell density of $\sim 10^3/\text{mm}^2$, areas up to 3×3 mm^2 and *a single output node*, they offer the possibility of measuring the intensity of the light field simply by counting the number of fired cells. The main features of SiPMs, due to their structural and operational characteristics are:

- high gain, granted by G-M operating mode, comparable to the values achieved by standard photomultipliers (PMTs);
- enhanced linearity, owing to cell structure, with deviations that become relevant when the average number of detected photons approaches the same order of magnitude of the number of cells of the device (Tarolli et al., 2010);
- large dynamic range, provided by the pixelated structure with a common output, spanning from the single photon regime up to high intensities;
- operability in magnetic fields, compactness and relatively low cost, granted by silicon-based technology.

On the other hand, since G-M avalanches that are triggered by electron-hole pairs extracted by the impinging photons are obviously indistinguishable from the ones originated from other processes, dealing with all possible sources of noise is far from being trivial:

- thermally extracted electron-hole pairs cause high Dark Count Rates (DCR), with values ranging from several hundreds of kHz up to the MHz level, depending on the total number of cells, the operating temperature and the overbias;
- spurious signals are also due to optical cross-talk (Sciacca et al., 2008): photons emitted by an avalanche can travel through silicon and reach the depleted region in a neighboring cell, thus triggering another avalanche;
- carriers extracted during an avalanche process may be trapped in a false potential minimum in the depleted region: escaping from that trap they can originate an *afterpulse* avalanche, so called because it typically happens shortly after (or even during) the recovery of the previous avalanche (Du et al., 2008; Eckert et al., 2010).

These apparently huge disadvantages can be overridden by means of an exhaustive characterization of the SiPM performances. An example of this procedure will be given in the following Section, where we present the characterization of the devices as a function of temperature.

Another aspect of the characterization of SiPMs will be addressed by providing a description of the G-M avalanches probability distribution functions, in order to reconstruct the statistics of the impinging light (Ramilli et al., 2010).

Modeling the SiPM response is also preliminary to application of these sensors as detectors in Fluorescence Fluctuation Spectroscopy biophysical experiments (Chen et al., 1999; Schwille, 2001), where the parameters describing the system under investigation are inferred by the deviations of the fluorescence intensity of suitable probes around its mean value. A feasibility study on this topic will be presented in the final part of the Chapter.

2. Characterization

Tests have been performed on existing devices with the main goal to define an exhaustive characterization protocol and to produce a comparative study. Three kinds of detectors from different manufacturers have been studied: SensL[1], Hamamatsu Photonics[2] and STMicroelectronics[3]. Detector characterization is a major task for all the applications of SiPMs and in particular for the identification of critical parameters: for example, DCR is an important parameter for low-rate-event applications, while thermal stability is essential for portable devices. For this reason, the following characterization protocol has been developed assessing:

- Geometrical parameters (number of cells, size of detectors and occupancy factor);
- I–V measurements;
- Noise measurements: DCR, optical cross-talk, dependence on the environmental parameters.
- Analysis of photon spectra: resolution power, gain, working point optimization (at low and large flux), electronic noise measurement taking into account cell-to-cell variations, dependence on the environmental parameters (temperature).
- Linearity and dynamic range.
- Spectral response measurement: photon detection efficiency (PDE).

[1] see http://sensl.com/.
[2] see http://www.hamamatsu.com/.
[3] see http://www.st.com/stonline/.

In particular, studies of the main SiPM parameters (such as gain, PDE, DCR and optical cross-talk) as functions of temperature will be described in details.

2.1 Experimental setup

In order to implement the characterization protocol, we have developed an experimental setup in which a green-emitting LED (λ = 510 nm), coupled to a fast pulse generator (PDL800-B PicoQuant), has been used as the light source. The SiPM output signal has been elaborated in the following way:

- SiPM was directly connected to a first stage amplification board:
  - in experimental situations where the SiPM was operated under a continuous light flux, a transimpedance pre-amplifier provided by SensL has been used: this device converts the raw current from the SiPM into a voltage, with an amplification of 470 V/A;
  - when the SiPM was operated in the pulsed regime, a different pre-amplifier, called Pulse Amplifier and also provided by SensL, has been used. It allowed the fast rise of the detector to be exploited, providing an amplification factor of 20;
- a leading edge discriminator (Lecroy 821) with a user-defined voltage threshold that the SiPM output has to exceed in order to provide a triggering signal has been then exploited at the output of the amplifier;
- in case of frequency measurements (e.g. DCR measurements), the discriminator output has been directly sent to a scaler;
- the output of the SiPM has been integrated by a CAEN QDC V792N board:
  - the board provides a charge measurement performing the integration of the input voltage signal, with a conversion time of 2.8 μs and a reset time of 4 μs;
  - the integration gate is generated by a NIM timing unit and must precede the analog input signal of at least 15 ns;
- data have been stored in a PC via a USB-VME Bridge (CAEN).

The system composed by the investigated SiPM and the first-stage amplification board has been located in a metal box sealed with grease, in which air has been replaced with helium. The cooling fluid has been pumped into the box through a copper pipe, allowing controlled temperature variations. The temperature of the system has been measured by a thermistor placed in contact with the external packaging of the sensor.

2.2 Studying temperature behavior

In order to quantitatively understand the effects of temperature changes on the SiPM main figures of merit, a suitable set of measurements must be accomplished. The procedure has been tested with SensL CSI 0747 015 A20 HD and the results regarding this sensor will be presented in more detail in the following. Once the procedure was tested, it has been repeated with different SiPM models: a SensL CSI 0740 001 A20 HD, a Hamamatsu S10362-11-100C and a ST Microelectronics TO-8 prototype.

Sensors have been placed in the cooling box described in Section 2.1 and several spectra have been acquired at different bias voltages and different temperatures in order to obtain the values of gain and PDE. Moreover, DCR measurements at different thresholds were used to evaluate both DCR and cross-talk contributions. To analyze the procedure in detail, a typical

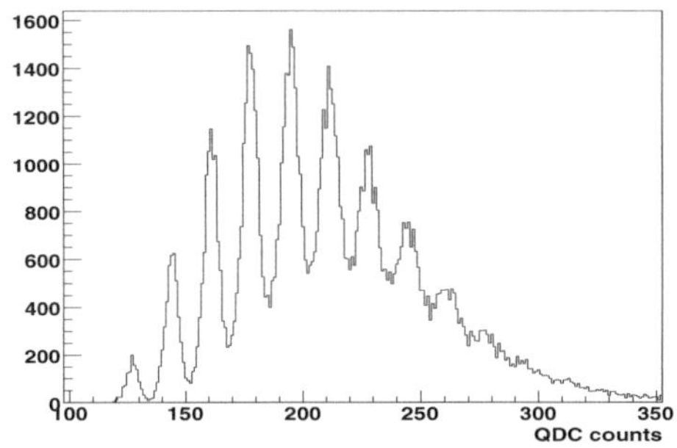

Fig. 1. Example of a low flux spectrum of a SiPM; each histogram bin represents a single QDC channel corresponding to 0.11 pC.

low-flux spectrum is presented in Fig. 1 that shows the charge measurements performed by the CAEN QDC: a waveform generator triggered both the light source and the *gate* needed to perform the charge measurement. The structure of the histogram reflects the characteristics of the SiPM, since the n-th peak position represents the most probable output value (in released charge) of n cells firing simultaneously and its Gaussian broadening is due to stochastic noise sources. In this experimental situation, the very first peak represents the *zero-photon* peak, i.e. the output of the QDC board with no SiPM signal. The broadening of the n-th peak σ_n can be described with good approximation as:

$$\sigma_n^2 = \sigma_0^2 + n\sigma_1^2, \tag{1}$$

where σ_0 is the variance of the zero-photon peak (giving the noise contribution due to the electronic chain), and σ_1 is the variance of the first photon peak, providing an estimation of the SiPM noise.

2.2.1 Gain

The gain has been evaluated by illuminating the SiPM with a low photon flux, thus obtaining a spectrum whose peaks are clearly recognizable (see Fig. 1). As the n-th peak corresponds to the mean charge released by n G-M avalanches, the gain G can be computed in the following way:

$$G = \frac{QDC_{cal}}{e^- K_{amp}} \Delta_{PP}, \tag{2}$$

where QDC_{cal} = 0.11 pC/channel is the charge corresponding to one QDC unit, e^- is the elementary charge, K_{amp} is the global amplification factor of the electronic setup, and Δ_{PP} is the distance in QDC units between two adjacent peaks of the collected spectra.

The gain behavior as a function of the bias voltage, at fixed temperatures, is shown in Fig. 2: a linear dependence in the range of interest is clearly observable with a slope independent of the temperature within the experimental errors. In Fig. 3 the same gain values are presented as functions of temperature, for fixed bias voltages: a linear behavior is still clearly recognizable; in this case as well the slope is not affected by the change of the applied bias.

This analysis suggests that the gain can be expressed as a linear function of a variable which can be re-scaled with temperature. Since it is well known that the breakdown voltage has, in

SiPM Model	m_{BD} (mV/degree)
SensL CSI 0747 015 A20 HD	23.2 ± 1.4
SensL CSI 0740 001 A20 HD	23.6 ± 0.9
Hamamatsu S10362-11-100C	61.9 ± 0.7
STM TO-8 prototype	31.5 ± 0.1

Table 1. Rate of change of the breakdown voltage with temperature.

our range of interest, a linear dependence on temperature (Goetzberger et al., 1963), the over voltage, defined as the difference between the applied bias voltage and the breakdown one, is thus a suitable candidate. The conditions can be summarized as follows

$$G(V,T) = m_V(V - V_{BD}(T)), \tag{3}$$
$$G(V,T) = m_T T + G(T_0, V), \tag{4}$$

where m_V and m_T are the slopes, $V_{BD}(T)$ is the breakdown voltage and T_0 is a reference temperature. Solving these equations for $V_{BD}(T)$, the rate of change of the breakdown voltage with temperature can be expressed as:

$$m_{BD} = -\frac{m_T}{m_V}. \tag{5}$$

The procedure has been applied to the different SiPMs. From the results in Table 1, we can see that the values of m_{BD} are technology dependent.

By using the breakdown voltage value measured at room temperature (RT) as the reference

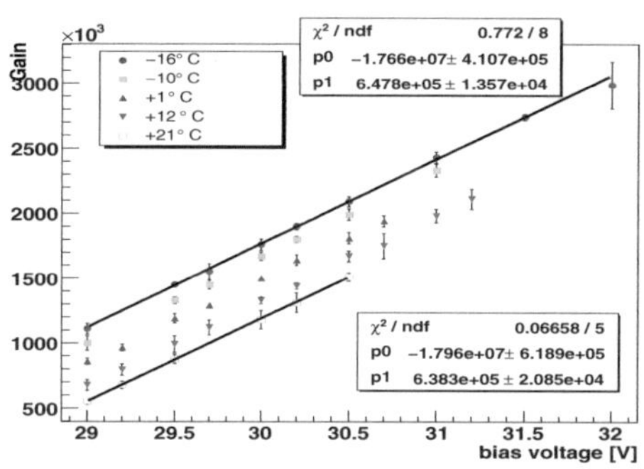

Fig. 2. Gain as a function of bias voltage, for different temperatures; a linear fit has been performed for each temperature set: the obtained fit slopes are in agreement.

value, the breakdown values for each temperature can be written as:

$$V_{BD}(T) = m_{BD}(T - T_{RT}) + V_{BD}(T_{RT}). \tag{6}$$

From Equation 6 it is possible to express the gain and all the other measured parameters as a function of the over voltage: in Fig. 4 the gain is plotted as a function of the over voltage, showing a global linear behavior independent of the temperature. Equation 6 can be used to fix the operational parameters of a SiPM regardless of the varying environmental conditions: if m_{BD} is known with enough precision and frequent measurements of the SiPM temperature are performed, the over voltage across each cell can be maintained fixed by simply continuously adjusting the applied bias according to the temperature variations. This

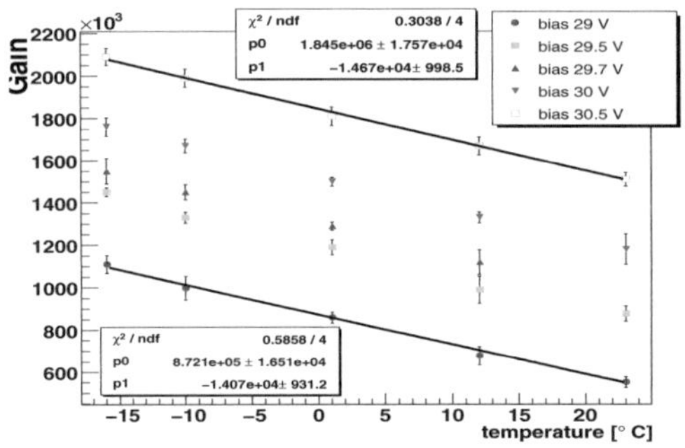

Fig. 3. Gain as a function of temperature (expressed in Celsius degrees), for fixed bias voltages; each set has been fitted using a linear law: the obtained slopes are in agreement.

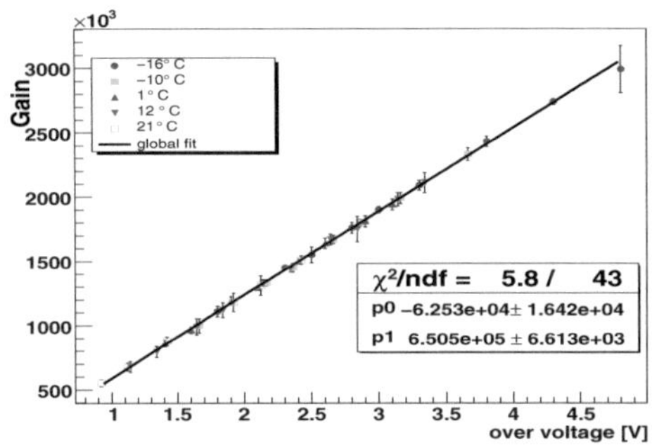

Fig. 4. Gain as a function of the over voltage: values are obtained by calculating the breakdown voltage for each temperature; all the data acquired at different temperatures have been fitted with the same linear law.

idea has led to a collaboration between Università dell'Insubria and CAEN for the realization of CAEN SP5600 General Purpose Power Supply and Amplification Unit module with an integrated threshold discriminator.

2.2.2 Dark count rate

As explained in Section 1, DCR is the frequency of the G-M avalanches triggered by thermally extracted carriers. A scan of the DCR at different thresholds can be done, and the resulting plot (an example is shown in Fig. 5) is usually referred to as *staircase function*.
DCR has been measured at different voltages and different temperatures, also setting different discrimination thresholds; in Fig. 6, for example, the results for a *"half photon threshold"* are shown. A clear and expected dependence on temperature is recognizable.
An exhaustive knowledge of DCR behavior is of utmost importance for a complete characterization of all the noise sources of the detector. It can be a fundamental figure of merit for low flux applications.

2.2.3 Optical cross-talk

An electron avalanche is modeled as a microplasma (Oldham et al., 1972): photons emitted by the accelerated carriers during this event have a certain probability to reach the neighboring

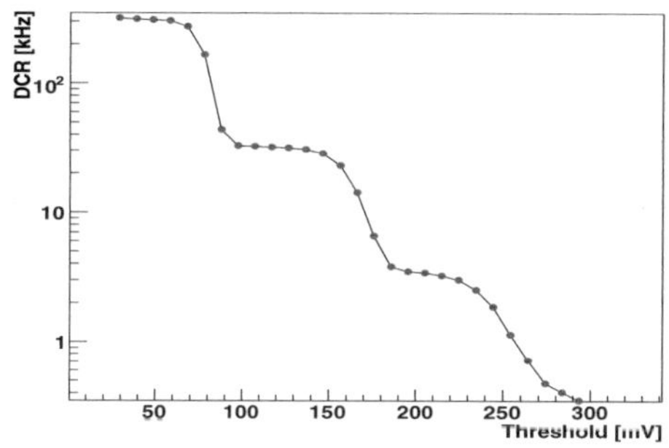

Fig. 5. Example of *staircase curve*, obtained after a three-stage amplification.

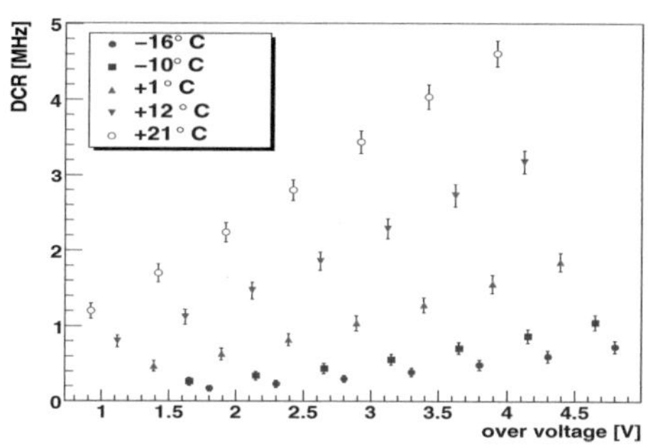

Fig. 6. DCR at different temperatures as a function of the over voltage, for a SensL SiPM.

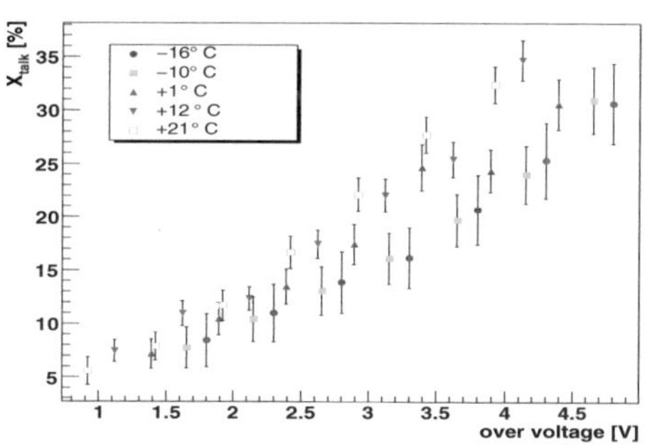

Fig. 7. Cross-talk as a function of the over voltage, at different temperatures, for a SensL SiPM.

cell diodes thus triggering a second avalanche. The quantity named optical cross-talk (in short cross-talk, X_T) is simply the percentage of avalanches triggered by such a mechanism.

Cross-talk has been calculated starting from DCR measurements by taking the ratio between the DCR frequencies with the discriminator threshold respectively at *"one-and-half photon"* and at *"half photon"*. This method is based on the assumption that the probability that two uncorrelated thermally-triggered avalanches are generated within the same rise time is negligible, so that all the second photon events are due to cross-talk.

In Fig. 7 the results of the cross-talk evaluation are presented: it is worth noting that, within the experimental errors, cross-talk does not seem to suffer a strong temperature dependence.

2.2.4 Photon detection efficiency

Photon detection efficiency (PDE) is a key parameter for every light detector and is defined as the product of three terms:
- quantum efficiency, that is the probability that the impinging photon produces a charge carrier, which is a function of the incoming light wavelength;
- the G-M probability that the extracted carrier generates an avalanche that depends on the over voltage;
- the so-called Fill Factor (FF), a geometrical factor that expresses the portion of the active area of the sensor with respect to the total area.

The PDE value thus represents the fraction of impinging photons that are actually detected.

To measure this quantity, SiPMs have been illuminated with a light intensity Φ which has been previously measured with a calibrated PMT HAMAMATSU H5783 and the resulting spectra have been acquired. PDE has been evaluated by estimating the mean number of detected photons $< n >_{meas}$ with respect to the incoming light intensity:

$$PDE = \frac{< n >_{meas}}{\Phi} \tag{7}$$

In the case of spectra in which each photon peak could be resolved, $< n >_{meas}$ has been obtained by fitting the peak positions with a Poissonian curve and by evaluating the mean value of the fit distribution. In the case in which the peaks could not be resolved, $< n >_{meas}$ has been estimated as

$$< n >_{meas} = \frac{QDC_{cal}}{e^{-}\, G\, K_{amp}} \Delta QDC \tag{8}$$

where G is the proper gain value, e^{-} is the elementary charge, $QDC_{cal} = 0.11$ pC is the charge per QDC unit, K_{amp} is the amplification factor of the electronics chain, and ΔQDC is the difference between the mean value of the obtained spectrum and the pedestal position.

However, this method provides a zero-order approximation of the PDE because it does not take into account cross-talk effects that can be as large as 40% (see Fig. 7 as a reference). To properly estimate the *"true"* number $< n >_0$ of impinging photons triggering an avalanche, the following relation has been used:

$$< n >_{meas} = \frac{< n >_o}{1 - X_T} \simeq < n >_o (1 + X_T). \tag{9}$$

The approximation is valid for small values of the cross-talk; the previously obtained PDE values have thus been corrected using $< n >_0$ as the mean number of detected photons as shown in Fig. 8. Again, PDE does not seem to have any remarkable temperature dependance, thus confirming the agreement of our results with the definition of PDE.

3. Photon-number statistics

Due to their good linear response, SiPMs can be considered as the ideal candidates for the reconstruction of the statistics of any light state (Afek et al., 2009). Obviously, the presence of DCR and cross-talk effects must be taken into account. A proper model of the detector

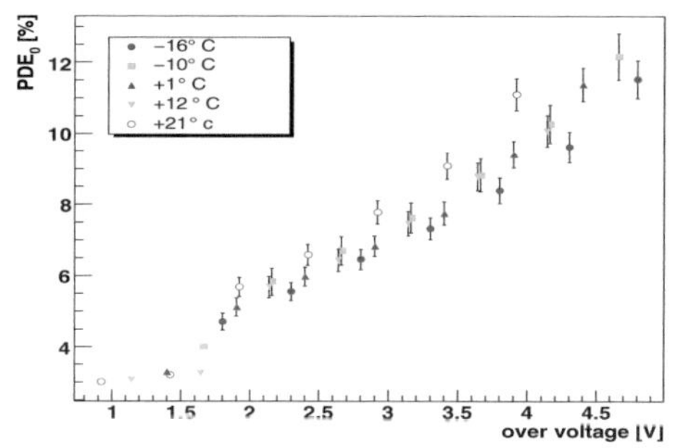

Fig. 8. PDE values corrected taking into account cross-talk effects, for a SensL SiPM. The wavelength of the impinging light is 510 nm.

	Hamamatsu MPPC S10362-11-100C
Number of Diodes:	100
Area:	1 mm × 1 mm
Diode dimension:	100 μm × 100 μm
Breakdown Voltage:	69.23 V
Dark Count Rate:	540 kHz at 70 V
Optical Cross-Talk:	25 % at 70 V
Gain:	$3.3 \cdot 10^6$ at 70 V
PDE (green):	15 % at 70 V

Table 2. Main features of the SiPM MPPC S10362-11-100C (Hamamatsu). The data refer to room temperature.

response is thus required in order to properly assess the statistical properties of the light field under investigation (Ramilli et al., 2010). In the following Section we present two different analysis methods, both validated by suitable measurements performed by using a SiPM produced by Hamamatsu Photonics. In particular, to keep DCR and cross-talk effects at reasonable levels and to optimize the PDE, a detector endowed with 100 cells was chosen (MPPC S10362-11-100C, see Table 2). To test the model, the sensor was illuminated by the pseudo-thermal light obtained by passing the second-harmonics (@ 523 nm, 5.4-ps pulse duration) of a mode-locked Nd:YLF laser amplified at 500 Hz (High Q Laser Production) through a rotating ground-glass diffuser (D in Fig.8). The light to be measured was delivered to the sensor by a multimode optical fiber (1 mm core diameter). The detector response was integrated by the charge digitizer V792 by Caen. The signal was typically integrated over a 200-ns-long time window synchronized with the laser pulse.

3.1 Detector response modeling

The response of an ideal detector to a light field can be described as a Bernoullian process (Mandel & Wolf, 1995):

$$B_{m,n}(\eta) = \binom{n}{m} \eta^m (1-\eta)^{n-m} \tag{10}$$

being n the number of impinging photons over the integration time, m the number of detected photons and $\eta < 1$ the PDE. Actually, η is a single parameter quantifying detector effects and losses (intentional or accidental) due to the optical system. As a consequence, the distribution $P_{m,\mathrm{det}}$ of the number of detected photons has to be linked to the distribution $P_{n,\mathrm{ph}}$ of the number of photons in the light under measurement by (Agliati et al., 2005; Mandel & Wolf, 1995; Zambra et al., 2004):

$$P_{m,\mathrm{det}} = \sum_{n=m}^{\infty} B_{m,n}(\eta) P_{n,\mathrm{ph}} \tag{11}$$

It can be demonstrated (Bondani et al., 2009a) that for a combination of classical light states the statistics is preserved by the primary detection process. This simple description has to be further developed to link $P_{m,\mathrm{det}}$ to the probability density distribution of the G-M avalanches of any origin. First we must take into account spurious hits and cross-talk effects, not negligible in the detectors under study. The DCR results in a Poissonian process that can be described as:

$$P_{m,\mathrm{dc}} = \overline{m}_{\mathrm{dc}}^{m} / m! \exp(-\overline{m}_{\mathrm{dc}}) \tag{12}$$

where $\overline{m}_{\mathrm{dc}}$ is the mean number of dark counts during the gate window and $\sigma^{(2)}_{m,\mathrm{dc}} = \sigma^{(3)}_{m,\mathrm{dc}} = \overline{m}_{\mathrm{dc}}$. As a consequence, the statistics of the recorded pulses may be described as:

$$P_{m,\mathrm{det+dc}} = \sum_{i=0}^{m} P_{i,\mathrm{dc}} P_{m-i,\mathrm{det}} \tag{13}$$

obviously shifting the mean value and increasing variance and third-order central moment in the following way: $\overline{m}_{\mathrm{det+dc}} = \overline{m}_{\mathrm{det}} + \overline{m}_{\mathrm{dc}}$, $\sigma^{(2)}_{m,\mathrm{det+dc}} = \sigma^{(2)}_{m,\mathrm{det}} + \overline{m}_{\mathrm{dc}}$ and $\sigma^{(3)}_{m,\mathrm{det+dc}} = \sigma^{(3)}_{m,\mathrm{det}} + \overline{m}_{\mathrm{dc}}$. As a further step, cross-talk effects must be taken into account. Cross-talk is a genuine cascade phenomenon that can be described at first order as (Afek et al., 2009)

$$C_{k,l}(x_t) = \binom{l}{k-l} x_t^{k-l} (1-x_t)^{2l-k}\,. \tag{14}$$

being x_t the (constant) probability that the G-M avalanche of a cell triggers a second cell (which becomes equivalent to the cross-talk probability X_T in the limit of $X_T \to 0$), l the number of dark counts and photo-triggered avalanches and k the actual light signal amplitude. Within this first-order approximation, the real sensor response is described by

$$P_{k,\mathrm{cross}} = \sum_{m=0}^{k} C_{k,m}(x_t) P_{m,\mathrm{det+dc}} \tag{15}$$

characterized by $\overline{k}_{\mathrm{cross}} = (1+x_t)\overline{m}_{\mathrm{det+dc}}$, $\sigma^{(2)}_{k,\mathrm{cross}} = (1+x_t)^2\sigma^{(2)}_{m,\mathrm{det+dc}} + x_t(1-x_t)\overline{m}_{m,\mathrm{det+dc}}$ and $\sigma^{(3)}_{k,\mathrm{cross}} = (1+x_t)^3\sigma^{(3)}_{m,\mathrm{det+dc}} + 3x_t(1-x_t^2)\sigma^{(2)}_{m,\mathrm{det+dc}} + x_t(1-3x_t+2x_t^2)\overline{m}_{m,\mathrm{det+dc}}$.

In the following we refer to this analytical model as *Method I* and compare it with a better refined model (*Method II*). This second method offers, in principle, an extended range of application even if it is limited to a numerical rather than an analytical solution. Irrespective of the model, the amplification and digitization processes that produce the output x can be simply described as a multiplicative parameter G:

$$P_{x,\mathrm{out}} = G P_{Gk,\mathrm{cross}} \tag{16}$$

whose moments are $\overline{x}_{\mathrm{out}} = G\overline{k}_{\mathrm{cross}}$, $\sigma^{(2)}_{x,\mathrm{out}} = G^2\sigma^{(2)}_{k,\mathrm{cross}}$ and $\sigma^{(3)}_{x,\mathrm{out}} = G^3\sigma^{(3)}_{k,\mathrm{cross}}$.

3.1.1 Method I: an analytical evaluation of the second and third order momenta

In Ref. (Bondani et al., 2009b) we presented a self-consistent method aimed at reconstructing the statistics of detected photons. According to this method, the analysis of the output of the detector is based on the assumptions that the detection process is described by a Bernoullian convolution and that the overall amplification-conversion process is given by a very precise constant factor G, which allows the shot-by-shot detector output to be converted into the number of detected photons. Experimentally, the value of G can be obtained by detecting a light field at different optical losses and keeping the detector parameters fixed. Once the amplification-conversion factor has been evaluated, the detected-photon distribution can be achieved by dividing the output values by G and re-binning the data in unitary bins. Here we present the extension of the method to detectors with a significant DCR and first order cross-talk effects. The second-order momentum of the recorded pulse distribution $P_{x,\text{out}}$ can be used to evaluate the Fano factor:

$$F_{x,\text{out}} = \frac{\sigma^{(2)}_{x,\text{out}}}{\overline{x}_{\text{out}}} = \frac{Q_{\text{det+dc}}}{\overline{m}_{\text{det+dc}}}\overline{x}_{\text{out}} + G\frac{1+3x_t}{1+x_t}, \tag{17}$$

where $Q_{\text{det+dc}} = \sigma^{(2)}_{m,\text{det+dc}}/\overline{m}_{\text{det+dc}} - 1$ is the Mandel factor of the primary charges. Note that, due to dark counts, the coefficient of $\overline{x}_{\text{out}}$ in Equation 17 cannot be written as $Q_{\text{ph}}/\overline{n}$ (Andreoni & Bondani, 2009; Bondani et al., 2009b), that is, the coefficient $Q_{\text{det+dc}}/\overline{m}_{\text{det+dc}}$ does not only depend on the light field to be measured. Similarly we can calculate a sort of symmetry parameter

$$S_{x,\text{out}} = \frac{\sigma^{(3)}_{x,\text{out}}}{\overline{x}_{\text{out}}} = \frac{Q_{s,\text{det+dc}} - 3Q_{\text{det+dc}}}{\overline{m}^2_{\text{det+dc}}}\overline{x}^2_{\text{out}} + G\frac{1+3x_t}{1+x_t}\frac{Q_{\text{det+dc}}}{\overline{m}_{\text{det+dc}}}\overline{x}_{\text{out}} + G^2\frac{1+7x_t}{1+x_t}, \tag{18}$$

where $Q_{s,\text{det+dc}} = \sigma^{(3)}_{m,\text{det+dc}}/\overline{m}_{\text{det+dc}} - 1$. In the presence of dark counts both coefficients $(Q_{s,\text{det+dc}} - 3Q_{\text{det+dc}})/\overline{m}^2_{\text{det+dc}}$ and $Q_{\text{det+dc}}/\overline{m}_{\text{det+dc}}$ depend on the light being measured (Bondani et al., 2009a).

3.1.2 Method II: a numerical evaluation based on the photon-number resolving properties of SiPMs

The above mentioned self-consistent method is very powerful, but requires the acquisition of several histograms at varying η, which might not always be easily performed in practical applications: from this point of view, the possibility to analyze each spectrum independently looks complementary to the self-consistent approach. This analysis has been performed with a two-step procedure:

- the areas of the spectrum peaks have been computed in order to obtain an estimation of the number of counts per peak;
- the obtained data points have been fitted with a theoretical function, which takes into account the statistics of the light, the detection and all the deviations of the detectors from ideality, such as DCR and cross-talk effects.

To evaluate the area of each peak, we performed a multi-peak fit of the spectrum histogram modeling each peak with a Gauss-Hermite function (Van Der Marel et al., 1993):

$$\text{GH} = Ne^{-w^2/2}\left[1 + {}^3hH_3(w) + {}^4hH_4(w)\right] \tag{19}$$

where $w = (x - \bar{x})/\sigma$, N is a normalization factor, $\bar{x}$ the peak position and σ the variance of the Gaussian function. $H_3(w)$ and $H_4(w)$ are the third and the fourth normalized Hermite polynomials and their contribution gives the asymmetry of the peak shape, whose entity is regulated by the pre-factors ^3h and ^4h, with values in the range $[-1, 1]$. The global fit function of the spectrum is a sum of as many Gauss-Hermite functions as the number of resolved peaks.

By using Equation 19 it is possible to calculate the area A_n of the n-th peak as:

$$A_n = N_n \, \sigma_n \, (\sqrt{2\pi} + {}^4h_n). \tag{20}$$

The error σ_{A_n} has been calculated by propagating the errors on the fit parameters.
This analysis is also useful to estimate the system gain G: in fact, from the fitted values of the peak positions $\bar{x}_n$, the peak-to-peak distance Δ for all the resolved peaks is:

$$\Delta_{n,n+1} = \bar{x}_{n+1} - \bar{x}_n. \tag{21}$$

The associated error $\sigma_{\Delta_{n,n+1}}$ can be once again obtained by propagating the fit errors; furthermore, the value of G can be derived as a weighted average on all the peak-to-peak values obtained from the analysis of the histograms.

The effect of detection, DCR and amplification have been modeled as described in the previous Sections (see Equations 10-13 and Equation 16).

The effect of cross-talk has been described by using a Bernoullian process, in a way analogue to what has been done in Equation 14. However, as the cross-talk process is intrinsically a cascade phenomenon, its contribution has been calculated by adding higher order effects:

$$P_{k,cross} = \sum_{m=0}^{k} \sum_{n=0}^{m} \sum_{j=0}^{n} P_{k-m-n-j,det+dc} \, B_{m,k-m-n-j}(x_t) \, B_{n,m}(x_t) \, B_{j,n}(x_t); \tag{22}$$

where the terms like $B_{j,n}(x_t)$ stand for the Bernoullian distribution

$$B_{j,n}(x_t) = \binom{n}{j} x_t^j (1 - x_t)^{n-j} \,. \tag{23}$$

Such a higher order expansion is not trivial to be achieved by the self-consistent approach of *Method I*, in which an explicit analytic expression of $P_{x,\text{out}}$ is needed in order to calculate its moments. Here, as all the elements of interest ($\overline{m}_{el}$, $\overline{m}_{dc}$, x_t, the number of modes μ) will be obtained as parameters of a fit, this is not necessary and therefore $P_{x,\text{out}}$ can be just numerically computed as the fitting function.

The major limit of this approach is obvious: as all the information on the statistics of the system is obtained from the peak areas, this method can only be applied to peak-resolving histograms with a number of peaks larger than the number of free parameters of the fitting function, which, in the present analysis, can rise up to five.

3.2 Experimental results

In Ref. (Ramilli et al., 2010), we presented the experimental validation of the two complementary methods by using both coherent and pseudo-thermal light. In particular, we have demonstrated that it is not possible to use coherent light to derive the DCR contribution,

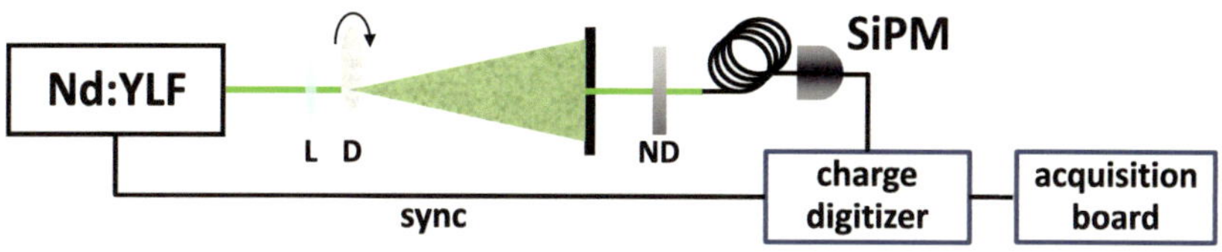

Fig. 9. Experimental setup. Nd:YLF: laser source, ND: variable neutral density filter, SiPM: detector.

which is described by a Poissonian distribution and is thus indistinguishable from the light under study. So we think that, within this context, it is more exhaustive and intriguing to show the results obtained with thermal light. The photon-number distribution of a field made of μ independent thermal modes is given by:

$$P_{n,\mathrm{ph}} = \frac{(n+\mu-1)!}{n!\,(\mu-1)!\,(N_{th}/\mu+1)^{\mu}\,(\mu/N_{th}+1)^{n}}\,, \qquad (24)$$

with $\overline{n} = N_{th}$, $\sigma_n^{(2)} = N_{th}\,(N_{th}/\mu+1)$ and $\sigma_n^{(3)} = N_{th}\,(N_{th}/\mu+1)\,(2N_{th}/\mu+1)$.

We reproduced a single mode of this pseudo-thermal light field by passing the coherent light of our laser source through an inhomogeneous diffusing medium and then selecting a single speckle with an aperture ($\sim$ 150 μm diameter), much smaller than the coherence area of the speckle pattern produced (Arecchi, 1965). By delivering this pseudo-thermal light to the SiPM sensor, we measured the values of the output, x, at 50000 subsequent laser shots and at 10 different mean values, obtained by means of a variable neutral-density filter (ND in Fig. 9). In Fig. 10 the plot of the experimental values of $F_{x,\mathrm{out}}$ and $S_{x,\mathrm{out}}$ is shown as a function of $\overline{x}_{\mathrm{out}}$, along with the fitting curves evaluated from Equations 17 and 18:

$$F_{x,\mathrm{out}} = \left(1-\frac{\overline{x}_{\mathrm{dc}}}{\overline{x}_{\mathrm{out}}}\right)^{2}\overline{x}_{\mathrm{out}} + B$$

$$S_{x,\mathrm{out}} = A\left(1-\frac{\overline{x}_{\mathrm{dc}}}{\overline{x}_{\mathrm{out}}}\right)^{3}\overline{x}^{2}_{\mathrm{out}} + 3B\left(1-\frac{\overline{x}_{\mathrm{dc}}}{\overline{x}_{\mathrm{out}}}\right)^{2}\overline{x}_{\mathrm{out}} + C\,, \qquad (25)$$

where μ, the number of independent thermal modes, has been set equal to one and, for simplicity of notation, the parameters $A = 2$, $B = G(1+3x_t)/(1+x_t)$ and $C = G^2(1+7x_t)/(1+x_t)$ have been introduced. We fitted the data to $F_{x,\mathrm{out}}$ thus obtaining the values of $\overline{x}_{\mathrm{dc}} = (5.82028 \pm 1.34015)$ and $B = (87.805 \pm 2.09009)$ ch.

Then the data for $S_{x,\mathrm{out}}$ have been fitted substituting the above values of $\overline{x}_{\mathrm{dc}}$ ch and B in order to obtain $A = (2.34754 \pm 0.091576)$ and $C = (8531.48 \pm 419.571)$ ch^2. These values were then used to evaluate G and x_t, obtaining $G = (74.2785 \pm 1.76982)$ ch and $x_t = (0.100174 \pm 0.050529)$. The x-values were then divided by G and re-binned in unitary bins (Bondani et al., 2009b) to obtain the $P_{k,\mathrm{cross}}$ distribution of the actual light signal amplitude measured in the presence of dark counts and cross-talk. In Fig. 11 six different $P_{k,\mathrm{cross}}$ distributions are plotted as bars at different mean values. Superimposed to the experimental values, two theoretical distributions are shown: the first one (white circles) is evaluated including the dark-count contribution that modifies the statistics of a single-mode thermal

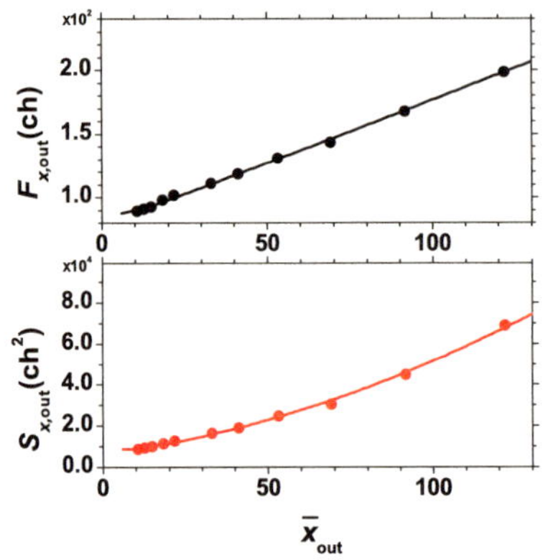

Fig. 10. Plot of $F_{x,\mathrm{out}}$ and $S_{x,\mathrm{out}}$ as functions of $\overline{x}_{\mathrm{out}}$ for pseudo-thermal light.

distribution (see Equation 24) into

$$\begin{aligned} P_{m,\mathrm{det+dc}} &= \sum_{k=0}^{m} P_{k,\mathrm{dc}} P_{m-k,\mathrm{det}} = \\ &= \frac{e^{-\overline{m}_{\mathrm{dc}}}}{(\mu-1)!}\left(1+\frac{\mu}{\overline{m}_{\mathrm{det}}}\right)^{-m}\left(1+\frac{\overline{m}_{\mathrm{det}}}{\mu}\right)^{-\mu} \\ & U\left[-m, 1-m-\mu, \overline{m}_{\mathrm{dc}}\left(1+\frac{\mu}{\overline{m}_{\mathrm{det}}}\right)\right], \end{aligned} \tag{26}$$

where $U(a,b,z)$ is the confluent hypergeometric function. The parameters are evaluated as $\overline{m}_{\mathrm{dc}} = \overline{x}_{\mathrm{dc}}/(G(1+x_t))$ and $\overline{m}_{\mathrm{det}} = (\overline{x}_{\mathrm{out}} - \overline{x}_{\mathrm{dc}})/(G(1+x_t))$. The second curve (full circles) is evaluated from Equation 15 to take into account the cross-talk. Unfortunately, the calculation does not yield an easy analytical result, and hence it has been evaluated numerically.

The comparison between the data and the theoretical functions can be estimated through the evaluation of the fidelity $f = \sum_{k=0}^{m} \sqrt{P_{k,\mathrm{exp}} P_{k,\mathrm{theo}}}$ (see Fig. 11).

Turning now to the other approach, it is worth noting that the number of fit parameters is large: the probability distribution is described by the expectation value $\overline{m}_{\mathrm{det}}$ of the avalanches generated by detection, the expectation value $\overline{m}_{\mathrm{dc}}$ of DCR contribution, the number of modes μ, the probability x_t of triggering a cross-talk event (up to three "iterations") and again a global normalization factor, for a total of 5 fit parameters: obviously, this puts a severe limit on the applicability of this method, needing at least 6 resolved peaks.

As it can be noted from the fit results in Fig. 12, the results obtained by using *Method II* are compatible within errors with what we found by applying *Method I*. However, even if the global fits present a very low χ^2 value for degree of freedom, the obtained fit parameters present large uncertainties, probably indicating the presence of very large off-diagonal elements in the minimization matrix and suggesting a strong correlation between the various parameters. This problem can be avoided by fixing some of the fit parameters (such as DCR or cross-talk), once their values have been retrieved from an accurate direct measurement (see Section 2).

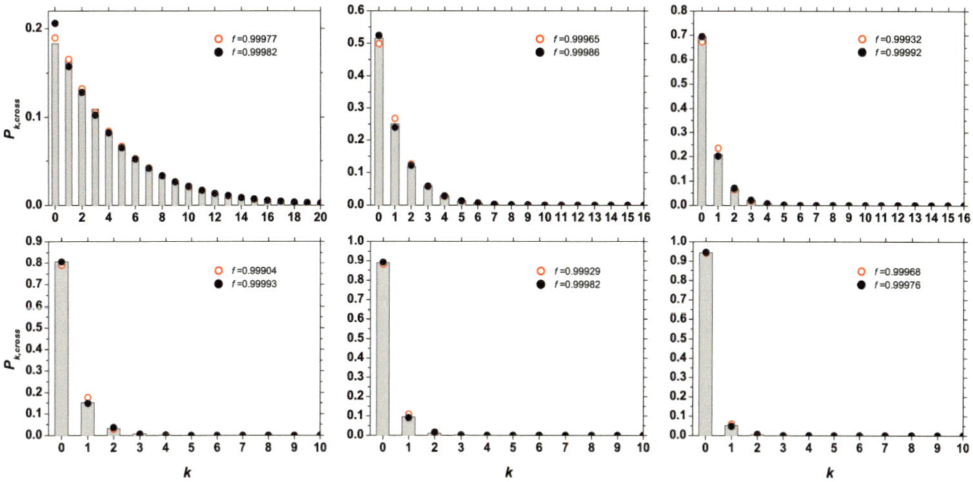

Fig. 11. Experimental $P_{k,\text{cross}}$ distributions at different mean values (bars) and theoretical distributions evaluated according to *Method I*: thermal modified by dark count distribution (white circles), thermal modified by dark counts and cross-talk effect (full circles). The corresponding fidelity values of the reconstruction are also shown.

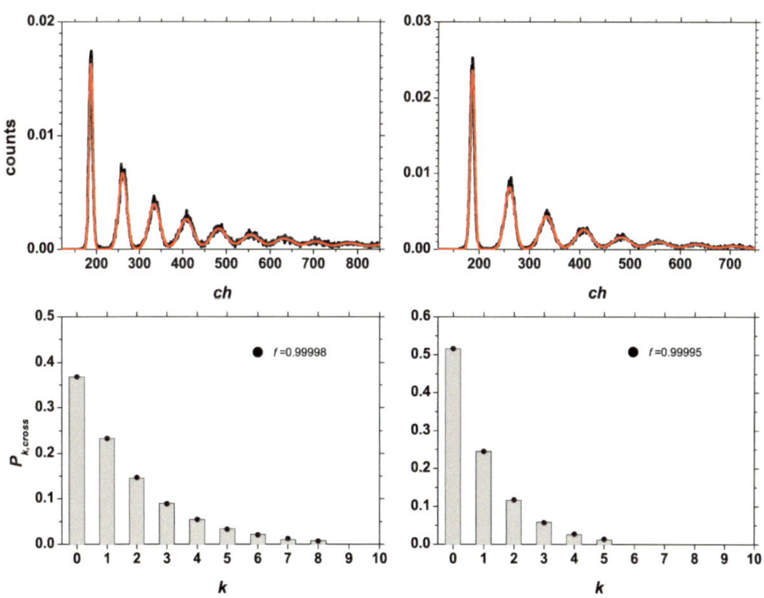

Fig. 12. Experimental results for *Method II* applied on two of the histograms acquired with thermal light. Upper panels: result of the multi-peak fit procedure; lower panels: fitted theoretical function. The corresponding fidelity values of the reconstruction are also shown.

3.3 Discussion

It is worth comparing the results of the two analysis methods on the same data sets (see Fig. 13). It is possible to notice that the agreement is better for the mean values of detected photons (panel (a)) and for the DCR (panel (b)), while the estimated values of x_t with the two *Methods* definitely disagree. This can be due to the different approximations adopted by the two *Methods* (first order *vs* third order).

In Table 3 we summarize the results of the two *Methods*. Both *Methods* work in a self-consistent way, even if they have two definitely different approaches. *Method I* does not need peak resolving capability, but requires the acquisition of several histograms at varying η. Once determined the parameters x_t and DCR, all the datasets in a series can be analyzed, independent of the number of distinguishable peaks in the pulse-height spectrum. *Method II* works analyzing each histogram independently, but, as the G-M-avalanches distribution is obtained with a fit of the data, it requires at least a number of resolved peaks larger than the number of free parameters.

The fact that the two *Methods* give very similar results for mean photon numbers is particularly important as in most applications this is the only important parameter. Merging the two *Methods* we can develop an optimal strategy based on a self-consistent calibration performed by measuring a known light and analyzing the data with *Method I*: once x_t and DCR are known, the determination of the mean photon number is independent of the specific light statistics. Hence the information on the mean photon number can be obtained from each single measurement, even when the fitting procedure of *Method II* cannot be applied.

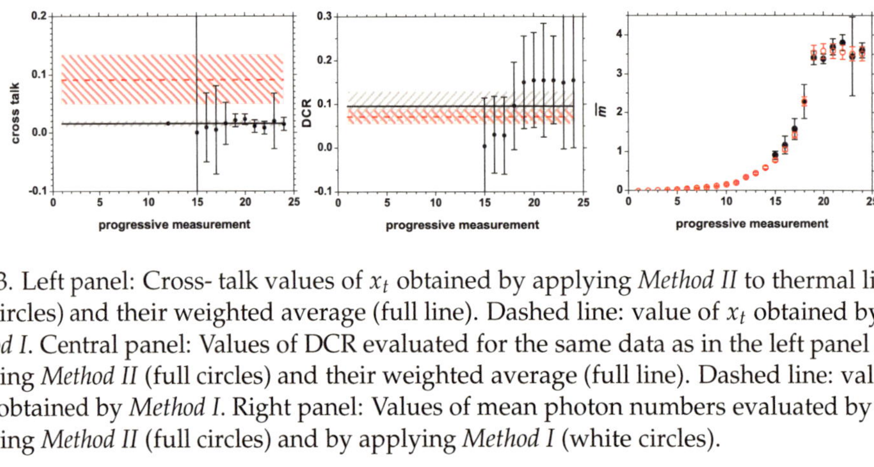

Fig. 13. Left panel: Cross- talk values of x_t obtained by applying *Method II* to thermal light (full circles) and their weighted average (full line). Dashed line: value of x_t obtained by *Method I*. Central panel: Values of DCR evaluated for the same data as in the left panel by applying *Method II* (full circles) and their weighted average (full line). Dashed line: value of DCR obtained by *Method I*. Right panel: Values of mean photon numbers evaluated by applying *Method II* (full circles) and by applying *Method I* (white circles).

	Pseudo-thermal	
	Method I	*Method II*
DCR	0.071 ± 0.017	0.094 ± 0.035
Cross-talk	0.091 ± 0.042	0.035 ± 0.004

Table 3. Comparison between the global DCR and cross-talk values obtained with *Method I* and weighted average of the values obtained with *Method II*.

4. Photon counting histogram with SiPM

Another instance in which the capability of resolving pulse-by-pulse the number of photons in high-repetition-rate periodic light signals is the analysis of fluorescence fluctuations. Traditionally, fluorescence techniques are applied to the study of systems constituted by huge ensembles of particles. For this reason, the probability distribution of the photon emission rate is peaked around its average value and fluctuations in the fluorescence signal are negligibly small. Nevertheless, if a microscopic system in which only few fluorophores are excited at a time is considered, the fluctuations in the fluorescence intensity become comparable to the average fluorescence intensity value, and can thus be measured. As these fluctuations are due to either the diffusion of the fluorophores in and out from the part of the specimen from which the fluorescence is collected and conveyed to the fluorescence detector, the so-called observation value, or to physical/chemical reactions inducing changes in the emission rate, their analysis can be applied both to characterize the diffusion kinetics and to study the physical/chemical reactions affecting emission. Such techniques are collectively called fluorescence fluctuation spectroscopy (FFS) techniques. This sort of analysis is particularly useful in probing biological systems both in vitro and in vivo. Indeed, many biomolecules (e.g. enzymes) and substances (such as minerals or ions) that are present in traces in cells have a determinant role in the cellular metabolism, and techniques allowing to study their action at physiological concentrations are most desirable. The most popular FFS technique is fluorescence correlation spectroscopy (FCS), which was developed in the early 1970′s (Gosch et al., 2005; Schwille, 2001), and consists in the analysis of the temporal correlations of fluorescence fluctuations in small molecular ensembles, combining maximum sensitivity with high statistical confidence. Among a multitude of physical parameters that are in principle accessible by FCS, it most conveniently allows to determine characteristic rate constants of fast reversible and slow irreversible reactions of fluorescently labelled biomolecules at very low (nanomolar and sub-nanomolar) concentrations. Moreover, the technique is a non-invasive one, which means that the above parameters can be assessed without perturbing the system equilibrium and, possibly, under physiological conditions. The technique also allows the determination of the mobility coefficients and local concentrations of fluorescent (or fluorescently labelled) molecules in their natural environments, provided that the exact value of the observation volume is known. More recently, at the end of the 1990′s, another FFS technique, called photon counting histogram (PCH), was developed by Chen et al. (Chen et al., 1999; 2005). This technique is based on the analysis of the statistical distribution of the fluctuations in amplitude of the fluorescence rate and is thus suited to yield steady state parameters of the system to be investigated, such as the number of different fluorescent species present in a solution, the average number of molecules of each species in the observation volume, and their molecular brightness, rather than dynamic parameters. This technique is thus to a certain extent complementary to FCS. Until now, both FCS and PCH measurements have been made by using APDs or other single-photon counters as the light detectors. Information on either the time-correlations or the statistical distribution of the fluorescence intensity fluctuations have been derived by counting the detection events in consecutive and equally long time intervals. As single-photon counters must be used only in the single photon regime (that is keeping the overall count rate very low in order to be sure that impingement on the detector sensitive area of photons during their dead time is a negligibly rare event), this sampling procedure is extremely time consuming and the sampling interval width must not be too small. Namely, in the most frequent case of pulsed excitation, it has to correspond to tens or even hundreds of pulse periods, which means tens of microseconds or more. Obviously, this puts an inferior limit on the sensibility to fluorescence fluctuations occurring on time scales comparable, or shorter than, the sampling time. Conversely, many biological reactions occur on time scales of hundreds of nanoseconds to few microseconds (Nettels et al., 2007), that is slightly beyond the detection limit with the current procedures. We believe that the usage, for FFS techniques, of detectors able to count how many photons are contained in each fluorescence pulse could allow avoiding sampling. This will significantly and inherently enhance the temporal resolution of both techniques, virtually pulling it on the scale of the excitation source pulse period. Another advantage connected to the release of the single-photon regime operation condition is the possibility of performing statistically reliable measurements in much shorter times, which is notably beneficial for in-vivo analyses, as the samples quickly deteriorate. Recently, we have addressed the task of performing PCH analysis with SiPM detectors. Firstly, the standard PCH model developed for fitting data acquired with APDs (Chen et al., 1999; 2005) is not suitable to properly analyze the SiPM output; thus, the phenomenological model for describing the SiPM response presented above has been used to modify the PCH equations so as to take into account the deviations for ideal detection caused by the presence of optical cros-talk and sizeable DCR, leading to the following fitting model:

$$\Pi^{\star}\left(k; \bar{N}, \epsilon, x_t, \langle n \rangle_{dcr}\right) =$$
$$= \sum_{N=0}^{\infty} p_{\#}(N; \bar{N}) \sum_{m=0}^{k} C_{k,m}(x_t) \sum_{n=0}^{\infty} P_n(\langle n \rangle_{dcr}) p^{(N)}(m-n; \epsilon) \tag{27}$$

where x_t is the parameter representing the probability for a cell of triggering a neighbour cell by cross-talk and $C_{k,m}(x_t)$ is the Bernoullian-like distribution defined in Section 3.1 describing the probability of measuring k G-M avalanches out of n previously triggered, with $k > m$. Secondly, a two-photon excitation setup has been commissioned and tested by performing standard PCH measurements on Rodamine B water solutions with a single-photon avalanche diode (PDM50, MPD) and a PC board allowing for real-time PCH reconstruction (SPC150, Becker & Hickl GmbH). The setup relied on fluorescence excitation by means of a continuous-wave SESAM mode-locked Ti:sapphire laser (Tiger-ps SHG, Time Bandwidth Product) delivering 3.9 ps pulses of 840 nm wavelength at a repetition rate of 48 Mhz. A 1.25 NA, 100X oil-immersion microscope objective (Nikon) was used both to focus the excitation beam onto the sample and to collect the emitted light in epifluorescence configuration. The fluorescence signal was separated from the back-scattered excitation light by using a dichroic mirror. Residual stray light was removed by using a short-wavelength-pass filter with cut at 700 nm. The fluorescence was delivered to the detector, through a 1 mm diameter multimode fiber. With this setup, appreciably superPoissonian PCH distributions were obtained for <300 nM dye solutions. Finally, a series of feasibility studies on the same dye were performed by using as the detector a Hamamatsu SiPM (MPPC S10362-11-100C). Given the 48 MHz repetition rate of the laser light source, the SiPM is expected to detect a fluorescence light pulse every 20 ns. In such experimental conditions, the usual detector signal processing seemed unfeasible with the instrumentation at hand by both the integration method and the free-time signal acquisition method. Indeed, in the first case the CAEN V792N QDC used for the quantum-light statistics measurements presented above requires an integration gate starting at least 15 ns before the rise of the signal, meaning that the board would not be able to process every pulse; moreover, with a detector recovery time of $\sim$ 60 ns, the output signal may exhibit pile-up features, which in turn would lead to difficulties in reconstructing the photon statistics. In the second case, the DT5720A Desktop Digitizer at our disposal samples the output signal every 4 ns, which means that the signal integration would be performed using only 5 points. For these reasons, a completely different approach to process the detector signal was followed. Namely, the coincidence frequency $\nu(x)$ between a threshold scan of the SiPM output and a trigger signal ν_T synchronous with the laser pulses (but with a threefold reduced frequency) has been acquired. The measured frequency can be described by the following relation:

$$\nu(x) = \nu_T \left(1 - \mathrm{e}^{-t_C \nu_S \int_x^{+\infty} \mathrm{d}y P(y)}\right) + \nu_T t_C \nu_S \int_x^{+\infty} \mathrm{d}y P(y); \tag{28}$$

where $P(y)$ is the probability of measuring a value y of the detector output signal, $\nu(x)$ is the threshold value, ν_S is the laser repetition rate and t_C is the coincidence time window. Equation 28 is based on the fact that a coincidence occurs both if after a trigger signal at least a SiPM signal over the threshold x is triggered within the coincidence time window or if the trigger signal rises within the coincidence time after the SiPM signal crosses the threshold. For small expectation values, Equation 28 can be approximated to

$$\nu(x) \approx 2\nu_T \nu_S t_C \int_x^{+\infty} \mathrm{d}y\, P(y). \tag{29}$$

Because of the Fundamental Theorem of Integration the detector output signal probability distribution can be written as

$$P(x) = \frac{-1}{2\nu_T \nu_S t_C} \frac{\mathrm{d}\nu(x)}{\mathrm{d}x} \tag{30}$$

which can be processed in the same way as the experimental detector output signal histograms experimentally obtained in the case of quantum-light statistics measurements to yield the probability distribution of the number of detected photons (see Section 3). In this set of measurements, the detector was controlled by the CAEN SP5600 General Purpose Power Supply and Amplification Unit module with an integrated threshold discriminator. The digital NIM output of the discriminator was put in coincidence with a synchronism signal triggered by the laser by using an EG&G CO4010 module. The coincidence frequency was sampled by a CAEN VME V1718. The fluorescence emission of two solutions of rhodamine B in water, with concentrations of 2 μM and 200 nM, respectively, have been analyzed with the threshold scan method described above. For the 2 μM concentrated solution a 1 optical density neutral filter was placed in front of the fiber, to reduce the mean detected fluorescence intensity to the value of the 200 nM concentrated solution. This makes the comparison between the two resulting distributions easier. Indeed, for equal expectation value, the 200 nM solution PCH should feature larger deviations from Poisson distribution, therefore showing a higher tail. Hereafter, the pertaining results are briefly summarized and their significance on the way towards implementation of PCH with SiPMs is outlined. Using Equation 30 the multi-peak spectra reported in Fig. 14 were obtained from the staircases for the 200 nM and the 2 μM solution, respectively. The PCH were derived from the spectra by applying the same procedure implemented in the quantum-light statistics data analysis. The zero-photons frequency was determined by imposing the normalization condition:

$$\Pi^\star(0) = 1 - \sum_{k=1}^{L} \Pi^\star(k) \tag{31}$$

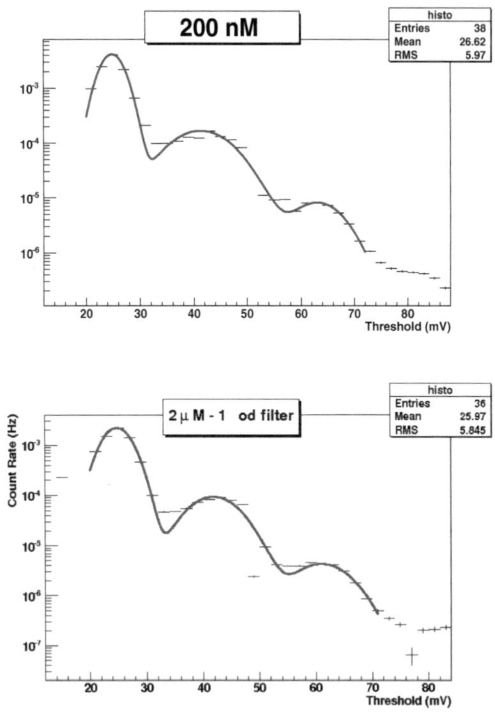

Fig. 14. Multipeak spectra of the fluorescence obtained from the threshold scans using Equation 30; in the upper panel spectrum from the 200 nM solution is shown, while in the lower panel the one obtained from the 2 μM solution is presented. Since the staircases has been acquired with a 2 mV step, histograms have 2 mV large bins. Spectra have been fitted with a multi-Gaussian function and the areas of each peak have been used as the PCH entries.

where L is the last detectable peak. The PCH are displayed in Figs. 15, for the 200 nM and the 2 μM solution, respectively. Fitting to Equation 27, with the average number of dark counts fixed to the experimental value preliminarily determined by performing a coincidence measurement of the dark counts, yields the fitting curves reported as dashed lines in Figs. 15.

The fitting parameters are summarized in Table 4.

Parameters	Concentrations (nM)	
	200	2000
$\bar{N}$	5.24 ± 0.03	71.70 ± 0.03
ϵ	$3.99 \times 10^{-3} \pm 2.4 \times 10^{-5}$	$1.536 \times 10^{-4} \pm 7 \times 10^{-7}$
x_t	$0.08085 \pm 5 \times 10^{-5}$	$0.0637 \pm 1 \times 10^{-4}$
$\langle n \rangle_{dcr}$	3.24×10^{-3}	3.24×10^{-3}

Table 4. Results for the fit of the PCH distributions. The errors listed are obtained from the fit procedure. The results for the mean number of particles and the molecular brightness are consistent with what expected, even though there is not exactly a factor 10 between the parameters of the two data sets.

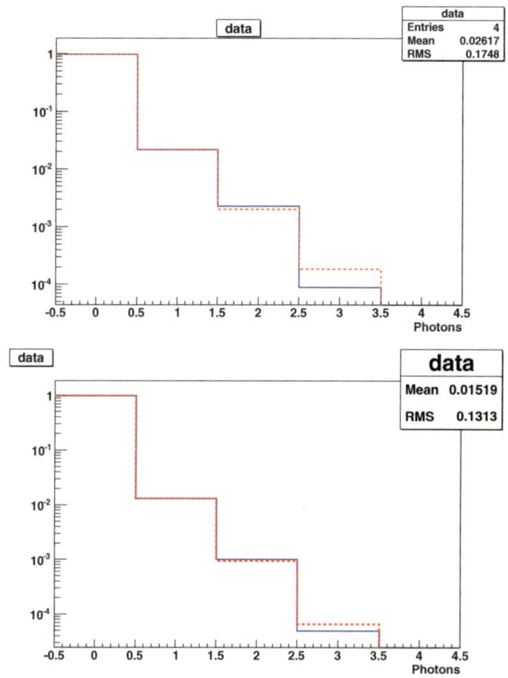

Fig. 15. Fit (red dashed line) of the experimental PCH data (blue continuous line) for 200 nM concentration (upper panel) and 2 μM concentration (lower panel).

The brightness and average particle number values are fairly consistent with the experimental conditions. Namely, the mean numbers of particles in the excitation volume differ by approximately one order of magnitude, reflecting the 10:1 concentration ratio between the two samples. To this purpose it should be noted that, while the concentration of the 2 μM solution could have been determined spectrophotometrically, 200 nM represents only a nominal value of the concentration of the less concentrated sample, obtained by dilution of the first, as the absorbance of it is below the sensitivity of our spectrophotometer (≈ 0.001) in this case. Also the difference in molecular brightness is roughly consistent with the insertion of the 1-optical-density filter. Indeed, the fact that the molecular brightness measured for the 2 μM concentrated sample is found to be more than twenty times lower could be ascribed to a slight misalignment of either the laser or the microscope objective during the measurement session. The latter lasted for many hours due to the necessity of determining with high statistical reliability the very low frequencies of coincidences occurring at a high threshold value. Even a slight degeneration in the quality of the optical setup alignment would determine a notable decrease in the two-photon absorption cross section, which scales as the square power of the excitation intensity. This would be reflected in the brightness value. A broader focus is also consistent with the mean number of molecules in the excitation volume detected for the 2 μM solution, which is slightly larger than expected. Finally, the difference between the cross-talk values is consistent with a lower over-biasing condition for the 2 μM set, suggesting that a worsening of the photon detection efficiency could also have contributed to the lower molecular brightness. It must be admitted that the above presented procedure is very far from ideal pulse-by-pulse PCH acquisition. However, these preliminary data demonstrate that despite the relatively high dark count rates with respect to APDs and the additional artifice constituted by optical cross-talk, SiPMs are detectors capable of discriminating the tiny deviations from Poisson distribution typically displayed by the PCH of diffusing fluorophores in solution at concentrations relevant for biophysical applications. Moreover, we notice that our fitting model yields consistent values of brightness and average fluorophore number. The above conclusions suggest that pulse-by-pulse reconstruction of PCH should be straightforwardly obtained with our detector and electronics apparatus for excitation rates <10 MHz, according to both the recovery time of the SiPM detector and the requirements of the CAEN integrator. An optical pulse-picker may be used to customize the Ti:sapphire repetition rate. Even working at such rates would allow to speed up PCH reconstruction by more than two orders of magnitude with respect to state-of-art APD-based methods. Alternatively, high-repetition rate measurements might be performed by reconstruction methods based on the analysis of the detector free-run output signal, as recorded by fast-sampling-rate digitizers. In such a case, the issues connected with detector pile-up effects would be overcome, thus confining the limits on the acquisition rate to the excitation rate and to the intrinsic decay time of the fluorophores under analysis.

5. Conclusions

We have presented the characterization of a class of photodetectors, namely the Silicon Photomultipliers, that has acquired great importance in the last decades because of its good photon-counting capability. Besides this feature, the compactness, robustness, low cost, low operating voltage and power consumption are added value that would make these detectors suitable for many applications. To this aim, we have implemented an exhaustive characterization protocol providing a quantitative evaluation of the main figures of merit these detectors are endowed with. In particular, we have studied the dependence of the photon detection efficiency, gain, dark count rate and optical cross-talk on temperature. The protocol has been applied to three SiPM detectors produced by different manufactures and the results have been then employed in a partnership with CAEN to develop the SP5600 General Purpose Power Supply and Amplification Unit module.

The versatility of SiPMs has been investigated in the reconstruction of the statistics of light states: two complementary models that take into account the DCR and the cross-talk effects have been developed in order to describe the output signal of the detectors. The correctness of these methods has been experimentally validated by illuminating the SiPM sensor with a single-mode pseudo-thermal light. However, further improvements could be achieved by including in the models the afterpulsing effect, which would become important in high-rate events.

The applicability of SiPMs in Fluorescence Fluctuation Spectroscopy studies has been preliminary evaluated by analyzing Rhodamine solutions by means of a Two-Photon Excitation setup: the capability of the sensors to reconstruct Photon Counting Histogram and in particular to provide the estimation of the mean number of fluorophores in the excitation volume and their molecular brightness has been proved.

6. References

Afek, I.; Natan, A.; Ambar, O.; & Silberberg, Y. (2009). Quantum state measurements using multipixel photon detectors. *Phys. Rev. A.*, Vol. 79, No. 4, April 2009, 043830, ISSN 1050-2947.

Agliati, A.; Bondani, M.; Andreoni, A.; De Cillis, G.; & Paris, M. G. A. (2005). Quantum and classical correlations of intense beams of light investigated via joint photodetection. *J. Opt. B Quantum Semiclass. Opt.*, Vol. 7, No. 12, November 2005, S652–S663, ISSN 1464-4266.

Akindinov, A. V.; Martemianov, A. N.; Polozov, P. A.; Golovin, V. M.; & Grigoriev, E. A. (1997). New results on MRS APDs. *Nuclear Instruments and Methods in Physics Research Section A: Accelerators, Spectrometers, Detectors and Associated Equipment*, Vol. 387, No. 1-2, March 1997, 231–234, ISSN 0168-9002.

Akindinov, A; Bondarenko, G.; Golovin, V.; Grigoriev, E.; Grishuk, Yu.; Malkevich, D.; Martemiyanov, A.; Ryabinin, M.; Smirnitskiy, A.; & Voloshin, K. (2005). Scintillation counter with MRS APD light readout. *Nuclear Instruments and Methods in Physics Research Section A: Accelerators, Spectrometers, Detectors and Associated Equipment*, Vol. 539, No. 1-2, February 2005, 172–176, ISSN 0168-9002.

Andreoni, A. & Bondani, M. (2009). Photon statistics in the macroscopic realm measured without photon counters. *Phys. Rev. A*, Vol. 80, No. 1, July 2009, 013819, ISSN 1050-2947.

Arecchi, F. T. (1965). Measurement of the Statistical Distribution of Gaussian and Laser Sources. *Phys. Rev. Lett.*, Vol. 15, No. 24, December 1965, 912–916, ISSN 0031-9007.

Bondani, M.; Allevi, A.; & Andreoni, A. (2009). Light Statistics by Non-Calibrated Linear Photodetectors. *Adv. Sci. Lett.*, Vol. 2, No. 4, December 2009, 463–468, ISSN 1936-6612.

Bondani, M.; Allevi, A.; Agliati, A.; & Andreoni, A. (2009). Self-consistent characterization of light statistics. *J. Mod. Opt.*, Vol. 56, No. 2, January 2009, 226–231, ISSN 0950-0340 print/ 1362-3044 online.

Chen, Y.; Müller, J. D.; So, P. T. C.; & Gratton, E. (1999). The Photon Counting Histogram in Fluorescence Fluctuation Spectroscopy. *Biophys. J.*, Vol. 77, No. 1, July 1999, 553–567, ISSN 0006-3495.

Chen, Y.; Tekmen, M.; Hillesheim, L.; Skinner, J.; Wu, B., & Muller, J. D. (2005). Dual-color photon counting histogram. *Biophys. J.*, Vol. 88, No. 3, March 2005, 2177–2192, ISSN 0006-3495.

Cova, S; Ghioni, M.; Lacaita, A.; Samori, C.; & Zappa, F. (1996). Avalanche photodiodes and quenching circuits for single-photon detection. *Appl. Opt.*, Vol. 35, No. 12, April 1996, 1956–1976, ISSN 1559-128X print/ 2155-3165 online.

Du Y. & Reti'ere, F. (2008). After-pulsing and cross-talk in multi-pixel photon counters. *Nuclear Instruments and Methods in Physics Research Section A: Accelerators, Spectrometers, Detectors and Associated Equipment*, Vol. 596, No. 3, November 2008, 396–401, ISSN 0168-9002.

Eckert, P.; Schultz-Coulon, H.-C.; Shen, W.; Stamen, R.; & Tadday, A. (2010). Characterisation studies of silicon photomultipliers. *Nuclear Instruments and Methods in Physics Research Section A: Accelerators, Spectrometers, Detectors and Associated Equipment*, Vol. 620, No. 2-3, August 2010, 217–226, ISSN 0168-9002.

Goetzberger, A.; McDonald, B.; Haitz, R. H.; & Scarlett, R. M. (1963). Avalanche Effects in Silicon p-n Junctions. II. Structurally Perfect Junctions. *Journal of Applied Physics*, Vol. 34, No. 6, June 1963, 1591–1600, ISSN 0021-8979.

Gösch, M. & Rigler, R. (2005). Fluorescence correlation spectroscopy of molecular motions and kinetics. *Adv. Drug. Del. Rev.*, Vol. 57, No. 1, January 2005, 169–190, ISSN 0169-409X.

Lutz, G. (1995). Silicon radiation detectors. *Nuclear Instruments and Methods in Physics Research Section A: Accelerators, Spectrometers, Detectors and Associated Equipment*, Vol. 367, No. 1-3, December 1995, 21–33, ISSN 0168-9002.

Mandel, L. & Wolf, E. (1995). *Optical Coherence and Quantum Optics*, Cambrige University Press, ISBN 0521417112, New York.

McKay, K. G. (1954). Avalanche Breakdown in Silicon. *Phys. Rev.*, Vol. 94, No. 4, May 1954, 877-884.

Nettels, D.; Gopich, I. V.; Hoffmann, A.; & Schuler, B. (2007). Ultrafast dynamics of protein collapse from single-molecule photon statistics. *Proc. Nat. Acad. Sci. USA*, Vol. 104, No. 8, February 2007, 2655–2660, ISSN 0027-8424.

Oldham, W. G.; Samuelson, R. R.; & Antognetti, P. (1972). Triggering phenomena in avalanche diodes. *IEEE Trans. Electron. Dev.*, Vol. 19, No. 9, September 1972, 1056-1060, ISSN 0018-9383.

Ramilli, M.; Allevi, A.; Chmill, V.; Bondani, M.; Caccia, M.; & Andreoni, A. (2010). Photon-number statistics with Silicon photomultipliers. *J. Opt. Soc. Am. B*, Vol. 27, No. 5, May 2010, 852–862, ISSN 0740-3224.

Schwille, P. (2001). Fluorescence correlation spectroscopy and its potential for intracellular applications. *Cell Biochem. Biophys.*, Vol. 34, No. 3, June 2001, 383–408, ISSN 1085-9195.

Sciacca, G.; Condorelli, E.; Aurite, S.; Lombardo, S.; Mazzillo, M.; Sanfilippo, D.; Fallica, G.; & Rimini, E. (2008). Crosstalk characterization in Geiger-mode avalanche photodiode arrays. *IEEE Electron. Dev. Lett.*, Vol. 29, No. 3, March 2008, 218-220, ISSN 0741-3106.

Tarolli, A.; Dalla Betta, G.-F.; Melchiorri, M.; Piazza, A.; Pancheri, L.; Piemonte, C.; & Zorzi, N. (2010). Characterization of FBK SiPMs under illumination with very fast light pulses. *Nuclear Instruments and Methods in Physics Research Section A: Accelerators, Spectrometers, Detectors and Associated Equipment*, Vol. 617, No. 1-3, May 2010, 430–431, ISSN 0168-9002.

Van Der Marel, R. P. & Franx, M. (1993). A new method for the identification of non-gaussian line profiles in elliptical galaxies. *Astrophysical Journal*, Vol. 407, No. 2, April 1993, 525–539, ISSN 0004-637X.

Zambra, G.; Bondani, M.; Spinelli, A. S.; Paleari, F.; & Andreoni, A. (2004). Counting photoelectrons in the response of a photomultiplier tube to single picosecond light pulses. *Rev. Sci. Instrum.*, Vol. 75, No. 8, August 2004, 2762–2765, ISSN 0034-6748 print/1089-7623 online.

5

High Sensitivity Uncooled InAsSb Photoconductors with Long Wavelength

Yu Zhu Gao
College of Electronics and Information Engineering, Tongji University, Shanghai China

1. Introduction

It is very attractive to research infrared (IR) materials and detectors in the 8-12 μm wavelength range for photodetection applications. Currently, HgCdTe is the dominant material system in this wavelength range. However, it suffers from chemical instability and nonuniformity due to the high Hg vapor pressure during its growth (Kim et al., 1996). Among III-V compound semiconductors, InAsSb alloy with a band gap as small as 0.1 eV has the advantages of high electron and hole mobilities, good operating characteristics at high temperatures, and high chemical stability. Therefore, the InAsSb system is a very promising alternative long-wavelength IR material to HgCdTe.

However, the lattice mismatch between InAsSb epilayers and the binary compound substrates is rather large (for InAs is larger than 6%, for GaAs is 7.2 ~ 14.6%). Thus it is very difficult to grow high-quality InAsSb single crystals with cutoff wavelengths of 8-12 μm using conventional technologies (Kumar et al., 2006). Narrow-gap InAsSb epilayers have been grown by molecular beam epitaxy (MBE) (Chyi et al., 1988), metalorganic chemical vapor deposition (MOCVD) (Kim et al., 1996), and liquid phase epitaxy (LPE) (Dixit et al., 2004). The thicknesses of these epilayers are about 2-10 μm. The dislocation densities observed in these thin films are as high as the order of 10^7 cm^{-2} that are caused by a large lattice mismatch (Kumar et al., 2006). It markedly lowers the terminal performance of the detectors.

A melt epitaxy (ME) method for growth of InAsSb single crystals on InAs and GaAs substrates with the wavelengths longer than 8 μm was proposed for a first time by Gao et al (Gao et al., 2002, 2006). The thickness of InAsSb epilayers reaches several decades ~ 100 μm. This thickness effectively eliminates the effect of lattice mismatch and results in a low dislocation density (the order of 10^4 cm^{-2}) in epilayers with a lattice mismatch larger than 6%. Based on the thick InAsSb epilayers grown by ME, high-sensitivity uncooled photoconductors with long wavelength were successfully fabricated (Gao et al., 2011). The IR photodetectors operating at room temperature need not coolers, thus have the important advantages of high speed, small volume, and good reliability. The response speed of them is more than three orders of magnitude faster than that of thermal detectors.

2. Melt epitaxy

We prepared InAsSb epilayers in a standard horizontal LPE growth system with a sliding fused silica boat in high-purity hydrogen ambient. Fig. 1 (a) shows the slideboat schematic

(Gao et al., 2002). The bottom of the tail part of the melt holder for the fused silica boat is flat. The original melt composition was $InAs_{0.04}Sb_{0.96}$. Seven Newtons of Sb, In, and non-doped InAs polycrystalline were employed as the source materials for the melt. The substrates were (100) oriented n-InAs and semi-insulating GaAs substrates. The growth melt of ME is not diluted, thus the composition of the solid is less different from that of the melt. We measured the composition of InAsSb solids by electron probe microanalysis (EPMA). Table 1 lists the composition and corresponding cutoff wavelength of InAsSb epilayers (Gao et al., 1999).

The growth process is as follows: Firstly, the InAs substrate and the source materials were set in the fused silica boat. The temperature of the furnace was increased to nearly 650°C and was kept at this temperature for 1 hour to mix the growth melt sufficiently. Next, the temperature was slowly decreased until 500°C with a cooling rate of 1°C/min. At this growth temperature, the melt contacts with the substrate, and the excess growth melt is immediately removed away from the substrate by pushing the melt holder. The key point is as follows: at the suitable growth temperature obtained by observation, some melt remains on the surface of substrate. Then the substrate was pushed under the flat part of the melt holder and cooled for 10°C at a cooling rate of 0.4°C/min to obtain an epilayer. Figs. 1(b)-(d) show the slideboat arrangement before contact, during contact with the melt, and during cooling for solidification, respectively. The thickness of the epilayer is dependent on the difference between the thickness of the substrate and the depth of the substrate well of the boat. Usually, the thickness of the grown layer reaches several decades ~ 100 μm. The samples were ground and polished using Al_2O_3 powder to obtain a flat and mirror-smooth surface for the characterization measurements.

We summarize the main differences between ME and LPE: (1) For the LPE, the growth process is controlled by solute diffusion. The solid composition is different from the melt composition. However, for the ME, the solute is not strongly distributed during the growth. The difference between the composition of the epilayer and that of the melt is small. (2) In the case of LPE, all of the growth melt is removed from the substrate after finishing growth. However, in the case of ME, a portion of the melt is aliquoted onto the substrate before growth, and this melt fraction is crystallized under the flat part of the melt holder.

x	Substrate	λ (μm)
0.946	InAs	9.5
0.948	InAs	10.5
0.95	InAs	11.0
0.96	InAs	12.5

Table 1. Compositional dependence of cutoff wavelengths of $InAs_{1-x}Sb_x$ epilayers (Gao et al., 1999)

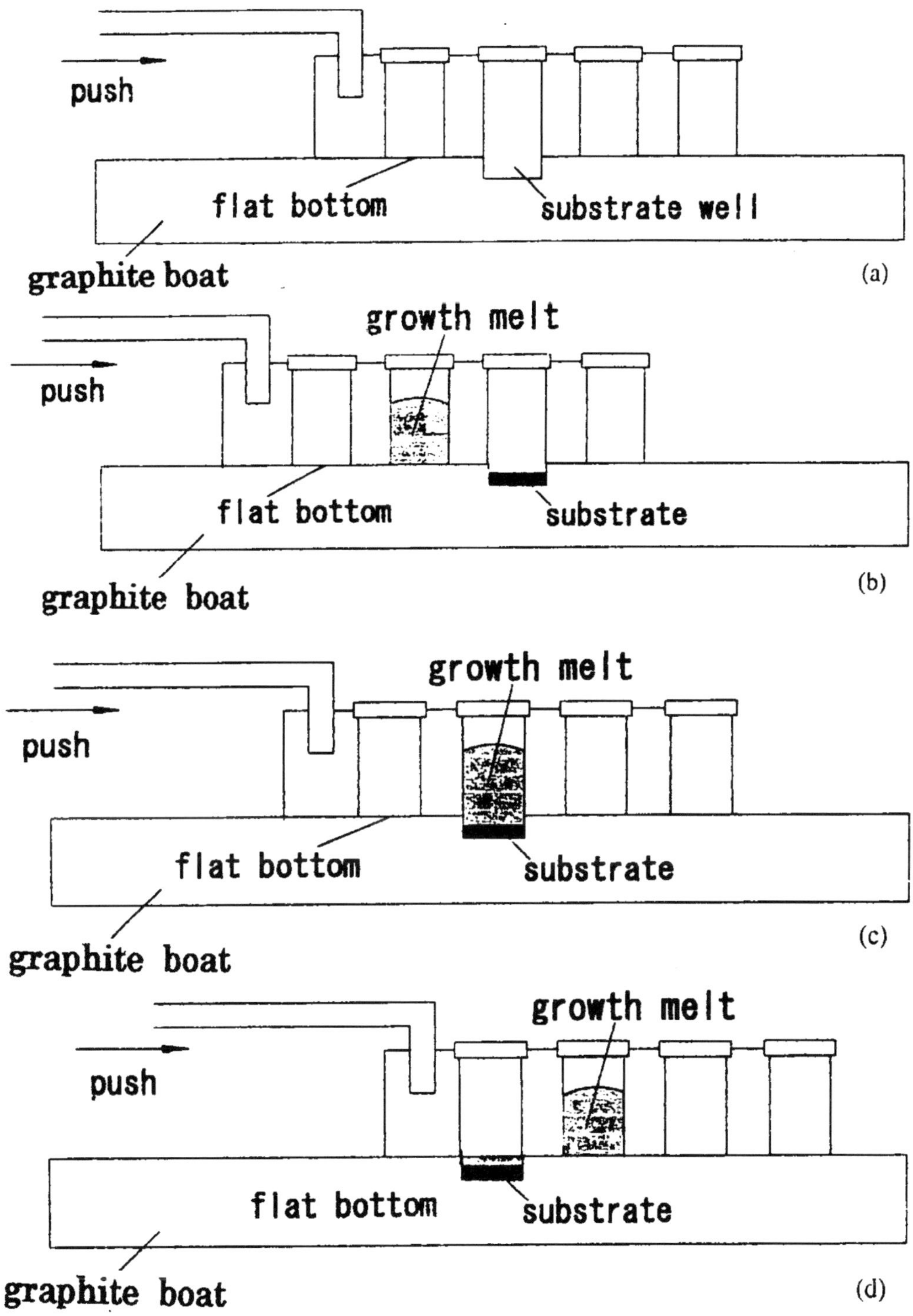

Fig. 1. Fused silica slideboat schematic (a) and the slide boat arrangement before contact (b), during contact with the melt (c) and during cooling for solidification (d) (Gao et al., 2002).

3. Characteristics of InAsSb single crystals grown by ME

3.1 Transmission spectra

Transmittance measurements for InAs/InAsSb samples were performed at 300 K using a Fourier transform infrared (FTIR) spectrophotometer (JIR-WINSPEC50, JEOL, Japan). In Fig. 2, the transmittance spectra for several InAsSb epilayers with different compositions are shown (Gao et al., 1999). We defined the cutoff wavelength as the mid-transmittance wavelength, and the results are summarized in Table 1. The band gap narrowing of ME-grown InAsSb is possibly caused by the increase in the bowing parameter, which may be induced by the lattice contraction of the ternary alloy. The lattice constant of $InAs_{0.051}Sb_{0.949}$ grown by ME is 6.4572 Å (see section 3.2), which is smaller than that of InSb (6.4789 Å). Fig. 2 shows that transmittances drop as the wavelength becomes longer than the intrinsic absorption region. This is attributed to the free carrier absorption in the epilayer. The absorption dip around 22-25 μm cannot be seen for InSb single crystals suggesting that it is related to InAs lattice vibrations. To and Lo Raman peaks for InAs are known to be 219 cm^{-1} (45.6 μm) and 236 cm^{-1} (42.4 μm) respectively (Gong et al., 1994). The observed dips are almost half the wavelength of these values, therefore they may be due to double frequency of lattice vibrations of InAs, i.e., the second harmonics.

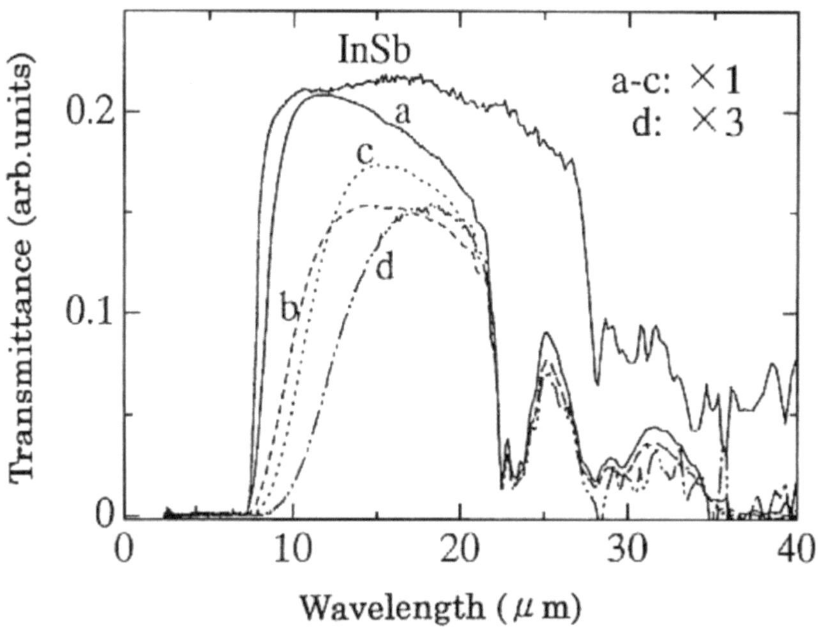

Fig. 2. Transmission spectra for several InAs/InAsSb samples with different compositions on InAs substrates. The cutoff wavelengths as defined by half transmittance are summarized in Table 1 (Gao et al., 1999).

In order to investigate the temperature dependence of the energy band gap, the temperature dependent absorption spectra were measured at 12, 77, 115 and 290 K for a $GaAs/InAs_{0.04}Sb_{0.96}$ sample with a total thickness of 600 μm. The energy band gaps of 0.147, 0.141, 0.136, and 0.112 eV were obtained respectively, at corresponding temperatures for this sample. To examine the temperature dependence of the energy band gap, we calculated the energy band gaps using a well-known Yarshni's empirical formula from 0 to 300 K (Yarshni, 1967):

$$E_g(T) = E_g(0) - \alpha T^2 / T + \beta \tag{1}$$

where $E_g(0)$ is the energy band gap at zero temperature, while a and β are two empirical parameters. Fig. 3 shows the calculated curve (solid line) and the date obtained from absorption measurements (opened squares) (Gao et al., 2006). Using least-square fits to equation (1), we obtained a = 0.147 meV/K, β = 60 K, and E_g (0) = 0.1475 eV, respectively.

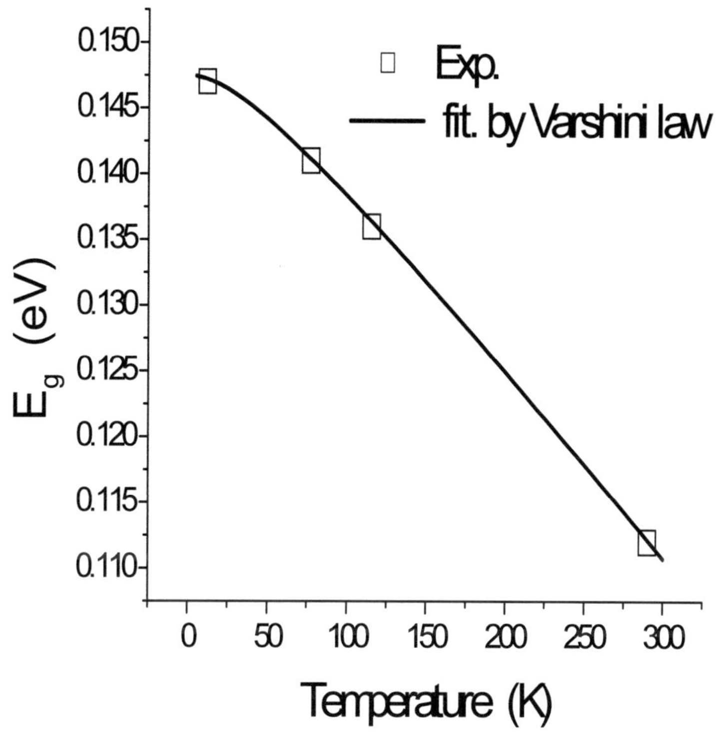

Fig. 3. Temperature dependence of energy band gaps for a GaAs/$InAs_{0.04}Sb_{0.96}$ sample (Gao et al., 2006).

3.2 Structural characteristics

The cross section of InAsSb epilayers was observed by scanning electron microscopy (SEM) (FEI Quanta 200F) at a magnification of 500. Fig. 4 shows a SEM cross-sectional image of an InAs/InAsSb sample. The boundary between the epilayer and substrate is flat and straight. It indicates the diffusion has not strongly occurred at the boundary for the growth temperature of about 500°C. This phenomenon may be benefitted from the special process of ME and the slow cooling rate of 0.4°C/min during the epitaxial growth. The cracks in the epilayer are caused by the cleavage. As is seen in Fig. 4, the thickness of the InAsSb epilayer grown by ME is about 40 μm, which is impossible to realize for MBE and MOCVD technologies. This thickness basically eliminates the effect of the lattice mismatch and results in the good crystal quality of the epilayer.

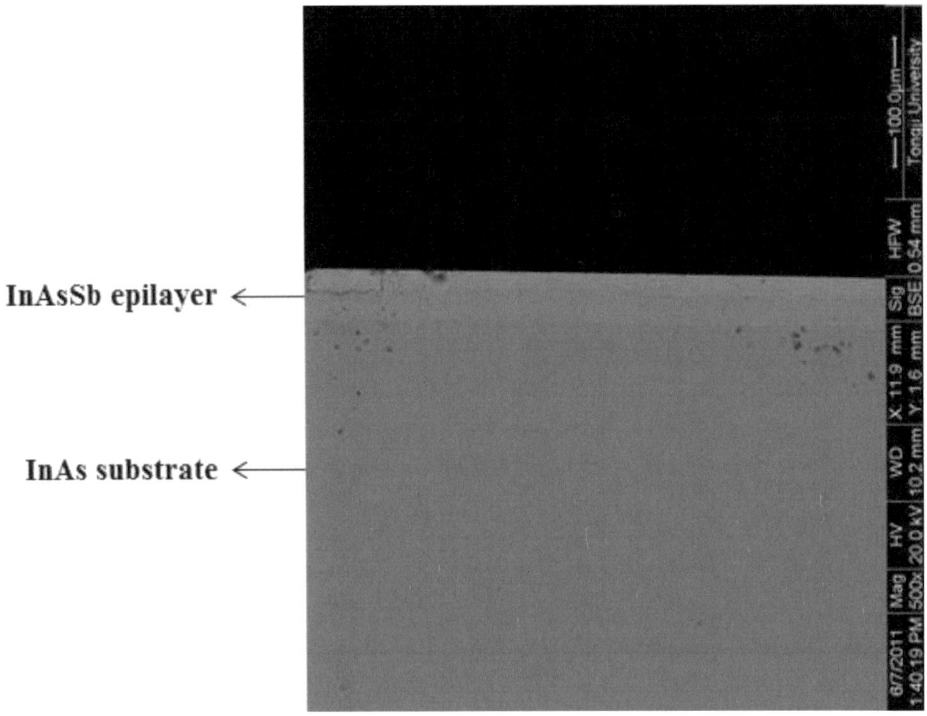

Fig. 4. Cross-sectional image of an InAs/InAsSb sample taken by SEM. The thickness of the InAsSb epilayer is about 40 μm.

Fig. 5 shows the X-ray diffraction (XRD) spectra of an InAs/InAsSb sample measured by an X-ray diffractometer (Rigaku D/MAX-2200PC, Cu barn) at a voltage of 40 kV, and a current of 40 mA (Gao et al., 2011). It is seen that (400) and (200) diffraction peaks of the InAsSb epilayer clearly appear, and no other crystal structures are observed. The growth direction of the epilayer is in agreement with the surface direction of the InAs substrate, i.e. the (100) orientation. This demonstrates that the InAsSb epilayer is indeed a single crystal. Since the thickness of the epilayer grown by ME reaches 40 μm, the diffraction peak of the InAs substrate does not appear. The sharpness and the full-width at half-maximum (FWHM) of 164 arcsec of the (400) diffraction peak indicate the high quality of InAsSb epilayers.

According to the Bragg diffraction equation, the lattice constant a for the $InAs_xSb_{1-x}$ sample shown in Fig. 5 is estimated to be 6.4572 Å. Based on the Vegard Law, the InAs mole fraction in epilayers can be calculated as:

$$x = (a_{\text{InAsSb}} - a_{\text{InSb}}) / (a_{\text{InAs}} - a_{\text{InSb}}) \tag{2}$$

where x is the InAs mole fraction in the epilayers. The InAs mole fraction is calculated to be 0.051. The calculated lattice mismatch between the $InAs_{0.051}Sb_{0.949}$ epilayer and the InAs substrate is as large as 6.58%.

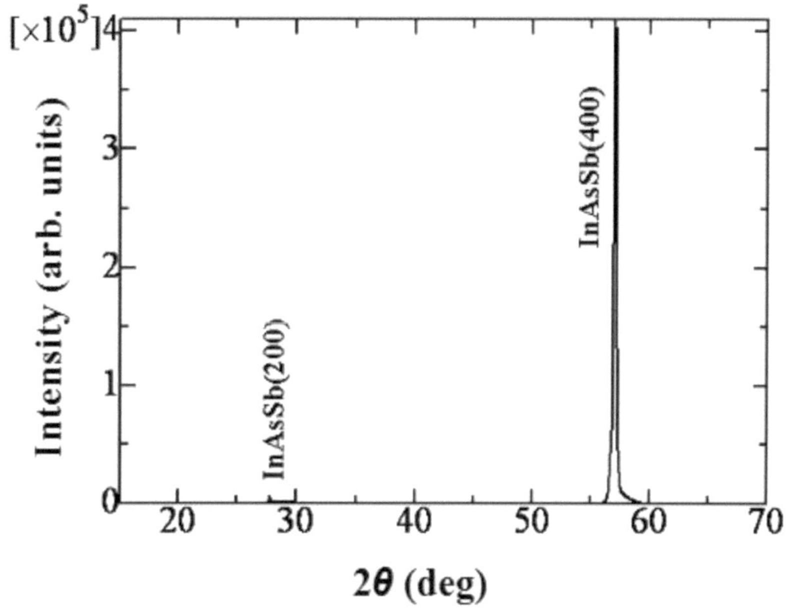

Fig. 5. XRD spectra of an InAs/InAsSb sample grown by ME. The lattice mismatch between the epilayer and substrate was estimated to be 6.58% (Gao et al., 2011).

3.3 Electrical properties

For measuring the electrical properties of the $InAs_{0.04}Sb_{0.96}$ epilayers, the surface of the samples should be ground and polished until mirror smooth, and the InAs substrate must be removed to eliminate the influence from the conductive InAs substrate. Moreover, an anodization treatment was performed on the surfaces of samples before measurements. Hence, the possibility of the presence of a surface conducting (accumulation) layer is also eliminated.

The carrier concentration and the electron mobility of InAsSb epilayers were measured using the Van der Pauw method with a standard Hall measurement system under a magnetic field of 2000 gauss. In was used as the contacts. To determine the temperature dependence of the electrical properties of the epilayers, the InAsSb sample holder was set in a cryostat. The temperature was changed from 300 to 50 K.

All $InAs_{0.04}Sb_{0.96}$ samples were n-type in the measurement temperature range. In Fig. 6 (Gao et al., 2004), measured carrier concentration as a function of temperature is shown for the two $InAs_{0.04}Sb_{0.96}$ epilayers with a thickness of 100 μm grown using a graphite boat and fused silica boat, respectively. It is seen that the carrier concentration of sample (a) grown using graphite boat has the same level of 0.2×10^{15} cm^{-3} at temperatures between 50 and 100 K, and then the carrier concentration gradually increases to 1.5×10^{15} cm^{-3} at 200 K. The carrier density of the $InAs_{0.04}Sb_{0.96}$ layer rapidly increases at the temperatures higher than 200 K due to the intrinsic generation of carriers in this temperature region. A carrier concentration of 2.3×10^{16} cm^{-3} was obtained at 300 K. This result is typical for narrow gap n-type InAsSb epilayers grown using a graphite boat by ME.

It is evident that except for the carrier density, which is about 3 times higher than that of sample (a), the tendency of the temperature dependence of the carrier concentrations for sample (b) grown using a fused silica boat is similar to that of sample (a).

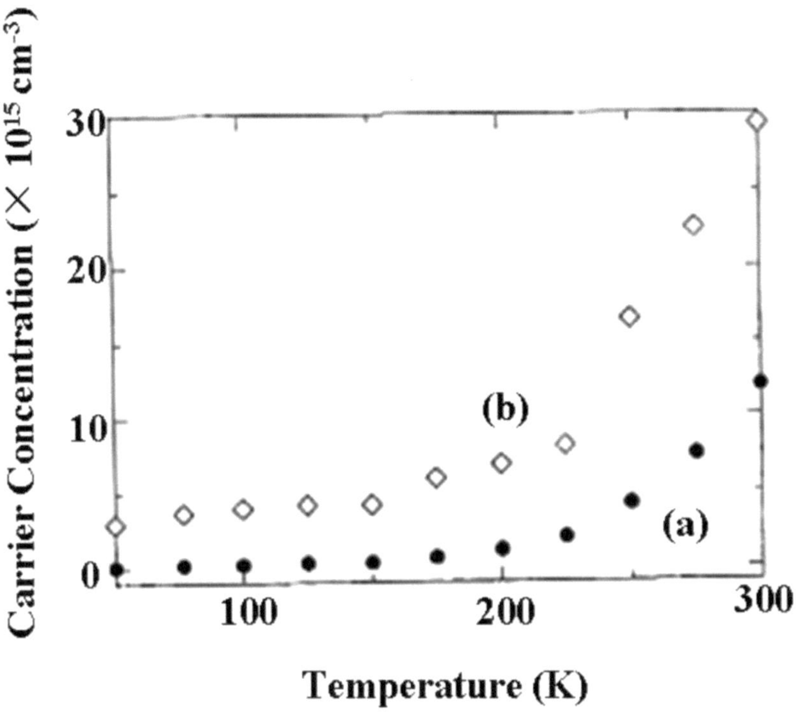

Fig. 6. Measured temperature dependence of carrier concentration for the two InAs/ $InAs_{0.04}Sb_{0.96}$ epilayers grown using (a) graphite boat; (b) fused silica boat (Gao et al., 2004).

Fig. 7 shows the measured temperature dependence of the electron mobility for the two InAs/$InAs_{0.04}Sb_{0.96}$ epilayers with a cutoff wavelength of 12 μm grown using graphite and fused silica boat (Gao et al., 2004).

It is seen that the electron mobility of sample (a) is lower than 2×10^4 cm^2/Vs at temperatures below 100 K, then quickly increases until 200 K. Electron mobility decreases again as temperature continuously increases to 300 K. Peak electron mobilities of 1×10^5 cm^2/Vs at 200 K, and 6×10^4 cm^2/Vs at 300 K have been obtained. This is the best result so far for InAsSb materials with cutoff wavelengths of 8-12 μm indicating the good quality of epilayers. Polar optical phonon scattering governs electron mobility at high temperatures. At low temperatures, alloy scattering is not important in these $InAs_{0.04}Sb_{0.96}$ epilayers because the As composition is only 0.04, and a dislocation density of the order of 10^4 cm^{-2} has been observed by counting the dislocation amount on the surface of the epilayers after chemical etching. Therefore, the rapid decrease in measured electron mobility with decreasing temperature below 175 K is probably due to impurity scattering.

The electron mobilities of sample (b) are all in the range of 4-5.27×10^4 cm^2/Vs at temperatures between 50 and 300 K. Only the peak points at 50 K and 225 K show mobilities higher than 5×10^4 cm^2/Vs. The most important phenomenon is that the electron mobility at 77 K is higher than that at 300 K. The purity improvement has possibly resulted from the reduction of carbon contamination in fused silica boat. This gives rise to a weaker impurity scattering in InAsSb epilayers grown with fused silica at 77 K.

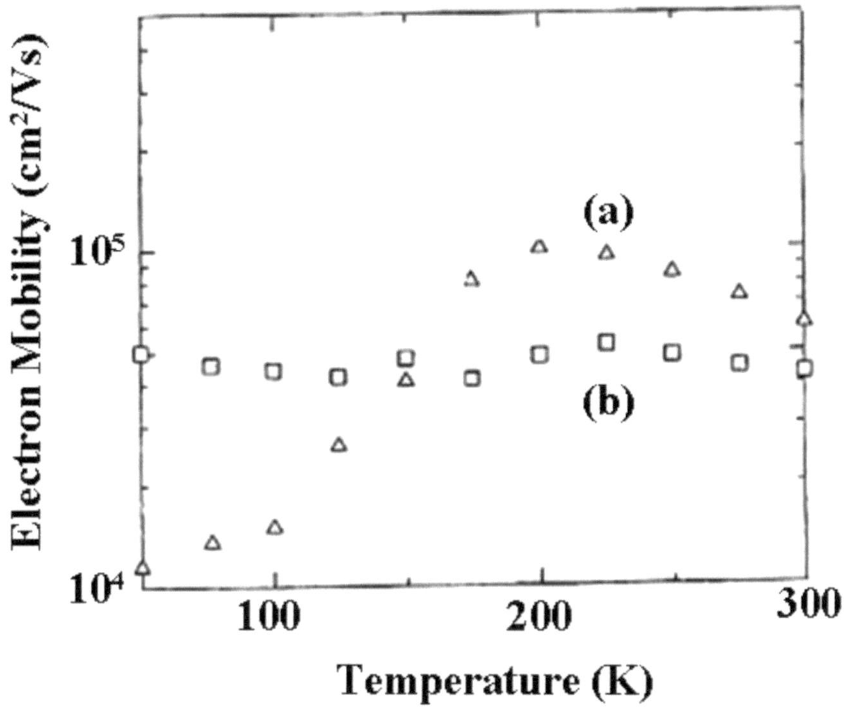

Fig. 7. Measured electron mobility as a function of temperature for the two InAs/$InAs_{0.04}Sb_{0.96}$ samples grown using (a) graphite boat; (b) fused silica boat (Gao et al., 2004).

The relation of ln n and T^{-1} of a GaAs/$InAs_{0.04}Sb_{0.96}$ sample is shown in Fig. 8, where ln n is the natural logarithm of electron concentration, and T^{-1} is the inverse of the temperature. As shown in Fig. 8, the saturation range appears between 40 K and 200 K. In this temperature range, donors are mainly ionized, i.e. most of electrons in donor level are excited to conduction band. Thus Fermi level is lower than donor level. However, most of electrons of valence band have not yet been excited to conduction band in this temperature region. When temperature increased to 200-300 K, the intrinsic region is observed as seen from Fig. 8. The electrons excited from valence band become the dominant conductive electrons due to intrinsic transition of carriers. Therefore, GaAs/$InAs_{0.04}Sb_{0.96}$ samples exhibit the intrinsic property in this temperature region. We obtained the relation between ln n and T^{-1} in the intrinsic region as follows (Yishida et al., 1989):

$$\ln n = (-E_g / 2k)T^{-1} + \ln(N_C N_V)^{1/2} \tag{3}$$

where N_C is effective density of state of conduction band, N_V is effective density of state of valence band, E_g is energy band gap, and k is Boltzmann's constant. In terms of equation (3), we calculated energy band gap E_g from the slope of intrinsic region in Fig. 8. The calculated energy band gap is 0.1055 eV, which is well consistent with the result obtained from room temperature transmittance measurements (0.1033 eV). This result demonstrates that the energy band gap of $GaAs/InAs_{0.04}Sb_{0.96}$ alloys grown by ME is strongly narrowed.

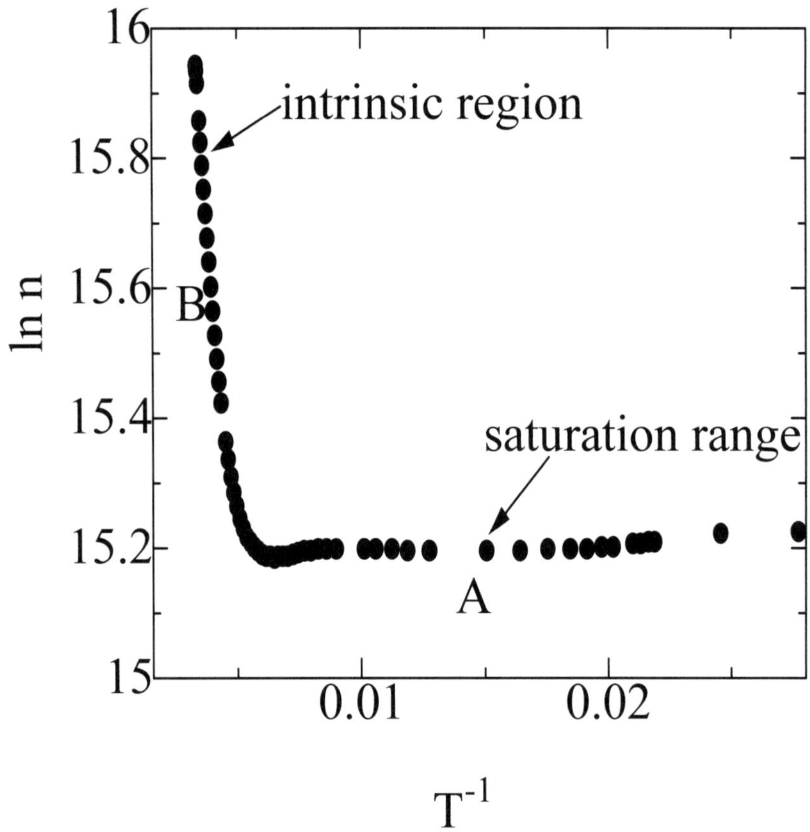

Fig. 8. Relation of ln n and T^{-1} for a $GaAs/InAs_{0.04}Sb_{0.96}$ sample (Gao et al., 2006).

4. Uncooled InAsSb photoconductors with long wavelength

4.1 Fabrication of immersion detectors

The immersion photoconductors were fabricated using InAs/InAsSb materials grown by ME. Since the thickness of the epilayers reached several decades ~ 100 μm, InAs substrates can be entirely removed away. Thus the effect of the lattice mismatch between epilayers and substrates were eliminated. In the device process, InAsSb epilayers were thinned to about 15 μm by grinding. Chemical polishing was carried out using a $C_3H_6O_3$ etchant after grinding to eliminate the mechanical damage to the wafers. Indium was used as the ohmic contact. The light-sensitive area of the photoconductors is 0.05 × 0.05 cm^2. The detector resistances are 20-110 Ω at 300 K. Ge optical lenses were set on the detectors. There are not any antireflective coatings deposited on the surfaces of the elements and lenses. Fig. 9 shows a photograph of uncooled InAsSb photoconductors.

Fig. 9. Photograph of uncooled InAsSb photoconductors.

4.2 Spectral photoresponse

The spectral photoresponse of InAsSb photoconductors were measured by a Fourier transform infrared (FTIR) spectrometer at room temperature, and the absolute responsivity was calibrated by a standard blackbody source at a temperature of 500 K and a modulation frequency of 1200 Hz. The bias current applied on the devices was 10 mA. Fig. 10 shows the detectivity D^* varying with the wavelength of uncooled $InAs_{0.051}Sb_{0.949}$ immersion photoconductors. At room temperature, the peak voltage responsivity is 164.3 V/W at the wavelength of 6.5 μm, resulting in the corresponding peak detectivity as high as 2.5×10^9 cm $Hz^{1/2} W^{-1}$. The peak detectivity of our detectors is more than one order of magnitude higher than that of long-wavelength type-II InAs-GaSb superlattice photodiodes (Mohseni et al., 2001). This may be caused by the following reasons: (1) the super-hemisphere immersion Ge lenses were set on our photoconductors. The incident IR radiation was focused by the lenses, thus the radiation energy density on the photosensitive surfaces was raised. It is well known that the super-hemisphere immersion component is able to increase the signal-to-noise ratio and detectivity of n_s^2 multiples, n_s = 4 is the refractive coefficient of Ge crystals. (2) ME-grown InAsSb epilayers with the thickness of several decades ~ 100 μm are the narrow gap materials, and have the properties of bulk single crystals. The narrow gap intrinsic semiconductors are particularly suitable for room temperature photon detector fabrication. The high density of states in the valence and conduction bands of them leads to the strong absorption of IR radiation (Piotrowski et al., 2004). As shown in Fig. 10 that D^* of 8.8×10^8 and 6.3×10^7 cm $Hz^{1/2} W^{-1}$ at the wavelength of 8.0 and 9.0 μm respectively, have been obtained.

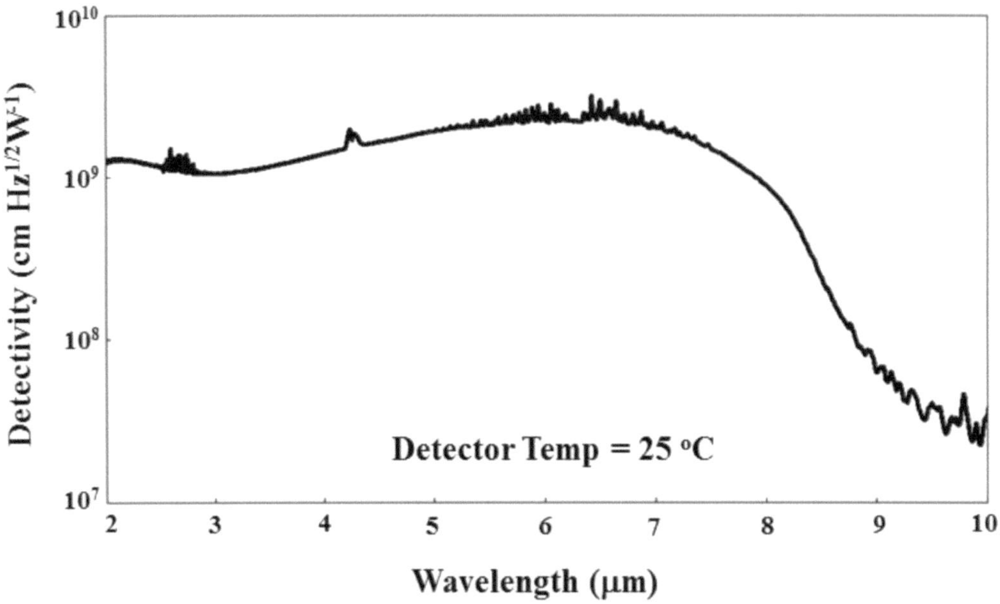

Fig. 10. Detectivity D^* versus wavelength of $InAs_{0.051}Sb_{0.949}$ immersion photoconductors at room temperature.

5. Conclusion

We provided high sensitivity uncooled InAsSb photoconductors with long wavelength. Ge optical lenses were set on the photoconductors without any antireflective coatings. The detectors are based on InAsSb epilayers with the thickness of 40 μm grown on InAs substrates by melt epitaxy. This thickness efficiently eliminates the effect of lattice mismatch and results in a low dislocation density (the order of 10^4 cm^{-2}) in epilayers, thus improves the terminal performance of detectors.

FTIR transmission spectra for $InAs_{1-x}Sb_x$ epilayers revealed the strongly band gap narrowing. The temperature dependence of energy band gap for $GaAs/InAs_{0.04}Sb_{0.96}$ was studied between 12 K and 290 K by measuring the absorption spectra. The structural properties of InAsSb materials were characterized by SEM observation and XRD spectroscopy. The measurement results of the materials exhibited the high quality.

The electrical properties were investigated by van der Pauw measurements. A peak electron mobility of 100,000 cm^2/Vs with a carrier density of 1×10^{15} cm^{-3} at 200 K, and an electron mobility of 60,000 cm^2/Vs with a carrier density of 2.3×10^{16} cm^{-3} at 300 K, have been obtained for an $InAs_{0.04}Sb_{0.96}$ epilayer. This is the best result so far, to our knowledge, for the InAsSb materials with cutoff wavelengths of 8-12 μm indicating the good quality of the epilayers. The purity of the epilayers grown with fused silica boat was improved. The purity improvement has possibly resulted from the reduction of carbon contamination in fused silica boat. A room temperature band gap of 0.1055 eV is demonstrated via analyzing the temperature dependence of the carrier density for the $GaAs/InAs_{0.04}Sb_{0.96}$ layers, which is in good agreement with the value obtained by transmittance measurements.

The spectral photoresponse of InAsSb photoconductors grown by ME were measured using a FTIR spectrometer at room temperature. The bias current applied on the devices was 10 mA. At room temperature, the peak detectivity $D_{\lambda p}$* (6.5 μm, 1200) reaches 2.5×10^9 cm $Hz^{1/2}$ W^{-1} for InAsSb immersion photoconductors. The detectivity D* at the wavelength of 8 μm is 8.8×10^8 cm $Hz^{1/2}$ W^{-1}, and that at 9 μm is 6.3×10^7 cm $Hz^{1/2}$ W^{-1}. The excellent performance of the detectors indicates the potential applications for IR detection and imaging.

6. References

[1] J. D. Kim, D. Wu, J. Wojkowski, J. Piotrowski, J. Xu, and M. Razeghi, "Long-wavelength InAsSb photoconductors operated at near room temperatures (200-300 K)", *Appl. Phys. Lett.*, 68, No. 1, 99 - 101 (1996).

[2] A. Kumar and P. S. Dutta, "Growth of long wavelength $In_xGa_{1-x}As_ySb_{1-y}$ layers on GaAs from liquid phase", *Appl. Phys. Lett.*, 89, No. 16, 162101-1 - 162101-3 (2006).

[3] J. -I. Chyi, S. Kalem, N. S. Kumar, C. W. Litton, and H. Morkoc, "Growth of InSb and $InAs_{1-x}Sb_x$ on GaAs by molecular beam epitaxy", *Appl. Phys. Lett.*, 53, No. 12, 1092 - 1094 (1988).

[4] V. K. Dixit, B. Bansal, V. Venkataraman, H. L. Bhat, K. S. Chandrasekharan, and B. M. Arora, "Studies on high resolution x-ray diffraction, optical and transport properties of $InAs_xSb_{1-x}$/GaAs ($x\leq0.06$) heterostructure grown using liquid phase epitaxy", *J. Appl. Phys.*, 96, No. 9, 4989 - 4997 (2004).

[5] Y. Z. Gao, H. Kan, F. S. Gao, X. Y. Gong, and T. Yamaguchi, "Improved purity of long-wavelength InAsSb epilayers grown by melt epitaxy in fused silica boats", *J. Cryst. Growth.*, 234, 85 - 90 (2002).

[6] Y. Z. Gao, X. Y. Gong, Y. H. Chen, and T. Yamaguchi, "High Quality $InAs_{0.04}Sb_{0.96}$/GaAs Single Crystals with a Cutoff Wavelength of 12 μm Grown by Melt Epitaxy", *Proc. of SPIE.*, 6029, 602911-1 - 602911-7 (2006).

[7] Y. Z. Gao, X. Y. Gong, G. H. Wu, Y. B. Feng, T. Makino, and H. Kan, "Uncooled InAsSb Photoconductors with Long Wavelength", *Jpn. J. Appl. Phys.*, 50, No. 6, 060206-1 - 060206-3 (2011).

[8] Y. Z. Gao, X. Y. Gong, H. Kan, M. Aoyama, and T. Yamaguchi, "$InAs_{1-y}Sb_y$ Single Crystals with Cutoff Wavelength of 8-12 μm Grown by a New Method", *Jpn. J. Appl. Phys.*, 38, No. 4A, 1939 – 1940 (1999).

[9] X. Y. Gong, H Kan, T. Yamaguchi, I. Suzuki, M. Aoyama, M. Kumagawa, N. L. Rowell, A. Wang, and R. Rinfret, "Optical Properties of High-Quality $Ga_{1-x}In_xAs_{1-y}Sb_y$/InAs Grown by Liquid-Phase Epitaxy", *Jpn. J. Appl. Phys.*, 33, No. 4A, 1740-1746 (1994).

[10] Y. P. Yarshni, "Temperature dependence of the energy gap in semiconductors", *Physica A.*, 34, 149 (1967).

[11] Y. Z. Gao, X. Y. Gong, Y. S. Gui, T. Yamaguchi, and N. Dai, "Electrical Properties of Melt-Epitaxy-Grown $InAs_{0.04}Sb_{0.96}$ Layers with Cutoff Wavelength of 12 μm", *Jpn. J. Appl. Phys.*, 43, No. 3, 1051 - 1054 (2004).

[12] T. Yishida and A. Shimizu, Semiconductor Devices, Corona Publishing Co., LTD, Tokyo, 1989. (In Japanese)

[13] H. Mohseni and M. Razeghi, "Long-Wavelength Type-II Photodiodes Operating at Room Temperature", IEEE PHOTONICS TECHNOLOGY LETTERS, 13, No. 5, 517 - 519 (2001).

[14] J. Piotrowski and A. Rogalski, "Uncooled long wavelength infrared photon detectors", *Infrared Phys. Technol.*, 46, 115 - 131 (2004).

6

Terahertz Emitters, Detectors and Sensors: Current Status and Future Prospects

M. Ghanashyam Krishna[1,2], Sachin D. Kshirsagar[1] and Surya P. Tewari[1,2]
[1]*Advanced Centre of Research in High Energy Materials*
[2]*School of Physics, University of Hyderabad, Hyderabad*
India

1. Introduction

The region of the electromagnetic spectrum from 0.3 to 20 THz (10 - 600 cm^{-1}, 1 mm - 15 μm wavelength) is an area for research that encompasses physics, chemistry, biology, materials science and medicine. The lack of sources of high quality radiation has limited developments in this area. However, there has been an upsurge of activity in the last five years due to the emergence of a wide range of new technologies. Terahertz radiation is now available in both continuous wave (cw) and pulsed form, down to single-cycles or less, with peak powers up to 10 MW [Ginzburg et al., 2011; Sizov& Rogalski, 2010; Williams, 2004; Tochitsky et al., 2005]. There is a wealth of information that can be extracted from a study of matter (solid, liquid or gaseous) at these frequencies. It is becoming evident that such directed basic research would also lead to a phenomenally large number of applications covering Physics, Chemistry, Life Sciences and Engineering [Jacobsen et al., 1996; Mueller, 2003; Chamberlain, 2004; Bolivar et al., 2004; Lewis, 2007; Tonouchi, 2007] .

As stated earlier, to achieve further understanding of physical phenomena in the THz region of the electromagnetic spectrum it is essential that there are reliable sources (including materials, devices and integrated systems) that emit THz radiation. Furthermore, there is need for materials, devices and systems that detect this radiation. Another aspect, which is central to all developments in THz science and technology is the development of reliable standards. Thus, THz metrology is another area that has to develop simultaneously with development of emitters and detectors. One of the applications, which is still in its infancy but has remarkable potential, is THz sensors for a variety of contexts in security, life sciences and medicine.

The aim of this review is to present the state-of-science and technology in the area of THz emitters, detectors and sensors. A simple search carried out on a general search engine with keywords "THz detectors or THz emitters" comes up with thousands of hits. Even a search on more focused science literature search engines and limited to the last five years, with the same keywords, generates over a thousand papers. This in essence, reveals the explosion of interest in the area of THz science as well as the ease of availability of sources, probes and tools for such studies. Therefore, the current review is limited to, what the authors consider, the most interesting and significant recent developments in the area of THz emitters, detectors and sensors. It is aimed at researchers intending to initiate work in these fields and therefore a considerable portion is focussed on basic principles. The rest of the review is organized in the following fashion: Section 2 will review some of the fundamental phenomena that occur in the THz region of the em spectrum; Section 3 will present a review on THz emitters; Section 4 is focused on THz detectors and sensors and; Section 5 will present an overview of the materials development work being carried out by the present authors. Section 6 will provide a summary and outlook for the future in this field

2. Significance of THz

The THz spectral range is characterized by several unique features [Jacobsen et al., 1996; Mueller, 2003; Chamberlain, 2004; Bolivar et al., 2004; Lewis, 2007; . Innocenzi et al., 2009; Roeser et al. 2010]. It is the highest frequency band where the field can be measured coherently without an interferometer. The THz region of the spectrum is sensitive to all of the thermally accessible excitations that determine the properties of correlated electron systems. Also, the ability to perform spectroscopy in the time as well as frequency-domain provides a unique window on the temporal evolution of optical response functions on time scales between 100 fs and 500 ps. Furthermore, THz methods are compatible with the study of materials under extreme conditions of temperature, electric, and magnetic field. THz probes are, in general, non-contact enabling polarization analysis and polarization control that may be of importance in situations where Ohmic contacts to certain materials are not possible due to the remote location of the material, or to avoid the introduction of extrinsic effects associated with leads. Terahertz probes can also provide detailed information on interface quality or on the presence of buried defects or structures.

Two, one and zero dimensionally quantum confined systems have electronic excitations in the THz regime. As the THz frequency exceeds various relaxation rates and broadening, it is possible to achieve well defined collective and single particle excitations. As a consequence many fundamental issues relating to control of carrier injection by doping, the relationship between geometry and energy level structure, line widths of transitions between states, excited state coherence and population lifetimes, superposition states of carriers, and spin excitations can be probed. These are issues not only of interest from a basic perspective but also critical for applications such as sources, detectors, novel ultrafast electro-optic devices and quantum information processing in semiconductors [Tonouchi, 2007].

Another area of interest is the coherent quantum control of the orbital and spin states of carriers in semiconductor nanostructures that occur at THz frequencies. The properties of semiconductor nanostructures at visible and near-IR wavelengths can also be controlled with THz-frequency electromagnetic radiation. In these structures, energy-level splittings are often in the THz range. At THz frequencies, electron interactions with acoustic and optical phonons are small leading to narrow line widths of electronic excitations. These small line widths and correspondingly long relaxation and dephasing times find application in a variety of devices, including light emitters, detector and quantum logic devices Magnetic level splittings in semiconductor nanostructures in magnetic fields of a few Tesla are frequently in the THz regime. Metals are typically very good reflectors of THz radiation. Majority of solids have optical phonons at THz frequencies. Most of these are infrared-active, which can be excited directly by resonant oscillating electric fields[Lewis, 2007]. Phonon features are particularly interesting near structural phase transitions, where they become "soft" and move towards zero frequency. Phonons interact strongly with electrons, a phenomenon that was central to explaining conventional superconductivity. A Josephson plasma resonance is produced by coherent propagation of Cooper pairs between superconducting "sheets" in layered superconductors. In a variety of layered superconductors this mode falls in THz frequency range[Tonouchi, 2007; Ferguson & Zhang, 2002; Sizov & Rogalski, 2010] Other similar strongly correlated electron systems have very complex phase diagrams with several transitions among them. Understanding the nature of these phases, as well as the transitions among them, requires detailed characterization of their excited states, or elementary excitations. The most important excitations are those with an energy of order $k_B T$ above the ground state and therefore in the THz region of the electromagnetic spectrum. Fundamental interactions between quasiparticles, phonons, spin-excitations and other constituents of correlated materials occur on ultrafast time scales [Kida et al., 2005]. It is worth noting that the phonon modes of inorganic and organic crystals fall in the THz region of the spectrum [Zheng et al., 2007].

The classical rotation period for a nitrogen molecule with one $\hbar$ of angular momentum is about a pico second. Similarly, the time of a molecular collision at room temperature is 0.1 to 1 ps. At very low temperatures the collisional energy of molecular collisions is comparable to the rotational energy level spacing. As a consequence, the collisions are quantum mechanical in nature and resonances in their cross sections occur in the THz region. The time scale of gas phase collisions is of the order 10^{-12} seconds which is ideally suited for THz studies. Rydberg atoms in the n = 20 to n = 60 range span the spectrum of sensitive response to THz fields. These are model systems for studying many aspects of few body quantum mechanics, including quantum chaos, quantum-classical correspondence, and quantum systems in the presence of strong external fields [Rangan & Bucksbaum, 2001] .

Interaction of high-intensity femtosecond lasers with solid matter generates plasmas that can be employed as sources of short x-ray pulses, coherent harmonic radiation, energetic electron and ion beams, plasma-based accelerators, ultra high magnetic fields. These plasmas can also serve as a source of intense THz emission [Li et al, 2011].

There are several processes in liquids that occur in the THz region and lend themselves to be probed using THz radiation.THz spectroscopy can also reveal information about protein structure and dynamics. It is possible to distinguish many amino acids based on their THz spectrum, particularly in the crystalline form. The THz spectra of individual DNA base pairs can be used to understand dynamics and conductivity. It could lead to label-free measurement of protein-protein interactions as cellular activity is occurring in live cells. THz radiation holds promise in new medical imaging techniques based on pulsed and cw THz sources. Differences in tissue water, architecture and chemical content can easily be detected using THz techniques since this radiation is strongly absorbed by water. This can lead to early detection of diseases by revealing features that are not apparent with other imaging techniques. [Nagel etal., 2003] .

It is evident that THz region is a region of the electromagnetic spectrum that is rich with information on all forms of matter. It is further obvious that significant developments in this area are technology limited rather than being limited by science.

The research and development in THz science and technology can be broadly divided in to the following categories based on the fundamental phenomena that occur at the THz region of the em spectrum. In all these cases there are issues relating to materials, processing technologies, devices, and system integration issues need to be addressed.

1. Sources
2. Emitters
3. Detectors and sensors

It is pertinent at this point to distinguish between detectors and sensors. For purposes of this review, detectors are used to signify components that provide only frequency information of the THz radiation. Sensors, in contrast, provide information on both the frequency and time domain as well as the source that causes variation in frequency of radiation incident on the detector. Detectors, thus, form a sub-set of the sensor. Sensors can, therefore be used for imaging, spectroscopy and sensing the nature of environment around the THz emitter or detector.

3. THz emitters

Terahertz radiation has potential in various short-wavelength communication devices and security applications. In order to attain the potential offered by terahertz technology, design of solid state terahertz emitters is indispensible. Terahertz sources based on femto second lasers pulses is one of the most promising techniques to generate THz radiation. Lasers are strong sources of electromagnetic energy where energy is stored both in the intensity and frequency. THz generation by ultrafast laser pulses can be classified as linear and non linear optical processes. The linear process involves injection of photocarriers by a laser pulse into the semiconductor connected to the antenna. The photocarriers are accelerated by dc bias applied to the antenna. The antenna radiates THz signal of a frequency that is determined by the pulse duration of laser pulses. In contrast, in the nonlinear optical process THz radiation is generated by optical rectification or difference frequency mixing techniques. The nonlinear optics methods are attractive for terahertz devices because of several properties, including (i) possibility of room-temperature operation (ii) easy operation. However, one of the major hurdles in operating terahertz emitters lies in poor power conversion efficiency, which is of the order of 10^{-6} to 10^{-5}. Terahertz sources can be divided according to their bandwidth, output power, and operating principle [Sakai & Tani, 2005]. In following sections we focus on the several kinds of terahertz emitters. Brief theories of the operating principle have been put forth with the explanation of their merits and demerits.

3.1 Photoconductive broadband THz antenna

An ultrashort laser pulse incident on the surface of a photoconductor at an angle of incidence θ, generates photocarriers, which then move under the influence of an electrical bias field. The bias field may be either externally applied (parallel or perpendicular to the surface) or internally generated by band bending from trapped surface charges (perpendicular to the surface). The resulting transient photocurrent gives rise to the emission of THz electromagnetic pulses in the reflected and transmitted directions. When a femto second (fs) laser pulse, with an arbitrary intensity profile, excites a biased semiconductor with photon energy greater than its bandgap, electrons and holes are produced at the illumination point in conduction and valence bands, respectively. The rapid changes of the density of photocarriers and their acceleration due to the applied DC bias produce an electromagnetic field radiation into free-space with the help of an antenna. The production of ultrashort currents with a FWHM lifetime of 1 ps or less strongly depends on the carrier's lifetime in the semiconductor. The space-charge screening and radiation-field screening are two factors preventing photoconductor from generating higher power THz radiation, which could affect all kinds of the photoconductive antenna. Based on the shape of antenna, their effects are different. Generally speaking, the space-charge screening can be important to small dipole antennas, but, for large area antennas, radiation-field screening is the major cause of saturation. Photoconducting switches are commonly used in time-domain terahertz spectroscopy, where, a generation and detection THz wave is carried out by use of biased photoconducting GaAs aperture antenna. The advantage of semiconductor photoconductive switches is that they can be used to efficiently generate signals and to generate and detect electrical transients in guided media or free space. However, the problem with photoconductive switches is that electrical pulse width is limited by a number of factors such as the laser pulse width, circuit parameters of the generation and detection site, and the carrier lifetime in the semiconductor. Furthermore, the carriers can be captured by traps at the grain boundaries, resulting in a rapid decay in number of carriers [Sizov & Rogalski, 2010].

Usually materials with reduced carrier lifetimes and high carrier mobility such as LT-GaAs or ion implanted semiconducting layers are utilized. Ga doped InAs based photo conductive switches are expected to yield higher terahertz output because of their higher electron mobility [Lloyd-Hughes et al., 2005]. Liu *et al.* report improvement in photoconductive response in shorter terahertz emission pulses from spiral antennas fabricated on multi-energy-implanted GaAs:As[Liu et al., 2005]. This is due to the uniformity of arsenic antisite defects (traps) distributed in multi-energy-implanted GaAs:As which contributes to the unchanged carrier trapping time as compared to single-energy-implanted GaAs:As. Miyadera *et al.* [Miyadera et al., 2004] used amorphous Ge thin films fabricated on (1 0 0) MgO substrates as a terahertz emitter by coupling with Au/Cr transmission lines. Amorphous Ge will be beneficial for optical to electrical signal converter, due to the potential advantages of fabrication without heating process and smaller bandgap of about 0.66 eV.

Light absorption process in the photoconductive switches which generates free carriers near the surface must also be considered. The absorption of terahertz pulses by free carrier can degrade overall efficiency of photoconductors. GaAs constituents of photoconductive

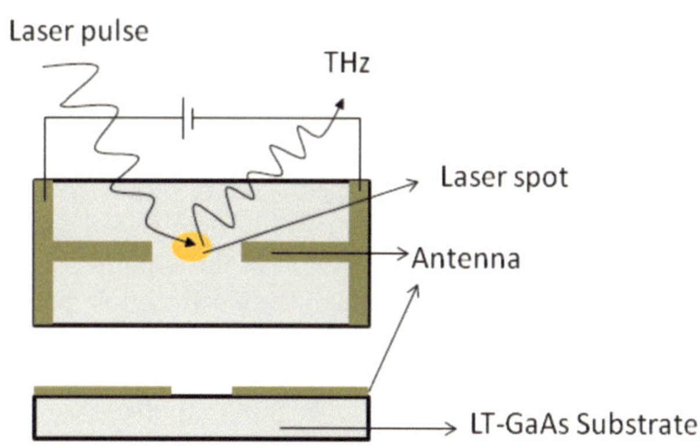

Fig. 1. Typical schematic of a photoconductive switch

switches have a phonon resonance around 8 THz that generates a broad absorption band between 7 and 10 THz. In order to reduce absorption caused by phonon resonance, Kasai *et al.* have fabricated LT-GaAs epitaxial layer which is transferred onto high resistivity Si substrates [Kasai et al., 2009]. These detectors showed higher sensitivity than conventional photoconducting switches at high frequencies. The typical schematic of a photoconductive switch is shown in fig 1.

3.1.1 Photo-Dember effect emitter

If surface energy bands of a semiconductor lie within its bulk band gap, then Fermi level pinning occurs, leading to band bending and the formation of a depletion/accumulation region at the surface. The resulting electric field will separate photo generated electrons and holes, forming a dipole perpendicular to the surface which emits THz radiation, this is the surface-field emission. By changing doping from *n* type to *p* type, the sign of the electric-field direction is reversed. This leads to a change in the polarity of the dipole, which is observed as a change in sign of the generated THz transient. A photo-Dember field can occur at the surface of a semiconductor after photo excitation. Two factors lead to this field: a difference in diffusion coefficients for electrons and holes, and a structural asymmetry. In a typical semiconductor, electrons have a larger diffusion coefficient than holes. Therefore, after photo excitation, the electron population diffuses more rapidly than the hole population. In the absence of a surface boundary, there would be no net dipole field, since the center of charge does not change. However, in the vicinity of the surface, reflection or capture of charges results in the center of charge for electrons and holes moving away from the surface. A dipole is thus formed perpendicular to the surface, leading to THz emission. In this case changing the semiconductor doping from *n* to *p* type has no effect on the sign of the emitted THz pulse. One of the main applications of photo-Dember effect is the generation of terahertz (THz) radiation pulses for terahertz time-domain spectroscopy. This effect is present in most semiconductors but it is particularly strong in narrow-gap semiconductors (mainly arsenides and antimonides) such as InAs and InSb owing to their high electron mobility.[Yi et al, 2010; Reklaitis, 2011; Klatt et al., 2010; Krotkus, 2010]

3.1.2 Current transient effect emitter

The current transient emitter or Auston switch is simply a gap in a uniform transmission line defined by thin-film metal fabrication on the top surface of an ultrafast photoconductor The transmission line is, generally, a coupled-strip line terminated in a planar antenna, within a planar strip dipole. The entire transmission line structure is biased with a voltage through an inductor. The principle of operation is that the gap in the transmission line initially creates an open-circuit condition. When a short pulse from a mode-locked laser arrives in the gap, a shower of electron-hole pairs are generated in the first few microns below the surface. The instantaneous concentration of these pairs is high enough to "short-out" the gap. When the laser pulse is terminated, the electron-hole pairs quickly recombine, creating an electrical impulse (or surge current) on the same time scale as the laser pulse. Due to the wide bandwidth of the planar transmission line, the electrical impulse propagates down the line to the antenna with minimal dispersion or droop. So upon reaching the antenna, the electrical pulse generates a significant amount of THz radiation that propagates primarily into the substrate.[Reimann, 2007; Suen et al., 2010;]

3.2 Optical rectification based emitters

One of the more popular means of generating THz frequency radiation is through the bulk second-order nonlinear effect of optical rectification. Optical rectification occurs if high-intensity light is directed onto an electro-optic material. If the excitation beam contains components of two or more frequencies then difference-frequency mixing, also known as optical rectification, may take place. Depending on the spectrum of the excitation beam, the resulting frequency may be in the THz range. The optical rectification process has been widely developed for THz generation, typically in materials with a completely nonresonant response. The advantage of such an approach is the inherent speed of the response, which may be viewed as essentially instantaneous. The disadvantage is the relative weakness of the nonlinear response. Thus, in order to obtain significant THz emission via the optical rectification mechanism it is desirable to use a medium of appreciable thickness which leads to the necessity of considering phase-matching constraints. The problem of broadband phase matching is a challenge. One possibility is exploiting the non-resonant second-order nonlinear response expected in any non-centrosymmetric semiconductor, such as the III–V materials. This effect can be identified by the distinctive dependence that it shows on the crystallographic orientation of the sample. Optical rectification using an ultra-short femto laser in the periodically poled structure can be considered as a special case of difference frequency mixing where terahertz radiation emitted from the polarization change that follows the transport of excited carriers in an applied or surface electric field. The physics of optical rectification leading to THz generation is well understood [Blanchard et. al, 2011]. The nonlinear coefficient of the second order susceptibility is an extremely important material property when high-power output THz emitters are considered. The output power is also predominantly determined by intensity of the excitation pulse and the phase matching conditions. The major disadvantage of this technique is low power output, broad line-width, velocity mismatch and requirement of expensive femto second laser sources. In addition, there are the restrictions implied by other nonlinear processes like two-photon absorption, competing with optical rectification and limiting its efficiency. The process of two-photon absorption not only influences efficiency of terahertz generation but also impacts the terahertz spectrum [Vidal et al., 2011]. THz generation via optical rectification of x-mode laser in a rippled density magnetized plasma has been demonstrated [Bhasin and Tripathi [Bhasin & Tripathi, 2009]. Fiber lasers are attractive as terahertz emitters because of their low cost, small size, high repetition rate, high pulse energy, and short pulse duration[Hoffmann et al., 2008].

Relativistic optical rectification, has opened up the potential realization of high frequency range terahertz emitter[Tsaur & Wang, 2009]. It is should be noted that bandwidth of emitter is sharply reduced in optical rectification due to the phase-matching condition. For EO sampling and optical rectification, the spectral response of the material is strongly sensitive to reststrahlen band absorption/reflection, phase matching, dispersion of refractive index and dispersion of EO coefficients, which can be expressed as

$$G(\Omega) = \frac{t(\Omega)(\exp(i2\pi\Omega\,\delta(\Omega)) - 1)}{\left[1 - r(\Omega)^2 \exp(\frac{i4\pi\Omega n(\Omega)d}{c})(i2\pi\Omega\delta(\Omega))\right]} \tag{1}$$

$$\delta(\Omega)=\frac{n_g(\lambda)-n(\Omega)}{c}d \tag{2}$$

$\delta(\Omega)$ is the velocity mismatch in time, $n_g(\lambda)$, is the optical group index, $n(\Omega)$, is the terahertz refractive index, d is crystal thickness, $t(\Omega)$ and $r(\Omega)$ are the Fresnel transmission and reflection coefficients and the multi-reflection of terahertz field inside the sensor material. The velocity mismatch between phase velocity of terahertz pulse and group velocity of laser beam hampers output power of these sources. This could be eliminated by placing nonlinear crystal in hollow metallic waveguide structure [Nikoghosyan, 2010]. The conversion efficiency in optical rectification depends primarily on the material nonlinear coefficient and the phase-matching conditions and various schemes for power enhancement have been reported [Radhanpura et al., 2010; Bugay & Sazonov, 2010; He et al., 2008; Reid et al., 2008]. To overcome the problem of phase mismatch, interaction of pump optical pulse with a two-level resonant impurities, for the terahertz wave is proposed [Bugay & Sazonov, 2010]. Alternatively, phase mismatching can also be effectively used to tune the operating frequency from 4.2 to 1.1 THz of a (110) oriented ZnTe crystal terahertz emitter. Materials with large drift velocity and short carrier lifetime such as GaAs and ZnTe are popularly used in optical rectification techniques. However, they are limited by large band gap which hampers their efficiency [Li & Ma, 2008]. In combination with second order susceptibility, effective second-order processes such as electric field induced optical rectification (EFIOR) has been shown to yield terahertz generation in centrosymmetric Ge crystals [Urbanowicz et al., 2007]. This effect has been termed EFIOR, because it is easily explained in the framework of third-order nonlinearity ($\chi^{(3)}$) contributed to effective second order process by DC surface electric field.

3.3 Quantum cascade laser emitters

Energy levels in semiconductor quantum wells can be designed and engineered to be of any value, with ease. In such systems, inter sub band transitions involving electrons that make lasing transitions between subband levels within the conduction band can be exploited to realize emissions in the far infrared and THz regions. Such lasers are called quantum cascade lasers (QCLs). Simple calculations show that energy level separation to achieve emission in the THz region from QCLs is in the meV region (1 THz = 4 meV). Design of quantum well structures for selective injection to the upper level and selective depopulation of electrons from the lower level is a challenge since, due to the narrow separation between subband levels, heating and electron-electron scattering have a great effect. Some of the other challenges that need to be overcome include detection and analysis of spontaneous emission; mode confinement at longer wavelengths; dielectric waveguide confinement (as the evanescent field penetration, which is proportional to the wavelength is of the order of several tens of microns>> the active gain medium of several microns. The physical origin of the THz emission of the quantum cascade laser, as stated above, can be traced to intersubband transitions. The quantum well thickness in QCL is one of the key parameters to control in order to finely tune laser energy. The background blackbody radiation and/or emissions from impurity levels may lead to poor reproducibility of the results. The micro-probe photoluminescence has been shown to provide a powerful tool for this purpose which can probe emission from individual states in the active region with the high spatial resolution [Freeman et al., 2011].

Unfortunately, it has been observed that QCL are not popular for THz source due to low energy (long wavelength) output. This is a consequence of the poor coupling between the small gain medium and the optical field. In fact, high optical losses are caused due to free electrons in the material at low energy. Recent developments in QCL technology have raised the hope for their application in terahertz devices. The nonequilibrium Green's function calculations performed by Yasuda *et al.* [Yasuda et al., 2009] predict that the 4L QCL has a larger terahertz gain than the conventional resonant phonon QCL. They have attributed it to the large number of electrons accumulate in the upper lasing level and contribute to lasing in the new scheme. They have found that the advantage of gain deteriorates at 200 K due to thermally activated phonon scattering.

Another major limitation of QCL is their low operation temperature. The maximum operating temperature of 178 K is currently the highest operating temperature achieved in a THz QCL. The need to use cryogenic cooling is considered to be a major obstacle to the introduction of room temperature QCL in terahertz devices. Ambient temperature operation can be achieved by integrating the optical nonlinearity for difference-frequency generation(DFG) into the active region of a dual-wavelength mid-IR QCL using band structure engineering of the QCL active region [Belkin et al., 2009]. However, the disadvantage of DFG QCLs is that they provide less terahertz power than normal QCL. Alternatively, it is also possible to improve temperature performance of QCL by implementation of a second type of waveguide (metal–metal (MM)) design which consists of metal films on both sides of the active region, and provides a mode confinement factor of nearly 100% [Belkin et al., 2009]. Optimized Second Harmonic Generation (SHG) in QC lasers, in which the QWs of the active regions simultaneously function as nonlinear oscillators, have led to high absolute SHG power levels of up to 2 μW and large linear-to-nonlinear power conversion efficiency of around 50–100 $\mu W/W^2$[Gmachl et al., 2003].

The present trend in the search of materials for QCL is to use silicon, which can offer the prospect of integrating coherent terahertz (THz) radiation sources with silicon microelectronics. Silicon is the only material that allows mature processing technology which may reduce costs and allow integration with conventional electronic devices. A recent report on silicon based QCL focused on a variety of *n*-type SiGe-based heterostructures as design candidates [Valavanis et al., 2011]. They have carried out theoretical studies on (001) Ge/GeSi, (111) Si/SiGe, and (001) Si/SiGe material configurations where they have found that (001) Ge/GeSi is the most promising system for development of a Si-based QCL

3.4 Photomixers

Modulated infrared radiation can cause the resonant excitation of plasma oscillations in quantum well diode and transistor structures with high electron mobility. This effect provides a mechanism for the generation of tunable terahertz radiation using photomixing of infrared signals [Preu et al., 2011] .

In terahertz photomixer, the outputs of two continuous-wave sources with frequency difference falling in the terahertz range mix in a photoconductors. The laser induced photocarriers short the gap of device producing photocurrent which is modulated at the laser difference frequency. The generated signal is usually fed to the antenna to transmit into free space. The photomixers are tuneable, coherent, compact, low cost and narrow-band

THz radiation sources. Photomixers have the greatest tuning range among all the coherent sources in terahertz region. Note that photomixing is fundamentally different from difference frequency mixing. Photomixing is more efficient than difference frequency generation at low THz frequencies. At higher THz frequencies, nonlinearities are more efficient due to parasitic impedances which limit the THz bandwidth of photomixers [Mittleman, 2003]. In difference frequency mixing, the outputs of two laser sources are mixed in nonlinear medium like ZnTe or $LiNbO_3$, where, an intense laser field can cause the polarization of the nonlinear medium to develop new frequency components. These resultant frequency components of polarization act as source of terahertz radiations.

The frequency and power of the generated terahertz signal can be tuned by tuning the central frequency of one of the lasers. Based on model calculations for LT-GaAs photomixer, it has been proposed [Pilla, 2007] that photomixing efficiency can be enhanced by reducing the transit time of majority of carriers in photomixers and photodetectors to < 1 ps. The model indicated that the output terahertz power, P_f is proportional to difference frequency of incident lasers (*f*) which is given by relation Pf α $f^{-2.77}$ in the 0.5–6.5 THz range.

In difference frequency mixing, an intense pulsed laser induces a significant polarization in non-centrosymmetric structured materials, which can be written as

$$P = \chi^{(1)}E + \chi^{(2)}E \tag{3}$$

where first term on the right hand side represents linear polarization while second term is responsible for SHG (second harmonic generation), dc rectification, Pockels effect, parametric generation etc

In nonlinear crystals, sum frequency generation (SFG) or difference frequency generation (DFG) can occur, where two pump beams generate another beam with the sum or difference of the optical frequencies of the pump beams. If the incident electric field contains two different frequency components ω_p and ω_s and amplitudes E_g and E_p respectively, the resultant intensity can written as

$$2\chi^{(2)}E_pE_s\ \{\cos(\omega_p + \omega_s)t + \cos(\omega_p - \omega_s)\, t\} \tag{4}$$

It is clearly seen from equation (4) that the second order nonlinear susceptibility will give rise to a nonlinear polarization and reemit radiation at sum ($\omega_p + \omega_s$) and difference ($\omega_p - \omega_s$) frequencies.

Advantages associated with a difference frequency generation include simple frequency control, narrow linewidth and wide tuning ranges based on low loss and phase matched NLO crystals and tunable pump lasers. Diode lasers are sometimes used in difference frequency generation sources for large band width THz signals. Narrow-band terahertz pulse generation via difference frequency mixing using fs laser irradiation of periodically poled lithium niobate (PPLN) and stoichiometric lithium tantalate (PPSLT) crystals[Yu et al., 2011]. The bandwidth of output terahertz radiation obtained was as low as 32 GHz for 1.38 and 0.65 THz for forward and backward THz pulses respectively. The phase mismatch is another critical factor affecting performance of difference frequency generation sources. The conversion efficiency in difference frequency generation process depends primarily on the material's nonlinear coefficient and the phase (velocity) matching conditions. The wave vector of the applied field and the generated field must have the same relation as their frequencies. Unfortunately in nonlinear crystals, the optical beam always travels faster than the THz beam, which makes the phase matching unachievable. Several phase matching techniques been explored to improve output power of terahertz emitters. The phase matching techniques can reduce decay of terahertz radiation in the crystals by techniques like reducing beam diameter [Shibuya et al., 2010] or artificially introducing reversed domain structures (quasi-phase matching, QPM) [Marandi et al., 2008]

4. THz detectors and sensors

4.1 Detectors

THz detectors have been developed based on a variety of principles as described below. However there are two generalized performance parameters that characterize THz detectors. These are Noise Equivalent Power (NEP) and detectivity.

Noise Equivalent Power (NEP) is defined the incident RMS optical power required in order to obtain signal-to-noise ratio of 1 in a bandwidth of 1 Hz. NEP is related to the minimum power that the photodetector can detect

$$NEP = \langle i_n^2 \rangle / R \tag{5}$$

Where i_n is the noise current, and R is the detector responsivity. The related figure of merit is detectivity defined as

$$D^* = \sqrt{SB}/NEP \tag{6}$$

where B is the bandwidth and S is the detector area. [Sizov & Rogalski, 2010]

4.2 Direct and heterodyne THz detection based on the principle of bolometer

Hot electron bolometers are very commonly used for THz detection. The term hot electron describes electrons which are not in thermal equilibrium with the lattice. In metals the heating of electrons does not change the electron mobility and therefore it does not affect the resistance value. The hot electron approach is very productive for semiconductors, where the mobility of electrons depends on their effective temperature. In a superconductor, the hot-electron phenomenon consists of the heating of electrons by radiation. A photon incident on a superconductor is absorbed by a Cooper pair. Due to the large coherence length only one of the electrons of the Cooper pair absorbs the photon. The Cooper pair breaks apart and one highly excited electron with energy close to the incident photon energy and one low-energy quasiparticle are created. Next the quasiparticle loses its energy via electron-electron scattering and creates a secondary excited electron. This process continues so long as the incident radiation is switched on. There are two different techniques to detect the power of submillimeter and far-infrared wavelengths. The first approach is to detect the radiation directly by creating charged carriers or converting the incoming energy to heat proportional to the flux of the incoming photons. The first method is direct detection which is also called an incoherent detection. Direct detectors do not have a fundamental limit of the sensitivity. A direct detector does not produce any noise power unless photons are absorbed. The second method is to shift the incoming radiation to a lower frequency band and then amplify and detect the power of the radiation. The method of shifting the incoming radiation to a lower frequency is called heterodyne principle or coherent detection. In a heterodyne receiver a locally generated frequency is mixed with a signal frequency to produce a signal at a much lower frequency. This frequency conversion is done by a mixer element. The locally generated frequency and the signal frequency are added together to produce a beat frequency, the difference between the signal and the locally generated frequency. The HEB can be used as a heterodyne mixer. The incident radiation and the local oscillator (LO) radiation excite a voltage across the bolometer. The dissipated power of the bolometer is a function of the average absorbed signal power and the average absorbed LO power. The output power consists of a component equal to the difference frequency of the electronic diode. Thus, a HEB can be used as a heterodyne mixer, if it is fast enough to follow the IF [Cherednichenko et al., 2002, Zmuidzinas & Richardds, 2004] .

Superconductor-insulator-superconductor (SIS) tunnel junctions are extremely sensitive heterodyne mixers and have been widely used for frequencies below 1 THz. The upper frequency limit of these devices is determined by the gap frequency of the superconductor defined by fgap = Δ/h with Δ the superconducting gap energy and h the Planck's constant. Drastic losses above the gap frequency of the superconductor (about 700 GHz for niobium) occur inside the superconducting tuning circuit which is used to match the impedance of the SIS device to the antenna impedance. Using superconductors (like NbN or NbTiN) having a higher gap frequency than Nb has increased the upper frequency limit of SISmixers to about 1.2 THz. The main advantage of SIS mixers is the wide Intermediate Frequency (IF) bandwidth provided by this type of mixer. HEB mixers do not have an upper frequency limit and provide a very high sensitivity (T_{rec} < 2000 K) and require very low LO power (< 1 μW). Although the IF bandwidth of HEB mixers is rather limited compared to Schottky and SIS mixers, HEB mixers are the most competitive devices for heterodyne detection in the THz range. When compared by characteristics like response time and dynamic range, SIS detectors are found to be much superior to their TES counterparts [Matsuo, 2006].

The use of low temperatures based on Nb, NbN and a number of related compounds have been reported . In these cases while the fabrication of the bolometer structures is routine since feature sizes are large, the very low operating temperatures (<20 K, in some cases) is a major drawback [Chen et al., 2009; Tretyakov *et al.* ; Stern et al., 2005; Schultz & Lichtenberger, 2007; Ilin et al., 2010; Liu et al., 2009; Kleinschmidt et al., 2007; Cherednichenko et al., 2007] . In addition, these bolometers work in the range < 5 THz, generally. The typical schematic view of a HEB structure is shown in figure 2.

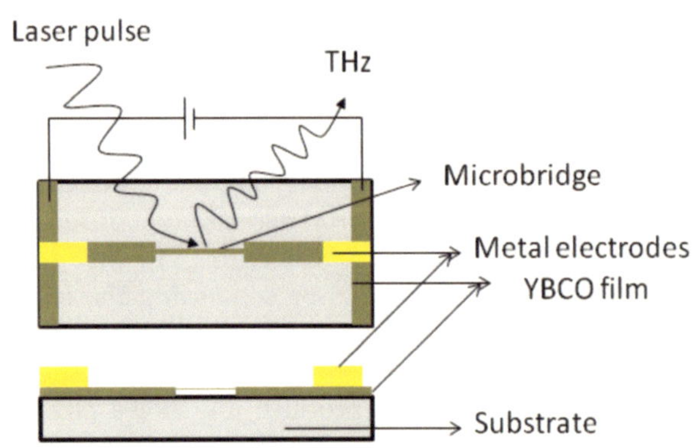

Fig. 2. The typical schematic view of a HEB structure

Bolometers have also shown potential for application in space and astronomy [Meledin et al., 2004; Caumes et al., 2009; Putz et al., 2011]. The discovery of high temperature superconductors has helped in increasing the operating temperature of the HEBs to liquid nitrogen temperatures. For example, an ultrathin PBCO/YBCO/PBCO HEB [Peroz et al., 2007] constructed by ultra-thin (10 nm thick) superconducting micro-bridge coupled with a planar antenna was demonstrated. However, the limitation on mixing operation of HEB is observed causing by aging of thinner layer and nano bridge structure. Tonouchi et al [Tonouchi et al, 2005] have reported extensively on the use of YBCO based antennae for THz detection. There are many other reports on the use of high Tc superconductors as bolometers, but only a few are listed [Du et al, 2011; Aurino et al., 2010; Laviano et al., 2010; Jagtap et al., 2009]

There are some disadvantages of bolometric type of detectors, such as they are extremely sensitive to background radiation, temperature fluctuation, mechanical vibration and electrical interference, and the performance deteriorates with increasing frequencies in the THz range. Further, bolometers are insensitive to phase which does not allow the reconstruction of the pulse shape in the time domain. It should be noted that background radiation defines the limiting condition of sensitivity of detector. The sensitivityof the bolometer is mainly determined by the system noise temperature. System noise includes noise from the input source and noise generated in the receiver.

4.3 Schottky barrier detectors

Schottky-diodes have been the only available technology for the detection of THz radiation for a long time {Roeser et al, 1994]. They do not require cooling to cryogenic temperatures and cover a wide frequency range up to several THz. The main disadvantages are the poor sensitivity and the high local oscillator (LO) power requirement. The LO is still a critical element for heterodyne detection in the THz frequency range. Solid state sources often do not provide enough power for the heterodyne operation in the THz range but are under development right now. One of the early reports on Schottky diodes for THz applications was by Suzuki etal [Suzuki et al 1999]. They presented results on submicrometer Pt/GaAs diode fabrication process that resulted in a significant reduction in low-frequency noise. Noise performance was optimized by establishing an even distribution of gallium and arsenic (stoichiometric surface) at the contact surface. Diodes fabricated with the optimized procedure exhibited a signal-to-noise ratio that was better than that of commercially available diodes at 0.9 THz. Rinzan et al [Riznan et al., 2005] demonstrated a heterojunction interfacial work function internal photoemission (HEIWIP) detector with a threshold frequency (f_0) of 2.3 THz (λ_0 = 128 μm). The threshold limit of ~ 3.3 THz (92 μm) was surpassed using AlGaAs emitters and GaAs barriers. The peak values of responsivity, quantum efficiency, and the specific detectivity at 9.6 THz and 4.8 K for a bias field of 2.0 kV/cm are 7.3 A/W, 29%, 5.3×10^{11}, respectively. Yasui et al [Yasui et al, 2006] developed a wide-bandwidth, high-sensitivity, continual terahertz-wave sensor that utilizes a quasioptical parabolic mirror and a Schottky barrier diode and successfully applied it at up to 7 THz range. This sensor utilized a parabolic cylindrical mirror, a long-wire antenna, and a Schottky barrier diode. The antenna, placed at the focal point of the parabolic mirror, quasioptically collects the terahertz signalleading to superior performance. Ryzhii et al., [Ryzhii et al., 2006] investigated the plasma oscillations in a two-dimensional electron channel with a reverse-biased Schottky junction. They show that the plasma instability can be used in a novel diode device - lateral Schottky junction tunneling transit-time terahertz oscillator. Magno et al grew [Magno et al., 2008] Antimonide-based p^{+}N junctions. The diodes had good rectification with ideality factors near 1, and high saturation current densities of 2.5×10^{-2} A/cm^2. A cutoff frequency of 6.5 THz is estimated from the *RC* product. The high saturation current indicates that this diode could be used as a terahertz mixer. Maestrini et al [Maestrini et al, 2010] reviewed some aspects of the technology of terahertz heterodyne receiver front-ends dedicated to astrophysics, planetary and atmospheric sciences with focus on frequency multipliers and Schottky mixers. It is clear from this review that there is much potential for research in the area of novel architectures of power-combined frequency multipliers at submillimeter-wavelengths, THz planar fundamental mixers, and integrated receivers and the fabrication of submillimeter-wave planar Schottky diodes. Chen et al. [Chen et al, 2010] demonstrated high performance nanometer $NiSi_2$-Si Schottky barrier diode arrays (SBDA) with various isolation designs, including poly Si gate (PSG) and resist protection oxide (RPO) for advanced radio frequency applications. The SBDAs could achieve a cutoff frequency of up to 4.6 THz.

It is evident from a brief survey of literature that Schottky diode structures have great potential for application in THz detectors impacting a variety of fields. However, there are several challenges in design, fabrication and operation of these device structures that remain unresolved

4.4 Field-effect transistor detectors

A channel of a Field Effect Transistor (FET) can act as a resonator for plasma waves. The plasma frequency of this resonator depends on its dimensions and for gate lengths of a micron and sub-micron (nanometer) size, can reach the Terahertz (THz) range {Dyakanov, 2008]. The interest in the applications of FETs for THz spectroscopy started at the beginning of 90s with the work of Dyakonov and Shur [Dyakanov and Shur, 1996] who predicted that plasma oscillations in a FET channel can lead to the THz emission. Recently, non-resonant plasma properties were successfully used for the room temperature broadband THz detection and imaging by silicon FETs. Both THz emission and detection, resonant and non-resonant, were observed experimentally at cryogenic, as well as at room temperatures, clearly demonstrating effects related to the excitation of plasma waves. The possibility of the detection is due to nonlinear properties of the transistor, which lead to the rectification of an ac current induced by the incoming radiation. As a result, a photoresponse appears in the form of dc voltage between source and drain which is proportional to the radiation power (photovoltaic effect). There are three distinct regions of operation of the FETs depending on gate lengths. It has been demonstrated that for gate lengths of the order of 0.1 μm the plasma oscillations will be in the low THZ region [Dyakanov, 2008]. Otherwise the FET can operate as a broad-band THz detector. There are a few other reports on the use of FETs as THz detectors, signifying again the potential of these structures as wellas the number of fabrication issues that need to be overcome [Knap et al., 2011]. The use of nitride based FETs for THz detection has also been demonstrated.[Shur, 2007]

4.5 High electron mobility transistor detectors

Semiconductors heterostructure devices such as high electron mobility transistor (HEMT) and heterojunction bipolar transistors have attracted increasing attention in THz devices

[Dyakonov 2005; Fatimy 2005; Knap 2004; Lusakowski 2005; Otsuji et al., 2010; Tonouchi 2007; Tredicucci 2005], which in fact, can be regarded as a new type of terahertz detectors. Otsuji *et al.* [Otsuji et al., 2010] reviewed recent advances in emission of terahertz radiation from 2D electron systems in semiconductor heterostructures. A self-consistent spatiotemporal variation of quantum confined electron gas with sufficiently high electron mobility in the channel of HEMT has been proved to be an efficient mechanism for the detection and emission of terahertz radiations. The main advantage of the HEMT over the non-laser based sources is in working temperature and hence hardware requirement for operation. Another major advantage of HEMT terahertz emitters are their high tunability and low fabrication cost. The introduction of closely-spaced heterojunction interfaces under the right conditions produces a two-dimensional electron gas (2 DEG). This low dimensional plasmon (electron gas) is favourable for solid-state far-infrared and THz sources. The strong confinement of charge carriers in the quantum well is believed to be the cause of large electron mobility. Further, the small channel length of the device, leading to an increase in transconductance with an increase in carrier mobility, is well known for HEMT. However, the high losses caused by free carrier absorption and small confinement factor of heterostructure are the factors which restricts use of HEMT at longer THz wavelengths. Monte Carlo simulation performed to study of transient electron transport in wurtzite GaN, InN, and AlN shown that, for ideal transistors with L = 0.1 μm, by replacing the GaN channel with InN, can be expected increase frequency of operation from 0.48 to 0.68 THz. Moreover, low m_e (effective electron mass) in the InN channels also gives a prospect for electron nonstationary dynamics, and operating frequency over 1 THz has been predicted for 0.1 μm gates in that case [Foutz et al., 1999]. Further increase in operating frequency (THz range) is, in fact, expected in ultrathin gate structured HEMT. Fatimy *et al.* [Fatimy et al., 2010] have demonstrated room temperature terahertz generation by a submicron size AlGaN/GaN-based high electron mobility transistor. They have showed that these transistors are tunable by the gate voltage between 0.75 and 2.1 THz. It is also possible to tune operating frequency by manipulating cap, window, and gate lengths and other structural parameters [Ryzhii et al., 2006].

Understanding the physics of plasma wave resonant line broadening is essential to improve performance of HEMT based plasma wave detectors of terahertz radiation. Properties of HEMT terahertz detectors are defined by plasma wave excitations of two dimensional electron gas [Tsaur & Wang, 2009]. Recently it has been demonstrated that narrow terahertz

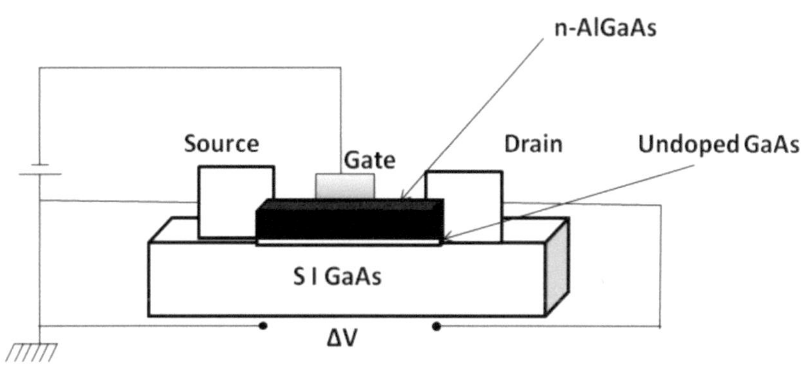

Fig. 3. The typical schematic view of a high electron mobility transistor structure

plasma wave resonant detection at low temperature in 200 nm gate length InGaAs/InAlAs multichannel HEMT. To achieve selective resonant and voltage tunable terahertz emission, they have maintained the gate width is in the order of the gate length [Shchepetov et al., 2008].The typical schematic of a High electron mobility transistor used for THz detection is shown in fig.3.

4.6 Sensors

The state of technology in THz sensors is briefly reviewed. A variety of THz sensors have been developed recently mainly for use in biological and medical applications . Federici et al have reviewed the developments in THz sensing [Federici et al, 2005] technology, spectroscopy and imaging for security applications. The typical view of a THz imaging set up is shown in fig.4. The authors state that since, (a) THz radiation can detect concealed weapons since many non-metallic, non-polar materials are transparent to THz radiation; (b) target compounds such as explosives and illicit drugs have characteristic THz spectra that can be used to identify these compounds and (c) THz radiation poses no health risk for scanning of people stand-off interferometric imaging and sensing for the detection of explosives, weapons and drugs is critical. Future prospects of THz technology are discussed. Nagel et al [Nagel et al, 2003] first reported a novel resonant THz sensor for the label-free analysis of DNA molecules. The sensor allows the direct detection of DNA-probe molecules at functionalized electrodes via hybridization. Subsequent time resolved photoconductive sampling of the THz transmission identifies the binding state between probe and target DNA. Integrating neighbouring sensors on a chip, this technique can be extended to a parallel analysis of multiple DNA sequences. Nagel et al [Nagel et al, 2006] then reviewed the state of biological THz technology applications and conclude that success in this area will strongly depend on the development of compact, low-cost and flexible systems. They have reported on different approaches for THz biosensor systems based on femtosecond lasers and discussed the technology for generation, transmission and detection of THz signals as well as their application formarker-free biomolecule detection on functionalized

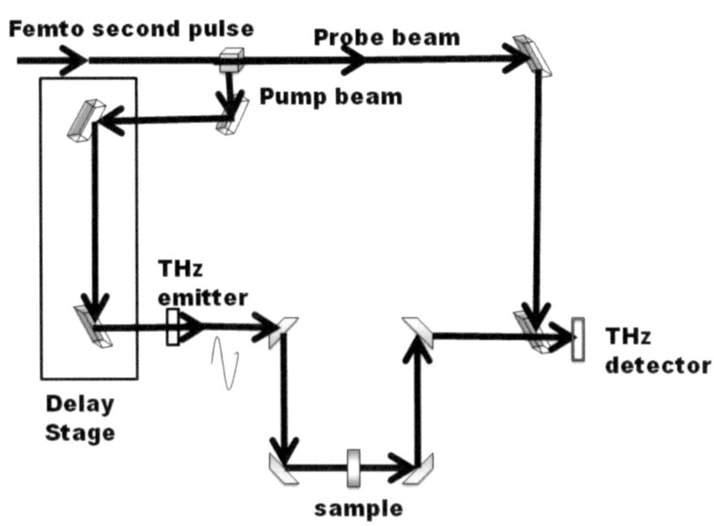

Fig. 4. Typical schematic view of a THz imaging set up.

surfaces in dry and fluid environments. Kitagawa et al [Kitagawa et al , 2006] demonstrated the THz spectroscopic performance of highly integrated sensor chips based on microstrip lines by measuring biomolecule/water systems. The concentration resolution of the chips reaches down to 0.05 g/ml. They have confirmed that the number of bound water molecules per biomolecule can be obtained with precision using solid state transmission lines. The chips are highly suitable for the inspection of small amounts of specimen and for the application to a wide range of water rich materials. McLaughlin et al [McLaughlin et al , 2008] employed poled polymer films 10–15 μm thick with electro-optic coefficients as high as 160 pm/V at 1300 nm are used to generate and sense subpicosecond pulses with continuous bandwidth up to 15 THz. The use of the poled polymer as the terahertz sensor for the identification of DAST phonons at 1.1, 3.0, 5.3, 8.5, and 12.5 THz was demonstrated. Kawase etal [Kawase etal , 2010] have suggested a wide range of real-life applications using novel terahertz imaging techniques. A high-resolution terahertz tomography has been demonstrated by ultra short terahertz pulses using optical fiber and a nonlinear organic crystal. They, further, describe a nondestructive inspection system that can monitor the soot distribution in a ceramic filter using millimeter-to-terahertz wave computed tomography. Another proposed application is in the area of thickness measurement of very thin films using the high-sensitivity metal mesh filter. Abbas et al [Abbas et al , 2009] describe the development, functionalization and functionality testing of a TeraHertz (THz) Bio-MicroElectroMechanical System (BioMEMS) dedicated to enzyme reaction analysis. The microdevice was fabricated by mixing clean room microfabrication with cold plasma deposition. The progression of the hydrolysis reaction over time was monitored by the THz sensor connected to a vectorial network analyzer. Preliminary results showed that sub-THz transmission measurements are able to discriminate different solid films, various aqueous media and exhibit specific transmission behavior for the enzyme hydrolysis reaction in the spectral range 0.06–0.11 THz. Rau et al [Rau et al., 2005] reported on a simple THz waveguide element used as a sensitive sensor for adsorbates on surfaces. The evanescent wave from total internal reflection off a silicon prism was used to couple pulsed THz radiation frequency selectively into the waveguide. The coupled frequencies were determined via time domain spectroscopy and react sensitively to any changes of thickness or phase shift upon reflection. In particular, the sensitivity to phase shifts makes this waveguide sensor attractive for the detection of very thin adsorbates. Xiao-li and Jiu-sheng [Xiao-li & Jiu-sheng,2011] used terahertz time-domain spectroscopy technique to test flour/talc powder mixture samples. They found that the two samples have distinct absorption peaks and different refractive index in the THz region leading to the possibility of THz applications to the field of food safety. Mendis et al [Mendis et al, 2009] describe a terahertz optical resonator suited for highly sensitive and noninvasive refractive-index monitoring. The resonator is formed by machining a rectangular groove into one plate of a parallel-plate waveguide, and is excited using the lowest-order transverse-electric (TE_1) waveguide mode. Since the resonator can act as a channel for fluid flow, it can be easily integrated into a microfluidics platform for real-time monitoring. Using this resonator with only a few microliters of liquid, a refractive-index sensitivity of 3.7×10^5nm/refractive-index-unit was demonstrated in the THz range. Lu et al [Lu et al., 2008] discuss broadband terahertz wave detection through field-induced second harmonic generation using selected gases. The dependences of the detected second harmonic intensity on probe pulse energy, bias field strength, gas pressure, and third order nonlinear susceptibility are systematically investigated with xenon, nitrogen, SF_6, and alkanes. Experiment results reveal that the detected second harmonic intensity quadratically depends on the third order nonlinear susceptibility of the gas. Two orders of magnitude enhancement in the dynamic range of broadband terahertz wave detection are observed with alkane gas (C_4H_{10}) sensor. Foltynowicz et al [Foltynowicz et al, 2006] reported the vapor-phase spectrum of 2,4-dinitrotoluene (DNT) from 0.05 THz to 2.7 THz utilizing pulsed terahertz time-domain spectroscopy. They observed a broad-band absorption profile from 50 GHz to 600 GHz, peaking at 240 GHz. This broad absorption profile corresponds to DNTs pure rotational spectrum, which was confirmed by asymmetric top model calculations. Superimposed on the broad absorption profile, we observed discrete structure that extended to the higher THz frequencies. Leahy-Hoppa et al [Leahy-Hoppa et al, 2007] have shown Time-domain terahertz spectroscopy (TDTS) to be a promising tool in detection of explosives and explosive related compounds. The THz absorption spectra over an extended frequency band from 0.5 to 6 THz were reported for four explosives: RDX (1,3,5-trinitroperhydro-1,3,5-triazine), HMX (1,3,5,7-tetranitroperhydro-1,3,5,7-tetrazocine), PETN (pentaerythritol tetranitrate), and TNT (2,4,6-trinitrotoluene). New distinctive spectral features are shown in these materials between 3 and 6 THz. Zhang et al [Zhang et al, 2008] present terahertz reference-free phase imaging for identification of three explosive materials (HMX, RDX, and DNT). They propose a feature extraction technique to locate the spectral position of an unknown material's absorption lines without using the reference signal. The samples are identified by their absorption peaks extracted from the negative first-order derivative of the sample signal phase divided by the frequency at each pixel. Hu etal [Hu et al, 2006] report experimental measurement and theoretical analysis of THz spectrum for five different explosives and related compounds are introduced. The refractive index and absorption coefficient of the samples are measured in the region of 0.2–2.6 terahertz by time-domain spectroscopy. The simulated spectrum of γ-HNIW is in agreement with the experimental data.

It is thus evident that there are many significant developments in the area of THz sensors that impact various aspects of biology, medicine and security.

4.7 New materials and device structures

A large variety of new types of materials and device structures have been proposed recently for THz applications. These include the use of carbon nanotubes [Costa et al., 2009], GaAs based nanotransistors [Łusakowski, 2007] and polar optical phonons in quantum wells [Liu et al., 2011].Sputter deposited zinc oxide photoconductive antenna have also been proposed for THz detection applications[Iwami et al., 2009]. Videlier et al have demonstrated Si MOSFET structures as THz detectors[Videlier et al].The possibility of utilizing two-dimensional plasmons in semiconductor heterostructures for THz detection has been also been proposed [Otsuji et al, 2010]. Cox et al [Cox et al., 2011} demonstrated a MEMS based uncooled THz detector. The development of novel single photon detectors that work in the range of 10-50um based on double quantum weel structures have been reported recently {Ueda& Komiyama, 2010]. There are many other reports on novel materials and structures, these are however beyond the scope of the current review.

5. ZnTe coatings for THz applications

In this section we present a review of the work being carried out by the present authors in the area of THz materials development. The ability to miniaturize THz photonics depends critically on two developments compact THz sources and elimination of the dependence on crystals as emitters and detectors. The focus is, therefore, mainly on the development of thick ZnTe coatings for use in THz emitter and detector applications, based on the principle of optical rectification. As stated in section 3.2, optical rectification can be controlled by manipulating the crystal structure and orientation. It is, however, a weak non-linear response. Hence, material thicknesses have to be high to generate a measurable response. Semiconductors from the II-VI group, especially ZnTe crystals [Turchinovich & Dijkhuis, 2007;], have drawn considerable interest in recent years for THz applications. ZnTe is intrinsically a p-type semiconductor from the II-VI family with a direct bandgap of 2.26 eV. The growth of *n*-type ZnTe is still a challenge, which must be overcome before *p-n* junction devices can be realized. Incorporation of excess Zn during growth may increase donor concentration favoring the formation of *n*-type ZnTe. The more commonly encountered polymorph of ZnTe is the zinc blende structure which has four asymmetric units in its unit cell as against two in the wurtzite form. The higher degree of asymmetric coordination in zinc blende ZnTe favors formation of non-centrosymmetric structure which is very useful in terahertz generation and detection applications [4]. The applications of wurtzite ZnTe, on the other hand, are less well reported. Information on properties like spectral transmission, refractive index, optical band gap, and electronic structure of wurtzite ZnTe thin films is very sparse.

Since most of the THz related work on ZnTe is on single crystals, the main objectives of our work is to fabricate ZnTe coatings that are 10-12 microns thick as well as stabilizing the wurtzite form of ZnTe and investigate its optical and electronic properties. The coatings were produced by either thermal or electron beam evaporation in high vacuum (10^{-6} Torr). The source material is zinc blende or wurtzite ZnTe. The substrates are borosilicate glass or fused silica and the films were deposited at room temperature. Spectral transmittance curves of the films in the wavelength range from 190 to 2500 nm were measured using a spectrophotometer and from these spectra the refractive index and optical band gaps were extracted. Crystal structure was determined suing a powder x-ray diffractometer and electron diffraction while microstructure and morphology was examined under an Atomic force microscope.

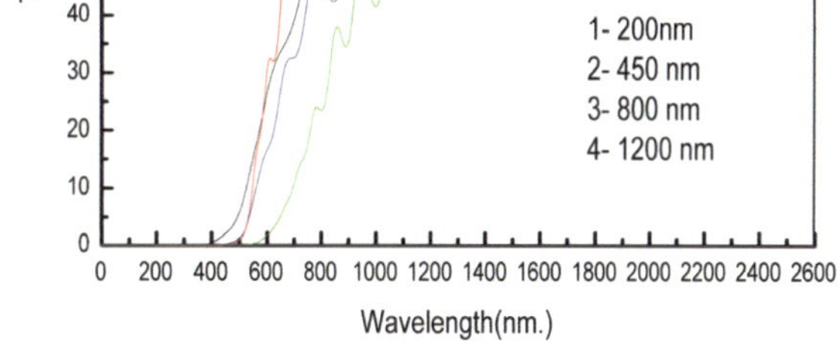

Fig. 5. Spectral transmission curves

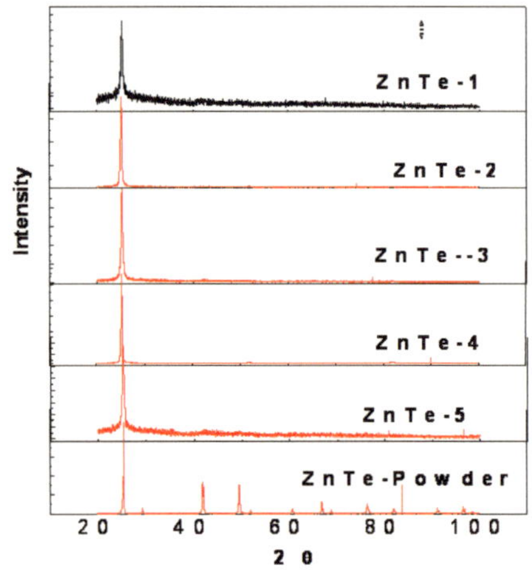

Fig. 6. X-ray diffraction patterns of the ZnTe films of different thicknesses showing the formation of zinc blende structure.

The measured spectral transmission curves for zinc blende ZnTe films of thicknesses upto 1200 nm and typical x-ray diffraction pattern of the 1200 nm film is shown in figs. 5 and 6 respectively. The transmission of the films is close to 90% in the near IR region and calculated refractive indices were between 2.0 and 2.6. The optical band gaps were between 2 and 2.2 eV. These values compared favourably with earlier reports.Significantly the 1200 nm thick film was nanocrystalline (with a crystallite size of 24 nm) and showed a preferred (111) orientation. The thickest films achieved using this technique had a thickness of 10-12 microns

The atomic force microscope image of a 2000 nm thick zinc blende ZnTe film is shown in fig. 7. The film is clearly very smooth with roughness less than 10 nm even at this thickness. The grain sizes are of the order of 100 nm.

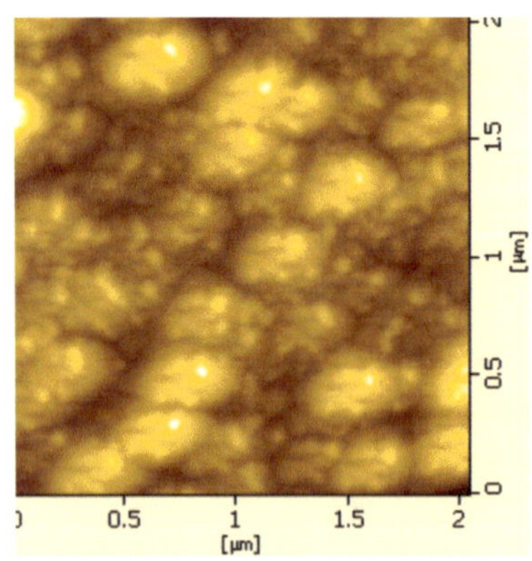

Fig. 7. Atomic force microscope image of a 2000 nm zinc blende ZnTe film.

ZnTe also crystallizes in the wurtzite structure wherein the stacking sequence of the cation and anion is ABABABAB. . . with each repeated period perpendicular to the basal plane, whereas in the zinc blende structure the sequence is ABCABCABC. Therefore, the distance between the third nearest-neighbor (NN) atoms is small in wurtzite ZnTe. In contrast, the distance between the first and second NN atoms is invariable in both structures. In addition, the Zn-Te bond in zinc blende ZnTe has greater ionic character than that in wurtzite ZnTe. Due to these differences it has been shown from ab-initio calculations that the band gaps of the structures should be different. Similarly, it is also expected that the electronic properties will be different. There are, however, very limited studies on the growth and stabilization of wurtzite ZnTe films. The present authors have grown wurtzite ZnTe films by electron beam evaporation of a wurtzite ZnTe source material. Typical x-ray diffraction patterns of the wurtzite films are shown in fig.8 for different thicknesses. The measured spectral transmittance for the zinc blende and wurtzite films are shown in fig.9. Detailed dispersion analysis of the films and comparison with zinc blende films has been carried out. ZnTe thin films exhibit either the wurtzite or zinc blende structure depending on the chemical nature of the source material. The formation of the structural phases is confirmed by a combination of x-ray diffraction and selected area electron diffraction. The films were nanocrystalline and randomly textured. The refractive index of the wurtzite phase (3.87 at 2000 nm) is higher than that for the zinc blende phase (3.03 at 2000 nm). The optical band gap, however, is lower for the wurtzite phase (1.1 eV) than the zinc blende phase (2.26 eV). The optical dispersion behavior has been investigated within the framework of the single effective oscillator model. It is shown that the oscillator strength, dispersion energy and optical charge carrier concentrations are all lower for the wurtzite phase than the zinc blende phase.

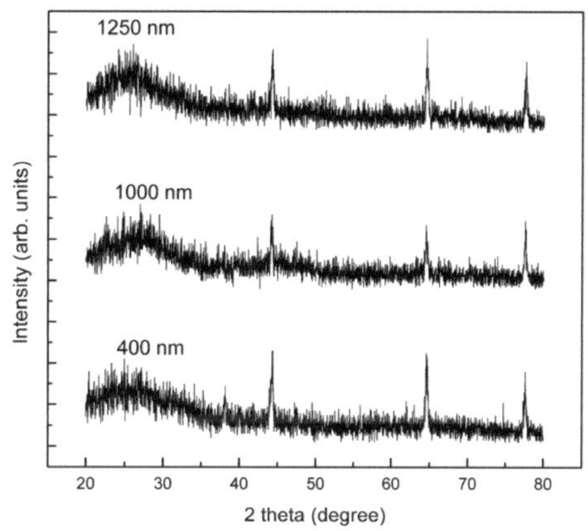

Fig. 8. XRD patterns of wurtzite ZnTe films of different thickness

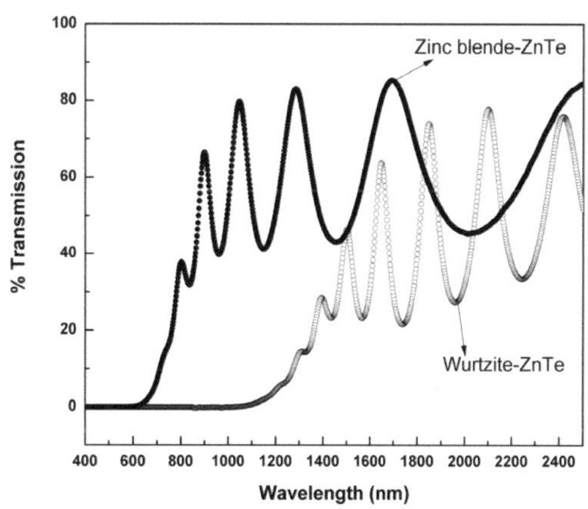

Fig. 9. Comparison of optical response of zinc blende and wurtzite ZnTe thin films.

The charge carrier concentration in wurtzite phase (5 x$10^{25}$$m^{-3}$) is lower than the zinc blende structure (4.17 x 10^{26} m^{-3}) This may make the wurtzite phase suitable for the development of *n*-type ZnTe, which has been a long standing challenge. Evidently the wurtzite phase has the potential to be employed as an n-type material leading to the formation of ZnTe p-n junctions for THz applications. Further developments in this direction are ongoing

Our work thus demonstrates the growth of thick ZnTe coatings in both the zinc blende and wurtzite polymorphs. Their optical properties are substantially different, signifying the possibility of applications in different frequencies regimes of the THz region [Kiran et al, 2010; Kshirsagar et al, 2011].

6. Conclusions and future prospects

In summary, a number of technologies for THz emitters, detectors and sensors have been reviewed. It is obvious that, while much progress has been made in all three areas, there is scope for new science and technology to emerge in all these areas. Significantly, the progress in the area of THz electronics has been much more rapid than that in the area of THz photonics. This is mainly due to the lack of cost effective so-called „table top" femto second laser sources. THz electronics, on the other hand, relies solely on existing mature technologies for materials and device fabrication. Much of the materials development has focused on single crystals. However, for miniaturization of THz technologies (Photonics and electronics) to succeed, the area of coatings, films and micro-nano devices has to be focused upon, in addition to development of compact THz sources. A report on the development of a high-resolution terahertz (THz) imaging beyond the diffraction limit by using a two-dimensional electron gas (2DEG) in a GaAs/AlGaAs heterostructure, points to effort in this direction[.Kawano & Ishibashi, 2010]. In the area of materials, strongly correlated electron systems have shown great potential[Ge et al, 2010]. In addition, novel device structures to exploit the plasma instability for THz emission and detection is another area that is receiving greater attention. Interestingly, access to fs laser sources has led to better understanding of physical and chemical phenomena in the THz region, that will eventually lead to new materials and device structures. Two examples exemplify this theme. It has recently been shown [Chan et al., 2011] that a nano mechanical oscillator coupled to single mode electromagnetic oscillator in a nano optical cavity can be designed to have oscillations in THz frequency domain. The physics of this domain of energy storage as envisaged by the Debye model has to be examined and the modes of vibrations at phase boundaries, and domain boundaries studied through THz experiments. These will reveal new materials and methods of generation and detection and sensing in the THz range. In another report [Chen et al, 2011], impulsive stimulated Raman scattering and Time-domain THz spectroscopy of coherent optical phonons in bismuth germinate ($Bi_4Ge_3O_{12}$) revealed more than 12 unique vibrational states ranging in frequency from 2 to 11 THz, each with coherent lifetimes ranging from 1 to 20 ps. Such studies on other systems can be expected have a profound impact on the developments in THz sources, emitters, detectors and sensors.

7. Acknowledgements

The authors acknowledge the facilities provided by the School of Physics under the UGC-UPE and CAS programs. A Dr D S Kothari Fellowship for SK is also acknowledged.

8. References

Abbas, A.; Treizebre, A.; Supiot, P.; Bourzgui, N.; Guillochon, D.; Vercaigne-Marko, D. & Bocquet, B. (2009). Cold plasma functionalized TeraHertz BioMEMS for enzyme reaction analysis, *Biosensors and Bioelectronics,* Vol.25, No.1, (September 2009), pp. 154-160, ISSN 09565663

Aurino, M., Kreisler, A.J., Türer, I., Martinez, A., Gensbittel, A., Dégardin, A.F. (2010) *Journal of Physics: Conference Series* Vol.234, No.4 (2010) art. no. 042002. ISSN: 1742-6588.

Belkin, M. A.; Jie, W. Q.; Pflugl, C.; Belyanin, A.; Khanna, S. P. Davies, A. G.; Linfield, E. H. Capasso, F. (2009). High-Temperature Operation of Terahertz Quantum Cascade Laser Sources, *IEEE Selected Topics in Quantum Electronics,* Vol.15, No.3 (June 2009) pp. 952-967, ISSN 1077-260X

Bhasin L. & Tripathi, V. K. (2009). Terahertz generation via optical rectification of x-mode laser in a rippled density magnetized plasma, *Phys. Plasmas* Vol.16, No. 10, (October 2009), pp. 103105- 1-6 ISSN 1089-7674

Blanchard,F.; Sharma, G.; Razzari, L.; Ropagnol, X.; Bandulet, H -C.; Vidal, F.; Morandotti, R.; Kieffer, J -C.; Ozaki, T.; Tiedje, H.; Haugen, H.; Reid, M. & Hegmann, F. (2011). Generation of Intense Terahertz Radiation via Optical Methods, IEEE Selected Topics in Quantum Electronics, Vol.17 No. 1, (June 2011), pp. 5-16, ISSN 1077-260X

Bolıvar P H, Nagel M, Richter F, Brucherseifer M, Kurz H, Bosserhoff A & Buttner R (2004), Label-free THz sensing of genetic sequences: towards 'THz biochips', *Philosophical Transactions of the Royal Society, London , A., Vol. 362* (2004) pp.323-335.

Bugay, A. N. & Sazonov, S. V. (2010). The generation of terahertz radiation via optical rectification in the self-induced transparency regime, *Phys. Lett. A,* Vol.374, No. 8 (February 2010), pp. 1093-1096, ISSN 0375-9601

Caumes, J. P.; Chassagne, B.; Coquillat, D.; Teppe, F. & Knap, W. (2009). Focal-plane micro-bolometer arrays for 0.5 THz spatial room-temperature imaging, *IEEE Electronics Letters,* Vol.45, No. 1, (December 2009), pp. 34-35, ISSN 0013-5194

Chan J., Mayer Alegre T.P., Safavi-Naeini A.H.,, Hill J.T., Krause A, Groblacher S., Aspelmeyer M & Painter O, (2011), Laser cooling of a nanomechanical oscillator into its quantum ground state, *Nature,* Vol. 478, (October, 2011), pp. 89-92. ISSN: 0028-0836.

Chamberlain J.M., (2004) Where optics meets electronics: recent progress in decreasing the terahertz gap, *Philosophical Transactions of the Royal Society, London , A., Vol. 362* (2004) pp.199–213. ISSN: 1471-2962.

Chen, J.; Liang, M.; Kang, L.; Jin, B. B.; Xu, W. W.; Wu, P. H.; Zhang, W.; Jiang, L.; Li, N. S. & Shi, C. (2009). Low Noise Receivers at 1.6 THz and 2.5 THz Based on Niobium Nitride Hot Electron Bolometer Mixers, *Appl. Superconductivity, IEEE Trans.* Vol.19, No. 3, (July 2009), pp. 278-271, ISSN 1051-8223

Chen, S.-M.; Fang, Y.-K.; Juang, F.R.; Yeh, W.-K.; Chao, C.-P. & Tseng, H.-C (2010). Terahertz Schottky barrier diodes with various isolation designs for advanced radio frequency applications, *Thin Solid Films,* Vol.519, No.1, (October 2010), pp. 471-474, ISSN 0040-6090.

Chen, Z.; Gao, Y.; & DeCamp, M. F. (2011). Retrieval of terahertz spectra through ultrafast electro-optic modulation, *Appl. Phys. Lett.,* Vol.99, No.1, (July 2011), pp. 011106-09 ISSN 1077-3118

Cherednichenko S., Khosropanah P., Kollberg E., Krong M., Merkel H., (2002), Terahertz superconducting hot electron bolometer mixers, *Physica C; Superconductivity,* Vol. 372-376, pt. 1, (August 2002), pp. 407-415.

Coquillat, D., Teppe, F., Videlier, H., Coquillat, D., Lusakowski, J., Skotnicki, T. Silicon field effect transistors for Terahertz detection and imaging Knap, W., Shuster, F., (2011) *Proceedings of the 5th European Conference on Antennas and Propagation, EUCAP 2011* , art. no. 5782258, (2011) pp. 3180-3182.

Costa M. R., Kibis O.V., & Portnoi M.E.(2009), Carbon nanotubes as a basis for terahertz emitters and detectors, *Microelectronics Journal* , Vol. 40 (2009) 776–778

Cox, J.A., Higashi, R., Nusseibeh, F., Zins, C. (2011) MEMS-based uncooled THz detectors for staring imagers *Proceedings of SPIE - The International Society for Optical Engineering* 8031, (2011) art. no. 80310D.

Du, J., Hellicar, A.D., Hanham, S.M., Li, L., MacFarlane, J.C., Leslie, K.E., Foley, C.P. (2011) YBCO hot-electron bolometers dedicated to THz detection and imaging: Embedding issues *Journal of Infrared, Millimeter, and Terahertz Waves* Vol. 32 No.5, (2011) pp. 681-690.

Dyakonov M. I,.(2010) Generation and detection of Terahertz radiation by field effect transistors *Comptes Rendus Physique*, Vol. 11 (2010) pp. 413–420

Dyakonov, M.; & Shur, M. (1996). Detection, mixing, and frequency multiplication of terahertz radiation by two-dimensional electronic fluid, (August 1996), *Electron Devices, IEEE Trans,* Vol.43, No.3, (1996) pp. 380-387, ISSN 0018-9383

Dyakonova, N.; Teppe, F.; Lusakowski, J.; Knap, W.; Levinshtein, M.; Dmitriev, A.P.; Shur, M.S.; Bollaert, S.; Cappy A. (2005). Magnetic field effect on the terahertz emission from nanometer InGaAs/AlInAs high electron mobility transistors, *J. Appl. Phys.* Vol.97, No.11, (May 2005), pp. 114313-1-3 ISSN 1089-7550

Fatimy, A. E.; Dyakonova, N.;, Meziani, Y.; Otsuji, T.; Knap, W.; Vandenbrouk, S.; Madjour, K.; Théron, D.; Gaquiere, C.; Poisson, M. A.; Delage, S.; Prystawko, P. & Skierbiszewski C.; (2010). AlGaN/GaN high electron mobility transistors as a voltage-tunable room temperature terahertz sources, *J. Appl. Phys.* Vol.107, No.2, (January 2010), pp-024504 1-4, ISSN 1089-7550

Federici, J. F.; Schulkin, B.; Huang, F.; Gary, D.; Barat, R.; Oliveira, F. & Zimdars, D. (2005). THz imaging and sensing for security applications—explosives, weapons and drugs, *Semicond. Sci. Technol.* Vol.20, No.7 (July 2005), pp. S266, ISSN 0268-1242

Ferguson B., Zhang X.C.,(2002) Materials for Terahertz Science and Technology, *Nature Materials* Vol. 1 (2002) 26-33. ISSN 1476-1122.

Foltynowicz, R. J.; Allman, R. E. & Zuckerman, E. (2006). Terahertz absorption measurement for gas-phase 2,4-dinitrotoluene from 0.05 THz to 2.7 THz, *Chemical Physics Letters,* Vol.431, No.1 (November 2006), pp 34-38, ISSN 0009-2614

Foutz, B. E.; O'Leary, S. K.; Shur, M. S. & Eastman, L. F. (1999). Transient electron transport in wurtzite GaN, InN, and AlN, *J. Appl. Phys.,* Vol.85, No.11 (February 1999), pp. 7727 1-8, ISSN 1089-7550

Freeman, J. R.; Brewer, A.; Beere, H. E. and Ritchie, D. A., (2011). Photo-luminescence study of heterogeneous terahertz quantum cascade lasers, *J. Appl. Phys.,* Vol.110, No.1, (July 2011) pp. 013103 1-6, ISSN 1089-7550

Ge C., Jin K., Lu H., Wang C., Zhao G., Zhang L., Yang G., (2010), Mechanisms for the enhancement of the lateral photovoltage in perovskite heterostructures, *Solid State Communications,* Vol. 150 (2010) pp. 2114–2117

Ginzburg N. S., Malkin A.M., Yu. Peskov N., Sergeev A. S., Yu. Zaslavsky V,& Zotova I.V.,(2011) Powerful terahertz free electron lasers with hybrid Bragg reflectors, *Physical Review Special Topics: Accelerators and Beams* Vol.14, No.4 (2011) 042001-1-9.

Gmachl, C.; Belyanin, A.; Sivco, D.L.; Peabody, M.L.; Owschimikow, N.; Sergent, A.M.; Capasso, F. & Cho, A.Y. (2003). Optimized second-harmonic generation in quantum cascade lasers, *Quantum Electronics, IEEE,* Vol.39 , No.11, (November 2003), pp. 1345-1355, ISSN 0018-9197

He, S.; Chen, X.; Wu, X.; Wang, G. & Zhao, F. (2008). Enhanced Terahertz Emission From ZnSe Nano-Grain Surface, *Journal of Lightwave Technology,* Vol.26, No.11, (June 2008), pp. 1519 - 1523, ISSN 0733-8724

Hoffmann, M. C.; Yeh, K.; Hwang, H. Y.; Sosnowski, T. S.; Prall, B. S.; Hebling, J. & Nelson, K. A. (2008). Fiber laser pumped high average power single-cycle terahertz pulse source, *Appl. Phys. Lett.* Vol.93, No. 14 (October 2008), pp. 141107 1-3 ISSN 1077-3118

Hu, Y.; Huang, P.; Guo, L.; Wang, X. & Zhang, C. (2006). Terahertz spectroscopic investigations of explosives, *Phys. Lett. A,* Vol.359, No.6, (December 2006), pp. 728-732, ISSN 0375-9601

Il'in, K.S.; Semenov, A.D.; Hübers, H.-W. & Siegel, M. (2010). Hot-electron bolometer mixers for terahertz radiation, *Electron. Lett.* Vol.46, No. 26, (January 2010), pp. S14-S16, ISSN 0013-5194

Innocenzi P, Malfatti L, Piccinini M, Sali D, Schade U,& Marcelli A (2009), Application of Terahertz Spectroscopy to Time-Dependent Chemical-Physical Phenomena, *Journal of Physical Chemistry A,* Vol. 113, No. 34, (2009) pp. 9418–9423.

Iwami, K., Ono, T., Esashi, M. 2009,Sputter deposited zinc oxide photoconductive antenna for terahertz time-domain spectroscopy *Proceedings of SPIE - The International Society for Optical Engineering* Vol. 7133, (2009), art. no. 713315

Jacobsen R. H., Mittleman D. M.,& M. C. Nuss, (1996) Chemical recognition of gases and gas mixtures with terahertz waves, *Optics Letters,* Vol. 21, No. 24 (December 1996) pp.2011-2013.

Jagtap V.S., Scheuring A., Longhin M., Kreisler A.J., Dégardin A.F., (2009) From superconducting to semiconducting YBCO thin film bolometers: Sensitivity and crosstalk investigations for future thz imagers *IEEE Transactions on Applied Superconductivity* Vol. 19, No. 3 (2009) pp. 287-292.

Kasai, S.; Katagiri, T.; Takayanagi, J.; Kawase, K. and Ouchi, T. (2009). Reduction of phonon resonant terahertz wave absorption in photoconductive switches using epitaxial layer transfer, *Appl. Phys. Lett.,* Vol.94, No.11, (2009) pp. 113505 1-3 ISSN 1077-3118

Kawano Y., Ishibashi K. (2010) Physica E 42 (2010) 1188–1191

Kawase, K.; Shibuya, T.; Hayashi, S. & Suizu, K. (2010). THz imaging techniques for nondestructive inspections, *Comptes Rendus Physique,* Vol.11, No.7, (October 2010), pp. 510-518, ISSN 1631-0705

Kiran M.S.R.N., Kshirsagar S.D., Krishna M.G., Tewari S.P. (2010), Structural, optical and nanomechanical properties of (111) oriented nanocrystalline ZnTe thin films, European Journal of Applied Physics: Applied Physics, vol. 51 (October 2010) pp.10502-9.

Kitagawa, J.; Ohkubo, T.; Onuma, M. & Kadoya, Y., (2006). THz spectroscopic characterization of biomolecule/water systems by compact sensor chips, *Appl. Phys. Lett.* Vol.89, No.4, (July 2006), pp. 041114 1-3, ISSN 1077-3118

Klatt G., Hilser F., Qiao W., Beck M., Gebs R., Bartels A., Huska K., Lemmer U., Bastian G., Johnston M.B., Fischer M., Faist J., & Dekorsy T., (2010) Terahertz emission from lateral photo-Dember currents, Optics Express, Vol. 18, Issue 5, (2010) pp. 4939-4947

Kleinschmidt, P.; Giblin, S. P.; Antonov, V.; Hashiba, H.; Kulik, L.; Tzalenchuk, A. & Komiyama, S. (2007). A Highly Sensitive Detector for Radiation in the Terahertz Region, *IEEE trans. Instru. measurement,* Vol.56, No. 2, (2007) pp. 463-467, ISSN 0018-9456

Kida N, Murakami H & Tonouchi M., (2005) Terahertz optics in strongly correlated electron systems, *Topics in Applied Physics,* Vol. 97 (2005) pp. 215-334.

Knap, W.; Lusakowski, J.; Parenty, T.; Bollaert, S.; Cappy, A.; Popov, V.V. & Shur, M.S. (2004). Terahertz emission by plasma waves in 60 nm gate high electron mobility transistors, *Appl. Phys. Lett.* Vol.84, No.13, (February 2004), pp. 2331 1-3, ISSN 1077-3118

Knap, W.; Shuster, F.; Coquillat, D.; Teppe, F.; Videlier, H.; Coquillat, D.; Lusakowski, J. & Skotnicki, T. (2011). Silicon field effect transistors for Terahertz detection and imaging, *Proceedings of the 5th European Conference on Antennas and Propagation, Antennas and propagation. european conference*, April, 2011, Rome, Italy

Krotkus A. (2010), Semiconductors for terahertz photonics applications, *Journal of Physics D ; Applied Physics, Vol.*43 (2010) pp.273001-273006. ISSN : 0022-3727

Kshirsagar S D, Krishna M.G., Tewari S.P., (2011), Morphological and Optical properties of Wurtzite ZnTe thin films, *AIP Conference proceedings,* Vol. 1349 (July 2011), pp. 1285-6.

Laviano, F., Gerbaldo, R., Ghigo, G., Gozzelino, L., Minetti, B., Rovelli, A., Mezzetti, E., (2010) Rugged superconducting detector for monitoring infrared energy sources in harsh environments, *Superconductor Science and Technology* Vol. 23, No.12 (2010), art. no. 125008 .

Leahy-Hoppa, M.R.; Fitch, M.J.; Zheng, X.; Hayden, L.M. & Osiander, R. (2007). Wideband terahertz spectroscopy of explosives, *Chem. Phys. Lett.,* Vol.434, No.4 (February 2007), pp. 227-230, 0009-2614

Lewis, R.A. (2007) Physical phenomena in electronic materials in the terahertz region *Proceedings of the IEEE* Vol. 95 No. 8, art. no. 4337844 (2007), pp. 1641-1645.

Li C., Zhou M., Ding W., Du F., Li Y., Wang W .M., Sheng Z.M., Ma J.L., Chen L M, Dong Q, & Zhang J, Effects of laser plasma interactions on THz radiation from solid targets irradiated by ultra short intense laser pulses, *Physical Review E,* Vol. 84 (2011) pp. 036405-036410

Li, D. & Ma, G. (2008). Pump-wavelength dependence of terahertz radiation via optical rectification in (110)-oriented ZnTe crystal, *J. Appl. Phys.* Vol.103, No. 12, (June 2008), pp. 123101 1-4, ISSN 1089-7550

Liu, H.C., Song, C.Y., Wasilewski, Z.R., Buchanan, M. (2011) Phonon and polaron enhanced IR-THz photodetectors *Proceedings of SPIE - The International Society for Optical Engineering* 7945, (2011) art. no. 79450X

Liu, L. Xu, H.; Percy, R. R.; Herald, D. L.; Lichtenberger, A. W.; Hesler, J. L. & Weikle, R. M. (2009). Development of Integrated Terahertz Broadband Detectors Utilizing Superconducting Hot-Electron Bolometers, *IEEE trans. Appl.Supercon.* Vol.19, No. 3, (July 2009), pp. 282-286, ISSN 1051-8223

Liu, T.; Lin, G.; Lee, Y.; Wang, S.; Tani, M.; Wu, H. & Pan, C. (2005). Dark current and trailing-edge suppression in ultrafast photoconductive switches and terahertz spiral antennas fabricated on multienergy arsenic-ion-implanted GaAs, *J. Appl. Phys.* Vol.98, No.1, (July 2005), pp. 013711 1-4, ISSN 1089-7550

Lloyd-Hughes, J.; Castro-Camus, E.; Johnston, M.B. (2005). Simulation and optimisation of terahertz emission from InGaAs and InP photoconductive switches, *Solid State Commun.,* Vol.136, No.11, (October 2005) pp. 595-600, ISSN 0038-1098

Lu, X.; Karpowicz, N.; Chen, Y. & Zhang, X.-C., (2008). Systematic study of broadband terahertz gas sensor, *Appl. Phys. Lett.* Vol.93, (December 2008), pp.261106, ISSN 1077-3118

Łusakowski J., (2007), Nanometer transistors for emission and detection of THz radiation, Thin Solid Films, Vol. 515 (2007) pp. 4327–4332.

Lusakowski, J.; Knap, W.; Dyakonova, N.; Varani, L.; Mateos, J.; Gonzalez, T.; Roelens, Y.; Bollaert, S.; Cappy, A. & Karpierz, K. (2005). Terahertz emission by plasma waves in 60 nm gate high electron mobility transistors, *J. Appl. Phys.* Vol.97, No. 13 (February 2005) pp. 064307 1-3, ISSN 1077-3118

Maestrini, A.; Thomas, B.; Wang, H.; Jung, C.; Treuttel, J.; Jin, Y.; Chattopadhyay, G.; Mehdi, I. & Beaudin, G. (2010). Schottky diode-based terahertz frequency multipliers and mixers , *Comptes Rendus Physique,* Vol.11, No.7, (2010). pp. 480-495, ISSN 1631-0705

Magno, R.; Champlain, J. G.; Newman, H. S.; Ancona, M. G.; Culbertson, J. C.; Bennett, B. R.; Boos, J. B. & Park D., (2008). Antimonide-based diodes for terahertz mixers, *Appl. Phys. Lett.* Vol.92, No. 24, (June, 2008) pp. 243502 1-3, ISSN 1077-3118

Marandi, A.; Darcie, T. E.; and So, P. P. M. (2008). Design of a continuous-wave tunable terahertz source using waveguide-phase-matched GaAs, *Optics Express,* Vol.16, No.14, (June 2008) pp 10427-10433, ISSN 1094-4087

Matsuo, H. (2006). Future prospects of superconducting direct detectors in terahertz frequency range, *Nuclear Instruments and Methods in Physics Research A,* Vol.559, (January 2006), pp. 748-750, ISSN 0168-9002

McLaughlin, C. V.; Hayden, L. M.; Polishak, B.; Huang, S.; Luo, J. Kim, T. & Jen, A. K. (2008). Wideband 15 THz response using organic electro-optic polymer emitter-sensor pairs at telecommunication wavelengths, *Appl. Phys. Lett.* Vol.92, No.15, (April 2008), pp. 151107 1-3, ISSN 1077-3118

Meledin, D. V.; Marrone, D. P.; Tong, C. Y.; Gibson, .H.; Blundell, R.; Paine, S. N.; Papa, D. C.; Smith, M.; Hunter, T. R.; Battat, J.; Voronov, B. & Gol'Tsman, G. A 1-THz superconducting hot-electron-bolometer receiver for astronomical observations, *IEEE Trans. on Microwave Theory and Techniques,* Vol.52, No. 10, (October 2004), pp. 2338-2343, ISSN 0018-9480

Mendis, R.; Astley, V.; Liu, J. & Mittleman, D. M. (2009). Terahertz microfluidic sensor based on a parallel-plate waveguide resonant cavity, *Appl. Phys. Lett.* Vol.95, No.17, (October 2009), pp.171113, ISSN 1077-3118

Mittleman, D. (2003). *in Sensing with Terahertz Radiation* Springer, ISBN 3-540-43110-1, New York

Miyadera, T.; Kiwa, T.; Kawayama, I.; Murakami, Tonouchi, H. M. (2004). Ultrafast optical study of amorphous Ge thin films for superconductor/semiconductor hybrid devices, *Physica C: Superconductivity,* Vol.412, No.2 (July 2004) pp.1602-1606, ISSN 0921-4534

Molis G., Adomavicus R., Krotkus A., (2008) Temperature-dependent terahertz radiation from the surfaces of narrow-gap semiconductors illuminated by femtosecond laser pulses, *Physica B,* Vol.403 (2008) pp. 3786–3788.

Mueller E.R., (2003) Terahertz Radiation: Applications and Sources, *The Industrial Physicist,* (Aug-Sept. 2003), pp 27-29.

Nagel, M.; Först, M & Kurz, H (2006). THz biosensing devices: fundamentals and technology, *J. Phys.: Condens. Matter,* Vol.18, No.18, (April 2006), pp. S601, ISSN 1361-648X

Nagel M, Richter F., Bolivar PH, & Kurz H (2003), A functionalized THz sensor for marker free DNA analysis *Phys. Med. Biol.* Vol.48 (2003) pp.3625-36. ISSN: 0031-9155

Nikoghosyan, A.S. (2010). Laser Driven Terahertz Dielectric Wedge Antenna laced in Free Space or in Hollow Metallic Waveguide, *35th International Conference on Infrared*

Millimeter and Terahertz Waves (IRMMW-THz), (September 2010) pp. 1-2, ISBN 978-1-4244-6655-9, Rome,

Otsuji, T., Tsuda, Y., Komori, T., El Fatimy, A., Suemitsu, T. (2009) Terahertz plasmon-resonant microship emitters and their possible sensing and spectroscopic applications *Proceedings of IEEE Sensors* , art. no. 5398309, (2009) pp. 1991-1996.

Otsuji, T.; Karasawa, H.; Watanabe, T.; Suemitsu, T.; Suemitsu, M.; Sano, E. Knap, W.; Ryzhii, V. (2010). Emission of terahertz radiation from two-dimensional electron systems in semiconductor nano-heterostructures, *Comptes Rendus Physique,* Vol.11, No.7, (July 2010), pp. 421-432, ISSN 1631-0705

Peroz, C.; Degardin, A. F.; Villegier, J. C. & Kreisler, A. (2007). Fabrication and Characterization of Ultrathin PBCO/YBCO/PBCO Constrictions for Hot Electron Bolometer THz Mixing Application, J. *Appl. Supercond., IEEE Trans.,* Vol.17 , No. (July 2007), pp. 637-640, ISSN 1051-8223

Pilla, S (2007). Enhancing the photomixing efficiency of optoelectronic devices in the terahertz regime, *Appl. Phys. Lett.,* Vol.90, No.16, (April 2007), pp. 161119-22, ISSN 1077-3118

Preu, S., Dhler, G.H., Malzer, S., Wang, L.J., Gossard, A.C. (2011) Tunable, continuous-wave Terahertz photomixer sources and applications *Journal of Applied Physics* Vol.109 No.6, (2011) art. no. 061301

Putz, P.; Jacobs, K.; Justen, M.; Schomaker, F.; Schultz, M.; Wulff, S. & Honingh, C. E (2011). NbTiN Hot Electron Bolometer Waveguide Mixers on ${rm Si}_{3}{rm N}_{4}$ Membranes at THz Frequencies, *Appl. Superconductivity, IEEE Trans.* Vol.21, (May 2011), pp. 636-639, ISSN 1051-8223

Radhanpura, K.; Hargreaves, S.; Lewis, R. A. & Henini, M. (2009). The role of optical rectification in the generation of terahertz radiation from GaBiAs, *Appl. Phys. Lett.* Vol.94, No. 25, (June 2009), pp. 251115-1-3, ISSN 1077-3118

Radhanpura, K.; Hargreaves, S.; Lewis, R. A.; Sirbu, L. & Tiginyanu, I. M. (2010), Heavy noble gas (Kr, Xe) irradiated (111) InP nanoporous honeycomb membranes with enhanced ultrafast all-optical terahertz emission, *Appl. Phys. Lett.* Vol.97, No. 18, (November 2010), pp. 181921-1-3, ISSN 1077-3118

Rangan C. & Bucksbaum P.H., Optimally shaped THz pulses for quantum algorithm on a Rydberg atom register, *Physical Review A,* Vol. 64 (2001) pp 37402- 37410.

Reklaitis A. , (2011) Crossover between surface field and photo-Dember effect induced terahertz emission, *Journal of Applied Physics,* Vol. 109, (2011) pp. 083108(*5 pages*)

Reimann K., (2007), Table-top sources of ultrashort THz pulses, Reports on Progress in Physics, Vol. *70 (2007) pp.1597.*

Rau, C.; Torosyan, G.; Beigang, R. & Nerkararyan, Kh., (2005). Prism coupled terahertz waveguide sensor, *Appl. Phys. Lett.* Vol.86, No.21, (2005), pp. 211119 1-3, ISSN 1077-3118

Reid, I.; Cravetchi, R.; Fedosejevs, I. M.; Tiginyanu, L.; Sirbu, & Robert W. Boyd, (2008). Enhanced nonlinear optical response of InP(100) membranes, *Phys. Rev. B* Vol.71, No. 8, (February 2005), pp. 081306, ISSN 1550-235x

Rinzan, M. B.; M. Perera, A. G. U.; Matsik, S. G.; Liu, H. C.; Wasilewski, Z. R. & Buchanan, M. (2004). AlGaAs emitter/GaAs barrier terahertz detector with a 2.3 THz threshold, *Appl. Phys. Lett.,* Vol.86, No.7 (Februvary 2005), pp. 071112 1-3, ISSN 1077-3118

Roser H.P., Hubers H.W., Crowe T.W., & Peatman W.C.B., Nanostructure GaAs Schottky diodes for far-infrared heterodyne receivers, *IR Physics* vol. 35 (1994) 451.

Roeser H.P., Haslam D.T., Hetfleisch F., Lopez J.S., vonSchoenermark M.F., Stepper M., Huber F.M., Nikoghosyan A.S., (2010) Electron transport in nanostructures: A key to high temperature superconductivity? *Acta Astronautica,* Vol. 67 (2010) pp. 546–552.

Ryzhii, V.; Satou, A.; Knap, W.; & Shur, M. S. (2006). Plasma oscillations in high-electron-mobility transistors with recessed gate, J. Appl. Phys. Vol.99, No.8, (May 2006), pp. 084507-1-5 ISSN 1089-7550

Sakai K., & Tani M., (2005) Introduction to Terahertz pulses, *Topics in Applied Physics,* Vol. 97 (2005), pp. 1-30.

Schultz, J. & Lichtenberger, A. (2007). Investigation of Novel Superconducting Hot Electron Bolometer Geometries Fabricated With Ultraviolet Lithography, *Applied Superconductivity, IEEE Trans.* Vol.17, (July 2007), No. 2, pp. 645-648, ISSN 1051-8223

Shchepetov, A.; Gardès, C.; Roelens, Y.; Cappy, A.; Bollaert, S.; Boubanga-Tombet, S.; Teppe, F.; Coquillat, D.; Nadar, S.; Dyakonova, N.; Videlier, H.; Knap, W.; Seliuta, D.; Vadoklis, R. and Valušis G. (2008). Oblique modes effect on terahertz plasma wave resonant detection in InGaAs/InAlAs multichannel transistors, *Appl. Phys. Lett.* Vol.92, No.24, (June 2008), pp. 242105-1-3, ISSN 1077-3118

Shibuya, T.; Suizu, K.; and Kawase, K.; (2010). Widely Tunable Monochromatic Cherenkov Phase-Matched Terahertz Wave Generation from Bulk Lithium Niobate, *Appl. Phys. Express,* Vol.3, (August 2010), pp. 082201-082204 ISSN 1882-0786

Sizov, F. & Rogalski A. (2010). THz detectors, *Progress in Quantum Electronics,* Vol.34, No.5, (September 2010), pp. 278-347, ISSN 0079-6727

Stern, J. A.; Bumble, B.; Kawamura, J. & Skalare, A. (2005). Fabrication of terahertz frequency phonon cooled HEB mixers, *IEEE Trans. Appl. Superconductivity,* Vol.15, No. (June 2005), pp. 499-502, ISSN 1051-8223

Suen, J. Y.; Li, W.; Taylor, Z. D. and Brown, E. R. (2010). Characterization and modeling of a terahertz photoconductive switch, *Appl. Phys., Lett.* Vol.96, No.14 (April 2010) pp. 141103-141106 ISSN 1077-3118

Suzuki, T.; Yasui, T.; Fujishima, H.; Nozokido, T.; Araki, M.; Boric-Lubecke, O.; Lubecke, V.M.; Warashina, H. & Mizuno, K. (1999). *Microwave Theory and Techniques, IEEE Trans.,* Vol.47 No.9, (September 1999), pp. 1649 - 1655, ISSN 0018-9480

Tochitsky S. Ya., Ralph J. E., Sung C., & Joshi C.(2005), Generation of megawatt-power terahertz pulses by noncollinear difference-frequency mixing in GaAs, *Journal of Applied Physics,* Vol. 98, No.2 (2005) 026101-1-3.

Tonouchi, M. (2007). Cutting-edge terahertz technology, *Nature Photoic.,* Vol.1, (February 2007), pp. 97-105, ISSN 1749-4885

Tredicucci, A.; Kohler, R.; Mahler, L.; Beere, H. E.; Linfield, E. H. & Ritchie, D.A. (2005). Terahertz quantum cascade lasers—first demonstration and novel concepts, *Semicond. Sci. Technol.* Vol.20, No.7, (June 2005) pp. S222–S227, ISSN 1361-6641

Tsaur G. & Wang, J. (2009). Relativistic optical rectification driven by a high-intensity pulsed Gaussian beam, *Phys. Rev. A,* Vol.80, No.2 (August 2009) pp. 023802-1-11 ISSN 1094-1622

Turchinovich D & Dijkhuis J.I. (2007) Performance of combined <100>–<110> ZnTe crystals in an amplified THz time-domain spectrometer, *Optics Communications* Vol. 270 (2007) pp. 96–99

Ueda T & Komiyama S (2010) Novel Ultra-Sensitive Detectors in the 10-50 mm wave length range , *Sensors,* Vol. 10, (2010) 8411-8423 ISSN 1424-8220

Urbanowicz, A.; Krotkus, A.; Adomavičius, R. & Malevich, V.L. (May 2007). Terahertz emission from femtosecond laser excited Ge surfaces due to the electrical field-induced optical rectification, *Physica B: Condensed Matter,* Vol.398, No. 98 (May 2007), pp. 98-101, ISSN 0921-4526

Valavanis, A.; Dinh, T. V.; Lever, L. J. M.; Ikonić, Z. and Kelsall, R. W. (2011). Material configurations for n-type silicon-based terahertz quantum cascade lasers *Phys. Rev. B,* Vol.83, No.19, (May 2011) pp. 195321-195329 ISSN 1550-235x

Vidal, S.; Degert, J.; Tondusson, M.; Oberlé, J. & Freysz, E. (2011). Impact of dispersion, free carriers, and two-photon absorption on the generation of intense terahertz pulses in ZnTe crystals, *Appl. Phys. Lett.* Vol.98, No. 19, (May 2011), pp. 191103-1-3, ISSN 1077-3118

Videlier, H., Nadar, S., Dyakonova, N., Sakowicz, M., Trinh Van Dam, T., Teppe, F., Coquillat, D.,&Lyonnet, J., (2009), Silicon MOSFETs as room temperature terahertz detectors, *Journal of Physics: Conference Series* Vol. 193, (2009) art. no. 012095 .

Williams G.P., High-power terahertz synchrotron sources, *Philosophical Transactions of the Royal Society London* A vol. 362 (2004), pp. 403–414. ISSN: 1471-2962.

Xiao-li, Z. & Jiu-sheng, L. (2011). Diagnostic techniques of talc powder in flour based on the THz spectroscopy, *J. Phys.: Conf. Ser.* Vol.276, No.1, (March 2011) pp.012234 ISSN 1742-6596

Yang, X.; Qi, S.; Zhang, C.; Chen, K.; Liang, X.; Yang, G.; Xu, T.; Han, Y. and Tian, J. (2011), The study of self-diffraction of mercury dithizonate in polymer film, *Optics Commun.* Vol.256, No. 4, (July 2011) pp. 414-421, ISSN 0030-4018

Yasuda, H.; Kubis, T.; Vogl, P.; Sekine, N.; Hosako, I. & K. Hirakawa, (2009). Nonequilibrium Green's function calculation for four-level scheme terahertz quantum cascade lasers, *Appl. Phys. Lett.* Vol.94, No.15, (April 2009), pp. 151109 1-3, ISSN 1077-3118

Yasui, T.; Nishimura, A.; Suzuki, T.; Nakayama, K. & Okajima S. (2006). Detection system operating at up to 7 THz using quasioptics and Schottky barrier diodes, *Rev. Sci. Instrum.* Vol.77, No.6, (2006) pp-066102 1-3, ISSN 1089-7623

Yi, M., Lee, K., Lim, J., Hong, Y., Jho, Y.-D., & Ahn J. (2010) Terahertz Waves Emitted from an Optical Fiber *Optics Express* vol. 18 No. 13, (2010) pp. 13693-13699.

Yu, N.E.; Lee, K.S.; Ko, D.-K. C. l.; Kang.; Takekawa, S. l. & Kitamura, K. (2011). Temperature dependent narrow-band terahertz pulse generation in periodically poled crystals via difference frequency generation, *Optics Commun.,* Vol.284 No.5, (November 2011), pp. 1395-1400, ISSN 0030-4018

Zhang, L.; Zhong, H.; Deng, C.; Zhang, C. & Zhao, Y. (2008). Terahertz wave reference-free phase imaging for identification of explosives, *Appl. Phys. Lett.* Vol.92, No.9, (March 2008), pp. 091117, ISSN 1077-3118

Zheng, X., McLaughlin, C.V., Cunningham, P., Hayden, L.M. (2007), Organic broadband terahertz sources and sensors, *Journal of Nanoelectronics and Optoelectronics* Vol. 2 No.1 (2007), pp. 58-76.

Zheng, X., McLaughlin, C.V., Cunningham, P., Hayden, L.M. (2007), Organic broadband terahertz sources and sensors, *Journal of Nanoelectronics and Optoelectronics* Vol. 2 No.1 (2007), pp. 58-76.

Zmuidzinas J., Richards P.L., (2004), Superconducting detectors and mixers for Millimeter and sub-millimeter Astrophysics, *Proceedings of IEEE,* vol.192, No.10, (October 2004) pp 1597-1616.

7

Far-Infrared Single-Photon Detectors Fabricated in Double-Quantum-Well Structures

Takeji Ueda and Susumu Komiyama
The University of Tokyo
Japan

1. Introduction

Sensitive infrared (IR) detectors are key elements for both fundamental research and applications, and development of sensitive IR detectors is one of intensive research subjects in resent years (Rogalski, 2002; Tidrow, 2000). The activities are motivated by both basic and applied research fields (Prochazka, 2005; Rogalski, 2002). For certain applications, an ultimately high sensitivity reaching a photon-counting level is indispensable; for instance, passive high-resolution microscopy, catching IR photons spontaneously emitted by a living cell or a small number of biomolecules, may be accessible only with such ultimately sensitive detectors.

Photon-counters are routinely used in visible and near infrared regions (Prochazka, 2005). In the long-wavelength range (λ > 10 μm, frequencies lower than 30 THz), called terahertz (THz) region, however, photon energies are far smaller ($h\nu$ < 124 meV for λ > 10 μm) and the single-photon detection is no longer trivial. Nevertheless, the THz region is one of the richest areas of spectroscopy of matters, encompassing the rotational spectra of molecules, vibrational spectra of molecules, liquids and solids, and the electron energy spectra in semiconductor nanostructures and superconducting energy gap in metals. Hence, sensitive microscopy of matters with high spatial resolution in this spectral region would provide a unique powerful tool for investigation of matters. Sensitive observation is also strongly demanded in astrophysics (Nakagawa et al, 2007).

In the last decade, a variety of novel detection schemes have been proposed (Astaviev et al., 2002; Day et al., 2003; Hashiba et al., 2006; Komiyama et al., 2000; Schoelkopf et al., 1999; Wei et al., 2008). Among them, only semiconductor quantum devices have demonstrated single-photon detection (Astaviev et al., 2002; Hashiba et al., 2006; Komiyama et al., 2000): The experimentally achieved noise equivalent power (NEP), less than 1×10^{-19} W/Hz$^{1/2}$, is by several orders of magnitude lower than those of any other detectors.

These Semiconductor photon-counters exploit a novel scheme, in which the photon absorption event leads to generation of a long-lived unit charge that is probed by a single-electron transistor (SET). The use of SETs, however, might restrict their application because the operation is limited to ultra-low temperatures (<1K). From the viewpoint of broader application, therefore, photon-counters usable at elevated temperatures are highly desirable. From the viewpoint of importance of the spectra of matters, it is also extremely important to fill the gap region between the near infrared and the far infrared regions; that is, the long-wavelength infrared (LWIR) and the mid-infrared (MIR) regions.

In this chapter, another type of charge-sensitive infrared phototransistors in 5-50 μm wavelength range is described. The detectors are called charge-sensitive infrared phototransistors (CSIPs). The detectors can have ultra-sensitivity reaching singe-photon detection level, as well as ultra-broad dynamic range (> 10^6, from attowatts to beyond picowatts)(Ueda et al., 2008). CSIPs can be operated reasonable temperatures (~25 K at 15 μm, depending on wavelength) (Ueda et al., 2009). The excellent noise equivalent power (NEP=6.8x10^{-19} W/$Hz^{1/2}$) and specific detectivity (D^*=1.2x10^{15} cm$Hz^{1/2}$/W) are demonstrated for λ=14.7 μm, which are by a few orders of magnitude superior to those of the other state-of-the-art detectors (Ueda et al., 2008). In addition, the simple planar structure is, similarly to CMOS sensors, feasible for array fabrication and will even make it possible to monolithically integrate with reading circuit.

2. Detection scheme and device structure

In CSIPs, an electrically isolated island of a QW is photoexcited to serve as a gate to a remote two-dimensional electron gas (2DEG) conducting channel. As schematically shown in Fig. 1 (a), photoexcited electrons escape the isolated QW island leaving holes behind. The photo electrons are driven to the 2DEG conducting channel yielding photocurrents. Another effect larger than this direct photocurrent arises from the positive charge left on the QW island, which, through capacitive coupling, increases the electron density in the 2DEG channel and thus its conductance. The effect persists until the excited electrons recombine with holes in the isolated island, serving as an amplification mechanism. CSIPs are thus charge-sensitive phototransistor, in which a QW island works as a photosensitive floating gate.

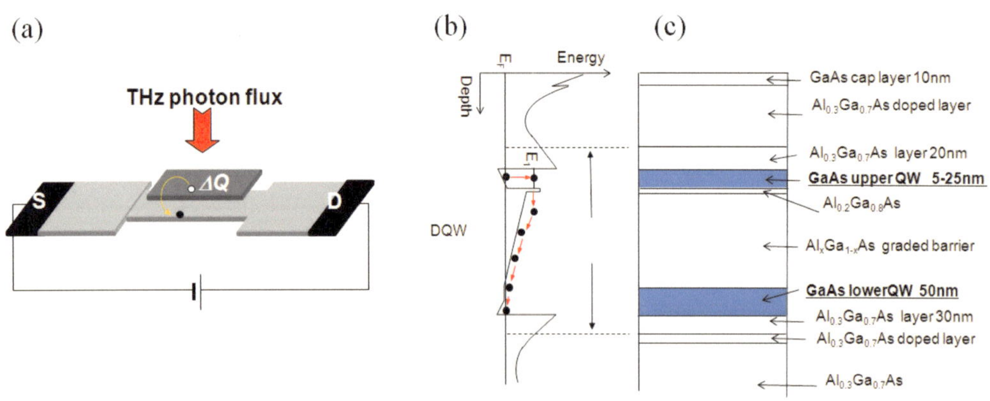

Fig. 1. (a) Schematic representation of the charge sensitive infrared phototransistor (CSIP) as a phpto-active field-effect transistor. (b) The energy diagram of double-quantum-well system. (c) Crystal strucuture for detection wavelength of 15 μm.

The upper QW is so designed that the energy spacing between the ground subband and the first excited subband is ΔE =84 meV (for detection wavelength of λ=15 μm). When radiation with photon energy of ΔE is incident on the isolated QW, electrons are excited to the first excited subband, where the thin tunnel barrier layer stands as schematically depicted in Fig. 1 (b). The electrons, having tunneled out of the QW, fall down the electrostatic potential slope in the graded barrier layer until they eventually reach the 2DEG channel to be absorbed there. This causes isolated QW island to be positively charged. Through capacitive coupling, the pile-up positive charge in the isolated island increases the electron density of the lower 2DEG channel leading to an increase in conductance.

The crystals are grown by molecular-beam epitaxy (MBE). The crystal structure is shown in Fig. 1 (c). Undoped GaAs (~400 nm) and 10 periods of an $Al_{0.3}Ga_{0.7}As$ (20 nm)/GaAs (2 nm) superlattice were first grown on the semi-insulating substrate. Then, a $1x10^{24}$ m^{-3} Si doped $Al_{0:3}Ga_{0:7}As$ layer (10 nm) and an $Al_{0.3}Ga_{0:7}As$ spacer layer (30 nm) were grown as a remote-doping layer to supply electrons to the successively grown undoped GaAs layer (20 nm), which serves as the lower QW in the DQW system. The 150-nm thick compositionally graded $Al_xGa_{1-x}As$ barrier (GB) and a 2-nm thick tunnel barrier (TB) are grown as the isolation barrier between the lower and upper QWs. Here, the growth was suspended between the GB and TB to change the Al beam. The GaAs layer as the upper QW was successively grown. Finally, an $Al_{0:3}Ga_{0:7}As$ spacer layer and an $Al_{0.3}Ga_{0.7}As$ uniform Si doping layer ($1x10^{24}$ m^{-3}, 40 nm) were grown to supply electrons to the upper QW. The surface is covered with a 10-nm-thick GaAs cap layer. The superlattice barrier below the lower QW improves the roughness of the GaAs/AlGaAs interface. The detection wavelength is controlled by changing the design width of upper QW (Ueda et al., 2011).

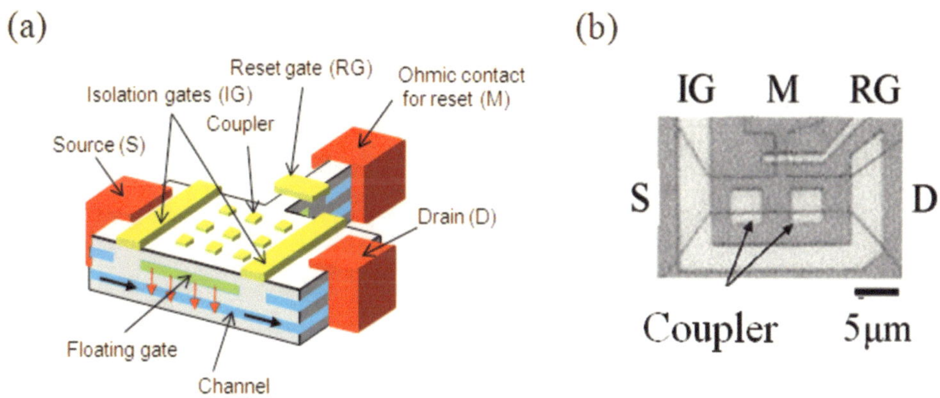

Fig. 2. (a) Schematic representation of QWs and ohmic contacts. The upper QW is electrically isolated by negative biasing of surface gates. (b) A microscope image of the device with a 16x4 μm² isolated island formed by the upper QW.

As shown in Fig. 2, the device consists of a wet-etched DQW mesa, alloyed AuGeNi ohmic contacts, Au/Ti Schottky gates (the isolation gate, IG, and reset gate, RG), and Au/Ti photo-coupler (antenna). The device is fabricated with standard electron-beam lithography technique. The 2DEG layer in the both of the QWs are normally connected by ohmic contacts, and can be electrically isolated by biasing metal isolation gates. The active areas of devices are freely defined in lithography. The coupler is used to cause intersubband transition by generating electric filed normal to the plane of the QW against the normally incident radiation. It should be mentioned that the simple planar structure of CSIPs is feasible for array fabrication (fabricated example is shown in Fig. 3). The FET structure would have an advantage of future monolithic integration with reading circuits.

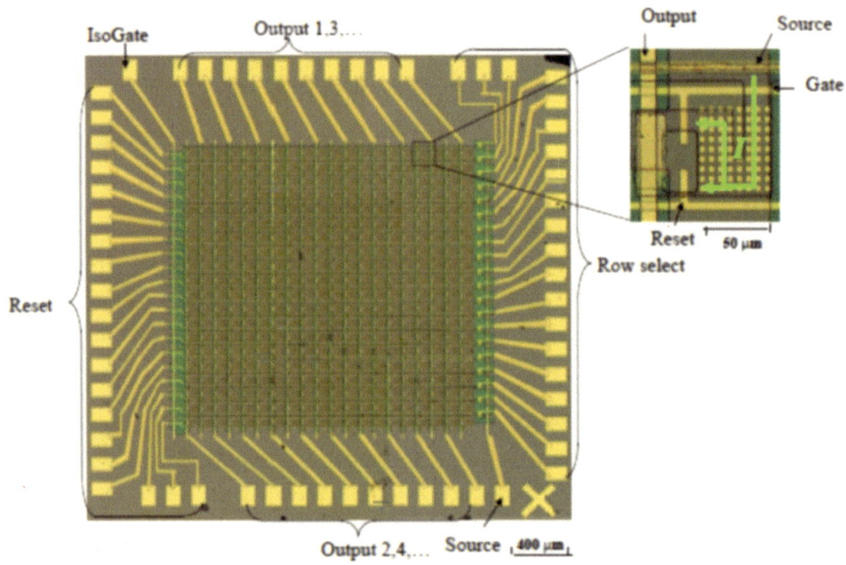

Fig. 3. 20x 20 CSIP focal plane array

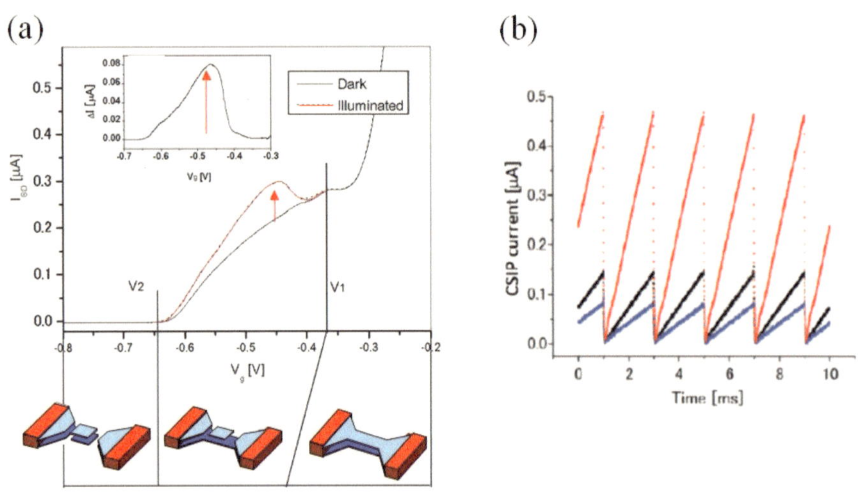

Fig. 4. (a) I-V measurement of a CSIP with scanning gates (IG and RG) bias. (b) Integration ramps of the measured photo-current against incident radiation with different intensities. Releasing the accumulated charge in the floating gate with reset gate pulses restore the detector to its original state.

In actual practice of operation, the optimized metal-gate bias is found in gate-bias-dependent IV curve where photosignal reaches maximum amplitude as shown in Fig. 4 (a). The signal appears when FET is formed: upper QW is electrically isolated and the accumulated photoholes induce larger current flowing in the lower QW.

As in Fig. 4 (b), to avoid the saturation of the signal, the accumulated charge is released to the reservoir by applying a brief positive pulse (e.g., +0.3V, 1 μs) to the RG, and the CSIP is reset to the original highly sensitive state (An et al., 2007). Conventional photoconductive detectors yield a certain amplitude of photocurrent that is proportional to the incident radiation intensity. The scheme of CSIPs is different: The source-drain current continues to increase with time under steady illumination, where the speed at which the current increases, $a=\Delta I/\Delta t$, is proportional to the incident radiation intensity.

This photoresponse can be interpreted by increase of electron density in lower QW induced by capacitively coupled photoholes stored in the isolated upper QW. The unit increment of current I_e induced by one photohole in the isolated upper QW (area of $L \times W$) is given by:

$$I_e = \frac{e\mu V_{SD}}{L^2} \tag{1}$$

where e is the unit charge, μ the electron mobility of lower QW, V_{SD} the souce-drain (SD) voltage, L the length of constricted channel. For example, unit increment I_e = 3 pA is given for μ = 1 m²/Vs, V_{SD} = 10 mV and L = 16 μm. The signal I_e persists as long as a photohole stays in upper QW. By setting the lifetime τ = 1 s, the amplification factor, or photoconductive gain, is given as $G = \tau I_e/e = 1.8 \times 10^7$. This value is comparable to that of

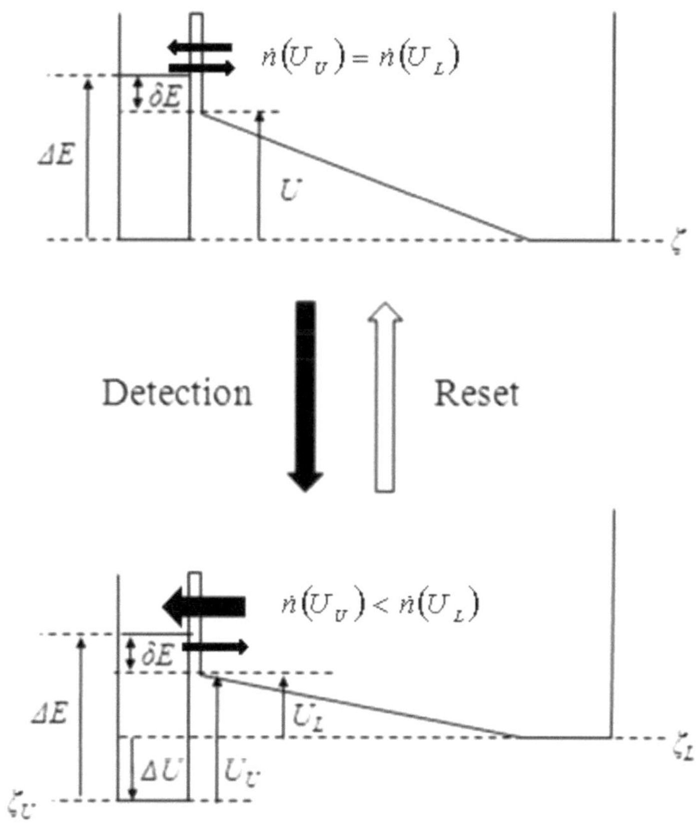

Fig. 5. Energy diagram illustrating photo-saturation.

photomultiplier tubes. The increase of the current is proportional to the number of accumulated photoholes p:

$$\Delta I = pI_e \tag{2}$$

Under steady illumination, p is a linearly increasing function of time if the lifetime of photoholes is longer than the relevant time of integration. The number of photoholes, of course, does not increase infinitely, but reach a saturated value. The saturation is caused by balance between generation and recombination speeds of photoholes which change with deformation of the potential profile due to accumulating positive charges in upper QW, as shown in Fig. 5. The potential drop is given by $\Delta U = pe^2d/\varepsilon LW$, where d = 150 nm is the distance between upper and lower QWs, $\varepsilon = 12 \times (8.85 \times 10^{-12})$ F/m is the electric permittivity of GaAs. The number of photoholes is described by the rate equation:

$$\frac{dp}{dt} = \eta\Phi - \frac{p}{\tau} \tag{3}$$

where η is quantum efficiency, Φ is the incident photon flux (photons/s), and τ is the lifetime of photoholes. The first and second term in the right-hand side of Eq. 3 refer to the generation and recombination speed of photoholes, respectively.

3. All-cryogenic spectrometer

It is not trivial to accurately characterize ultrasensitive detectors in far infrared region. In usual setup, a monochromator system is placed at room temperature. Then, in case of ultrasensitive detection in far-IR and THz region, the measurements suffer from the thermal background blackbody radiation (BBR) at 300K that can be much stronger than the light guided from the external source. This caused a serious problem for the study of extremely sensitive CSIPs, the intrinsic characteristics of which may be significantly affected by relatively weak background radiation. It was also difficult to determine accurate excitation spectrum.

We characterize CSIPs in reliable manners by developing a home-made all-cryogenic spectrometer (4.2K) (Ueda, 2007). As illustrated in Fig. 6 (a), the all-cryogenic spectrometer consists of an emitter, a parabolic mirror, a rotating diffraction grating, and the detector. The whole system is assembled on a 1cm-thick copper base plate and inserted in a 60mmϕ metal pipe.

The emitter is a 1kΩ thin metal-film chip resistor (Fig. 6 (b)), consisting of a 1x2mm^2 resistive metal film coated on a 0.5mm-thick alumina substrate and covered with a ~20μm-thick glass overcoat layer. The resistance (1kΩ) does not change appreciably in the range of temperature studied (4.2K up to 244K). The resistor is sealed in a small metal chamber (8 mm-length and 4 mm-diameter, not shown in Fig.6 (b)) and suspended in vacuum (~3x10^{-4} Pa) by current leads (0.2mmϕ Manganin wires). We suppose that the resistor serves approximately as a black body when heated with current, since the emissivity of the overcoat glass layer is beyond 90% in the LWIR range. The emitted radiation is transmitted through a 625μm-thick GaAs window (optical transmittance of 50-60% in a 5-15 μm wavelength region) of the small chamber. The resistance of the current leads is ~30 Ω over the total length (2 m) in the cryostat. Hence the electrical power is substantially dissipated by the chip resistor. The emitter temperature, $T_{emitter}$, is monitored with a thermocouple, Chromel-Au/Fe (0.07 %) glued to the backside of the resistor (Fig.6 (b)). The values of $T_{emitter}$ will be shown as a function of the input electrical power P_{in} in the inset of Fig.9 (b).

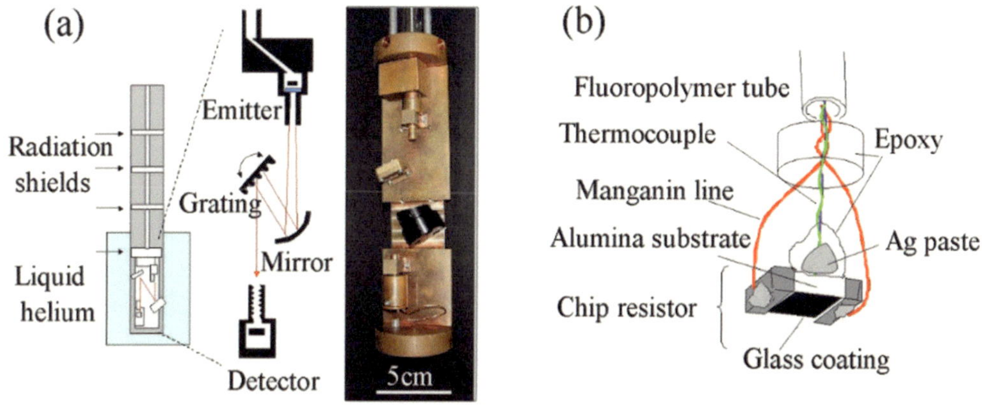

Fig. 6. (a) All cryogenic spectrometer; schematic representation and a photograph. (b) A 1kΩ thin metal film chip resistor serving as a black-body radiator. The emitter is packaged in a small vacuum metal chanber with a GaAs window.

The parabolic mirror collimates broadband BBR from the emitter and directs it to the diffraction grating, which will be described in the next paragraph. The angle of diffraction grating is controlled mechanically from the outside of the cryostat. The radiation diffracted at the grating is guided to the detector. The detector is installed in a small copper box with a 4 mmϕ and 20 mm-long metal pipe at the entrance. The inner wall of the entrance pipe is screw patterned so as to prevent stray lights from entering the detector box.

A home-made diffraction grating is used. Commercially available diffraction gratings are usually composed of materials of different thermal expansion coefficients, like an epoxy substrate coated with aluminum films. They are hence unreliable for the use at cryogenic temperatures. We fabricated a diffraction grating using a GaAs crystal. A GaAs crystal surface was patterned into 7.5μm-wide and 7.5μm-deep trenches at 15 μm-period via mesa etching with a solution of $H_2PO_4 : H_2O_2 : H_2O = 4 : 1 : 40$ (etching rate: 40 nm/min) at 25 °C. The etching was made for 4.5 hours. The slow rate and the long time of etching were chosen so as to round the grove edges through under-etching the resist mask. The rounding of grove edges was necessary for obtaining a reasonable amplitude for the 1st-order diffraction by reducing the 0th-order diffraction (specular reflection). Finally the trenched GaAs surface was coated with a 20nm-Ti/300nm-Au layer via vacuum evaporation.

The whole spectrometer system is immersed in liquid helium (T=4.2K). The 300K BBR from the warm part of the cryostat is blocked by placing copper plates at a 10 cm-interval inside the metal pipe. The emitter chamber is made vacuum by pumping through a thin, 4-mm diameter, metal pipe. The pipe is narrowed to 2mmϕ at the entry to the chamber and bended at two places for minimizing the 300K-BBR entering through the vacuum pipe.

The wavelength resolution of the present optical setup was evaluated to be $\Delta\lambda_g$=1.0 μm (full width at half maximum, FWHM) by using a Fourier Transform Infrared Spectroscopy (FTIR) system.

4. Spectral response

The spectral response of the detector was studied by varying the angle of the diffraction grating of All cyogenic spectrometer at a step of $\Delta\theta = 0.4$ degree (corresponding change in the wavelength of 0.2 μm). The emitter was fed with an electrical power of P_{in}=0.1 W, where the thermocouple indicated $T_{emitter}$=130K.

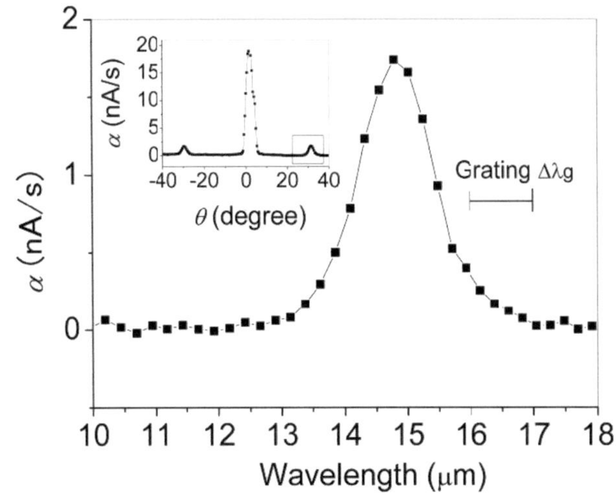

Fig. 7. The detection spectrogram of the CSIP studied by the all cryogenic spectrometer, where the instrumental resolution is $\Delta\lambda_g$ = 1μm as marked by the bar. The inset displays the original spectral chart.

The inset of Fig. 7 shows the current increases $a=\Delta I/\Delta t$ against the angle of grating θ, where the central peak (θ=0) corresponds to the 0th order diffraction and the two symmetrically located side peaks are the ±1st order diffractions. Figure 7 elucidates the spectrogram of the right-hand side peak, where the horizontal axis is scaled by the wavelength. The peak wavelength is λ=14.7 μm, and the apparent spectral bandwidth (FWHM) is $\Delta\lambda_{app}$= 1.6 μm. By noting the instrumental resolution of $\Delta\lambda_g$=1.0 μm, we evaluate the true bandwidth to be $\Delta\lambda_d$~1.0 μm (Ueda et al, 2007).

Similarly, spectral responses for CSIPs with different upper QW widths, i.e. subband energy, have been studied (Ueda et al., 2011). As shown in Fig. 8, CSIPs cover the wavelength range of 10-50 μm, where there was no available ultrasensitive detector, reaching single-photon counting level.

5. Single-photon signal and dynamic range of CSIPs

Because the incident radiation intensity to the detector is not accurately determined by the experimental setup shown in Fig. 6 (a), the detector box was placed immediately in front of the emitter chamber as illustrated in the inset of Fig. 9 (a). Knowing the geometry, we can derive absolute values of the incident radiation intensity as will be discussed later. (Though wavelength selection is not made, only a narrow detection band (Fig. 7) contributes to the photo-response.)

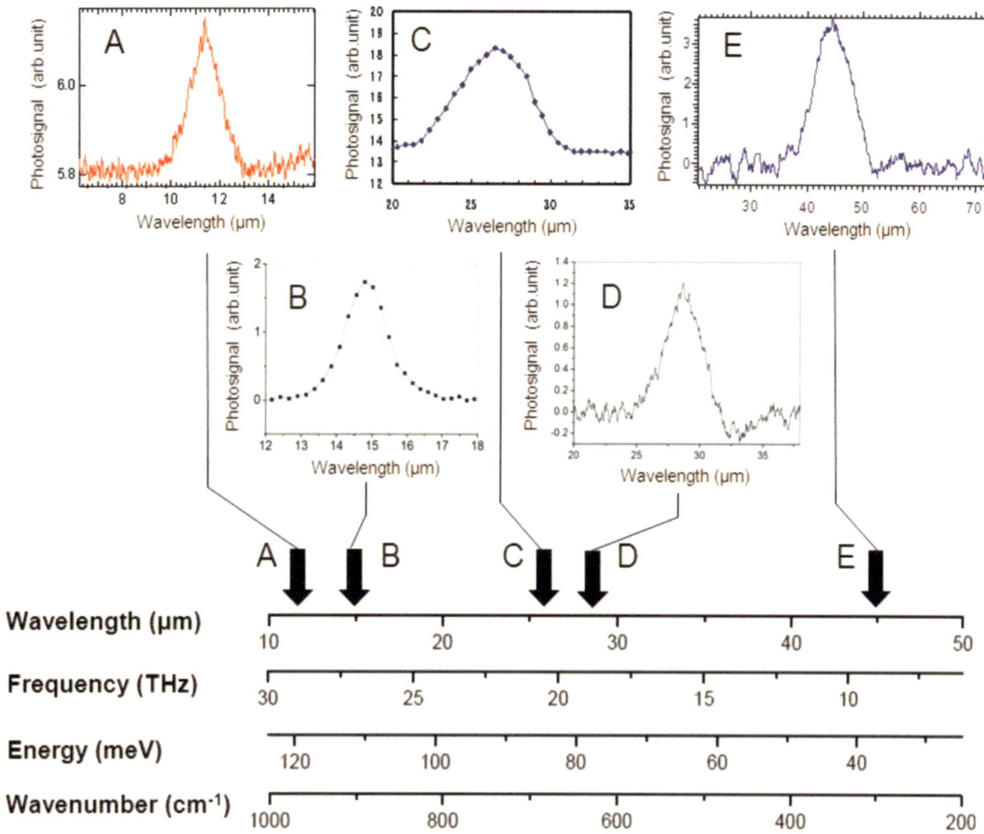

Fig. 8. Spectral response of CSIPs with different intersuuband energies.

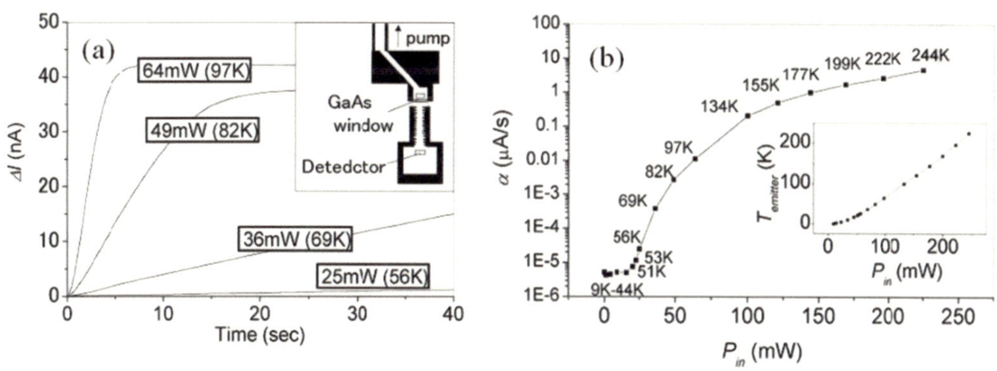

Fig. 9. (a) Time traces of the photo current, ΔI, obtained in different incident-radiation intensities, where the emitter directly faces the detector as shown in the inset. The electric power fed to the emitter, P_{in}, and the emitter temperature, $T_{emitter}$, are indicated for the respective curves. (b) The rate of photo-current increase ($a=\Delta I/\Delta t$) vs. the input electric power fed to the emitter (P_{in}). The emitter temperature $T_{emitter}$ is also indicated. The inset shows $T_{emitter}$ as a function of P_{in}, studied by the thermocouple attached to the emitter.

Figure 9 (a) displays time traces of the photo-induced change in the source-drain current, ΔI, obtained after the detector has been reset at t=0 sec with a pulse of 1μsec duration. The curves are taken with four different radiation intensities ($T_{emitter}$=56K, 69K 82K and 97K) for P_{in}= 25mW~64mW. It is clearly seen that the increase of current levels off deviating from the linear increase when ΔI exceeds around 20 nA. The current increases linearly in a small range of ΔI <20 nA. The slope of the linear increase, $a=\Delta I/\Delta t$, is the signal intensity indicating the rate at which photo-generated holes are accumulated in the isolated upper QW. Figure 9 (b) shows $a=\Delta I/\Delta t$ as a function of the input electric power P_{in} fed to the emitter, where $T_{emitter}$ is marked on each data point. Bachground blackbody radiation corresponding to 50 K still remains, but fortunately, the analysis is not afected by the radiation.

Though not visible in Fig.9 (a), closer look at the time traces makes it probable that the current does not increase smoothly but it increases stepwise, as exemplified in Fig. 10. The data of Fig.10 are taken with a time constant of measurements of Δt=1sec for P_{in}=1mW ($T_{emitter}$=14K).

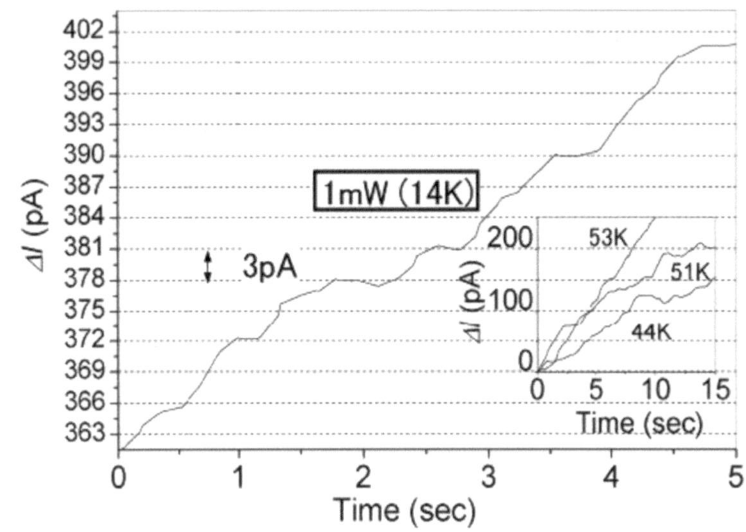

Fig. 10. Time trace of _I obtained with Temitter=14 K, which is represented in a magnified scale.

Figure 11 shows the histogram of the frequency of step-wise change ($\Delta I_{plateau}$) between clear plateaus (omitting small winding in continuous slope), occurring in the curves obtained at different low illumination intensities ($T_{emitter}$<50K). It is likely that the histogram consist of the Gaussian distributions with the mean values of $<\Delta I>=\mu$= 3pA, 6pA, 9pA and 12pA. It is hence suggested that the increase of ΔI takes place with a step of ΔI_{Step} =3pA.

In the present scheme of photo detection (Figs. 1 and 2), the current through the lower QW layer is expected to increase by Eq. (1). The values of the mobility of the lower QW layer μ = 4.3x10^{3}cm^2/V, source-drain bias voltage V_{SD}=10mV, channel length L=16μm yield ΔI_e =2.7pA, which is close to the experimentally observed amplitude of the unit step ΔI_{Step} =3pA in the above.We note in Fig. 9 (a) that the detector exhibits linear response for the excited holes less than 20 nA/3 pA = 7x10^3.

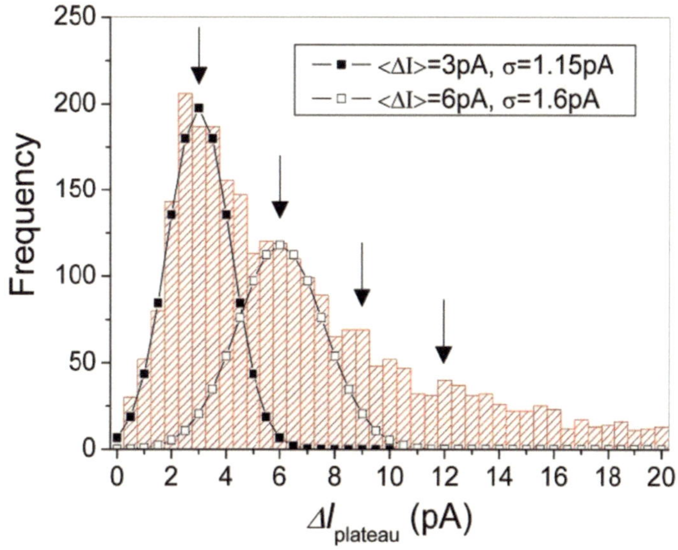

Fig. 11. The histogram of the stepwise change of current ($\Delta I_{plateau}$), occurring in ΔI versus t curves. Fitting lines represent the Gaussian distribution with $<\Delta I>$ the mean value and σ the standard deviation.

The quantum efficiency, $\eta = \{a/\Delta I_{Step}\}/\Phi$, is an important parameter defining the ratio of the rate of detected photons, $a/\Delta I_{Step}$, to that of the total incident photons, Φ. Here, the total incident photon flux is given by

$$\Phi = (\varepsilon f_{att})\left[\frac{B(\lambda,T)\Omega\Delta\lambda_d}{hc/\lambda}\right]S_{em} \tag{4}$$

where $B(\lambda,T) = (2hc^2/\lambda^5)\{\exp(hc/\lambda k_B T)\text{-}1\}^{-1}$ is the Plank's formula with h the Planck constant, k_B the Boltzmann constant and c the light velocity, $\Delta\lambda_d$=1μm is the detector bandwidth and $\Omega = S_{iso}/d^2$ = 1.2x10^{-7}sr is the solid angle with the active detector area of $S_{iso}=WL_{iso}$=64μm^2 with an emitter/detector distance of d=2.3cm. Here, S_{em}=8mm^2 is the effective emitter area, ε ~1 is the emissivity of the emitter (glass surface) , f_{att} =0.45 is the attenuation due to reflection at the GaAs window and $hc/\lambda = h\nu$ = 83 meV is the photon energy (λ=14.7μm). Knowing these parameter values, we can derive the photon flux, Φ, from $T=T_{emitter}$.

The curve of a versus P_{in} ($T_{emitter}$) in Fig. 9 (b) is re-plotted in terms of the count rate, $a/\Delta I_{Step}$, and the incident photon flux, Φ, in Fig.12. The linear relationship between $a/\Delta I_{Step}$ and Φ assures the validity of our analysis, and indicates an efficiency of η =2± 0.5 %. The dynamic range of linear response is ~10^7 (Φ=4x10/sec~1x10^8/sec, or $P=\Phi h\nu$=5.3x10^{-19}W ~ 1.3x10^{-12}W). Note that the maximum radiation power in Fig. 6 (Φ = 1x10^8/sec or $P=\Phi h\nu$=1.3x10^{-12}W) is not determined by the capability of the detector but is restricted by the emitter ($T_{emitter}$ =244K with P_{in}=225mW). As discussed later, the detector is expected to show linear response up to much higher power levels (P=10^{-6}W) in an appropriate condition of reset pulses, suggesting a much wider dynamic range reaching 10^{13}.

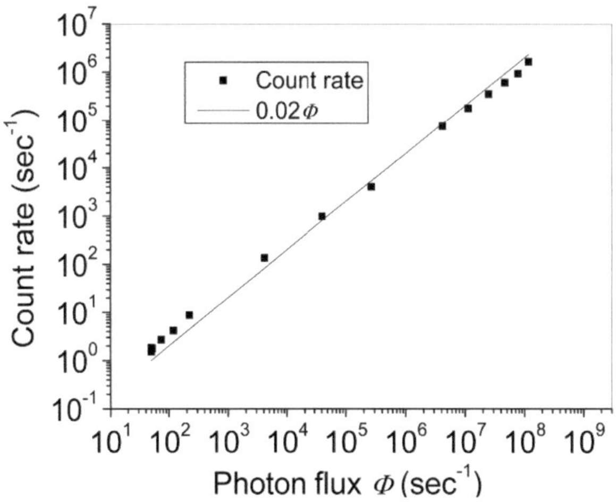

Fig. 12. Count rate of the photosignal vs. incident photon flux Φ. Photon fluxes of Φ=10^2 photons/sec corresponds to $P_{in} = \Phi h\nu = 1.3\text{x}10^{-18}$W.

6. Figures of merit

The current responsivity of the detector, $R=\Delta I/(h\nu\Phi)$, is the photocurrent amplitude ΔI divided by the incident radiation power. In the CSIP, $\Delta I=a\Delta t_I$ is proportional to the integration time, Δt_I, which can be arbitrarily chosen. If we take a typical reset interval, in a range of Δt_I=10msec~1sec, we derive $R=\eta(\Delta I_{Step}/h\nu)\Delta t_I$ = (4x10^4~4x10^6) A/W by noting $\eta=\{a/\Delta I_{Step}\}/\Phi$ with ΔI_{Step} =3pA and η=0.02. The responsivity is thus by many orders of magnitude larger than that of well known QWIPs, for which R is typically ~0.5A/W (Levine, 1993; Yao, 2000). We can also write the responsivity as $R=\eta(e/h\nu)$g by using the "photoconductive gain" g, which refers to the effective number of electrons generated by a single photon. We mention that the large value of R comes from the extremely high photoconductive gain, g=$(\Delta I_{Step}/e)\,\Delta t_I$ = 1x10^5~1x10^7, in the CSIP.

Noise equivalent power (NEP) is defined by

$$NEP = \frac{P}{SNR\sqrt{\Delta f}} \tag{5}$$

where P is the power of incident radiation, SNR is the signal to noise ratio, and Δf=1/(2Δt) is the frequency bandwidth of measurements with Δt the averaging time. We evaluate Eq.(5) by taking three curves of ΔI versus t shown in the inset of Fig.7 as an example. These curves are taken for T_{emitter}=44K (P_{in}= 2.8aW), 51K (4.5aW) and 55K (5.0aW). Considering the time constant τ =1sec of measurements, we evaluate the noise from $\sqrt{\langle(\alpha-\bar{\alpha})^2\rangle}$, where $\bar{\alpha}$ is the average slope (count rate) of each curve for a long time (>15 sec) and $a=\Delta I/\Delta t$ is the slope at each point over Δt=1sec. For each curve, we derive SNR = 2.1, 1.9 and 2.2 for Δt = 1sec and obtain NEP = 1.9x10^{-18}, 3.4x10^{-18} and 3.3x10^{-18} W/$\text{Hz}^{1/2}$.

It is interesting if the noise in the present measurements arises from detector specific electrical noise or is ascribed to the photon noise. If the detector noise is ignored, random events of photon arrival give rise to the photon noise, like the shot noise in the electron current. On the average, $N\Delta t$ photons are counted over integration time of Δt when the photon count rate is $N=\eta\Phi=a/\Delta I_{Step}$. Since the standard deviation from the average is given by $(N\Delta t)^{1/2}$ in the Poisson distribution, *SNR* of an ideal detector (free from any detector-specific noises) is $(N\Delta t)^{1/2}$. It follows that *NEP* is given by (Knuse et al., 1962)

$$NEP_{BLIP} = h\nu\sqrt{\frac{2\Phi}{\eta}} = h\nu\frac{\sqrt{2N}}{\eta} \tag{6}$$

in the condition when the noise is dominated by the photon noise; i.e., in the background limited performance (BLIP). We have NEP_{BLIP} = 1.9x10^{-18}, 2.5 x10^{-18} and 2.6 x10^{-18} W/Hz$^{1/2}$, respectively, for the three conditions (N=4.2, 6.7 and 7.4) of the data in the inset of Fig.10. The fact that these values are close to the *NEP* values derived from Eq.(5) indicates that the fluctuation in the present measurements comes from the randomness in the photon arrival events and that the detector-specific noise is indiscernibly small.

The true *NEP* is obtained by replacing *N* in Eq.(6) by the dark count rate Γ, (Brule Technologies, Inc., 1980)

$$NEP = h\nu\frac{\sqrt{2\Gamma}}{\eta} \tag{7}$$

In the present spectrometer, weak background radiation makes hinders direct determination of Γ because of background blackbody corresponding at 50 K. In another optical system, we studied the detector in the complete dark condition, and obtained Γ=0.5 sec^{-1}, which yields *NEP*=6.8 x10^{-19} W/Hz$^{1/2}$.

The fact hat the noise in the present measurements arises from the photon noise is also supported by the inter-step interval histograms. A time trace of ΔI for P_{in}=1mW was studied for 700 seconds in the present spectrometer (partially shown in Fig. 10). The local slope of the curve, $a(t)=\partial I/\partial t$, was numerically calculated at each time point t of the curve, where the value is averaged over 1sec at each point. Figure 13 (a) represents the frequency of different slopes as a function of the steps per seconds or $a(t)/\Delta I_{Step}$ with ΔI_{Step} =3pA. In the trace of ΔI, upward stepwise change as well as downward change are recognized, which were separately analyzed in Fig.13 (b) (Gain) and (c) (Loss). As expected for single-photon detection, all histograms in Figs. 10 are fairly well described by the Poisson distribution, $f(k,N)=e^{-\nu}\nu^{k}/k!$, where k is the number of occurrence of an event and ν is the variance. We mention that the average rate of Loss events ν=0.5 in Fig.13 (c) is close to the dark count rate.

Specific detectivity, D^*, is the reciprocal value of NEP normalized by the detector active area S_{iso},

$$D^* = \frac{\sqrt{S_{iso}}}{NEP} \tag{8}$$

With S_{iso}=64x10^{-8}cm^2 for the present detector, we have D^*_{BLIP} = 4.1x10^{14}, 2.3x10^{14} and 2.5x10^{14}W/Hz$^{1/2}$ in the background limited conditions (*NEP* = 1.9x10^{-18}, 3.4x10^{-18} and 3.3x10^{-18} W/Hz$^{1/2}$). As for the true specific detectivity of the detector at 4.2K, we have

$D^*\cong1.2\times10^{15}$cmHz$^{1/2}$/W from $NEP\cong6.8\times10^{-19}W/Hz^{1/2}$. CSIPs are thus highly sensitive detectors with values of D^* much higher than those of conventional detectors like QWIPs ($D^*=10^{10}$-10^{13}cmHz$^{1/2}$/W at 4.2~80K) (Levine, 1993; Yao et al., 2000) and mercury-cadmium telluride (MCT) detectors ($D^*\sim10^{10}$cmHz$^{1/2}$/W at ~80K) (Rogalski, 2002).

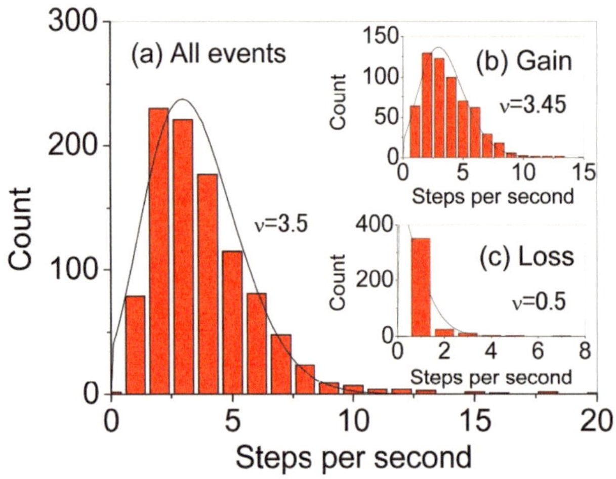

Fig. 13. Histogram of the photo-current ΔI per second. The photocurrent is measured in unit of ΔI_{Step} =3pA so that the horizontal axis is scaled by the number of steps per second, $(\Delta I/\Delta I_{Step})/\Delta t = a/\Delta I_{Step}$. The data are derived by analyzing a time trace of ΔI for 700 sec at P_{in}=1mW: (a) All events of both gain signal (generation of holes) and loss signal (loss of holes due to recombination) are included. In (b) and (c), the gain signal and the loss signal are separately shown.

The intrinsic speed of detection will be limited by the transistor operation, which may be faster than 1 ns. In actual practice of detector operation, the detection speed may be determined by the frequency of applicable rest pulse, which may be restricted to about 300 MHz (~3 ns interval). The maximum incident radiation power will be hence $P_{max}\sim$ (7×10^3 $h\nu/3$ $ns)/\eta\sim2$ μW, which makes us to expect that the dynamic range can be expanded to 10^{13} by assuming the minimum detectable power to be $P_{min}\sim1\times10^{-19}$W.

7. Improvement of quantum efficiency

Quantum efficiency η is an important parameter for the detector. High values of η (>30%) are reported in conventional QW IR photodetectors, where photons are absorbed by more than 30 multiple QWs. High quantum efficiency is not trivial in CSIPs, where photons are absorbed by only one QW. In most experimental conditions of CSIP for λ = 12-15 μm, the efficiency $\eta = \eta_1\eta_2$ is primarily determined by the coupling strength of electrons to the incident radiation, namely, $\eta\approx\eta_1$ ($\eta_2\approx1$).

Recently we proposed and demonstrated efficient photo-couplers for CSIPs (λ=15 μm) by exploiting surface-plasmon-polariton (SPP) resonance occurring in aperture metal sheets coated on top of the crystal surface (Fig. 14 (a)) (Nickels et al, 2010). The SPP resonance induces wavelength-selective strong electric confined near the surface of the metal sheets intensifying the subband transition in the QW 100 nm below the surface. The quantum efficiency has been experimentally studied by applying different metal photocouplers as displayed in Fig. 14 (b) and compared with simulation results in Fig. 14 (c).

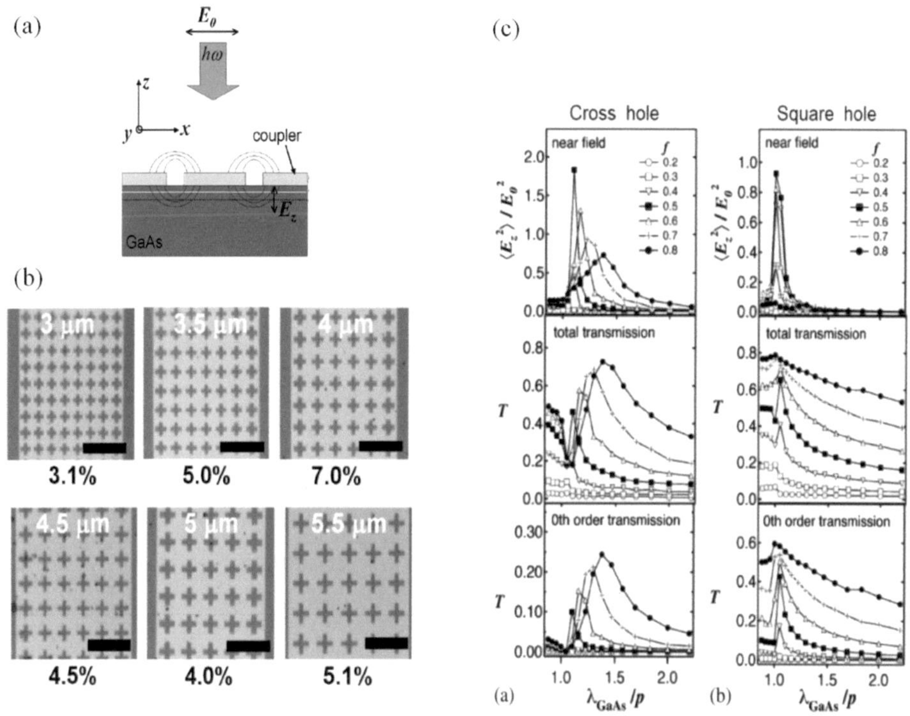

Fig. 14. (a) Schematic representation of photocoupler. (b) Metal meshes studied. White and black characters indicate, respectively, the lattice period p and the experimentally derived quantum efficiency η. (c)Results of simulation calculation are shown for (a) cross-hole arrays and (b) square-hole arrays: from the top, normalized field intensity $<E_z^2>/<E_{z0}^2>$, the total power transmission, and the zeroth-order transmission, against the ratio λ_{GaAs}/p with λ_{GaAs} = 4.3 μm.E_z is the electric field normal to the QW plane at the position of upper QW(100 nm below the surface) and E_{x0} is the electric field of incident radiation. Theoretically, $<E_z^2>/<E_{z0}^2>$,= 2 gives roughly η = 7%, which agrees with the experiment

In the experiments, an efficiency of $\eta \approx 7\%$ has been achieved by utilizing the surface plasmon excitation in a 2-D metal hole array (inductive metal mesh) with cross-shaped holes lining up at a lattice period of 4 μm (~ wavelength in GaAs). Other coupler geometries such as 2-D metal hole arrays with square-shaped holes, 2-D metal patch arrays (capacitive metal mesh), and patch antennas (microstrip antennas) have yielded lower values of $\eta \approx$ 1%–3% (Ueda et al., 2008; Nickels et al, 2010). The device used for Fig. 12, for instance, utilizes a patch antenna coupler yielding $\eta \approx 2\%$ (Ueda et al, 2008). Simulation suggests that $\eta \approx 7\%$ is not a highest achievable value but can be improved by a factor of ~3 ($\eta \approx 20\%$) when a resonant cavity is formed by the metal hole array and the n-type substrate. Values of the current responsivity and the NEP given in the section 6 have been so far derived by assuming η = 2%. The figure-of-merits of the CSIP will be improved in accordance to η.

8. Temperature dependence of the performance

Higher temperature operation is desired for practical applications. There is in general, however, a trade-off between high sensitivity of an IR detector and operation temperature. The external constraint is that sensitive devices are saturated by strong background blackbody radiation from surrounding materials. The internal constraint is that the thermionic emission inside the device becomes equivalent to photo-emission. The former may be determined by optical setup, as well as relative intensity of signal radiation to the background. Here, we discuss the latter intrinsic constraint.

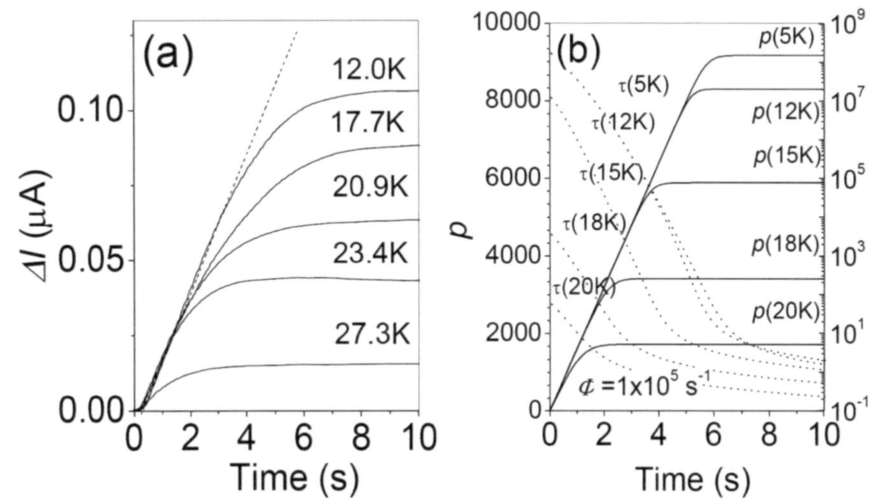

Fig. 15. (a) Time traces of the photocurrent with Φ=1x10^5 s^{-1} at different temperatures. (b) Theoretical time traces of photo current in terms of p at different T (solid line), and lifetime change (dotted line)

In Fig. 15 (a), time traces of photo-current at different temperatures are displayed under the fixed photon flux Φ=1x10^5 s^{-1}(Ueda et al., 2009). The temperature effect appears as the lower amplitude of photo-current saturation. It should be noted that the slope, $\alpha=\Delta I/\Delta t=\eta\Phi I_e$, in the initial stage of each trace is independent of T, assuring that ηI_e is independent of T. This means higher frequency reset operation is required, i.e., the integration time is shortened, in the elevated temperatures. The photo-signal is discernible up to 30K for the CSIP of λ= 15 μm. The derived NEP and D^* up to T=23 K with integration time of 1s are given as NEP=8.3x10^{-19} W/Hz$^{1/2}$, and D^*=9.6x10^{14} cm Hz$^{1/2}$ /W, which are not very different from the 4.2 K values mentioned above (Ueda et al., 2009).

The understanding of photo-current saturation directly leads to temperature dependent physics inside the device. The potential profile changing in the detection is shown in Fig. 5. The potential height of the triangular barrier in equilibrium is $U=\Delta E-\delta U$, i.e., barrier hight is lower, by $\delta U\approx$15 meV, than subband energy splitting of ΔE. Here the barrier hight $U=U_U=U_L$ is measured from electrochemical potential of upper or lower QW, $\zeta=\zeta_L=\zeta_U$.

Under illumination, the energy of the floating upper QW with p photoholes decreases against the grounded lower QW by $\Delta U= |\zeta_L-\zeta_U| = pe^2d / \kappa LW$, where e is the charge of electron, d is the distance between the upper and the lower QWs, and κ=12 x (8.85x10^{-12}) F/m is the electronic permittivity of the crystal. The net potential barrier height measured from ζ_L , then decreases to be U_L= (ΔE-δE)-ΔU by p photoholes, while U_U =ΔE-δE remains constant. The barrier reduction assists the electrons in the lower QW to recombine with photoholes in the upper QW. The photo-current saturation occurs when the recombination process becomes equivalent to the photohole generation process. Obviously, more electrons in the lower QW contribute to the recombination process at higher temperature. Finally at around T=30 K, the number of traveling electron across the barrier exceeds that of photo-emitted electrons even with no band-deformation.

Temperature dependence shown in Fig. 15 (a) arises from decrease of the lifetime of photoholes. The number of photoholes is determined by the rate equation (3). The recombination rate, p/τ, is given by difference between $\dot{n}(U_L)$ and $\dot{n}(U_U)$,

$$\frac{p}{\tau} = \dot{n}(U_L) - \dot{n}(U_U) \tag{9}$$

where $\dot{n}(U_U)$ is the rate from the upper QW to the lower QW and $\dot{n}(U_L)$ the rate of the opposite flow. Here again, U=U_L and U=U_U are the potential heights of triangular barrier measured from each electrochemical potential ζ of the lower QW, ζ=ζ_L, and the upper QW, ζ=ζ_U, respectively (Fig. 5).

The thermal electron emission is given as

$$\dot{n}(U) = \frac{LW}{d} D \int_0^{\infty} v \Theta f d\varepsilon \tag{10}$$

where D=m*/$\pi\hbar^2$ is the two-dimensional density of states with m*=0.0665x(9.1x10^{-31})kg the effective mass of conduction electrons in GaAs, and v=($2\varepsilon/m^*$)$^{1/2}$ the electron velocity with ε measured from the bottom of the conduction band, Θ is the transmission probability, and f= [1+exp{(ε-ζ)/k_BT}]$^{-1}$ is the Fermi distribution function with k_B the the Boltzmann constant,. Here, we use WKB approximation of Θ, in which Θ=1 (at ε>U), and

$$\Theta = \exp\left[-\frac{4}{3eF} \left(\frac{2m^*}{\hbar^2} \right)^{1/2} (U - \varepsilon)^{3/2} \right] \quad \text{(at } \varepsilon < U\text{)}, \tag{11}$$

where F= U_L/ed is the electric field given by the slope of the deformed triangular barrier.

As shown in Fig. 15 (b), the predicted time traces of p (photo-signal) at different T's substantially reproduce experimental results. It should be noted that adjustable parameters are not involved in the calculation. The temperature dependence is hence concluded to arise from those thermally activated electrons that are transmitted across the potential barrier to be recombined with holes in the isolated QW.

At the temperature limit, T_{lim}, Eqs. 3 and 9 becomes $2\eta\Phi$= $\dot{n}(U_L)$,by assuming $\dot{n}(U_U)$≈$\eta\Phi$ under dp/dt=0 and ΔU=0. The relationship between subband energy the ΔE and temperature limit T_{lim} can be obtained:

$$\Delta E - \delta E \approx -k_B T_{\lim} \ln\left(\frac{2\eta\Phi d}{LWDvk_B T_{\lim}} \right) = \gamma T_{\lim} \tag{12}$$

where $v=(2E_F/m^*)^{1/2}$ is Fermi velocity. Here, integration is made only for $E > U$ with $\Theta=1$ and $f \approx \exp(-E/k_BT)$, i.e., only thermionic emission is considered. Since contribution of T_{lim} in the logarithm is not significant, the temperature limit shows almost linear dependence on the subband energy with quasi-constant $\gamma > 0$. For the λ=15 µm, the parameter values yields T_{lim} =29 K, which is close to the experimentally observed limit temperature.

9. Conclusion

We have developed the novel ultrasensitive detectors, named charge-sensisve infrared phototransistors (CSIPs). We demonstrated the single-photon detection in the wavelength range of 10 - 50 µm for the first time. The accurately determined specific detectivity as $D^*=9.6\text{x}10^{14}$ cm Hz$^{1/2}$/W for λ=14.7 µm is by a few orders of magnitude higher than those of other conventional detectors. In addition, the planar FET structure of CSIPs is feasible for developing large scale array including monolithic integration with readout circuit. The quantum efficiency is improved up to 7% by using the surface-plasmon polariton couplers. The detectors can keep the sensitivity up to T=23 K for λ=15 µm, and the temperature dependence of the detector performance is reasonably interpreted by a simple theoretical model. We developed CSIPs for several wavelengths in 10-50µm, covering attractive spectral range for passive observation of spontaneous emission from RT object. Recently CSIPs have been applied to the construction of a highly sensitive passive microscope (Kajihara et al., 2009; 2010; 2011).

10. Acknowledgment

This work was supported by CREST and SORST projects of Japan Science and Technology Agency (JST). The reserach work has been carrierd with number of collaborators. Particularly T. Ueda would like to thank Z. An, P. Nickels, N. Nagai, Z. Wang and S. Matsuda for their valuable contributions.

11. References

An, Z., Ueda, T., Komiyama, S. & Hirakawa, K. (2007). Reset Operation of Qwantum-Well Infrared Phototransistors, *IEEE Transaction on Electron devices*, Vol. 75, pp. 1776-1780, ISSN 0018-9383

Andersson, J. & Lundqvist, L. (1991). Near-Unity Quantum Efficiency of AlGaAs/GaAs Quantum Well Infrared Detectors Using a Waveguide with a Doubly Periodic Grating Coupler, *Applied Physics Letters*, Vol. 59, pp. 857 1-3, ISSN 0003-6951

Astaviev. O., Komiyama, S., Kutsuwa T., Antonov, V. & Kawaguchi, Y. (2002). Single-Photon Detector in the Microwave range, *Applied Physics Letters*, Vol. 80, pp. 4250 1-3, ISSN 0003-6951

Burle Technologies, Inc. (1980). *Photomultiplier Handbook – Theory Design Application*, Lancaster, PA

Cwik, T. & Yeh, C. (1999). Higjly Sensitive Quantum Well Infrared Photodetectors, *Journal of Applied Physics*, Vol. 86, pp. 2779 1-6, ISSN 0021-8979

Day, P., LeDuc, L., Mazin, B., Vayonakis, A. & Zmuidzinas, J. (2003). A Broadband Superconducting Detector Suitable for Use in Large Arrays, *Nature*, Vol. 425, pp. 817-821, ISSN 0028-0836

Hashiba, H., Antonov, V., Kulik, L., Tzalenchuk, A., Kleindschmid, P., Giblin, S. & Komiyama, S. (2006). Isolated quantum dot in application to terahertz photon counting, *Physical Review B*, Vol. 73, pp. 081310 1-4, ISSN 0613-1829

Ikushma, K., Yoshimura, Y., Hasegawa, H., Komiyama, S., Ueda, T. & Hirakawa, K. (2006). Photon-Counting Microscopy of Terahertz Radiation, *Applied Physics Letters*, Vol. 88, pp. 152110 1-3, ISSN 0003-6951

Komiyama, S., Astaviev. O., Antonov, V., Kutsuwa T. & Hirai, H. (2000). A Single-Photon Detector in the Far-Infrared Range, *Nature*, Vol. 403, pp. 405-407, ISSN 0028-0836

Kajihara, Y., Komiyama, S., Nickels, P. & Ueda, T. (2009). A Passive Long-Wavelength infrared Microscope with a Highly sensitive phototransistor, *Review of Scientific Instruments*, Vol. 80, pp. 063702 1-4, ISSN 0034-6748

Kajihara, Y., Kosaka, K., Komiyama, S., Kutsuwa T. & Hirai, H. (2010). A Sensitive Near-Field Microscope for Thermal Radiation, *Review of Scientific Instruments*, Vol. 81, pp. 033706 1-4, ISSN 0034-6748

Kajihara, Y., Kosaka, K.& Komiyama, S. (2011). Thermally excited Near-Field Radiation and Far-Field Interfarence, *Optics Express*, Vol. 19, pp. 7695-7704, ISSN 1094-4087

Knuse, P., McGlauchlin, L. & McQuistan, R. (1962). *Elements of Infrared Technology*, Wiley, ISBN 978-047-1508-86-1, New York

Levine, B. (1993). Qwantum-Well Infrared Photodetectors, *Journal of Applied Physics*, Vol. 74, pp. R1-81, ISSN 0021-8979

Nakagawa, T. & Murakami, H. (2007). Mid- and far-infrared astoronomy mission SPICA, *Advances in Space Research*, Vol. 40, pp. 679-683, ISSN 0273-1177

Prochazka, I. (2005). Semiconducting Single Photon Detectors: The State of the Art, *Physica Status Solidi (c)*, Vol. 2, pp. 1524-1532, ISSN 1610-1634

Rogalski, A. & Chrzanowski, K. (2002). Infrared Devices and Techniques, *Opt-Electrics Review*, Vol. 10, pp. 111-136, ISSN 1230-3402

Schoelkopf, R., Moseley, S., Stahle, C., Wahlgren, P. & Delsing, P. (1999). A concept for a submillimeter-wave single-photon counter, *IEEE Transactions on Applied Superconductivity*, Vol. 9, pp. 2935-2939, ISSN 1051-8223

Tidrow, M. (2000). Device Physics and State-of-the-art of Quantum Well Infrared Photodetectors and Arrays, *Materials Science and Engineering: B*, Vol. 74, pp. 45-51, ISSN 0921-5107

Ueda, T., An, Z., Hirakawa, K. & Komiyama, S. (2008). Charge-Sensitive Infrared Phototransistors: Characterization by an All-Cryogenic Spectrometer, *Journal of Applied Physics*, Vol. 103, pp. 093109 1-7, ISSN 0021-8979

Ueda, T., An, Z., Hirakawa, K., Nagai, N. & Komiyama, S. (2009). Temperature Dependence of the Performance of Charge-Sensitive Infrared Phototransistors, *Journal of Applied Physics*, Vol. 105, pp. 064517 1-8, ISSN 0021-8979

Ueda, T., Soh, Y., Nagai, N., Komiyama, S. & Kubota, H (2011). Charge-Sensitive Infrared Phototransistors Developped in the Wavelength Range of 10-50μm *Japanese Journal of Applied Physics*, Vol. 50, pp. 020208 1-3, ISSN 0021-4922

Wei, J., Olaya, D., Karasik, B., Pereverzev, S., Sergeev, V. & Gershenson, M. (2008). Ultrasensitive hot-electron nanobolometer for Terahertz Astrophysics, *Nature Nanotechnology*, Vol. 3, pp. 495-500, ISSN 1748-3387

Yao, J., Tsui, D. & Choi, K. (2000). Noise characteristics of quantum-well infrared photodetectors at low temperatures, *Applied Physics Letters*, Vol. 76, pp. 206 1-3, ISSN 0003-6951

Nickels, P., Matsuda, S., Ueda, T., An, Z. & Komiyama, S. (2010). Metal Hole Arrays as a Resonant Photo-Coupler for Charge-Sensitive Infrared Phototransistors, *IEEE journal of Quantum Electronics*, Vol.46, pp. 384-3906, ISSN 0018-9197

8

Broadband Photodetectors Based on *c*-Axis Tilted Layered Cobalt Oxide Thin Films

Shufang Wang* and Guangsheng Fu
Hebei Key Lab of Optic-Electronic Information and Materials, Hebei University, Baoding PR China

1. Introduction

Laser-induced voltage (LIV) effects in *c*-axis tilted thin films of $YBa_2Cu_3O_{7-\delta}$ (YBCO) and $La_{1-x}Ca_xMnO_3$ (LCMO) have been extensively studied in the past decades due to their potential applications in photodetectors [1-9]. Compared to the commonly used photodiodes-based photodetectors, this new type of photodetectors has the advantage of broad spectrum response ranging from ultra-violet (UV) to infrared (IR). Other advantages of this type of detectors consist in that they can be operated without cryogenic cooling and bias voltage.

The origin of the LIV signal was explained as result of the transverse thermoelectric effect which becomes effective when the *c* axis of YBCO or LCMO films are tilted by an angle α with respect to the film surface normal [8]. The induced voltage can be quantitatively described by the equation

$$U = \frac{l}{2d}\sin(2\alpha)\Delta S \Delta T \tag{1}$$

Where ΔT is the temperature difference between the top and bottom of the film, which is generated by heating the film surface due to the absorption of the incident laser radiation; $\Delta S = S_{ab} - S_c$ is the difference of the seebeck coefficient in the *ab*-plane and along the *c*-axis of the film; α is the tilted angle of the film with respect to the surface normal; d is the thickness of the tilted film and ι is the laser spot diameter [1]. According to this equation, searching for materials having large anisotropy in seebeck coefficient, that is large ΔS, is a key factor for developing this new type photodetectors.

Recently, layered cobalt oxides including Na_xCoO_2, $Ca_3Co_4O_9$, $Bi_2Sr_2Co_2O_y$ and etc. have attracted great attention as promising thermoelectric materials due to their good thermoelectric performance as well as the good thermal stability, lack of sensitivity to the air and non-toxicity [10-14]. The crystal structure of these cobalt oxides consists of the conducting CoO_2 layer and the insulating Na, Ca_2CoO_3 or $Bi_2Sr_2O_4$ layer, which are alternately stacked along the *c*-axis. This layered structure results in a large anisotropy of the

* Corresponding Author

seebeck coefficient in the ab-plane and along the c-axis of the film, realizing $\Delta S = S_{ab} - S_c$ of tens of μV/K, which is about several times larger than that of YBCO (~ 10 μV/K) and hundreds times larger than that of LCMO (~ 0.22 μV/K) [3, 4]. This fact indicates the layered cobalt oxides might have potential applications in the field of high-sensitive broadband photodetectors. In this chapter, we present our investigation of LIV effects in *c*-axis tilted cobalt oxide thin films (Na_xCoO_2, $Ca_3Co_4O_9$ and $Bi_2Sr_2Co_2O_y$) with different pulsed laser sources with wavelength ranging from UV to NIR. The open-circuit voltage signals were detected in these films when their surfaces were irradiated by the laser light. The results demonstrate that the *c*-axis tilted cobalt oxide thin films have great potential applications in the broadband photodetectors.

2. Sample preparation and LIV measurements

2.1 Fabrication of *c*-aixs tilted Na_xCoO_2, $Ca_3Co_4O_9$ and $Bi_2Sr_2Co_2O_y$ thin films

The *c*-axis inclined Na_xCoO_2 thin film was obtained by epitaxially growing a layer of Na_xCoO_2 (x~0.7) on a tilted Al_2O_3 (0001) single crystal substrate by topotaxially converting an epitaxial CoO film to Na_xCoO_2 with annealing in Na vapor. A CoO film was first epitaxial grown on the tilted Al_2O_3 (0001) by pulsed laser deposition. The CoO film was then sealed in an alumina crucible with $NaHCO_3$ powder and heated to 700-750°C for 60 min to form the Na_xCoO_2 film. The *c*-axis tilted $Ca_3Co_4O_9$ and $Bi_2Sr_2Co_2O_y$ thin films can be grown on the tilted $LaAlO_3$ (001) or Al_2O_3 (0001) substrates with the pulsed laser deposition (PLD) or chemical solution deposition (CSD) methods. The detailed PLD and CSD fabrication parameters can be found in Ref. 15-17. Transport measurements on these films reveal that they have the room temperature seebeck coefficient comparable to that of the single crystals, suggesting good quality of these films.

Fig. 1a-c presents the x-ray diffraction (XRD) θ-2θ scans of the *c*-axis tilted cobalt oxide thin films on 10° tilted substrates. The offset angle ω is the angle between the *c*-axis direction and the substrate surface-normal direction and it is set as 10° (See the inset of Fig. 1a). Apart from the substrate peak, all peaks in these patterns can be indexed to the (00*l*) diffractions of the corresponding layered cobalt oxides, indicating that phase-pure *c*-axis tilted Na_xCoO_2, $Ca_3Co_4O_9$ and $Bi_2Sr_2Co_2O_y$ thin films are obtained and the tilted angle is 10°.

2.2 LIV measurements

Fig.2 presents the schematic illustration of the LIV measurements. Two indium or gold electrodes with the diameter of ~ 0.4 mm were symmetrically deposited on the film surface along the inclined direction and they were separated by 4 mm. To prevent the generation of any electric contact effect, the electrodes were always kept in the dark. A XeCl excimer pulsed laser (λ=308 nm, t_p~20 ns) and an Nd:YAG pulsed laser (λ=532 and 1064 nm, t_p~25 ps) were used as the light sources. The incident laser beam, adjusting to 2 mm in diameter using an aperture, was directed perpendicular to the film surface at the middle position between the two electrodes. The induced lateral voltage signals were recorded using a digital oscilloscope of 500 MHz bandwidth terminated into 1 MΩ (Tektronix, TDS 3052).

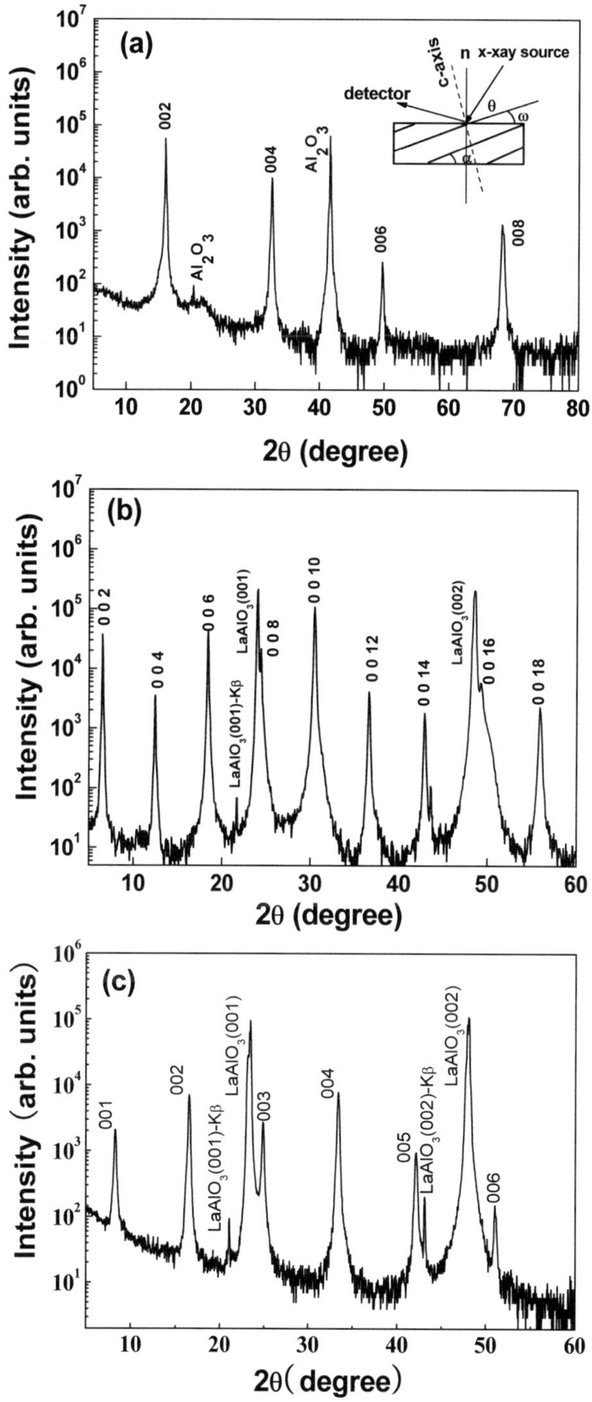

Fig. 1. XRD θ-2θ scans for the c-axis tilted (a) Na_xCoO_2, (b) $Bi_2Sr_2Co_2O_y$ and (c) $Ca_3Co_4O_9$ thin film on 10° tilted single crystal substrates. The inset of Fig. 1a is the sketch map of the XRD θ-2θ measurement and the offset angle ω is set as 10°.

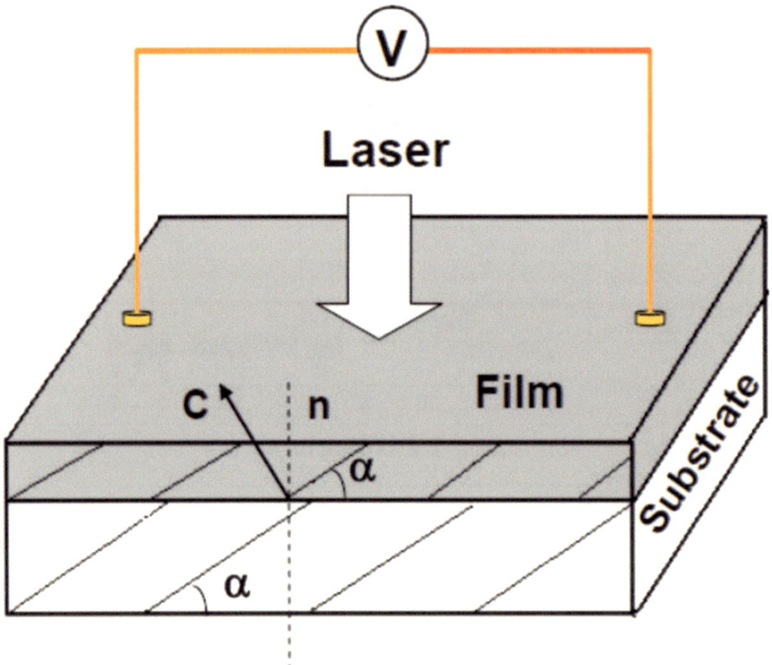

Fig. 2. Schematic illustration of the LIV measurements.

3. Results and discussion

3.1 LIV response of the c-axis tilted $Bi_2Sr_2Co_2O_y$ thin films to different laser pulses

Fig. 3a-c presents the response of a 10° tilted $Bi_2Sr_2Co_2O_y$ thin film (~ 100 nm) to the laser pulses of 308 nm, 532 nm and 1064 nm, respectively. Large open-circuit signals with the response time in the order of several hundred nanoseconds are detected for these three wavelengths. The voltage responsivity is calculated to be about 440 mV/mJ, 348 mV/mJ and 65 mV/mJ for 308 nm, 532 nm and 1064 nm pulsed radiations respectively. The stronger LIV signals obtained for the UV 308 nm and visible 532 nm lasers than that for the NIR 1064 nm laser can be explained by considering the different optical absorption and penetration depth of the radiations with different wavelength in $Bi_2Sr_2Co_2O_y$ film. The higher absorption and smaller penetration depth of 308 nm and 532 nm radiation in comparison with these of the 1064 nm radiation lead to a larger ΔT and thus a larger induced voltage.

To reduce the influence of the RC measurement circuit on the response time, a 2 Ω load resistance is connected parallel with the tilted film while keeping other experimental conditions unchanged. As shown in Fig. 4, the rise time is dramatically decreased from 100 ns shown in Fig. 2a to about 6 ns and the FWHM is also decreased from 470 ns to be about 20 ns. It should be mentioned here that the response time has a pronounced dependence on the pulse width of the incident lasers. The smaller pulse width usually leads to a faster response time. For example, the FWHM of the induced voltage signal is reduced to 1-2 ns when the same film is irradiated by the Nd:YAG picosecond laser pulse. The nanosecond-scale response of the tilted $Bi_2Sr_2Co_2O_y$ thin film to different laser pulses ranging from UV to NIR reveals that it has a potential application in broadband photodetectors with fast response.

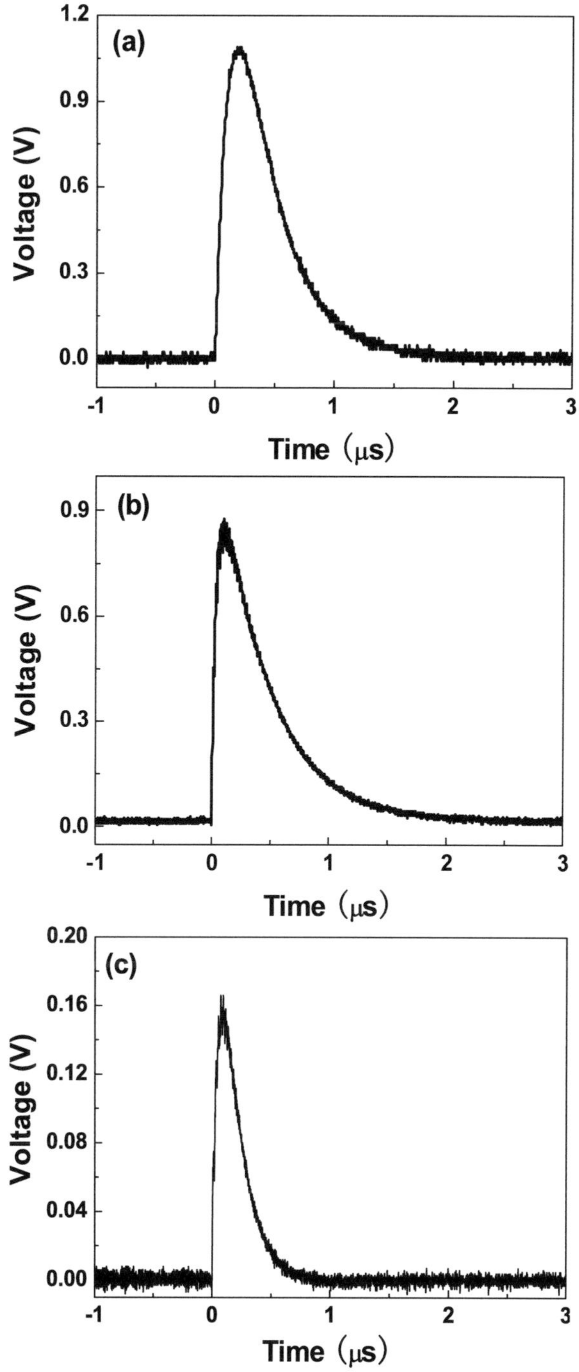

Fig. 3. LIV signals of a 10° *c*-axis tilted $Bi_2Sr_2Co_2O_y$ thin film (~ 100 nm) when its surface is irradiated by the pulsed lasers with different wavelength of (a) 308 nm, (b) 532 nm and (c) 1064 nm. The input impedance of an oscilloscope is 1 MΩ and the laser energy on the sample is about 2.5 mJ.

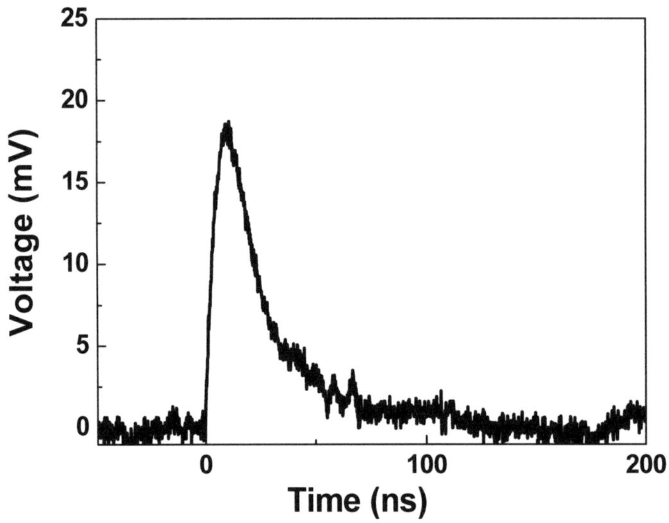

Fig. 4. LIV signal of a 10° *c*-axis tilted $Bi_2Sr_2Co_2O_y$ thin film under the 308 nm radiation after connecting a 2 Ω load resistance in parallel with the tilted film.

3.2 Dependence of the peak voltage of the LIV response on the laser energy, tilted angle and film thickness

Fig. 5a shows the dependence of the peak value of the open-circuit voltage signal (V_p) in the tilted $Bi_2Sr_2Co_2O_y$ thin film on the 308 nm laser energy on the film surface. A linear dependence is obtained for laser energy below the destruction limit of the film. The dependence of the LIV signal of the tilted $Bi_2Sr_2Co_2O_y$ film on the tilted angle α is also investigated and an almost linear relationship between the V_p value and α is obtained, seen in Fig. 5b. This $U \propto \sin 2\alpha$ dependence again demonstrates that the LIV effect in the *c*-axis tilted layered cobalt oxides thin films mainly originates from the transverse thermoelectric effect since all other light-induced effects do not shown such tilt angle dependence.

Fig. 5c presents V_p as a function of the film thickness under 308 nm laser radiation. The V_P increases with the decrease of film thickness and reaches a maximum value when the film thickness is about 100 nm, and then V_p turns to a reduction with further decreasing film thickness from 100 nm to 60 nm. This is inconsistent with Eq. (1) where V_p increases monotonically with decreasing d. Similar V_p-d dependence was observed in the LIV measurements for *c*-axis tilted YBCO and LCMO films. An improved equation based on the plane heat source and cascade power net model was proposed to explain this abnormal behavior. Calculations based on this improved equation revealed that V_p was no more monotonic variation with the thickness *d* and there existed an optimum thickness corresponding to a maximum peak of the induced signal [9].

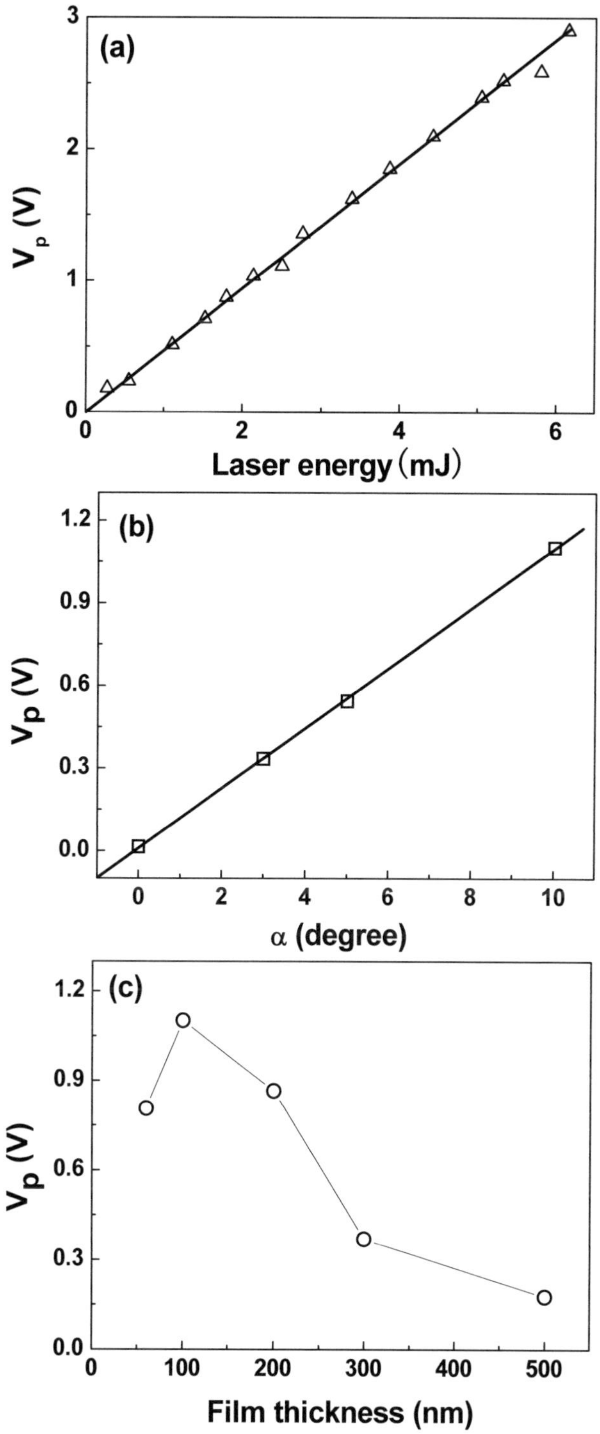

Fig. 5. Dependence of the peak value of the induced open-circuit voltage (V_P) on the (a) laser energy on the film surface, (b) tilted angle α and (c) film thickness. Solid lines are guide for eyes.

3.3 LIV signal in c-axis $Ca_3Co_4O_9$ and Na_xCoO_2 thin films

Similar results are also observed in the *c*-axis tilted $Ca_3Co_4O_9$ thin films when their surface is irradiated by the above lasers. Fig. 6 presents a typical laser-induced open-circuit voltage signal of the 10° tilted $Ca_3Co_4O_9$ thin film (~ 100 nm) under the 308 nm pulsed illumination with the laser energy on the sample of 1 mJ. Both the responsivity (~ 230 mV/mJ) and the response time (the rise time~60 ns and the FWHM~700 ns) of the LIV signal in this tilted $Ca_3Co_4O_9$ thin film is in the same order as that of the tilted $Bi_2Sr_2Co_2O_y$ thin film with the same thickness.

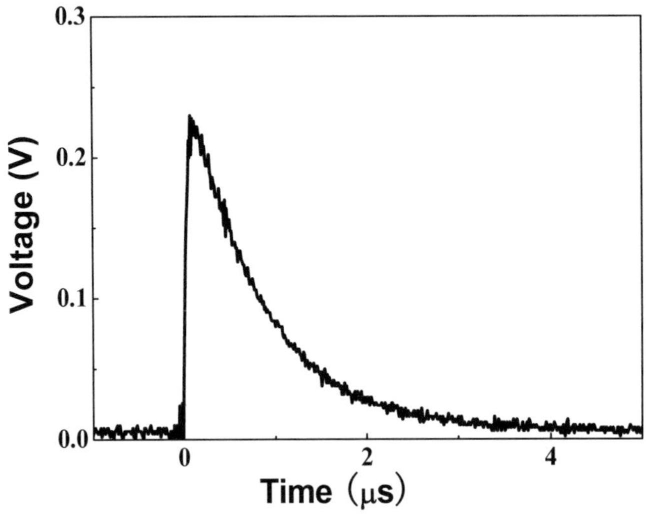

Fig. 6. A typical LIV signal of a 10° *c*-axis tilted $Ca_3Co_4O_9$ thin film (~ 100 nm) under the 308 nm laser radiation. The laser energy on the sample is about 1 mJ.

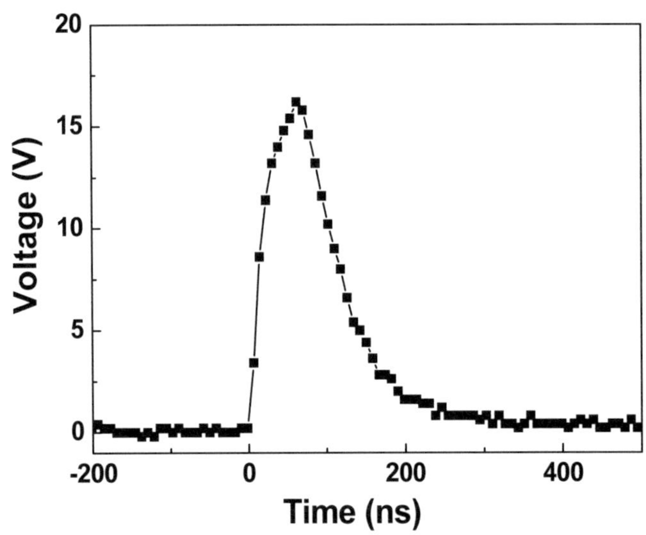

Fig. 7. A typical LIV signal of a 10° *c*-axis tilted Na_xCoO_2 thin film (~140 nm) under the 308 nm laser radiation. The laser energy on the sample is about 1.5 mJ.

The performed LIV measurements on the *c*-axis tilted Na_xCoO_2 thin films show similar dependence of V_p on the laser energy, tilted angle α and film thickness. However, the *c*-axis tilted Na_xCoO_2 thin films have a much larger induced voltage signal than that of the $Bi_2Sr_2Co_2O_y$ and $Ca_3Co_4O_9$ thin films. Fig. 7 illustrates the laser-induced open-circuit voltage signal of the 10° tilted Na_xCoO_2 film irradiated by the 308 nm pulsed laser with the laser energy on the sample of 1.5 mJ. A giant open-circuit voltage signal with V_p of 16.3 V is observed. The responsivity is calculated to be about 11 V/mJ. It is much larger than the responsivity of YBCO, LCMO and other two cobalt oxide thin films [1-4, 18, 19]. This might be due to the Na_xCoO_2 film has better crystal quality which corresponds to a larger ΔS, and the exact reason needs to be clarified in our future work.

4. Conclusion

In conclusion, the performed investigations of the LIV effect in *c*-axis tilted Na_xCoO_2, $Bi_2Sr_2Co_2O_y$ and $Ca_3Co_4O_9$ thin films at room temperature show that large open-circuit lateral voltage signals with the rise time in the order of several nanoseconds are measured when the surface of these films is irradiated by different laser sources with the wavelength ranging from UV to NIR. The obtained results suggest that *c*-axis tilted cobalt oxide thin films can be used for broadband photodetectors with fast response.

5. Acknowledgments

This work was partially supported by NSFC of China under Grant No. 10904030, SFC of Hebei Province under Grant No. A2009000144 and E 2006001006.

6. References

[1] H. Lengfellner, G. Kremb, A. Schnellbbgl, J. Betz, K. F. Renk, W. Prettl, Appl. Phys. Lett. 60, 601(1992)
[2] S. Zeuner, W. Prettl, H. Lengfellner, Appl. Phys. Lett. 66, 1833 (1995)
[3] Th. Zahner, R. Stierstorfer, S. Reindl, T. Schauer, A. Penzkofer, H. Lengfellner, Physica C 313, 37(1999)
[4] P.X. Zhang, W.K. Lee, G.Y. Zhang, Appl. Phys. Lett. 81, 4026(2002)
[5] H.-U. Habermeiera, X.H. Li, P.X. Zhang, B. Leibold, Sol. State Commun. 110, 473(1999)
[6] K. Zhao, M. He, G.Z. Liu, H.B. Lu, J. Phys. D: Appl. Phys. 40, 5703(2007)
[7] S. Zeuner, H. Lengfellner, W. Prettl, Phy. Rev. B 51, 11903 (1995)
[8] H. Lengfellner, S. Zeuner, W. Prettl, K. F. Renk, Europhys. Lett. 25, 375 (1994)
[9] P.X. Zhang and H.-U. Habermeier, J. Nanometers 2008, 329601 (2008)
[10] I. Terasaki, Y. Sasago, K. Uchinokura, Phys. Rev. B 56, R12685(1997)
[11] Y.Y. Wang, N.S. Rogado, R.J. Cava, N.P. Ong, Nature (London) 423, 425 (2003)
[12] M. Shikanoa, R. Funahashi, Appl. Phys. Lett. 82, 1857(2003)
[13] R. Funahashia, M. Shikano, Appl. Phys. Lett. 81, 1459(2001)
[14] K. Koumoto, I. Terasaki, R. Funahashi, MRS Bulletin 31, 206 (2006)
[15] S.F. Wang, Z.C. Zhang, L.P. He, M.J. Chen, W. Yu, G.S. Fu, Appl. Phys. Lett. 94, 162108 (2009)
[16] S.F. Wang, A. Venimadhav, S.M. Guo, K. Chen, X. X. Xi, Appl. Phys. Lett. 94, 022110 (2009)

[17] S.F. Wang, M.J. Cheng, L.P. He, W. Yu, G.S. Fu, J. Phys. D: Appl. Phys. 42, 045410 (2009)
[18] S.F. Wang, J.C. Chen, X.H. Zhao, S.Q. Zhao, L.P. He, M.J. Chen, W. Yu, J.L. Wang, G.S. Fu, Appl. Sur. Sci. 257, 157(2010)
[19] S.F. Wang, J.C. Chen, S.R. Zhao, L.P. He, M.J. Chen, W. Yu, J.L. Wang, G.S. Fu, Chin. Phys. B 19, 107201 (2010)

9

Geiger-Mode Avalanche Photodiodes in Standard CMOS Technologies

Anna Vilà, Anna Arbat, Eva Vilella and Angel Dieguez
Electronics Department, University of Barcelona
Spain

1. Introduction

Photodiodes are the simplest but most versatile semiconductor optoelectronic devices. They can be used for direct detection of light, of soft X and gamma rays, and of particles such as electrons or neutrons. For many years, the sensors of choice for most research and industrial applications needing photon counting or timing have been vacuum-based devices such as Photo-Multiplier Tubes, PMT, and Micro-Channel Plates, MCP (Renker, 2004). Although these photodetectors provide good sensitivity, noise and timing characteristics, they still suffer from limitations owing to their large power consumption, high operation voltages and sensitivity to magnetic fields, as well as they are still bulky, fragile and expensive. New approaches to high-sensitivity imagers tend to use CCD cameras coupled with either MCP Image Intensifiers, I-CCDs, or Electron Multipliers, EM-CCDs (Dussault & Hoess, 2004), but they still have limited performances in extreme time-resolved measurements.

A fully solid-state solution can improve design flexibility, cost, miniaturization, integration density, reliability and signal processing capabilities in photodetectors. In particular, Single-Photon Avalanche Diodes, SPADs, fabricated by conventional planar technology on silicon can be used as particle (Stapels et al., 2007) and photon (Ghioni et al., 2007) detectors with high intrinsic gain and speed. These SPAD are silicon Avalanche PhotoDiodes biased above breakdown. This operation regime, known as Geiger mode, gives excellent single-photon sensitivity thanks to the avalanche caused by impact ionization of the photogenerated carriers (Cova et al., 1996). The number of carriers generated as a result of the absorption of a single photon determines the optical gain of the device, which in the case of SPADs may be virtually infinite.

The basic concepts concerning the behaviour of G-APDs and the physical processes taking place during their operation will be reviewed next, as well as the main performance parameters and noise sources.

1.1 Basic concepts for Geiger-mode avalanche photodiodes, G-APDs

APDs can be obtained by implementing two possible approximations that produce two different structures with differentiated capabilities. On one hand, thin silicon APDs (Lacaita et al., 1989) are devices with a depletion layer of few micrometres and low breakdown voltages. They also present good detection efficiency and time resolution. As in planar

APDs it is important to avoid the possibility of edge breakdown of the sensor, many different terminations have been proposed, generating a variety of planar CMOS compatible devices. On the other hand, thick APDs (Cova et al., 2004) are devices with a depletion layer of some tens of micrometres that work at high breakdown voltage; they have good detection efficiency but moderate time resolution. These components are fabricated in dedicated technologies, increasing their cost.

An example of these two different structures is shown in Fig. 1.A: a thick APD at the left side (Alexander, 1997) and a thin APD at the right side. The structures are composed of different sections of semiconductor with different doping profiles shown in B. In C, the charge density of the structures is represented. The p region develops a net negative charge, while the n region develops a net positive charge, generating the energy diagrams visible in D and an electric field profile such as in E.

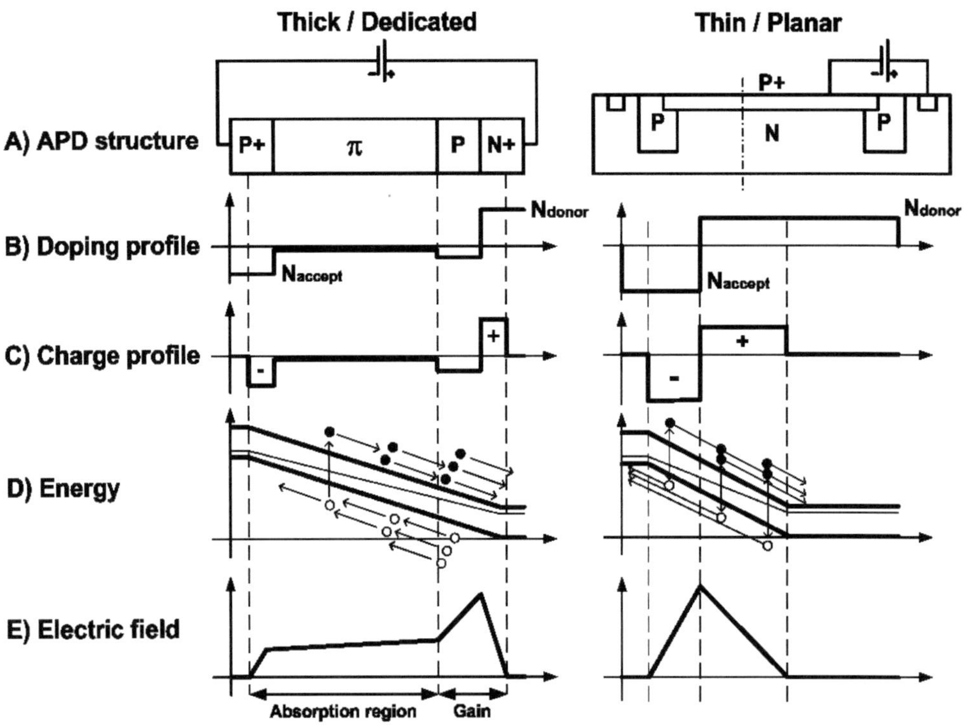

Fig. 1. Comparison between thick (left) and thin (right) APD structures. A) structure, B) doping profile across the dash-dotted line in A, C) charge profile, D) energy-band diagram, E) electric field profile.

An electron-hole pair, EHP, can be generated due to photon absorption, minimum-ionizing particle crossing the detector or by thermal generation. Due to the electric field in the sensing area, carriers are accelerated to the collection zone. The high electric field give to the carriers enough kinetic energy to produce additional EHP by impact ionization. This additional charge is also accelerated and eventually produces new EHP in their way towards the electrodes. The whole effect generates a multiplication of the electric signal from a single photon.

There are two possible operation modes depending on the biasing of the APD: linear (Fig. 2.A) and Geiger (Fig. 2.B). In the linear mode, only the electrons achieve enough kinetic energy to produce new EHPs by impact ionization, producing gains around 1000 (Kindt, 1994). The gain is then moderate and affected by strong excess noise (McIntyre, 1972), making difficult the detection of single photons. The gain is strongly dependent on the biasing voltage, requiring a uniform field over the sensible area to have a constant gain. When the device is biased well above the breakdown voltage, Geiger mode, the electric field in the depletion layer is so high that also holes produce new EHPs, generating a self-sustaining avalanche current. In this case, the avalanche can be triggered by a single photon or a single EHP generated by a MIP, and the device becomes a SPAD. The current rises swiftly (sub-nanosecond rise-time) to a macroscopic steady level in the milliampere range, irrespective to the quantity of photons detected (essential effect for silicon photomultipliers, where the output signal is proportional to the number of APDs in avalanche).

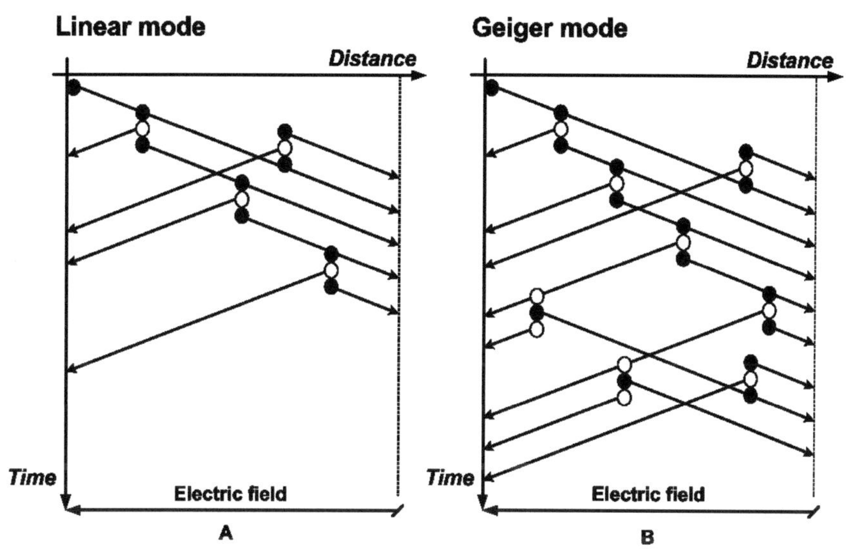

Fig. 2. Avalanche phenomena due to impact ionization. A) In linear mode, only electrons contribute to EHP generation. B) In Geiger mode, both electrons and holes produce newEHPs, promoting a self-sustained avalanche.

In Geiger mode, the current generated during the avalanche continues to flow until lowering the bias voltage down to the breakdown quenches it. Then, the lower electric field is no longer able to accelerate carriers to impact-ionize lattice atoms, therefore current ceases. In order to be able to detect another photon, the biasing needs to be restored to the initial value in a recharge process to leave the detector ready to work again.

Consequently, the detection cycle includes the quenching and the recharge periods, and usually specific circuitry helps to do these actions. The basic goal is to limit the avalanche charge flowing through the APD at every ignition, in order to reduce three detrimental effects:

- Afterpulsing: avalanche events produced by the release of charges trapped into deep levels (Cova et al., 1996).

- Self-heating: more dark counts are generated, but also breakdown voltage is increased and excess voltage decreased, reducing the detection efficiency.
- Crosstalk: interference between neighbouring sensors produced by emission of secondary photons (Zappa et al., 1997).

1.1.1 Performance parameters

In this context, the main figures of merit that have been used to describe the capabilities of G-APDs as photodetectors are shortly reviewed next. Some of them are specific of the case under study and have no sense related to other types of technologies. However, all they are of crucial importance in G-APDs.

Gain

First of all, the gain or multiplication factor, M, is defined as the ratio between the observable photocurrent at the APD terminals and the internal photocurrent before multiplication. This primary photocurrent depends on the detector quantum efficiency, defined as the number of electrons collected in a pixel over the number of photons penetrating it, the photon energy and the received optical signal power. This effect is similar to that achieved with a photomultiplier.

Ionization coefficients

The ionization coefficients for electrons and holes, α_e and α_h respectively, are defined as the probability that the carrier causes some ionization within a unit length. These coefficients increase with the field, as the carrier achieves a higher energy, while decreases with temperature, as the atoms are thermally excited, reducing the collision probability. High-quality APDs can present a ratio between α_h and α_e of between 0.003 and 0.01.

Time delay

The time delay of an avalanche corresponds to the time between the photodiode generates an EHP and the avalanche arrives to the measuring circuit, and it represents a fundamental parameter of an APD because it, together with the RC time-constant and the transit time, determines its bandwith. The time delay is determined by the transition times for the electrons/holes to reach the avalanche region (i.e. the time required for charges to go through an absorption region and arrive to the avalanche region) and the avalanche build-up time.

Timing resolution or jitter

The timing resolution or timing jitter of the sensor is defined as the statistical fluctuation of the time interval between the arrival of the photons at the sensor and the output-pulse leading edge. It can be described as the full-width half-maximum measure of the temporal variation in the avalanche breakdown pulses resulting from an incident photon beam. Among the timing-resolution components are the variation caused by the generated-carrier transit time from the depletion layer to the multiplication region, which is dependent on the depth of absorption of the incident photon and, more important, the statistical build up of the avalanche current itself (Ghioni et al., 1988), which timing improves with the excess bias voltage.

Photon detection efficiency, PDE, or photon detection probability, PDP

The photon detection efficiency, PDE, is the probability of avalanche generation for each detected photon. For a photon to be detected, it must be absorbed in the detector active volume and generate a primary carrier, but it is also necessary for this primary carrier to succeed in triggering an avalanche. The photon detection probability, PDP, is related to the PDE, and it is defined as the probability of a photon to generate an avalanche. Both concepts are wavelength dependent and increase with the excess bias voltage, as the higher electric field enhances the triggering probability.

Sensitivity or dynamic range

The dynamic range, DR, quantifies the ability of a sensor to adequately image both clear lights and dark shadows in the same scene. It is defined as the maximum signal divided by the noise. In the case of APDs, the DR is given by the maximum number of counts generated by the sensor and the dark count noise. The sensitivity increases with the biasing voltage and depends on the technology.

Fill factor

The fill factor is the ratio between the light-sensitive area of a pixel and its total area. The electric field strength should be consistent across the whole active area (the central photon-sensitive portion of the detector), so as to yield a homogeneous breakdown probability. However, electric field concentration at the edges and corners can cause higher photon-detection probability and dark counts. Such zones should be avoided by proper design solutions such as guard rings or shallow trenches, which limit the active detector volume. This is a main drawback for present G-APDs, especially for large-area high-precision applications such as particle detectors in colliders or telescopes.

Dead time

The dead time is the time during which the sensor is not responsive to further incoming photons, due primarily to accumulation of charge in the active region. It involves the periods comprising the avalanche quenching and the device recharge, i.e. the reset of the final bias conditions. However, in the case of a passively quenched SPAD this is not strictly the case, because, as the device is recharged via the quenching resistor, it becomes increasingly biased beyond its breakdown voltage, so it is able to detect the next photon arrival prior to being fully reset. This behaviour is coupled with a significant fluctuation in the reset waveform and can increase the afterpulsing probability due to inadequate trap flushing time.

1.1.2 Noise sources

The inherent noise sources of APDs are dark counts and afterpulsing events. There is also an external noise source between neighbouring sensors due to electrical and optical crosstalk.

Dark count rate, DCR

A dark count is an avalanche event caused by non-photogenerated carriers, which can be originated from four factors (Haitz, 1965): diffusion from neutral regions, thermal generation, band-to-band tunnelling or by release from a charge trap (see also afterpulsing).

This generation is critical when operating in Geiger mode, as it can produce an output pulse indistinguishable from the one originated by a detection event. The per-second rate at which dark counts occur is the DCR in Hertz.

Dark-count noise increases with temperature, as the thermal carrier generation increases and so the probability of avalanche. It also increases with the excess voltage because of two effects, namely the field-assisted enhancement of the emission rate from generation centres and the increase of the avalanche-triggering probability with the higher electric field at the depletion layer, which reduces the recombination probability. Of course, thermal generation is strongly dependent on the fabrication process, which determines the concentration of traps, the breakdown voltage and the avalanche probability. On the contrary, tunnelling generation depends on the doping profile and the bias voltage, and its temperature dependence is weak.

Afterpulsing

Charge traps due to unintentional impurities and crystal defects can result in generation-recombination centres. The high current peak through the junction during an avalanche breakdown introduces a probability that the trap is filled by a carrier that is later released, initiating a second, follow-on Geiger 'after-pulse' (Haitz, 1965). Afterpulsing is then a false detection caused by secondary dark pulses correlated to a previous primary one. This phenomenon suffers from a statistical fluctuating delay whose mean value depends on the deep levels actually involved.

The number of carriers captured during an avalanche pulse increases with the total number of carriers crossing the junction, that is, with the total charge involved in the avalanche pulse, and with the number of traps in the multiplication area. Therefore, afterpulsing increases with the delay of avalanche quenching and with the current intensity, which is proportional to the excess bias voltage. This overvoltage also enlarges the depletion zone, and so the number of available traps, increasing the trapping probability. Finally, released charges can retrigger avalanche.

Crosstalk

A crosstalk is a signal interference between neighbouring pixels produced by electrical or optical interaction between them. The electrical crosstalk corresponds to the migration of carriers from one pixel to another, and can be probably reduced by a potential barrier designed properly to collect these charges. On the other hand, hot carriers in an avalanche emit 2.9 photons per 10^5 electrons (Lacaita et al., 1993) that can produce an avalanche in the neighbouring pixel generating an optical crosstalk. The probability of this phenomenon depends on many factors, such as pixel size, dead space between active areas, sensitive volume and gain, but it can be diminished by reducing the number of carriers generated in the avalanche.

2. Technology considerations for GAPDs

This section consists on a review of the fundamental technology considerations that deserve to be taken into account referred to the actual G-APDs, including the state-of-the-art of avalanche photodiodes fabricated in CMOS compatible technologies.

2.1 Historical review of APDs technologies

The switching effect observed in reverse-biased silicon p-n junctions by McKay (McKay, 1954), marked the beginning on the study of the structures in avalanche. Imperfect junctions with a large number of impurities were used in initial investigations. In 1961 Shockley (Shockley, 1961) proposed the ideal reverse junction characteristics which were confirmed in the early 1960s (Goetzberg et al., 1963), (Haitz, 1964) and (Haitz, 1965). With initial progresses of silicon technology, it was possible to fabricate better junctions nearly free from defects (called microplasmas by then). But practical devices were large (about $1mm^2$) and they did not attract much attention due to their high cost low efficiency.

In the early 70s, McIntyre (McIntyre, 1972) and Webb (Webb et al., 1974) developed new APD structures that were the precursors of commercial devices implemented in the Single Photon Counting Modules, SPCM. These new devices were fabricated on a sophisticate dedicated process which was incompatible with the planar technology (thick-APDs), but improved the photodetection efficiency. This was the beginning of dedicated technologies for APDs, able to reduce the dark count noise and improve the features of the detector.

Evolution has allowed APDs exhibiting excellent quantum efficiency, with values around 80% in the near ultra-violet range, dropping to about 40% in the blue region, which is to be compared to typical values of 5-8 % in the blue for standard photomultipliers. Although in the late 80s the planar technology to fabricate APDs was improved by Ghioni when they introduced the first epitaxial device structure (Ghioni et al., 1988), nowadays mainly all commercial APDs are produced by dedicated technologies not compatible with CMOS planar technology, what increases the difficulty to integrate an array of detectors together with the control electronics.

At the beginning of the XXI century the research on APDs was reoriented to the integration of the sensors together with the electronics thanks to the improvement on the CMOS technologies. Main objectives were the fabrication of SPAD integrated cameras (Charbon, 2007) and of large SPADs, which were restricted to the dedicated technologies (Ghioni et al., 2007). The first APDs in CMOS compatible technologies were published by research centers (Spinelli et al., 1998, Jackson et al., 2001 and Gulinatti et al., 2005). They were developed using the high voltage CMOS, HV-CMOS, technologies (Rochas et al., 2002 and Zappa et al., 2004), as they provide a relatively low-doped deep n-well that allows up to 50 V of isolation from the substrate (Rochas et al., 2003a) and low noise detectors.

The first SPAD-based pixel array in CMOS technology has been demonstrated a few years ago (Rochas et al., 2003b), consisting on a 4 by 8 array of SPADs with a passive quenching circuit and a simple comparator (inverter). Further developments followed quickly, facilitated by the availability of commercial HV-CMOS technologies and of specially tailored imaging processes boosted by the huge market of mobile phone cameras. Designs using a 0.8 μm CMOS technology have also been presented. In (Niclass & Charbon, 2005) a 64x64 image sensor with single pixel readout is employed for 3D imaging. Another APD array based on event-driven reading was also developed using the same technology (Niclass et al., 2006a). The evolution of the technologies also produces an evolution on the sensors, and some arrays were presented using 0.35 μm technology (Sergio et al., 2007). An array of 128x2 APDs was developed including latches to preserve the timing information. Finally the evolution of APDs arrived to 130 nm, with an array of 32x32 sensors (Guerrieri et al., 2009).

In the last years, the use of submicronic technologies has also included the study of APD structures, and some comparison between technologies has been reported recently (Arbat et al., 2010a), concluding that high-voltage technology has a low dark-count rate related to the low trap concentration but the sensor is slow, while high-integration technology generates fast sensors with high dark count rates. In 2006 Finkelstein presented a new structure for an APD based on shallow-trench isolation, STI (Finkelstein et al., 2006a). Although a reduction of size, the structure suffered from many dark counts that limited the device performances. In 2007, an n-well-based structure was designed on a 130nm technology (Niclass et al., 2007a and b), and the modifications needed for the 130nm-process to improve the detector features were presented (Gersbach et al., 2008). With the same 130nm technology, p-guard STI structures were also proposed (Gersbach et al., 2009), as well as a lower p-doping well in a deep n-well (Richardson et al., 2009). Going further, the first APD structures obtained using 90nm technology have been reported (Karami et al., 2010). In a parallel way, the use of nanostructures of germanium deposited in silicon to obtain small APDs able to work at voltages of only 1.5V and speeds of 30GHz have been proposed (Assefa et al., 2010). Even, a novel effort to use APDs to construct detectors for charged particles has recently been reported (Graugés et al., 2010, Vilà et al., 2011). Results suggest that, through control of the doping concentration, devices with a much improved fill factor could be achieved. Consequently, submicron CMOS technologies with ever decreasing minimum feature size seem to be useful for SPAD fabrication.

2.2 Dedicated vs. standard technologies

The present situation of SPADs may be described in terms of the manufacturing processes and construction details. In particular, two main process categories can be distinguished:

a. Dedicated technologies, which improve performances by optimizing individual technological parameters. They generate low-noise high-quantum-efficiency sensors via thick depletion layers (> 30μm), but integration of electronics with sensor is impossible.
b. CMOS compatible technologies, which can be only compatible (i.e. full-custom processes optimized to yield the best possible performing single detector element) or standard CMOS processes, without any modifications to the layers normally available to the designers. Both single detectors and arrays have been implemented with HV-CMOS processes (Rochas et al., 2003a and b, Stoppa et al., 2007 and Vilà et al., 2011), demonstrating their suitability for SPADs.

Both types of APD processes offer different features. In Table 1, a comparative of the main characteristics of each type of sensors is given. As APDs are mainly used as photon detectors, the detection efficiency is given for wavelengths. When high-energy particles are considered, the detection efficiency is related to the number of electrons generated by ionization in the multiplication region due to the crossing particle.

A method to build a thick APD using dedicated technologies consists on building the sensor on a p substrate with an n+ layer on one side and a p+ diffused layer to improve the ohmic contact on the other side (Webb et al. 1974). This produces a reach-through structure formed by four layers in a vertical structure, generating a p+-π-p-n+ stack (see figure 1) that can be improved through processes covered by patents (McIntyre, 1990 and 1996). As its operation is based on the complete depletion of the device (30 to 100μm deep), the required voltages are high, typically over 100V. This produces sensors with low noise and high quantum efficiency (Cova et al., 2004, Lacaita et al., 1988 and Ghioni & Ripamonti, 1991). This technology allows the possibility to produce wide active areas, up to 500 µm in diameter, with variable depletion-layer thickness (Dautet et al., 1993).

Main features	Planar - Thin	Dedicated - Thick
Breakdown	10 – 50 V	100 – 500 V
Multiplication thickness	1 – 2 µm	30 – 100 µm
Active area diameter	5 – 150 µm	100 – 500 µm
Photon Detection Efficiency (PDE)	~ 45 % @ 500 nm ~ 32 % @ 630 nm ~ 15 % @ 730 nm ~ 10 % @ 830 nm 0.1 % @ 1064 nm	> 50 % @ 540 - 850 nm ~ 3 % @ 1064 nm
Resolution in photon timing	<100 ps FWHM	< 350 ps reach-through ~ 150 ps smoother field profile
Dark count rate	Technology dependent	Very low noise
Power dissipation	Low, cooling system not required	High, cooling system required
Array fabrication	Compatible	Not compatible
Robustness	Good	Delicated
Cost	Low	High

Table 1. Dedicated vs. planar technologies comparison.

There are some commercial compact modules that include the bias and the quenching circuitry (Spinelli et al., 1996 and EG&G, 1996). Some performances of commercial APDs based in reach-through structures are given in Table 2. Although the reach-through structures have good performance, they also have a number of practical drawbacks. The main problem is the high power dissipation due to the high biasing voltage, arriving to values of 5 to 10 W. Another negative aspect is the high fabrication cost of the sensors, due to a low fabrication yield. Finally, the produced devices are delicate and degradable, and of course not integrable with circuitry.

Feature	InGaAs	Si	Array 4x8
Active area (mm)	0.04	0.2 - 5	1.6 x 1.6 Pitch 2.3mm
Wavelength (nm)	950 - 1700	200 - 1100	320 - 1000
λ peak (nm)	1550	620 – 940	600
Sensibility @ λp (A/W)	0.8 – 0.9	0.42 -0.5	-
Quantum efficiency (%)	-	75 - 80	70
V breakdown (V)	40 - 60	150 - 300	400 - 500

Table 2. Hammamatsu commercial APD features.

APD devices developed in planar technologies generate a thin depletion layer of few micrometers, being called thin APDs. These devices are generated by epitaxy over silicon wafers (Spinelli et al., 1998). Over the last years, improvements of the technology to fabricate integrated circuits have allowed the development of APDs integrated in commercial CMOS

Reference	Technolgy (termination)	Diameter (μm)	V_{BD} (V)	DCR (c/s)	PDE (%) @500nm	FWHM (ps)
(Spinelli et al., 1998)	Double junction	10	27	-	35	35
(Jackson et al., 2001) (Jackson et al., 2002)	1.5μmNMRC (1)	15/20/30 40/50	30	200/1k/3k 9k/30k	26	-
(Gulinatti et al., 2005) (Ghioni et al., 2006)	CNR-IMM	20/50 100/200	24	300/700 4k/40k	38	30/30 31/35
(Rochas et al., 2002)	HV0.8μm (4)	80	19.5	-	-	-
(Rochas et al., 2003a) (Stapels et al., 2006)	HV 0.8μm (2)	7	25	900	30	60
(Zappa et al., 2004) (Zappa et al., 2005)	HV 0.8μm (2)	12	16	600	38	36
(Panchieri & Stoppa, 2007)	HV 0.7μm (2)	10x10 20x20 40x40	20.5 - 21	<10k <60k <200k	26	144
(Dandin et al., 2007)	0.5μm (5)	-	16.85	16k	-	-
(Mosconi et al., 2006)	HV 0.35μm (2 + IC)	20x20	28.3	3 – 5k	10	80
(Xiao et al., 2007)	HV 0.35μm (3)	-	45 - 50	<50	33	80
(Tisa et al., 2008)	HV 0.35μm (2 + IC)	20	24	<5k	37	39
(Niclass et al., 2006b) (Niclass et al., 2009)	0.35μm (2)	4/10	-	6/750	36	80
(Arbat et al., 2010a)	HV 0.35μm (2)	20x20	17.3	3.4k	-	-
(Finkelstein et al., 2006a) (Finkelstein et al., 2006b)	180nm IBM (6)	2x2 14x14	10	185k	12 - 20	-
(Faramarzpour et al., 2008)	180nm (2)	10/20	10.2	60k(10) 240k (20)	45	-
(Niclass et al., 2007a) (Niclass et al., 2007b)	130nm mod (2)	10	10	100k	41	144
(Gersbach et al., 2008) (Gersbach et al., 2009)	130nm mod (6)	4.3	10/17.2	1M/90k	36	125
(Richardson et al., 2009)	130nm STM (3)	8	14.4	<20	28	200
(Arbat et al., 2010b)	130nm STM (2)	20x20	10.5	3.5k	-	-
(Arbat et al., 2010b)	130nm STM (6)	20x20	10.55	1200k	-	-
(Karami et al., 2010)	90nm mod (2)	8x8	10	8.1k (0.13V)	8 - 14	398

Table 3. CMOS planar APDs. Termination (see section 2.3): 1-N+diffusion over p-well. 2-P-guard ring. 3-Deep p-well. 4-P-diffusion. 5-P-guard and polysilicon control gate. 6-Shallow trench isolation

technologies. Table 3 contains a summary of the published structures fabricated using planar technologies. See section 2-3 for details about termination. Some performances are associated to these devices, such as the possibility of integrating the electronics of the detector, of developing arrays of sensors, and of using low voltages and reducing the power consumption of the devices. One of the main disadvantages of the utilization of standard fabrication processes is the impossibility of modifying the fabrication process without the foundry interaction, inhibiting the possibility of device optimization.

2.3 Termination techniques for planar APDs

As commented before, the working principle of APDs requires a uniform electric field in the whole active area, what is only achieved in the central part of the sensor. Electrical-field concentration at the edges of the active area disturbs this uniformity and can cause premature edge breakdown, PEB. Different techniques have been developed in order to avoid PEB and consequently the device failure (Charbon, 2008), with the objective of limiting the electric field at the edges to be weaker than at the central multiplication region. Some of the most used options are summarized next.

Adding an n+ layer between the n substrate and the p surface confines the high electric field at the central part of the sensing area (Fig. 3), owing to the higher doping at the central junction (p-n+ instead of p-n at the edges). The concept was used with a p enrichment below the n+ diffusion to confine the electric field (Spinelli et al., 1998 and Jackson et al., 2000).

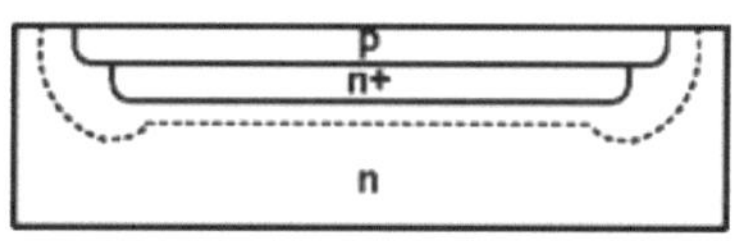

Fig. 3. N+ layer added between the n substrate and the p surface to avoid PEB. Termination technique (1) in table 3.

A second option largely used consists on the implementation of a p guard ring with lower doping level around the structure (Niclass et al., 2006b and Stoppa et al., 2005). This structure allows the possibility to share n-tubs between different sensors, reducing the pitch (Pancheri & Stoppa, 2007), and can be used in triple-well technologies which provide low-doping layers, as in the case of HV-CMOS (Xiao et al., 2007). This last approximation avoids heavily doped layers which are prone to have generation and trapping centres. Fig. 4 shows the initial and triple-well structures. Sensor arrays have been demonstrated with this technology (Tisa et al., 2008 and Arbat et al., 2010a) and it suitability has also been demonstrated in submicron technologies (Faramarzpour et al., 2008), but it is always important to avoid the full depletion of the n well, which would lead to punch through between the well and the substrate.

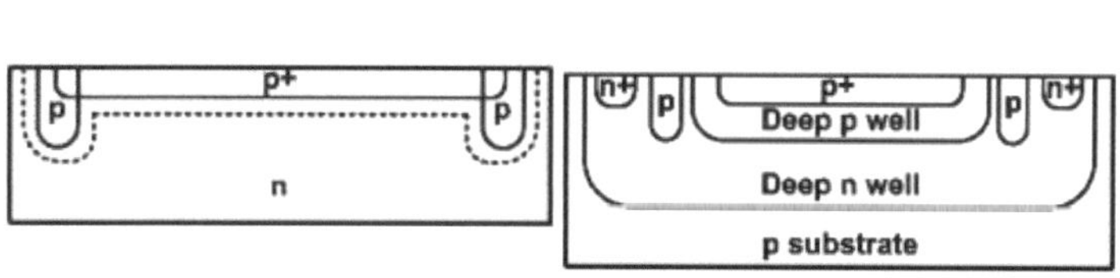

Fig. 4. Structures for avoiding PEB based on p-well guard ring at left, termination technique (2) in table 3, and deep tubs available in HV-CMOS technologies at right, termination (3) in table 3.

A third option frequently used consists on a p diffusion close to the edge of the p well at the sensing area (Rochas et al., 2002). In this case, the ion lateral diffusion during the processing generates a low-doped n region between p zones which can be completely depleted thanks to the biasing of an additional gate located on it (fig. 5, left). When using a p-substrate technology with doble well, it is possible to build similar guard-ring structures (fig. 5, right) with adjacent n wells placed at the minimum distance allowed by the technology (Dandin et al., 2007).

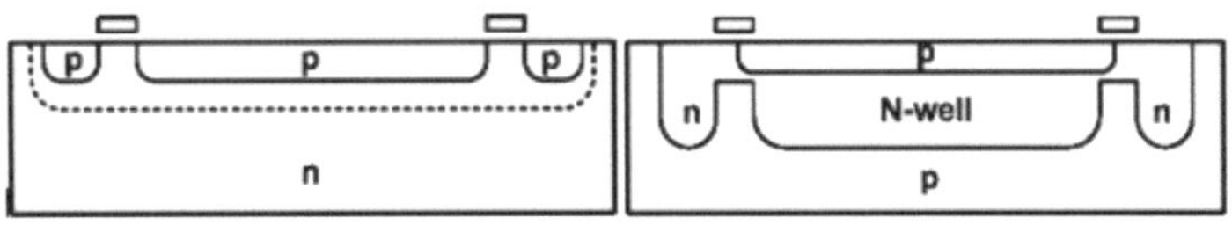

Fig. 5. Double low-doped n-well structure with gates (left) and guard rings and control gate signal (right). Termination techniques (4) and (5) in table 3, respectively.

Finally, for technologies beyond 250nm, the use of shallow trench isolation, STI, becomes a technological solution to reduce the negative effects of the birds-beak-shape of the SiO_2 (Finkelstein et al., 2006a). The STI prevents punch-through and latch-up, and it confines the electric field because the dielectric strength of SiO_2 is much higher than that of Si (fig. 6).

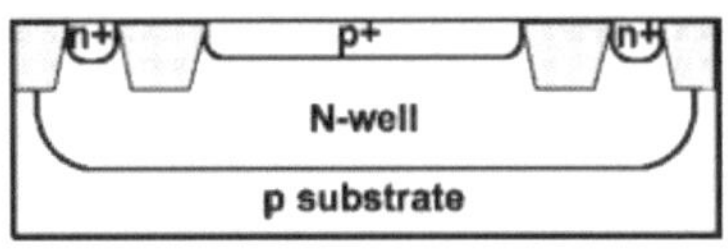

Fig. 6. Structure to avoid PEB based on STI elements. Termination technique (6) in table 3.

All these strategies have demonstrated suitability for preventing PEB, but not all of them are possible in a specific technology. The better choice depends on the possible structures.

3. G-APD simulation and characterization

Characterization of the individual sensing elements is needed for detector design, using different test benches and a black Faraday box. This section analyzes firstly the possibility of physical description of the device, to better understand its optical and electrical characteristics. Next, the characterization techniques usual for APDs are introduced: the I-V characteristic of the detector gives information about the breakdown voltage, the dark current below breakdown (linear mode) and the avalanche current. A 4-wire setup directly connected to the sensor terminals is recommended. Finally, the noise figures need to be characterized: dark count rate, afterpulsing and optical and electrical crosstalk. These parameters are not evident to minimize, and a definitive reading of actual detection requires new strategies based mainly on the accompanying electronics, as exposed in next section.

3.1 Physical simulation

Three main aspects have to be taken into account to calculate the electrical behaviour of the G-APDs: firstly, the charges generated in the semiconductor during the breakdown phenomenon; secondly, the movement of these charges inside the device towards its electrodes; and thirdly, the influence of the external circuitry on the recorded signal after detection, which is the only result that can be compared with experimental data.

Interaction between radiation and matter is a statistical phenomenon that can be calculated via Monte Carlo models able to describe a particle or photon beam as it interacts with the different layers in a device. The conventional model that explains the generation-recombination processes via traps is the well known Shockley-Read-Hall model (Shockley & Read, 1952), in which four basic mechanisms are supposed to be involved: electrons and holes capture and electrons and holes escape from a trap. However, for high electric field, tunnelling effects are dominant, and newer models are introducing corrections to take them into account.

So long, Monte Carlo calculations have then allowed a detailed description of a beam inside a device, discriminating the contribution of tunnelling to the dark-count rate, and explaining most of the inner effects of absorption, backscattering, etc. referred to a given structure (Vilà et al., 2011). After that, the evolution of the generated charges must be described, and classical drift-diffusion models can give some macroscopic parameters related to interactions. However, the avalanche is difficult to be described via these semi-classical models, as the impact-ionization process itself depends at nearly atomic level on the generated fields, which in turn depend on the generated charge. In particular, the ionization coefficients for electrons and holes can be calculated for biasing up to the breakdown voltage, but for higher polarizations the charges generated by impact ionization modify locally the electric field, so that the coefficients get stabilized and no further progress can be calculated for any overvoltage. Solving this situation needs further advances in models.

Consequently, proprietary software needs to be developed and applied to properly describe the avalanche, and then, well established electrical models will help describing the device inside the whole detector system. The complete simulation process is complex, but necessary to understand the device behaviour in operating systems as for characterization.

3.2 I-V characteristics

The breakdown voltage is evidenced by a sudden current increase of several orders of magnitude in I-V characteristics for reverse bias. It is strongly technology dependent, because it is related to concentration and depth of the doping profiles, and can even be sensitive to the particular run and chip, as illustrated in fig. 7 (Arbat et al., 2010b). The analysis of the temperature effect on the breakdown voltage indicates that both parameters increase together. All these variations can be managed with adequate readout electronics.

3.3 Dark count rate

The dark-count rate increases exponentially with biasing, as shown in figure 8, and approximately linearly with the detector area. However, important differences can be found depending on the technology used. In particular, the trench-based structures present much more dark counts than the structure based on an n well. This suggests that the fabrication of these trenches introduces traps in their walls, increasing the pulse-generation probability (Hamamoto, 1991). Consequently, trench-based technologies should be treated with care when a low noise rate is required. The influence of other factors such as degradation due to radiation is still to be studied.

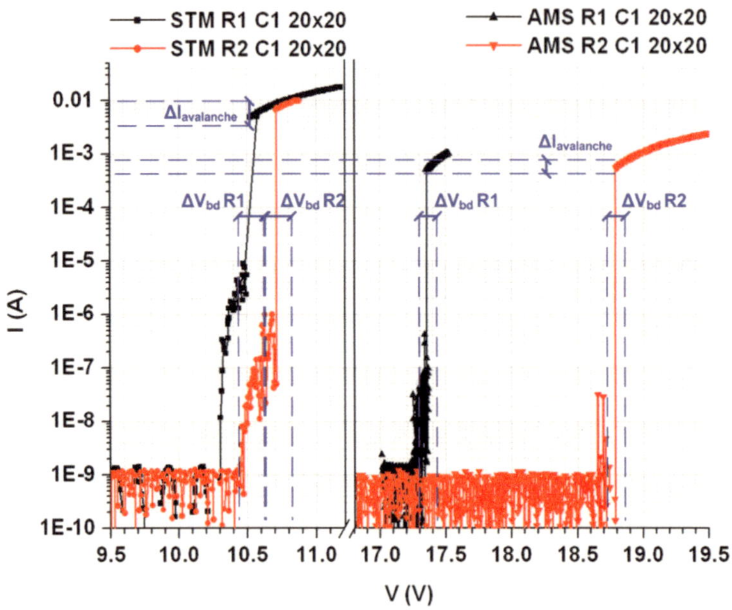

Fig. 7. I-V characteristics of 20μmx20μm APDs made using STM (left) and HV-AMS (right) technologies, for two different runs (R1 and R2) each.

An analysis with the temperature allows observing that the number of dark-count pulses increases exponentially with the temperature, as expected due to thermal generation of electron-hole pairs. The dynamic-range analysis of these types of structures gives sensibilities with similar behaviour independently from the technology, incrementing the counts linearly with illumination, until saturation is reached. On the other hand, increasing biasing raises the sensor sensibility.

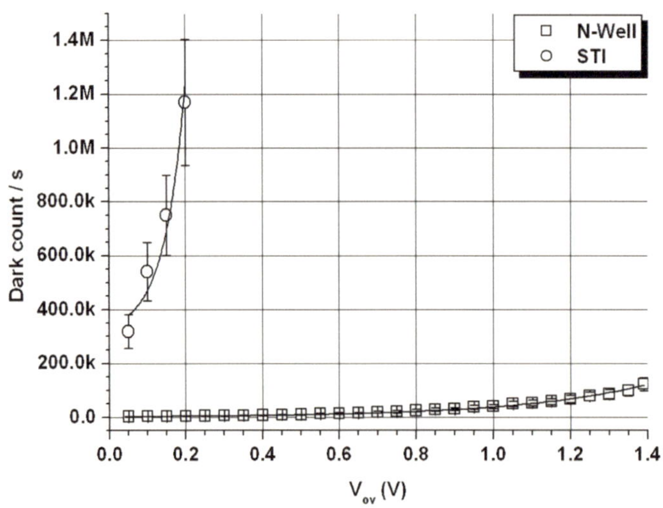

Fig. 8. Dark-count rate for two detector structures based on n-well and on shallow-trench isolation, STI, respectively, vs. operation voltage, V_{op}.

3.4 Time response and afterpulsing

Direct displaying of the voltage measurements vs. time indicates the afterpulsing behaviour of a sensor. Figure 9 shows the noise counts shortly after the pulse generation. Although the probes capacitance introduces some deformation of the signal, this type of graph also allows analyzing in some detail the time response of the sensor. For different technologies there are clear different timing behaviours, which can be summarized in a longer quenching time for n-well structures as compared to trench based ones, what favours the probability of charge trapping and increases afterpulsing. When the quenching and readout circuits are introduced into the ASIC, this quenching time can be reduced, as well as the probability of afterpulsing. Another noticeable remark is the longer dead time for n-well than for trench structures, which should be reduced by implanting active-recharge systems.

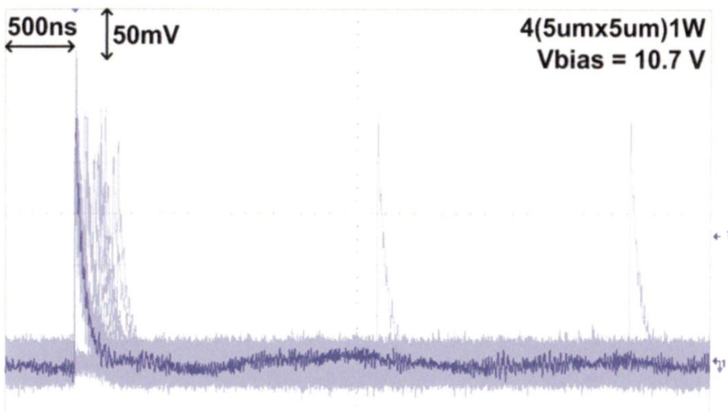

Fig. 9. Afterpulsing and noise visible in direct displaying vs. time.

3.5 Sensitive areas and cross-talk in APD arrays

A novel method has been recently reported to characterize APD-arrays response with high space resolution, whose results suggest that the fill factor can be drastically improved via good control of the doping profiles (Vilà et al., 2011). This research represents a novel effort to use APDs to construct an active pixel detector for charged particles in colliders. The key issue consist on using a focused ion beam/scanning electron microscopy, FIB/SEM setup to precisely characterize the sensitive areas and the quantum efficiency. The beams provided by FIB/SEM setup are nanometric, in contrast to usual test beams that lighten the detectors with lasers providing spots of several tens of microns -comparable to detector dimensions.

A FIB/SEM setup allows a beam to be accelerated, collimated into a nanometric spot and focused with nanometric precision onto one pixel, as well as to perform a detection experiment without trigger setup, thanks to the controlled amount of particles injected per measuring time. The beam is scanned from a pixel to its neighbour one, as shown in figure 10-left, and the agreement between the mathematical description of the detection events and the measurement corroborates both. The measured detection profiles are as shown in figure 10-right, where different applied bias are represented. The position and separation between adjacent pixels is marked onto the profiles taking onto account their distance on the layout, the width of the guard rings and their lateral diffusion.

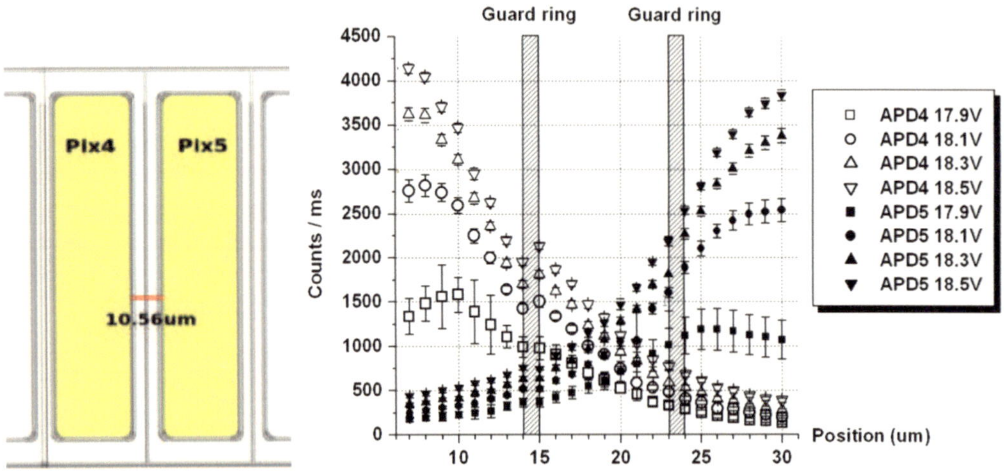

Fig. 10. Layout of the pixels used for sensitive-area and crosstalk demonstration (left). Detection frequencies measured by scanning a 30kV electron beam onto these two neighbouring pixels (right).

It is important to remark that this new characterization is only useful with a detailed functional simulation. The qualitative agreement between Monte Carlo calculations and experimental detection frequencies under a low-flow electron beam (Vilà et al., 2011) avails the correctness of both and suggests good efficiency for the APD prototypes. The nanometric dimension of the beam, as well as its accurate positioning onto the APD array, allows a detailed study of the dead areas around pixels, where the high detection rate observed suggests a noticeable efficiency of the guard ring as detector. This situation could improve drastically the fill factor of the devices, which could virtually arrive to 100%.

4. Circuitry involved in the G-APD operation

The improvement of the sensor performances has been strongly influenced by the development of the related electronics. In the case of dedicated technologies, the quenching electronics is implemented externally to the sensor. In the case of monolithic technologies, the quenching and any additional signal processing can be done close to the sensor, allowing an increase of functionality while reducing area, consumption and power requirements. In this section, basic circuits will be revised, starting from some SPICE models for describing the behaviour of the APDs.

4.1 Models for APD detectors

In order to develop a complete electronic system involving the sensor, an electrical model will be necessary for it. Few models are already available to reproduce partially the behaviour of the APD, but they do not offer a complete vision of the sensor. Recent developments are summarized in this section.

Since the initial investigations of the p-n junction characteristics, developed after the Shockley prediction (Shockley, 1961) and confirmed on the early '60s (Goetzberg et al., 1963), the multiplication behaviour of such structures working in Geiger mode generated a great expectation. The first electrical model was developed as a consequence of these studies (Haitz, 1965) and has been largely used. The model contemplates the parasitic capacitances of the device and reproduces the junction behaviour when a photon is detected. Recently, a second model was presented, using programmable voltage switches (Dalla Mora et al., 2007). In this model, a voltage programmable source is used to reproduce the different behaviours of the APD current depending on the bias. It also contemplates the self-sustaining process reached in APDs when the current rises to a certain threshold. An alternative to this model includes a voltage-controlled switch to reproduce both the static and dynamic behaviours (Mita et al., 2008). The switch is defined in Verilog-A language, and the model also deals with the dependence of the device capacitance on the reverse bias. A newer modification includes the current-voltge behaviour of a diode in the whole polarization range. It adds forward behaviour and the second breakdown (Zappa et al., 2009). The model is presented for both SPICE and Virtuoso Spectre simulators. All these three fundamental models are presented in figure 11.

However, all the models presented so far lack to consider uncorrelated (dark counts) and pulse-correlated (afterpulsing) noises. The electronics used after the sensor has to be designed considering that these false hits must be stored for postprocessing. The first model including diode behaviour for all the polarization range and noise reproduction has been presented recently (Arbat et al., 2010b). Its validity is demonstrated by adjusting the measurements obtained after the characterization of two different APD structures fabricated with a standard 130nm CMOS processing.

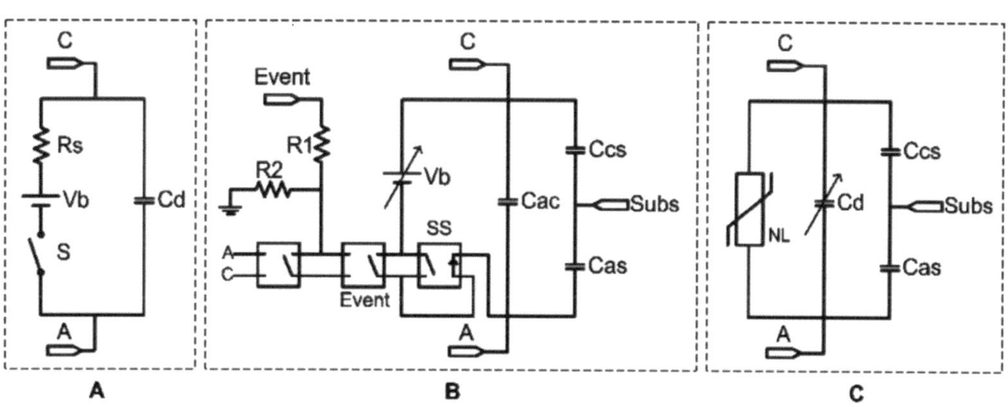

Fig. 11. Models for APDs operation: A) (Haitz et al., 1965), B) (Dalla Mora et al., 2007) and C) (Mita et al., 2008).

This new model uses a source with a noise pattern, together with other active and passive components from Spectre (Cadence™), to reproduce the noise behaviour in the time domain. (Arbat, 2010b). The noise transitory analysis available from the Cadence™ software since version 2008 (Cadence, 2008) allows the possibility of reproducing the actual sensor behaviour. Including the noise behaviour in the model it is possible to obtain information about the probability of information losses due to the dead time related to the noise generation. It also allows studying the strategies that can be used for avoiding the noise interference into the signal, such as gated-mode acquisition or active quenching and recharge circuits. Finally, the analysis of architectures for array readout considering the generated noise, indistinguishable from the signal, is also possible.

The model is represented in figure 12. It contains the elements presented in the previous models, such as anode-cathode capacitor, C_{ac}, the anode-bulk capacitor, C_{ab}, and the cathode-bulk capacitor, C_{cb}. However, the different I-V regions are reproduced with three different groups of resistor, voltage source and voltage controlled switch. The first branch models the forward behaviour of the diode, V_f, allowing the current flow when the forward bias is higher than the bandgap. The second branch represents the second breakdown (Vb2) related to the breakdown of the lateral sensor junctions.

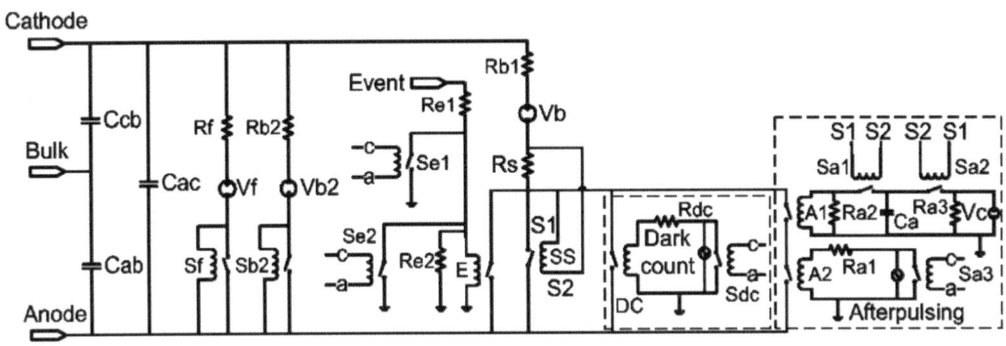

Fig. 12. APD model containing I-V behaviour and noise sources for dark counts and afterpulsing.

Finally, the third branch corresponds to the Geiger-mode region (V_b), and has been modified from the previous models to include noise effects. It offers four possible ways to allow conductivity between both terminals: photon/particle arrival (E), self-sustaining avalanche (SS), dark-count event (DC) and afterpulsing generation (A1-A2). To reproduce the spurious nature of dark-count pulses, the source of the model generates a random noise signal with the frequencies obtained during the sensor characterization. Similarly, to reproduce afterpulsing events, the sensing resistor Rs is used to sense the current flowing between anode and cathode during an avalanche.

The novelty of this model is to add these sources of noise to an electrical model of the sensor, and its validity has been tested by comparison with experimental measurements obtained after characterization of two APD structures fabricated with a standard 130nm CMOS process. Figures 13 A, B and C summarize the obtained results, demonstrating the validity of this newest model for APDs.

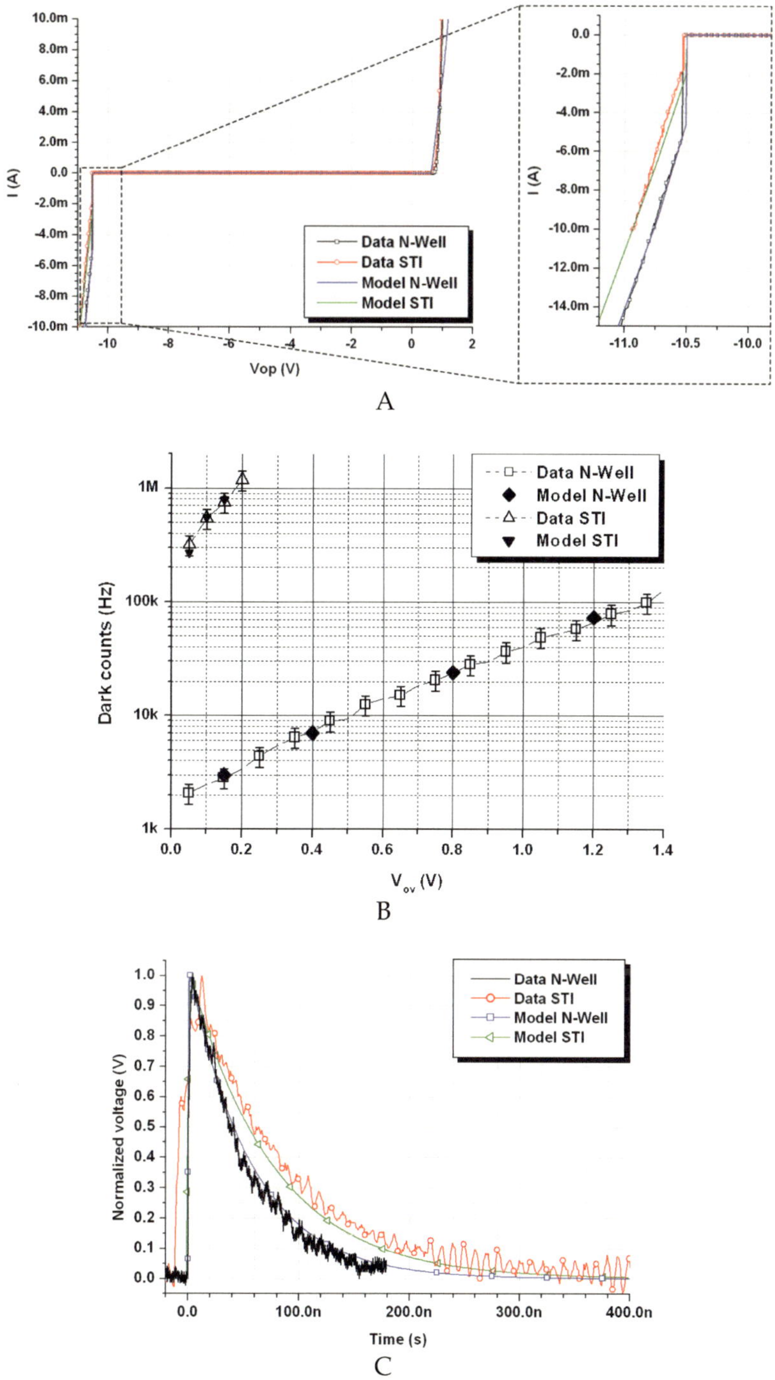

Fig. 13. A) I-V measurements and simulation using the models for an N-well and a STI structures. B) Dark counts measured for the N-well and the STI structures with a 10 kΩ quenching resistance. C) Time response of the two structures and the models represented with a normalized voltage and a quenching resistance of 10 kΩ.

4.2 Quenching and recharge circuits

The dead time, i.e. the total time between the start of a pulse and the resetting of the bias voltage after quenching and recharge, limits the device speed. Many different circuits have been developed to minimize this time, which can be divided into active, passive or combination of both. In all cases, the avalanche reading is done by converting analogue signal to digital by means of a comparator.

As the simplest approximation, a passive quenching circuit (Cova et al., 1996) can be obtained by connecting a high-valued quenching resistance, R_q, in series to the APD (fig. 14). As the avalanche current starts to flow, the resistance forces the operating voltage, V_{op}, to drop down to the breakdown voltage, V_{bd}, quenching the avalanche. This process is not instantaneous because of the parasitic capacitance of the diode, but follows an exponential decay determined by R_qC_{APD} and the avalanche current. Also the circuit recovery after quenching is exponential, but slower owing to the recharge current is much lower than the avalanche current. The sum of the quenching, T_q, and recharge, T_r, times corresponds to the dead time, T_{dead}, which in this case is quite long.

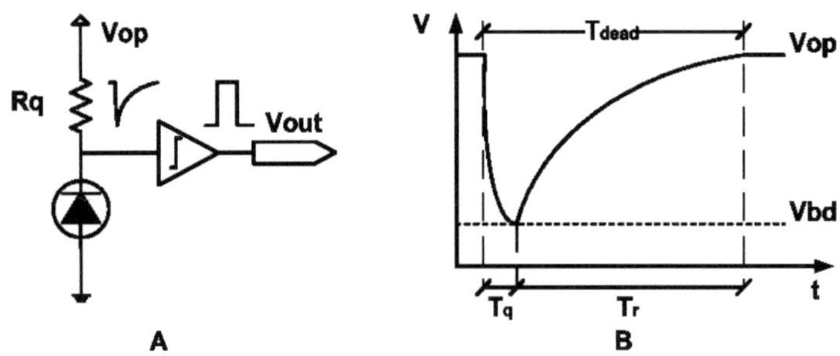

Fig. 14. A) Both passive quenching and recharge circuits. B) Sketch of the timing behaviour.

The next development of the passive quenching for integrated APDs corresponds to the use of a saturated MOS transistor as quenching resistance (figure 15). During the recharge period, the transistor works as a variable resistor, thanks to the change in the biasing.

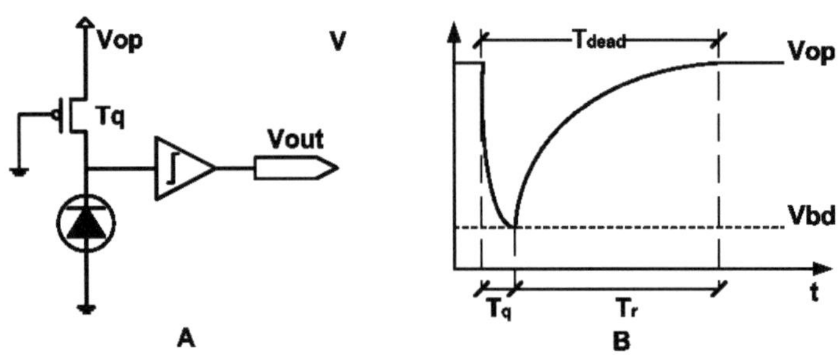

Fig. 15. A) Passive quenching by a transistor and detecting circuits. B) Sketch of the timing behaviour.

As the main contribution to the dead time is due to the recharge stage, a first improvement can be obtained by using active recharge. A simple circuit to provide it consists of an additional MOSFET with low capacitance as switch in parallel to the quenching resistor.

The circuit in figure 16 includes a 50Ω resistor R_s to match impedances with the measuring circuit. When the avalanche is detected, the comparator generates signal, the switch is closed and the sensor recovers the operating voltage, V_{op}. When the process is finished, the switch is opened again and the detector becomes ready for a new detection. This implementation reduces the dead time, but it can increase the afterpulsing noise, as the traps filled during the avalanche may not be released.

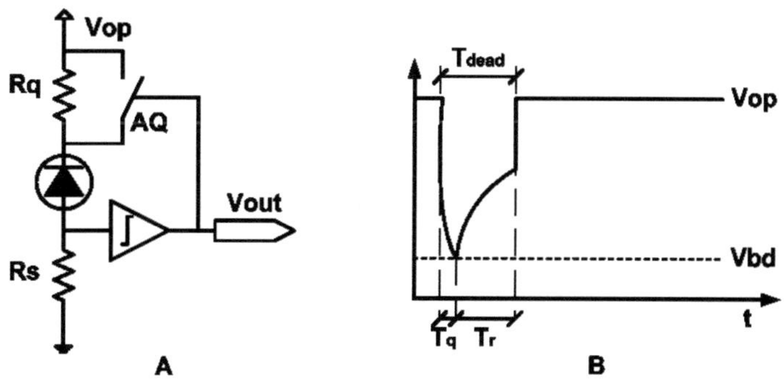

Fig. 16. A) Passive quenching and active recharge by a switch AQ. B) Sketch of the timing behaviour.

Using an active quenching can reduce afterpulsing noise, as the probability for a charge to be trapped is proportional to the duration of the avalanche event. A simple way to generate an active quench is to connect a switch between the APD and a voltage source below the breakdown (figure 17). When the avalanche is detected, the switch is closed and the sensor is quickly biased to this low voltage, producing fast avalanche quenching and limiting the charge through the detector.

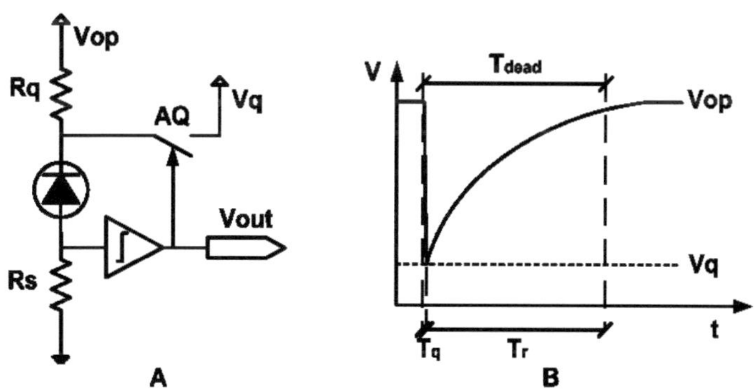

Fig. 17. A) Active quenching with passive recharge. B) Sketch of the timing behaviour.

The most complete approach to improve timing consists of both active quenching and recharge (figure 18). When an avalanche pulse is sensed, the comparator activates the voltage driver, which switches AQ and lowers the biasing down to the breakdown voltage. After a hold-off time (not represented in the figure), the bias voltage is switched back to the operating level by using AR. This circuit reduces the sensor dead time allowing to reach rates of million counts per second while maintaining the high resolution in photon timing.

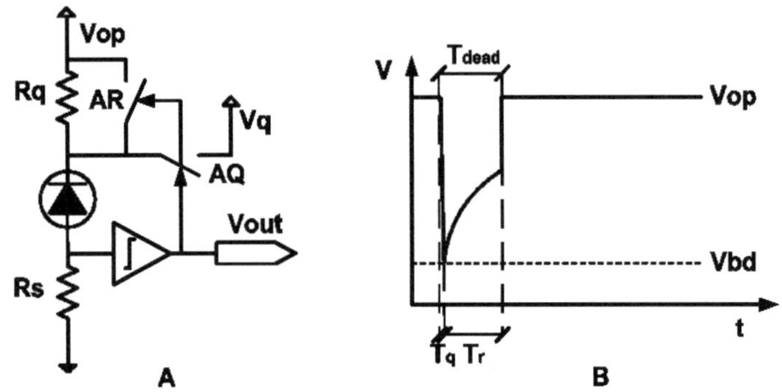

Fig. 18. A) Active quenching by switch AQ and recharge by switch AR. B) Sketch of the timing behaviour.

The development of improved active quenching and recharge circuits has been important. Integrated circuits including improvements of this simple approach have been reported (Zappa et al., 2000 and 2002) involving mixed passive-active circuit configurations that combined integrated circuit with external components (figure 19).

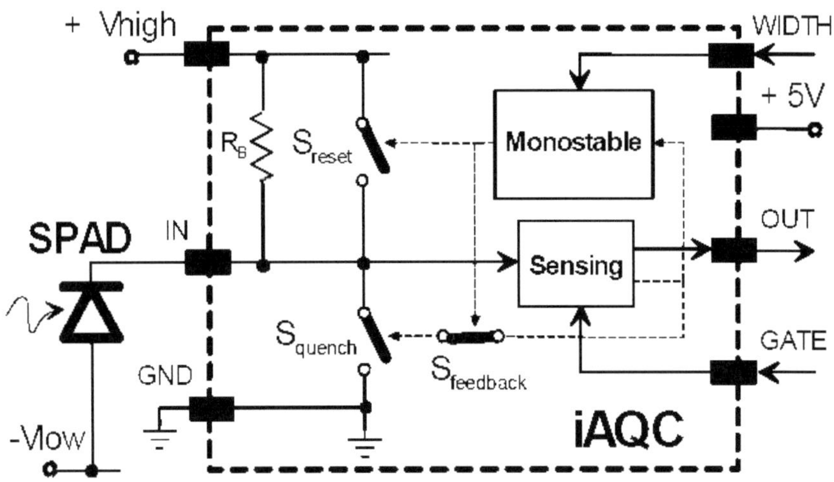

Fig. 19. Integrated active quenching and recharge structure (Zappa et al., 2002).

4.3 Operation modes and readout methods

APDs can be always reverse-biased at a fixed voltage above V_{BD}, so that they are always ready to detect (free running). However, in those applications where the signal arrival time is known, the sensor can also be activated only in certain periods (gated acquisition). In contrast with the free-running operation mode, in the gated acquisition mode the reverse bias voltage swings from over to under V_{BD} to periodically enable and disable the photodiode. In this way, the sensor is kept active only for short periods of time that can be synchronized with the expected signal arrival. Consequently, the probability to detect dark counts interfering with the signal-triggered ones is linearly reduced with the width of the active period of the sensor without missing any photon counts, and hence the signal/noise ratio is improved (Vilella et al., 2011a and b). In addition, long enough non-active periods, longer than the lifetime of the trapping levels, allow to completely eliminate the afterpulsing probability. The basic circuits for running in gated mode can be seen in figure 20.

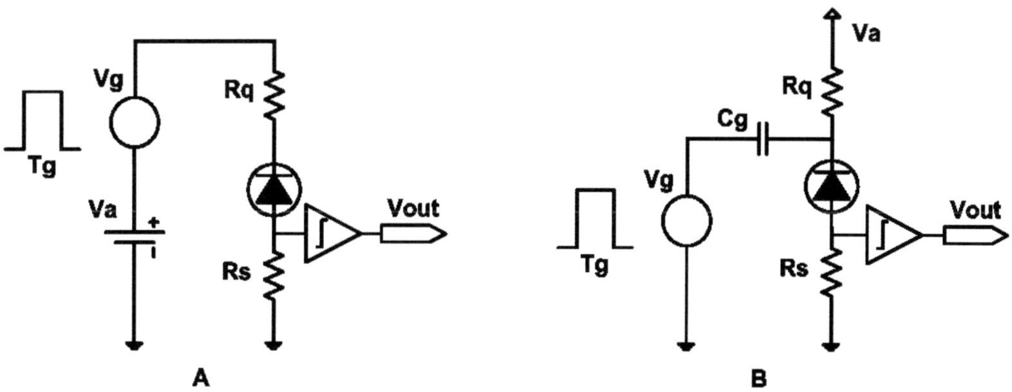

Fig. 20. Circuits for gated-mode operation. A) DC coupling and B) AC coupling.

Recently, a new prototype having the front-end electronics for gated acquisition monolithically integrated with the G-APD has been demonstrated (Vilella et al., 2011a and b). The pixel detector, based on a gated G-APD with passive quenching and active recharge (through RST transistor), is shown in figure 21. The operation mode, together with suitable readout electronics that allows low reverse overvoltage, reduces both the dark-count rate and probability, and eliminates afterpulsing at the pixel. Also, gating with shorter observation periods allows a considerable increase in the sensor dynamic range.

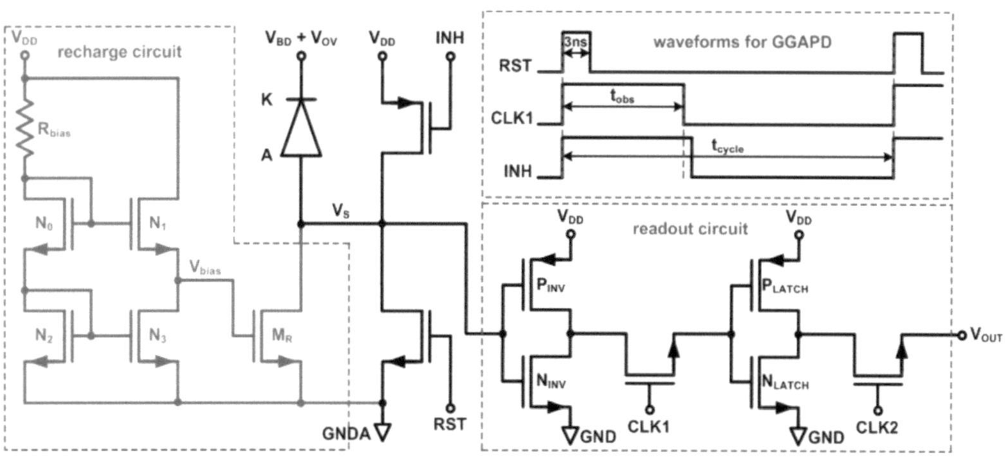

Fig. 21. Schematic diagram of the pixel detector integrated with gated-acquisition electronics, with the waveforms of the gated acquisition, in the inset up right (Vilella et al., 2011b).

Finally, several readout circuits for CMOS G-APDs allowing low-noise operation have been also reported (figure 22), to complete the previous sensor with gated acquisition (Vilella et al., 2010 and 2011b). By using a front-end circuit based on a floating ground, avalanche detections were possible for very low overvoltage, which also reduced the dark-count rate. The characterization of this pixel showed that it could provide acceptable performance even for high-speed applications.

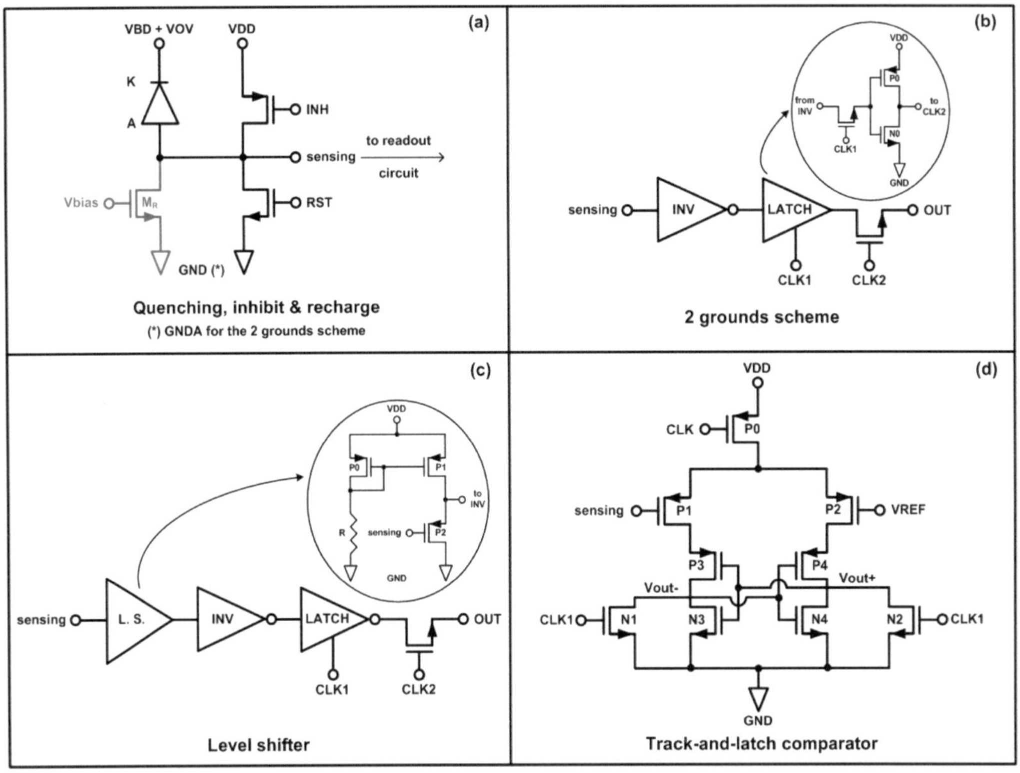

Fig. 22. A) Generic schematic diagram of the structure with gated acquisition. B) Readout circuit based on a 2-grounds scheme, C) on a level shifter and D) in a track-and-latch comparator (Vilella et al., 2011b).

4.4 Readout options for G-APDs arrays

Finally, G-APDs are usually used in large area applications that require their organization in sensor arrays or cameras. The reading options of a camera are mainly related to its final application: for imaging every pixel must be read, while for an even-driven information, such as the position of a particle arrival, only the information of the hit pixels is needed. Consequently, a detector array can be read sequentially, i.e. all the elements are read in an ordered way, or event-driven, what means reading only the pixels in which some signal has been detected. The sequential readout can also be performed with one or several output lines (figure 23 A and B), being possible to adjust the data-transfer speed to the application-required speed. In this case, every pixel contains the basic quenching and recharging circuits together with the elements for selecting row and column (Niclass & Charbon, 2005). On the contrary, in event-driven readout only the pixels that have received a pulse are read (figure 23 C), and this is optimal when only a small part of the array receives information. In this case, the electronics becomes more complicated than for sequential readout, but the timing is noticeably reduced (Niclass et al., 2006a).

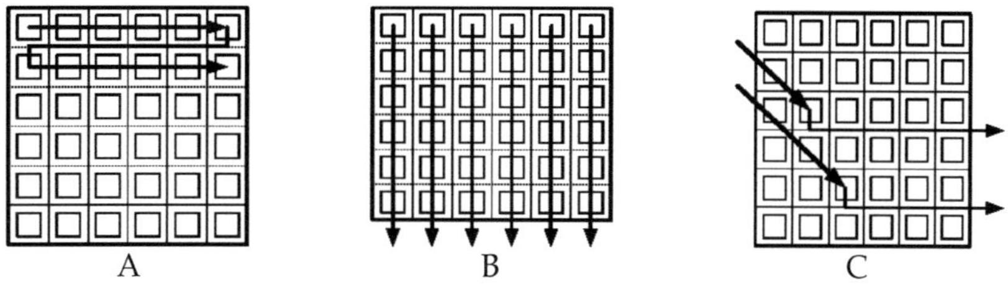

Fig. 23. Sequential readout for a detector array, with only one (A) or several (B) outputs. In C, event-driven readout mechanism.

After reading, signal must be processed, and, for some applications, the implementation of a part of this processing in the pixel can be very interesting (figure 24). In this case, the pixel-contained electronics becomes complicated, and readout of the whole camera after the complete measurement is required. A simplest circuit could be a pulse counter able to provide the image gray levels (Tisa et al., 2008 and 2009).

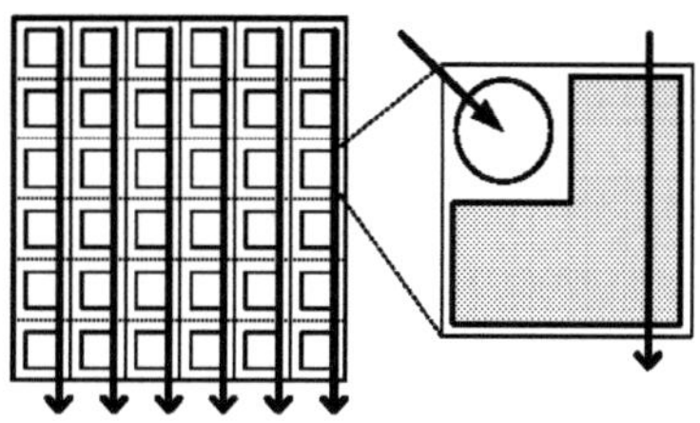

Fig. 24. Electronic processing inside the pixel.

5. Summary and conclusions

A general overview of the main aspects related to Geiger-mode avalanche photodiodes has been exposed in this chapter. Those sensors have been largely used for single photon detection, and nowadays are also used for X- and gamma-rays detection and high-energy particles. After the revision of the main concepts related to their operation, dedicated and standard technologies have been considered from the historical review of their respective evolution. The different aspects corresponding to the different types of sensors, the related circuits and the measurement and definition of the principal figures of merit of the sensors have been presented. Recent developments of the scientific community related to the use of G-APDs have also been exposed here.

6. Acknowledgment

This work has been partially supported by the Spanish projects "Desarrollo de nuevas tecnologías en aceleradores y detectores para los futuros colisionadores de física de partículas", coded FPA2008-05979-C04-02 and "Desarrollo de nuevos detectores para los futuros colisionadores en Física de Partículas", coded FPA2010-21549-C04-01.

7. References

Alexander, S.B. (1997). *Optical communication receiver design (SPIE tutorial texts in optical engineering Vol. TT22)*, SPIE publications, IEE Telecommunications Series Vol. 37, ISBN 0-8194-2023-9, Bellingham WA

Arbat, A., Trenado, J., Gascon, D., Vilà, A., Comerma, A., Garrido, L. & Dieguez, A. (2010a). High voltage vs. high integration: a comparison between CMOS technologies for SPAD cameras. *Proc. of SPIE Optics and Photonics 2010*, Vol. 7780, pp. 77801G, ISSN 0277-786X, San Diego CA, August 2010

Arbat, A. (2010b). Towards a forward tracker detector based on Geiger mode avalanche photodiodes for future linear colliders. PhD dissertation, December 2010

Assefa, S., Xia, F. & Vlasov, Y.A. (2010). Reinventing germanium avalanche photodetector for nanophotonic on-chip optical interconnects. *Nature*, Vol. 464, (March 2010), pp. 80-85, ISSN 0028-0836

Cadence. (2008). Application notes on direct time-domain noise analysis using virtuoso spectre, version 1.0

Charbon, E. (2007). Will avalanche photodiode arrays ever reach 1 megapixel?. *Proc. 2007 International Image Sensor Workshop*, pp. 246-249, Ogunquit ME, 2007

Charbon, E. (2008). Towards large scale CMOS single-photon detector arrays for lab-on-chip applications. *Journal of Physics D: Applied Physics* Vol. 41, (2008), pp. 094010-1:9, ISSN 0022-3727

Cova, S., Ghioni, M., Lacaita, A., Samori, C. & Zappa, F. (1996). Avalanche Photodiodes and Quenching Circuits For Single-Photon Detection. *Applied Optics*, Vol. 35, No. 12 (April 1996), pp. 1956-1976, ISSN 1943-8206

Cova, S., Ghioni, M., Lotito, A., Rech, I. & Zappa, F. (2004). Evolution and prospects for single-photon avalanche diodes and quenching circuits. *Journal of Modern Optics*, Vol. 51, No. 9-10 (June-July 2004), pp. 1267-1288, ISSN 0950–0340

Dalla Mora, A., Tosi, A., Tisa, S. & Zappa, F. (2007). Single-Photon Avalanche Diode Model for Circuit Simulations. *IEEE Photonics Technology Letters*, Vol. 19, No. 23 (December 2007), pp. 1922-1924, ISSN 1041-1135

Dandin, M., Nelson, N., Saveliev, V., Ji, H., Abshire, P. & Weinberg, I. (2007). Single photon avalanche detectors in standard CMOS. *IEEE Sensors '07*, ISSN 1930-0395, pp. 585-588, Atlanta, October 2007

Dautet, H., Deschamps, P., Dion, B., MacGregor, A.D., MacSween, D., McIntyre, R.J., Trottier, C. & Webb, P.P. (1993). Photon counting techniques with silicon avalanche photodiodes. *Appl. Opt.*, Vol. 32, No. 21 (1993), pp. 3894-3900, ISSN 2155-3165

Dussault, D. & Hoess, P. (2004). Noise Performance Comparison of ICCD with CCD and EMCCD Cameras, *Proceedings of SPIE*, Vol. 5563, pp. 195-204, ISBN 0-8194-5501-6, Denver, August 2004

EG&G (1996) Canada, Vaudreuil, P.Q., Canada, SPCM-AQ Series Data Sheet ED-0043/11/96, 1996

Faramarzpour, N., Deen, M.J., Shirani, S. & Fang, Q. (2008). Fully integrated single-photon avalanche diode detector in standard CMOS 0.18 μm technology. *IEEE Transactions on electron devices*, Vol. 55, No. 3 (March 2008), pp. 760-767, ISSN 0018-9383

Finkelstein, H., Hsu, M.J. & Esener, S.C. (2006a). STI-Bounded single-photon avalanche diode in a deep-submicrometer CMOS technology. *IEEE Electron Device Letters*, Vol. 27, No. 11 (November 2006), pp. 887-889, ISSN 0741-3106

Finkelstein, H., Hsu, M.J. & Esener, S. (2006b). An ultrafast Geiger-mode single photon avalanche diode in 0.18μm CMOS technology. *Advanced photon counting techniques, Proc. of SPIE*, Vol. 6372, ISBN 9780819464705, pp. 63720W-1:10, Boston MA, October 2006

Gersbach, M., Niclass, C., Charbon, E., Richardson, J., Henderson, R. & GRANT, L. (2008). A single photon detector implemented in a 130 nm CMOS imaging process. *Proc. of the 38th European Solid-State Device Research Conference, ESDERC 2008*, pp. 270-273, ISSN 1930-8876, Edinburgh, September 2008

Gersbach, M., Richardson, J., Mazaleyrat, E., Hardillier, S., Niclass, C., Henderson, R., Grant, L. & Charbon, E. (2009). A low-noise single-photon detector implemented in a 130 nm CMOS imaging process. *Solid-State Electronics,* Vol. 53, (May 2009), pp 803-808, ISSN 0038-1101

Ghioni, M., Cova, S., Lacaita, A. & Ripamonti, G. (1988). New silicon epitaxial avalanche diode for single-photon timing at room temperature, *Electron. Lett.*, Vol. 24, No. 24 (Nov. 1988), pp 1476-1477, ISSN 0013-5194

Ghioni, M. & Ripamonti, G. (1991). Improving the performance of commercially available Geiger-mode avalanche photodiodes. *Rev. Sci. Instrum.*, Vol. 62, No. 1 (January 1991), pp. 163-167, ISSN 0034-6748

Ghioni, M., Gulinatti, A., Rech, I. & Cova, S. (2006). Recent advances in silicon single photon avalanche diodes and their applications. *Annual Meeting of the IEEE Lasers and Electro-Optics Society, LEOS 2006*, ISBN 0-7803-9555-7, pp. 719-720, Montreal, October 2006

Ghioni, M., Gulinatti, A., Rech, I., Zappa, F. & Cova, S. (2007). Progress in silicon single-photon avalanche diodes. *IEEE Journal of selected topics in quantum electronics,* Vol. 13, No. 4 (July-August 2007), pp. 852-862, ISSN 1077-260X

Goetzberg, A., McDonald, B., Haitz, R. & Scarlett, R. (1963). Avalanche effects in silicon p-n junctions. II. Structurally perfect junctions. *Journal of App. Phys.*, Vol. 34, (1963), pp. 1591-1600 ISSN 0021-8979

Graugés, E., Comerma, A., Garrido, L., Gascón, D., Trenado, J., Diéguez, A., Vilà, A., Arbat, A., Freixas, L., Hidalgo, S., Fernández, P., Flores, D. & Lozano, M. (2010). Study of Geiger avalanche Photo-diodes (Galds) applications to pixel Trucking detectors. *Nucl. Inst. Met. Phys. Res. A,* Vol. 617, (2010), pp. 541-542, ISSN 0168-9002

Guerrieri, F., Tisa, S. & Zappa, F. (2009). Fast Single-Photon imager acquires 1024 pixels at 100 kframe/s. *Proceedings of SPIE,* Vol. 7249, pp. 72490U-1 - 72490U-11, ISSN 0277-786X, San José CA, January 2009

Gulinatti, A., Maccagnani, P., Rech, I., Ghioni, M. & Cova, S. (2005). 35 ps time resolution at room temperature with large area single photon avalanche diodes. *Electronic letters,* Vol. 41, No. 5 (March 2005), pp. 272-274, ISSN 0013-5194

Haitz, R. (1964). Model for the electrical behavior of a microplasma. J. Appl. Phys., Vol. 35, (1964), pp. 1370-1376 ISSN 0021-8979

Haitz, R. (1965). Mechanisms contributing to the noise pulse rate of avalanche diodes. J. Appl. Phys., Vol. 36, (1965), pp. 3123-3131, ISSN 0021-8979

Hamamoto, T. (1991). Sidewall damage in a silicon substrate caused by trench etching. *Applied physics letters,* Vol. 58, No. 25 (June 1991), pp. 2942-2944, ISSN 0003-6951

Jackson, J.C., Morrison, A.P. & Lane, B. (2000). Characterization of large area SPAD detectors operated in avalanche photodiode mode. *IEEE 13th Annual Meeting Laser and Electro-Optics Society, LEOS 2000,* Vol. 1, pp. 17-18, ISBN 0-7803-5947-X, Rio Grande, November 2000

Jackson, J.C., Morrison, A.P., Hurley, P., Harrell, W.R., Damjanovic, D., Lane, B. & Mathewson, A. (2001). Process monitoring and defect characterization of single photon avalanche diodes. *Proc. of the 2001 International Conference on Microeletronic Test Structures,* Vol. 14, pp. 165-170, ISBN 0-7803-6511-9, Kobe, March 2001

Jackson, J.C., Morrison, A.P., Phelan, D. & Mathewson, A, (2002). A novel silicon Geiger-mode avalanche photodiode. *Digest. International Electron Devices Meeting, IEDM'02,* ISBN 0-7803-7462-2, pp. 797-800, San Francisco CA, December 2002

Karami, M.A., Gersbach, M., Yoon, H.J. & Charbon, E. (2010). A new single-photon avalanche diode in 90nm standard CMOS technology. *Optics Express*, Vol. 18, No. 21 (October 2010), pp. 22158-22166, ISSN 0146-9592

Kindt, W.J. (1994). A novel avalanche photodiode array. *Nuclear Science Symposium and Medical Imaging Conference,* 1994 IEEE Conference Record, Vol. 1, (Oct-Nov 1994), pp. 164-167 ISBN 0-7803-2544-3, Norfolk, VA, Oct-Nov 1994

Lacaita, A., Cova, S. & Ghioni, M. (1988). Four-hundred-picosecond single-photon timing with commercially available avalanche photodiodes. *Rev. Sci. Instrum.,* Vol. 59, no. 7 (July 1988), pp. 1115-1121, ISSN 0034-6748

Lacaita, A., Ghioni, M. & Cova, S. (1989) Double epitaxy improves single-photon avalanche diode performance. *Electron. Lett.,* Vol. 25, No. 13 (June 1989), pp. 841-843, ISSN 0013-5194

Lacaita, A.L., Zappa, F., Bibliardi, S. & Manfredi, M. (1993). On the bremsstrahlung origin of hot-carrier-induced photons in silicon devices. *IEEE Transactions on Electron Devices,* Vol. 40, No. 3 (March 1993), pp. 577-582, ISSN 0018-9383

McIntyre, R. (1972). The distribution of gains in uniformly multiplying avalanche photodiodes: Theory, *IEEE trans. Electron Devices,* Vol. 19, No. 6 (June 1972), pp 703-713, ISSN 0018-9383

McIntyre, R. (1990). Silicon avalanche photodiode with low multiplication noise. *US patent* n° 4,972,242"

McIntyre, R. & Webb, P. (1996). Low-noise, reach-through, avalanche photodiodes. *US Patent* n° 5,583,352"

McKay, K.G. (1954). Avalanche breakdown in silicon. *Phys. Rev.,* Vol. 94, No. 4 (1954), pp. 877–884, ISSN 1943-2879

Mita, R., Palumbo G. & Fallica, P.G. (2008). Accurate model for single-photon avalanche diodes. *IET circuits devices systems,* Vol. 2, No. 2 (April 2008), pp. 207–212, ISSN 1751-858X

Mosconi, D., Stoppa, D., Pancheri, L., Gonzo, L. & Simoni, A. (2006). CMOS single-photon avalanche diode array for time-resolved fluorescence detection. *Proc. of the 32nd European Solid-State Circuits Conference, ESSCIRC'06,* ISSN 1930-8833, pp. 564-567, Montreux, September 2006

Niclass, C. & Charbon, E. (2005). A single photon detector array with 64x64 resolution and millimetric depth accuracy for 3D imaging. *IEEE International Solid-State Circuits Conference, ISSCC 2005,* pp. 364-366, ISSN 0193-6530, San Francisco, February 2005

Niclass, C., Sergio, M. & Charbon, E. (2006a). A CMOS 64x48 single photon avalanche diode array with event-driven readout. *Proc. of the 32th European SolidState Circuits Conference, SSCIRC'06,* pp. 556-559, ISBN 1-4244-0303-4, Montreux, September 2006

Niclass, C., Sergio, M. & Charbon, E. (2006b). A single photon avalanche diode array fabricated in 0.35μm CMOS and based on an event-driven readout for TCSPC experiments. *Proc. of SPIE, Advanced Photon Counting Techniques,* Vol. 6372, ISBN 9780819464705, pp. 63720S-1:12, Boston MA, October 2006

Niclass, C., Gersbach, M., Henderson, R., Grant, L. & Charbon, E. (2007a). A single photon avalanche diode implemented in 130 nm CMOS technology, *IEEE Journal of selected topics in quantum electronics,* Vol. 13, No. 4 (July/August 2007), pp. 863-869, ISSN 1077-260X

Niclass, N., Gersbach, M., Henderson, R.K., Grant, L. & Charbon, E. (2007b). A 130 nm CMOS single photon avalanche diode. *Proceedings of SPIE,* Vol. 6766 (2007), pp. 676606, ISSN 0277-786X, Boston, September 2007

Niclass, C., Favi, C., Kluter, T., Monnier, F. & Charbon, E. (2009). Single-photon synchronous detection. *IEEE Journal of Solid-State Circuits,* Vol. 44, No. 7 (July 2009), pp. 1977-1989, ISSN 0018-9200

Pancheri, L. & Stoppa, D. (2007). Low-noise CMOS single-photon avalanche diodes with 32 ns dead time. *37th European Solid State Device Research Conference, ESSDERC 2007,* ISSN 1930-8876, pp. 362-365, Munich, September 2007

Renker, D. (2004). Photosensors. *Nucl. Inst. and Meth. in Phys. Res. A,* Vol. 527, No. 1-2, (July 2004), pp. 15-20, ISSN 0168-9002

Richardson, J.A., Grant, L.A. & Henderson, R.K. (2009). Low dark count single-photon avalanche diode structure compatible with standard nanometer scale CMOS technology. *IEEE Photonics Technology Letters,* Vol. 21, No. 14 (July 2009), pp. 1020-1022, ISSN 1041-1135

Rochas, A., Pauchard, A.R., Besse, P.A., Pantic, D., Prijic, Z. & Popovic, R.S. (2002). Low-Noise silicon avalanche photodiodes fabricated in conventional CMOS technologies. *IEEE Transactions on Electron Devices,* Vol. 49, No. 3 (March 2002), pp. 387-394, ISSN 0018-9383

Rochas, A., Gani, M., Furrer, B., Besse, P.A., Popovic, R.S., Riborby, G. & Gisin, N. (2003a). Single photon detector fabricated in a complementary metal-oxide-semiconductor high-voltage technology. *Rev. Sci. Instrum.,* Vol. 74, No. 7 (July 2003), pp. 3263-3270, ISSN 0034-6748

Rochas, A., Gösch, M., Serov, A., Besse, P.A., Popovic, R.S., Lasser, T. & Rigler, R. (2003b). First Fully Integrated 2-D Array of Single-Photon Detectors in Standard CMOS Technology. *IEEE Photonics Technology Letters,* Vol. 15, No. 7 (July 2003), pp. 963-965, ISSN 1041-1135

Sergio, M., Niclass, C. & Charbon, E. (2007). A 128x2 CMOS single-photon streak camera with timing-preserving latchless pipeline readout. *IEEE International Solid-State Circuits Conference, ISSCC 2007,* pp. 394-396, ISSN 0193-6530, San Francisco, February 2007

Shockley, W. & Read, W.T. (1952). Statistics of the recombinations of holes and electrons. *Phys. Rev.,* Vol. 87, No. 5 (September 1952), pp. 835-842

Shockley, W. (1961). Problems related to p-n junctions in silicon. *Solid-State Electron.* Vol. 2, No. 1 (January 1961), pp. 35-60, ISSN 0038-1101

Spinelli, A., Davis, L.M. & Dautet, H. (1996). Actively quenched single photon avalanche diode for high repetition rate time-gated photon counting. *Rev. Sci. Instrum.,* Vol. 67, No. 1 (January 1996), pp. 55-61, ISSN 0034-6748

Spinelli, A., Ghioni, M.A., Cova, S.D. & Davis, L.M. (1998). Avalanche detector with ultraclean response for time-resolved photon counting. *IEEE Journal of Quantum Electronics,* Vol. 34, No. 5 (May 1998), pp. 817-821, ISSN 0018-9197

Stapels, C.J., Lawrence, W.G., Augustine, F.L. & Christian, J.F. (2006). Characterization of a CMOS Geiger photodiode pixel. *IEEE Transactions on electron devices,* Vol. 53, No. 4 (April 2006), pp. 631-635, ISSN 0018-9383

Stapels, C.J., Squillante, M.R., Lawrence, W.G., Augustin, F.L. & Christian, J.F. (2007). CMOS-based avalanche photodiodes for direct particle detection. *Nucl. Inst. and Meth. In Phys. Res. A,* Vol. 579, No. 1 (2007), pp. 94-98, ISSN 0168-9002

Stoppa, D., Pancheri, L., Scandiuzzo, M., Simoni, A., Viarani, L. & Dalla Betta, G.F. (2005). A CMOS sensor based on signle photon avalanche diode for distance measurement applications. *Proc. of the IEEE. Instrumentation and measurement technology conference,* Vol. 2, ISBN 0-7803-8879-8, pp. 1162-1165, Ottawa, May 2005

Stoppa, D., Pancheri, L., Scandiuzzo, M., Gonzo, L., Dalla Betta, G.F. & Simoni, A. (2007). A CMOS 3-D imager based on single photon avalanche diode. *IEEE Transactions on Circuits and Systems - I,* Vol. 54, No. 1 (January 2007), pp. 4–12, ISSN 1057-7122

Tisa, S., Guerrieri, F. & Zappa, F. (2008). Variable-load quenching circuit for single-photon avalanche diodes. *Optics express,* Vol. 16, No. 3 (February 2008), pp. 2232-2244, ISSN 0146-9592

Tisa, S., Guerrieri, F. & Zappa, F. (2009). Monolithic array of 32 SPAD pixel for single-photon imaging at high frame rates. (2009). *Nucl. Instr. Meth. Phys. Res. A:*

Accelerators, Spectrometers, Detectors and Associated Equipment, Vol. 610, No. 1 (October 2009), pp. 24-27, ISSN 0168-9002

Vilà, A., Trenado, J., Arbat, A., Comerma, A., Gascon, D., Garrido, Ll. & Dieguez, A. (2011). Characterization and simulation of avalanche photodiodes for next-generation colliders. *Sensors and Actuators A: Physical,* (2011) doi:10.1016/j.sna.2011.05.011

Vilella, E., Arbat, A., Comerma, A., Trenado, J., Alonso, O., Gascon, D., Vilà, A., Garrido, L & Dieguez, A. (2010). Readout electronics for low dark count pixel detectors based on Geiger mode avalanche photodiodes fabricated in conventional CMOS Technologies for future linear colliders. *Nucl. Instr. Meth. Phys. Res. A,* Vol. 650, (2010), pp. 120-124, ISSN 0168-9002

Vilella, E., Comerma, A., Alonso, O. & Dieguez, A. (2011a). Low-noise pixel detectors based on gated Geiger mode avalanche photodiodes. *Electronics Letters,* Vol. 47, No. 6 (March 2011), ISSN 0013-5194

Vilella E., Arbat, A., Comerma, A., Trenado, J., Alonso, O., Gascon, D., Vilà, A., Garrido, L. & Dieguez, A. (2011b). Readout electronics for low dark count Geiger mode avalanche photodiodes fabricated in conventional HV-CMOS technologies for future linear colliders. *Topical Workshop on Electronics for Particle Physics 2010,* pp. C01015, Aachen, September 2011

Webb, P., McIntyre, R. & Conradi, J. (1974). Properties of avalanche photodiodes. *RCA Rev.,* Vol. 35, (June 1974), pp. 234-278, ISSN 0033-6831

Xiao, Z., Pantic, D. & Popovic, R.S. (2007). A new single photon avalanche diode in CMOS high-voltage technology. *International solid-state sensors, actuators and microsystems conference, Transducers 2007,* ISBN 1-4244-0842-3, pp. 1365-1368, Lyon, June 2007

Zappa, F., Ghioni, M., Cova, S., Varisco, L., Sinnis, B., Morrison, A. & Mathewson, A. (1997). Integrated array of avalanche photodiodes for single-photon counting. *Proc. 27th Eur. Solid-State Device Research Conf.,* pp. 600-603, ISBN 2-86332-221-4, Stuttgart, September 1997

Zappa, F., Ghioni, M., Cova, S., Samori, C. & Giudice, A. (2000). An integrated active quenching circuit for single-photon avalanche diodes. *IEEE Transactions on instrumentation and measurement,* Vol. 49, No. 6 (December 2000), pp. 1167-1175, ISSN 0018-9456

Zappa, F., Giudice, A., Ghioni, M. & Cova, S. (2002). Fully-Integrated Active-Quenching Circuit for Single-Photon Detection. *Proceedings of the 28th European Solid State Circuit Conference – ESSCIRC'02,* ISBN 88-900847-9-0, pp. 355-358, Firenze, September 2002

Zappa, F., Tisa, S., Gulinatti, A., Gallivanoni, A. & Cova, S. (2004). Monolithic CMOS detector module for photon counting and picosecond timing. *Proc. of the 34th European Solid-State Device Research Conference, ESSDERC,* ISBN 0-7803-8478, pp. 341-344, Leuven, November 2004

Zappa, F., Tisa, S., Gulinatti, A., Gallivanoni, A. & Cova, S. (2005). Complete single-photon counting and timing module in a microchip. *Optics Letters,* Vol. 30, No. 11 (June 2005), pp. 1327-1329, ISSN 0146-9592

Zappa, Z., Tosi, A., Dalla Mora A. & Tisa, S. (2009). SPICE modeling of single-photon avalanche diodes. *Sensors and Actuators A,* Vol 153, No. 2 (August 2009), pp. 197-204, ISSN 0924-4247

Part 2

Theoretical Modelling and Simulations

10

Avalanche Process in Semiconductor Photo Detectors in the Context of the Feedback Theory

Vasily Kushpil
Nuclear Physics Institute of Academy Science of the Czech Republic
Czech Republic

1. Introduction

The feedback theory is a quite advantageous tool in the process of description of complex systems' behaviour or the procedure of complex processes. We can characterize the complexity of the system from its structural properties when there is a large amount of bounds between many elements of system. We can also characterize the complexity of the system as its functional complexity when the current state of system is defined as a result of many self-consistent states originating during the evolution of the system. It can be said in most general words that the feedback theory (FBT) describes the behaviour of a system or a process when the current status is defined as a result of achieving the self-consistency of the main system parameters. The feedback theory is most fruitfully applied to the description of those complex systems for which, due to their statistical character it is impossible to construct the full physical-mathematical model with a simple solution, but for which it is easy to select statistical parameters of the system and define their connections with the system properties that are of our interest. This approach allows applying FBT to describe the evolution of complex systems in chemistry, physics, biology, sociology and economics [1].

In this paper, we discuss the avalanche process in semiconductor avalanche photo detectors (APD) proceeding from three different points of view on this process. Then, we sum up the conclusions made in relation to the avalanche process in connection with avalanche photo detectors. We present APD as a converter of photo radiation to electric current with further current amplification. As a result, we can interpret the APD operation from various points that will allow us, finally, both to make several interesting conclusions on their practical application and analyze a possibility to choose optimal parameters in APD manufacturing.

To describe efficiently the processes in the frames of the feedback theory, it is necessary to fetch out the main physical parameters that determine the process procedure and establish quantitative ratios between the character of the process procedure and the change of these parameters. Further, we discuss the main characteristic parameter - the multiplication factor of the avalanche photo detectors M as an applied biasing voltage function (of the material parameters and topology) and attempt to describe the conditions for the achievement of self-consistency with the choice of the feedback coefficient function. It should be mentioned that at present FBT is most widely used in the control theory [2], i.e. in the description of the behaviour of the systems for which it is necessary to achieve the desired character of behaviour. It can be usually attained by the choice of the corresponding system feedback function if the system transfer function without feedback is known. From this point of view, the application of FBT should allow us to approach the designing of avalanche photo detectors from the taken characteristics. We will obtain an opportunity to build the necessary APD transfer function and define the necessary feedback function with it. We have to learn to connect the topology and technology with the parameters of the established Feedback Function (FBF) and then, proceeding from the given FBF, we will be able to obtain topological and technological parameters. The practical application of the described physical-technological algorithm is not discussed in this paper. Our task is to show how it is possible to come from different APD models to their generalized description using FBT.

2. Feedback theory basics

Let us regard the main notions of the feedback theory in radio technique. Hereinafter, we will have to describe an avalanche photodiode as an amplifying device with feedback and present it in the form of the corresponding equivalent scheme.

In radio technique the behaviour of a system with feedback is described by the function of a special type - the system function (in a complex form, as the transfer function H(s)). Let us discuss a simple scheme of the signal amplifier shown in Fig.2.1 to whose input a signal is given that depends on time x(t), and the signal y(t) is received in the output.

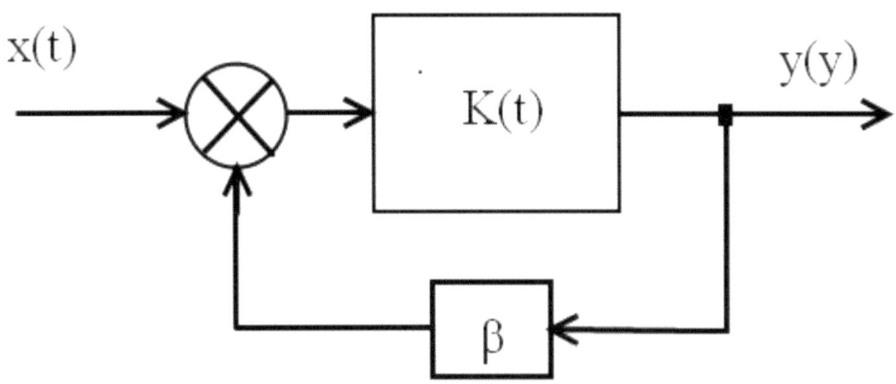

Fig. 2.1. Feedback in amplifier.

The system coefficient of gain function is defined as a ration $K(t)=y(t)/x(t)$. If a part if the signal from the system output is transmitted to the input it is supposed that the feedback is active in the system and the feedback coefficient is defined as $\beta(t)=((x'(t)-x(t))/y(t))$, where $x'(t)$ - is the signal in the system input in the action of the feedback in the system. At $\beta(t)>0$ the feedback is positive; at $\beta(t)<0$ the feedback is negative; at $\beta(t)=0$ there is no feedback in the system. The system gain is written in the system with acting feedback in the form (2.1)

$$K=K_o/(1-\beta^*K_o) \quad (2.1)$$

where K_o is gain of the system without the feedback. Negative feedback (NFB) is most often used in electronic systems as in this case the system stability increases. Besides, providing

K_o>>1 the gain is defined mainly with the feedback coefficient $K \approx 1/\beta$ that allows a desired coefficient of gain of the system be set by the choice of only one parameter β.

It should be stressed that the transfer function (1.1) in radio technology is given in frequency representation [3]. Speaking more accurately, it is regarded as a result of Laplace transformations on the complex plane. In this case the transfer function is equivalent to the frequency characteristic of the chain if active sources are absent.

$$H(s)=Y(s)/X(s)=L[y(t)]/L[x(t)] \tag{2.2}$$

The frequency representation for radio technical amplifying systems is more informative and suitable for their designing and study. We will discuss here only the time representation of the system coefficient of gain (amplifying) for avalanche detectors. The reason is the suitability and simplicity of this representation for the description of the avalanche photo detectors' operation and sufficiently simple procedure of measuring of the APD pulse characteristic. The time representation of the system gain is easily obtained if the APD pulse characteristic is known. In this stage, we do not consider the transfer function for APD, not to make the description and comprehension of the idea to describe the process of avalanche multiplication in the FBT frames too complicated.

3. Probability model of the avalanche multiplication

Let us consider an idealized model of avalanche multiplication of electrons in a homogeneous semiconductor with the following suppositions: electron multiplication takes place in a limited region with the length L_a with constant electron field voltage E_a=const in this region. The probability of multiplication along the L_f electron free flight is P<1. See Figure 3.1.

If there are N_0 electrons at the beginning of multiplication in the output of this region there will be $N >> N_0$ electrons, as a result of multiplication. It can be expressed quantitatively in the form (3.1).

The multiplication factor is defined as $M=N/N_0$

$$N=N_0*(1+P+P^2+P^3+...+P^n) \text{ where } n=L_a/L_f \tag{3.1}$$

For n>>1 this expression can be written in the form

$$N=N_0(1-P)^{-1} \tag{3.2}$$

Formally, let us re-write (2.1) in the form

$$N_x=N_0*(1+(P-ß)+(P-ß)^2+(P-ß)^3+...+(P-ß)^n)=N_0*(1-(P-ß))^{-1} \tag{3.3}$$

In this expression ß<1 and it describes a probability of electron multiplication suppression along its free flight. From expressions (3.1) and (3.3) we get

$$M=(N_x/N_0)=M_0/(1+ß*M_0), \text{ where } \qquad M_0=(1-P)^{-1} \tag{3.4}$$

Therefore, we see that with this simple model the physical sense of the feedback coefficient is obvious. It describes a probability of electron multiplication suppression in its free flight.

It is to imagine a situation when to account for the multiplication probability of not only electrons but also holes, we upgrade the expression (3.3) so as $P=P_e+P_h$ and $ß= ß_e+ß_h$, while P_e, P_h are multiplicity probabilities for electrons and holes in their free flight length. In this case $N_x=N_e+N_h$.

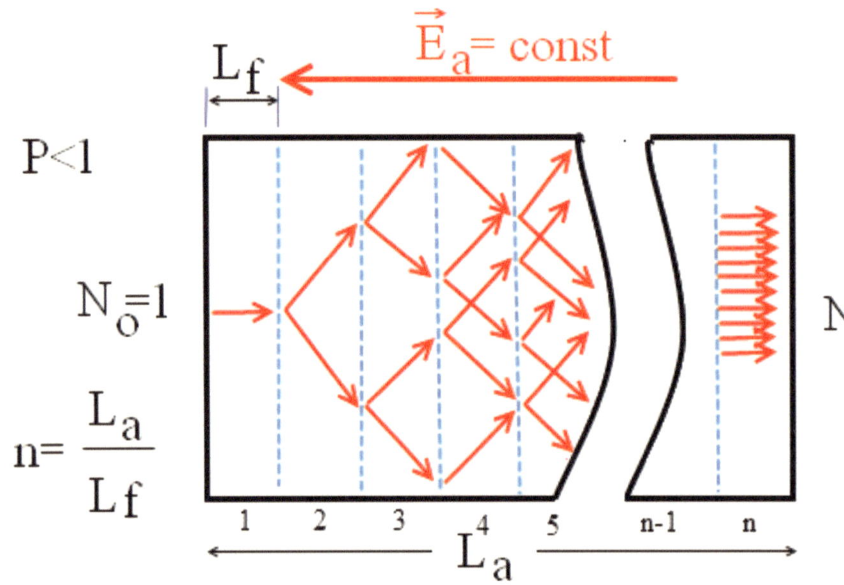

Fig. 3.1. Illustration of Probability model for avalanche process.

This simple model presupposes independence of multiplication processes of electrons and holes. It is not easy to implement it in practice.

$$N_x=N_e+N_h.=N_{0e}*(1-(P_e-ß_e))^{-1}+ N_{0h}*(1-(P_h-ß_h))^{-1} \quad (3.5)$$

We would like to note that it is the first illustration of moving from a certain definite (in this case, simple probability) model for avalanche photo-detectors to further interpretation of this model in the frames of FBT.

We shall discuss in more detail some not very obvious statements of the suggested model. First, we suppose that all electron multiplication events are independent as a whole. The second supposition comes from the sum row notion in the form of the algebraic expression (3.2). This expression has uncertainty at P=1. It leads to a natural restriction P<1 and shows that the multiplication probability can be as close to the entity but cannot be equal to it. Both statements are not significant for the main conclusion – the result of the mathematical description of the avalanche generation process can be presented in the form of a formula for the coefficient of gain of the amplifier with feedback after some identical transformations, and, consequently, we can apply it for the analysis of the operation of FBT avalanche photo-detectors. To illustrate some advantages of the interpretation, we consider (3.4) in the case $M_0>>1$. Then $M\sim1/ß$ and, consequently, the multiplication factor is definitely determined by the feedback coefficient ß. In other words, the avalanche suppression processes in APD define the multiplication factor. As such requirements correspond to the operation principle by convention Geiger Mode Avalanche Photo Diode

(GAPD) [4], the probability interpretation of multiplication coefficient has quite a definite meaning. Namely, as GAPD operates in the region where the field intensity is critical and P~1, the avalanche suppressing should be defined with a certain internal process whose probability should be practically constant to provide for low noise of the detector. The cell structure of GAPD is the simplest and most natural way to develop them, as the cell dimensions and the intensity of the electric field in it define the maximal amplifying. The smaller is the dimension of the cell the better are noise properties of GAPD due to electron fluctuations decrease in it. Further, we will continue the discussion of this subject in the phase of more accurate mathematical calculations.

4. Physical model of the avalanche multiplication

Let us discuss a classical one-dimensional system of continuity equations for P-N transition shown in Fig. 4.1 [5]

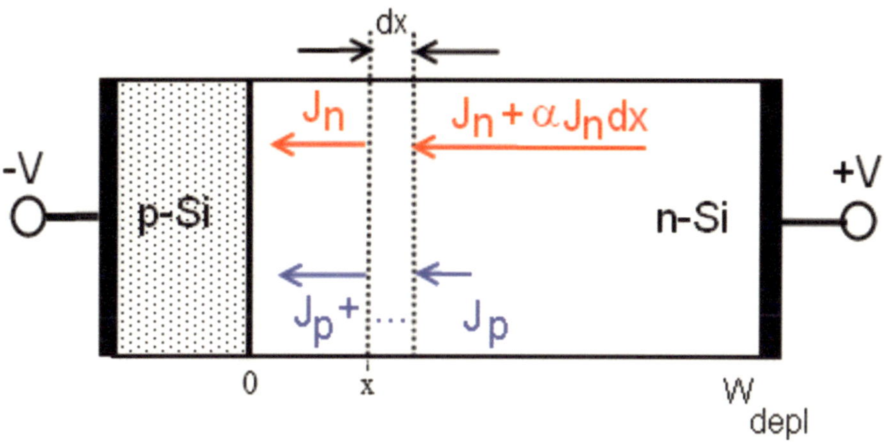

Fig. 4.1. Avalanche multiplications of electrons in high electrical field region of P-N junction.

$$\frac{dJ_n}{dx} = \frac{dJ_p}{dx} = \alpha_n \cdot J_n + \alpha_p \cdot J_p \tag{4.1}$$

at boundary conditions

$$J_n(0)=J_{no} \qquad J_p(0)=J-J_{po}$$

$$J_n(Z)=J-J_{po} \qquad J_p(W_{depl})=J_{po}$$

$$J(x)=J_n(x)+J_p(x)=\text{const}$$

The solution of this system will be interesting for us in the form of multiplication factor Mn=Jn/Jno for the case of purely electronic injection.

$$M_n = \frac{1}{1-\int\left(\alpha_n \cdot \exp\left(-\int\left(\alpha_n - \alpha_p\right)dx\right)\right)dx'} \tag{4.2}$$

if M_{no} is written in the form

$$M_{no} = \frac{1}{1 - \int\left(\alpha_n \cdot \exp\left(-\int\left(\alpha_n\right)dx\right)\right)dx'} \tag{4.3}$$

It is possible to represent (4.2) in the following form

$$M_n = \frac{M_{no}}{1 + \beta_n \cdot M_{no}} \tag{4.4}$$

In this expression the feedback coefficient is in the form

$$\beta_n = \int \alpha_n \cdot e^{-A(x)} \cdot \left(1 - e^{-B(x)} dx\right) \tag{4.5}$$

where

$$A(x) = \int \alpha_n dx \quad \text{and} \quad B(x) = \int \alpha_p dx$$

We can easily obtain the expression for the avalanche multiplication factor in the case of purely "holes" injection, simply changing n to p in formula (4.4) and vice versa. This formal transformation, as it may seem, of a well-known expression makes it possible for us to regard electron and hole multiplication from the FBT point of view.

Let us study some limit cases for formula (4.5).

Firstly, the trivial case $e^{B(x)} = 1$ means that there is no feedback or $B(x) = \int \alpha_p dx = 0$ - there is no multiplication of the hole component. On the other hand, the presence of the holes multiplication process is a factor that means the presence of positive feedback. This fact easily shows the case $e^{B(x)} >> 1$ when we get from formula (4.5)

$$\beta_n = -\int \alpha_n \cdot e^{-(A(x) - B(x))} dx \tag{4.6}$$

In the obtained expression, the minus mark indicates the feedback type – namely positive feedback, and the ratio of two components of the exponent shows the FB depth.

Concerning the restrictions on the transfer from representation (4.2) to representation (4.4), they are connected with the necessity to follow the implementation of requirement

$$\int(\alpha_p \exp(-\int \alpha_n) dx) dx' \neq 1 \tag{4.7}$$

For FB coefficient the following condition should be implemented

$$\beta_n \cdot M_n > 1 \qquad \text{or}$$

$$0 \leq \int \alpha_n \cdot e^{-(A(x) - B(x))} dx < \frac{1}{M_{no}}$$

Let us consider the character of possible behaviour of the feedback coefficient in the region of avalanche multiplication. We differentiate (4.6) in coordinate. As a result we get three variants of behaviour.

Case A(x)>B(x) dβn/dx - the rate of hole multiplication in the direction of differentiation gradually decreases.

Case A(x)=B(x) dβn/dx - the rate of hole multiplication in the direction of differentiation is constant.

Case A(x)<B(x) dβn/dx - the rate of hole multiplication in the direction of differentiation gradually increases.

Thus, having done a formally identical transformation of the initial equation, we get an opportunity to mark three characteristic cases for positive FB. The method discussed above will be further applied and allow us to make a clear and generalizing classification of various APD types.

There two reasons that show that there is no need to analyze deeper the obtained expressions. Firstly, our main aim is an acquaintance with a possibility itself to interpret an avalanche process on various levels of abstraction and for various models. Secondly, further studies will have sense if we use a specific APD topology. There exists a simpler model of FBT for avalanche detectors to make general description of some common properties of APD. This model allows one to use a more illustrative and simpler approach, from the point of practical application of APD.

5. Miller's formula

It is well-known in APD applications that in practice the use of the physical model to describe a specific APD will be too complex and far from informative. That is why various empiric or semi-empiric models are used. Miller suggested the first model of this type in 1955 [6]. It turns out that the new approach based on the identical transformation of the main formula of this model allows one to obtain sufficiently interesting results.

In the first variant of formula (5.1) the influence on avalanche amplification was not considered of the APD internal base and the contact resistance (the equivalent APD scheme is in Fig. 5.1a).

$$M_{no} = \frac{1}{1 - \left(\frac{V}{V_{br}}\right)^n} \tag{5.1}$$

Later, the formula was modified and presented in the form (5.2) for equivalent APD scheme shown in Fig. 5.1в.

$$M_n = \frac{1}{1 - \left(\frac{V - i \cdot R_{fb}}{V_{br}}\right)^n} \tag{5.2}$$

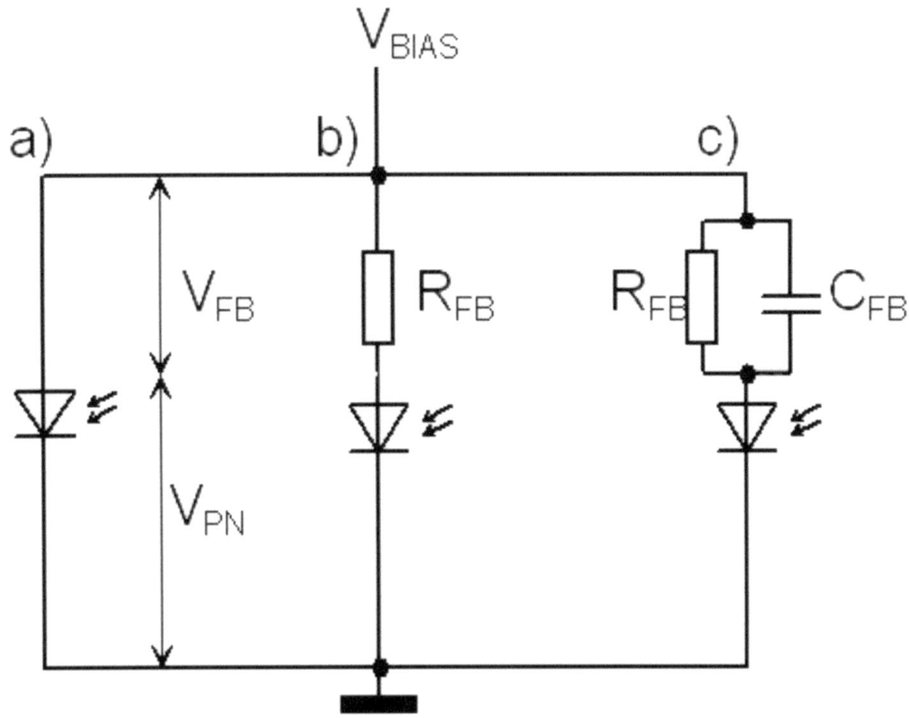

$$V_{BIAS} = V_{PN} + V_{FB}$$

Fig. 5.1. The evolution of equivalent circuit in Miller's formula.

It turns out that the Miller formula can be transformed to the form of expression (2.0). Let us do transformation for formula (5.3) (the equivalent APD scheme is suggested in [7] and shown in Fig. 5.1c).

$$M_n = \frac{1}{1-\left(\dfrac{V-V_{fb}}{V_{br}}\right)^n} \tag{5.3}$$

In the case under consideration the transformation has sense as the model exactly describes the APD operation with an internal FB. In (5.4) we show the feedback coefficient in the general form to indicate that the feedback value is defined by the change of the voltage V_{fb}. On the other hand, in this expression a significant FB difference is clearly seen if the FB resistance consists of purely active component and if it contains the reactive component [8].

$$M(V) = \frac{M_o(V)}{1-\beta \cdot M_o(V)} \tag{5.4}$$

here $\beta = \frac{V_{fb}}{V_{br}} = \frac{i \cdot R_{fb}}{V_{br}}$

In the first case external chains of APD connection define the rate of the avalanche multiplication suppression; in the second case, the influence of the reactive component on the rate of the avalanche multiplication suppression is strong. Due to this, the suggested model allows one to describe accurately the operation of APD with internal local negative FB. The FB locality implies that the process of the avalanche suppressing is much quicker than the changes in the parameters of external chains of APD power supply. Thus, we can choose parameters C_{fb} and R_{fb} in such a way that we can accurately describe the APD behaviour with negative local FB. A step further is more complicated as it is necessary to connect the obtained distributed parameters with topological and technological parameters of the APD under study.

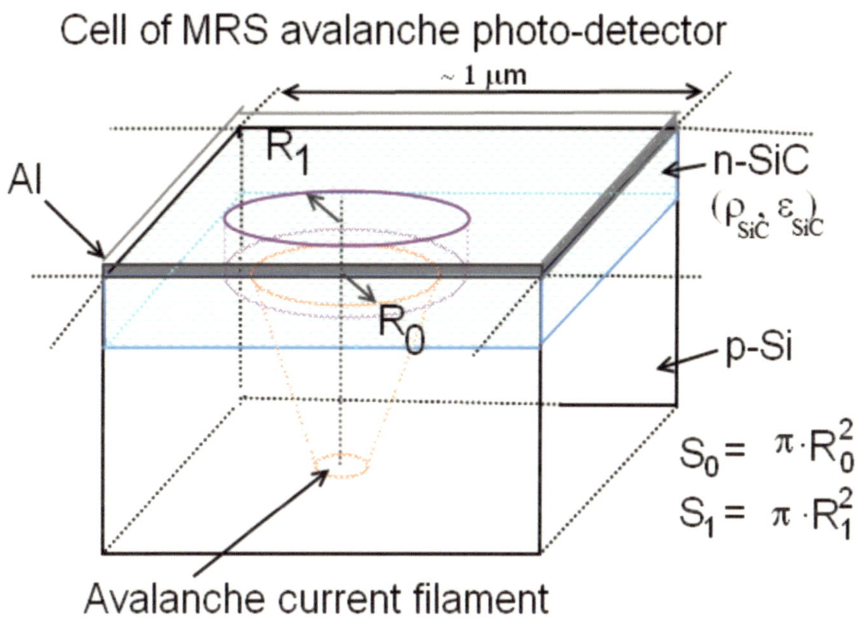

Fig. 5.2. Metal - Resistive Layer - Silicon (MRS APD) cell. The diameter of current filament must be change on boundary n-SiC – p-Si. (ρ_{SiC}/ $\rho_{p\text{-}Si}$)~10

In the discussed example (Figure 5.2) we can proceed from the fact that we have already known the parameters of the resistive layer (resistivity) ρ_{SiC} and area N^+ region S_N. The value C_{fb} is more significant for the description of the dynamic mode, while for the static one R_{fb} is more important, so we can expect that $R_{fb} = \rho_{SiC} \cdot \frac{d_{SiC}}{S_o}$ where S_0 determines the cross section area of the avalanche in stationary mode (avalanche current is constant).

To describe the dynamic mode, the value C_{fb} is more significant, while for the static mode R_{fb} is more significant. We can expect here that $R_{fb} = \rho_{SiC} \cdot \frac{d_{SiC}}{S_1}$ where S_1 determines the dimension of the spreading area of the filament of the current on the boundary N_{SiC}-P_{Si}, proceeding from the term of achieving the current stationary when M(t)=const, and $C_{fb} = \frac{\varepsilon_o \cdot \varepsilon_{SiC} \cdot S_1}{d_{SiC}}$

where S_1 defines the maximum dimension of the avalanche current from which the avalanche suppression starts.

We can define a characteristic constant τ_{fb} (5.5) that we will call the time of feedback of the process of avalanche multiplication suppression. This value is determined by the APD technology and topology. It should be noted that the constant does not depend on the resistive layer thickness, but does depend on the spreading area cross-section and the thickness of the avalanche current area cross-section.

$$\tau_{fb} = \varepsilon_o \cdot \varepsilon_{SiC} \cdot \rho_{SiC} \cdot \frac{S_1}{S_0} = \tau_{SiC} \cdot \frac{S_1}{S_0} \tag{5.5}$$

It seems obvious that for local negative FB it is necessary to implement the condition $\tau_{fb} \geq \tau_{av}$, namely, the time of FB suppression must be a little longer than the time of the avalanche generation τ_{av}. If we consider the model of the abrupt P-N junction and presuppose that the avalanche region W_{av} ~0.1 of the thickness of the depletion region W_{depl} and the velocity of the electron movement in the multiplication region V_{sat} we get an evaluation in the form (5.6).

$$\varepsilon_o \cdot \varepsilon_{SiC} \cdot \rho_{SiC} \cdot \frac{S_1}{S_0} \geq \frac{W_{av}}{V_{sat}} \tag{5.6}$$

As V_{br} is depend on the silicon parameters [9], and V_{sat}=const, we get the evaluation (for the n-SiC/p-Si structure) of the minimal concentration of acceptors in p-Si in the form (5.7).

$$W_{av} \leq V_{sat} \cdot \tau_{SiC} \frac{S_1^2}{S_0^2} \tag{5.7}$$

$\tau_{SiC} = \varepsilon_o \cdot \varepsilon_{SiC} \cdot \rho_{SiC}$ – Maxwell time for n-SiC.

This implies (for simplicity we suppose that S_1=S_0) that at ρ_{SiC}=156 [ohm/cm] the acceptors concentration in silicon must be lower $N_p < 2*10^{15}$ 1/cm^3. But at the same time the concentration cannot be too low. Otherwise FB will not have time to suppress locally the avalanche production process. If we take into account that an agreement of this kind is necessary for each new value ρ_{SiC} and, besides, that in the general case $S_1 \neq S_0$ depends on the impurity concentration and conditions on the border n-SiC/p-Si, it becomes clear why the suggested FB mechanism was implemented by only one research group [10].

The presented derivations are one more example of FBT application for the description of APD operation. In the Chapter 6 that considers examples of application of the models based on FBT ideas we discuss the model of the n-SiC/p-Si structure in more detail.

Let us regard the general analysis of formula (5.4)[11]. We mark that in the first approximation we thought that n=1. If we take the general case n>1, the feedback coefficient must be presented in the form (5.8)

$$\beta = \frac{V^n - (V - V_{br})^n}{V_{br}^n} \tag{5.8}$$

We study the character of behaviour of the FB coefficient depending on the voltage changes V.

If we define the avalanche gain efficiency coefficient in the form

$G_{ef} = \frac{dM_0}{dV} - M_0^2 \cdot \frac{d\beta}{dV}$ the derivative on V from (4.4) will be in the form (5.9)

$$\frac{dM}{dV} = \frac{G_{ef}}{(1 - \beta \cdot M)^2} \tag{5.9}$$

If we define the FB efficiency parameter in the form (5.10) it is possible to outline three characteristic cases of FB demonstration in APD in Fig. 5.3.

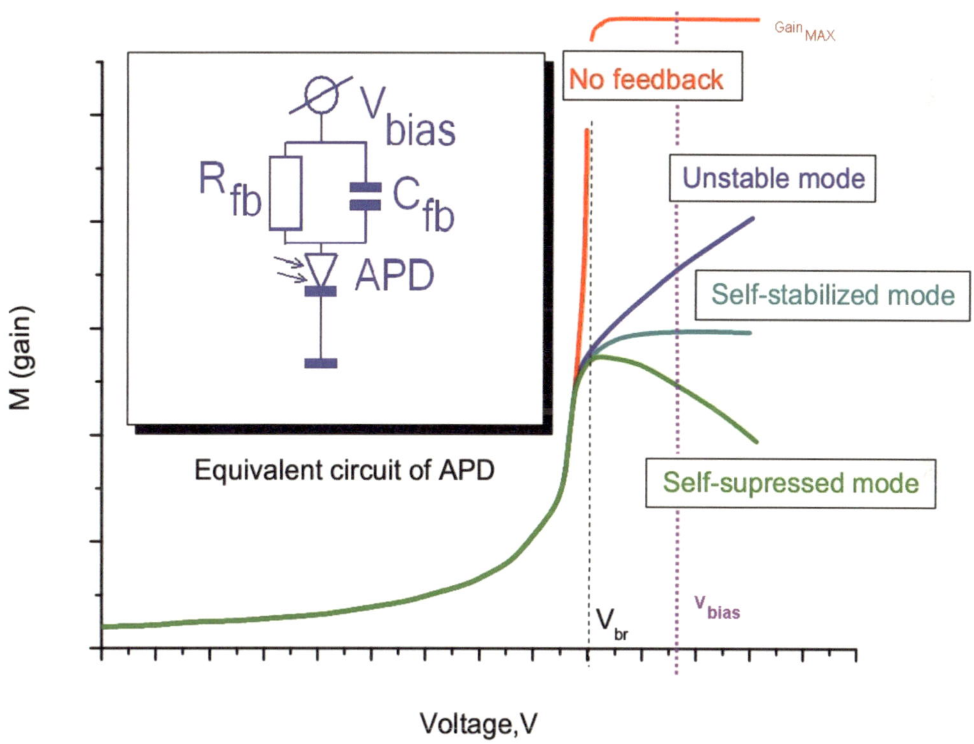

Fig. 5.3. Classification of APD using model of FB.

$$K_{ef} = \frac{\frac{d\beta}{dV}}{\frac{1}{M_0} \cdot \frac{dM_0}{dV}} \tag{5.10}$$

$K_{ef} < 1$ - with the growth of voltage the feedback increases slower than the avalanche gain that corresponds to the APD operation in the unstable mode;

$K_{ef} = 1$ - with the growth of voltage the feedback increases as quickly as the avalanche gain that corresponds to the APD operation in the self stabilization mode;

$K_{ef} > 1$ - with the growth of voltage the feedback increases quicker than the avalanche gain that corresponds to APD operation in the self-suppressing mode.

A classification of this type allows us to define the character of the FB behavior. For some particular cases $\frac{d\beta}{dV} \approx \frac{1}{M_0} \cdot \frac{dM_0}{dV}$ for example we can state:

$\frac{d\beta}{dV} > 0$ FB coefficient monotonously grows with the voltage increase; thus, theoretically, there must exist voltage when, if higher than its value, relative changes of the avalanche gain $\frac{1}{M_0} \cdot \frac{dM_0}{dV}$ are lower than the FB introduced suppressing. Initially, the condition $\frac{d\beta}{dV} < \frac{1}{M_0} \cdot \frac{dM_0}{dV}$ is implemented.

$\frac{d\beta}{dV} = 0$ The FB coefficient is constant with the voltage growth; therefore, theoretically there must be voltage for which relative changes in avalanche gain $\frac{1}{M_0} \cdot \frac{dM_0}{dV}$ correspond to FB introduced suppressing. Initially, the condition $\frac{d\beta}{dV} < \frac{1}{M_0} \cdot \frac{dM_0}{dV}$ is implemented.

$\frac{d\beta}{dV} < 0$ The FB coefficient monotonously decreases with the voltage growth; thus, theoretically there must exist voltage when, if high than its value, relative changes of the avalanche gain $\frac{1}{M_0} \cdot \frac{dM_0}{dV}$ become higher than the FB introduced suppressing. Initially, the condition $\frac{d\beta}{dV} > \frac{1}{M_0} \cdot \frac{dM_0}{dV}$ is implemented.

All the above stated, despite the fact that only cases of quite particular character were considered, can further assist in sorting out correctly the main mechanism of FB for each specific case. And this will be the first and basic step how to connect experimental data with the APD physics, topology and technology.

6. Examples of FBT application for APD

6.1 Logistic model of front of avalanche in GAPD

In this chapter we consider model that describes the process of avalanche generating in GAPD. The peculiarity of the model is in the fact that we will not solve the Fundamental Equation System for semiconductors and will not actually use the Poisson equation. The dynamics of an avalanche process in GAPD is determined by two important characteristics of this detector. The first is the difference of the bias voltage and breakdown voltage and the second is the dimension and structure of the GAPD cell. Let us discuss a GAPD cell as it is shown in Fig. 6.1.

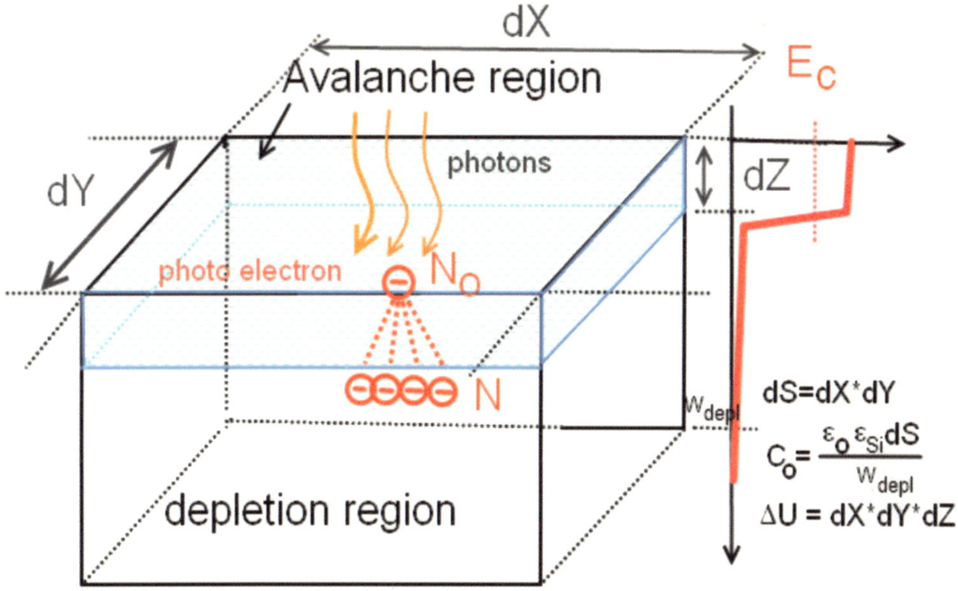

Fig. 6.1.1. Structure of GAPD for model.

The electric field in the cell is close or more to the critical one; therefore, the probability of the avalanche generation of a secondary carrier P~1. Moreover, there exists a certain threshold value of photo-generated electrons N_o necessary for the start of the avalanche process. Thus, to suppress the avalanche in cell from thermal noises for the thermo-generated carriers N_T the $N_T < N_o$ inequation should be implemented. We can evaluate N_T supposing that the noise generates only by the detector dark current. This supposition will allow us evaluate theoretically the dependence of the size of cell on the concentration of the doping applied for the APD material. It is obvious that the higher is the doping concentration the lower is the maximal cell volume $\Delta\bar{U}$ at the same cell area. At backward bias voltage V_{bias}, the dark current J_d and equals to N_T will determine the concentration of minor carriers generated per time unit in the depletion region. For APD with a long base, presupposing that the dark current is defined by the volume component from [12] it follows $J_d \approx \frac{e \cdot N_i \cdot \Delta U}{2 \cdot \tau_{bulk}}$ Let us consider to be definite a cell of the S area in the form of a abrupt P-N junction on silicon of the N type with the donor concentration N_d and depletion region size $W_{depl}(N_d, V_{br})$ at break-down voltage V_{br}.

$$N_T \approx \frac{N_i \cdot \Delta U}{2 \cdot \tau_{bulk} \cdot U_{sat}} \tag{6.1.1}$$

Consequently $N_0 > \frac{N_i \cdot \Delta U}{2 \cdot \tau_{bulk} \cdot U_{sat}}$ or $\Delta U = S \cdot W_{depl}(N_d, V_{br}) < \frac{2 \cdot \tau_{bulk} \cdot U_{sat}}{N_i}$

It should be noted that, if the threshold value N_o is large, the avalanche probability is small and the front time increases. Big lifetime of the minor charge carriers in SCR (Space-Charge Region) makes it possible to produce GAPD cells of a larger area. After we have evaluated the maximal cell area let us consider the following model.

Let the maximal number of electrons in the cell available for multiplication be equal to N. At the moment of time t after the start of the avalanche multiplication process there are already x electrons; therefore, the number of electrons potentially available for multiplication will be (N-x). If the length of the avalanche gain region is L_{av} and the velocity of electron in the electric field U_{sat}=const, (the time of front of avalanche in the order T_{av}~L_{av}/U_{sat}). The differential equation for the described situation is given in the form (6.1.2).

$$\frac{dx}{dt} = k \cdot x \cdot (N - x) \tag{6.1.2}$$

with the initial requirement $x(t = 0) = N_0$

The solution of this equation is well known as a logistic equation [13].

$$x(t) = \frac{N}{(1 - a \cdot \exp(-k \cdot N \cdot t))} \tag{6.1.3}$$

where $a = \frac{N - N_0}{N_0}$

The constant 1/k must have time dimensionality, so we input a certain constant of the time of the avalanche front and define it on the basis of the dimension analysis. $\tau_0 = \frac{1}{k} = T_{av} \cdot \frac{N_0}{N} = \frac{L_{av}}{V_{sat}} \cdot \frac{N_0}{N}$ (at N_o=1 τ_o=T_{av}/N has a simple physical interpretation - it is the time of one multiplication act, and if the threshold number N_o grows the time of one multiplication act becomes longer). Now we write (5.1.3) in the form

$$x(t) = \frac{N}{(1 - a \cdot \exp(-\frac{t}{\tau_0}))} \tag{6.1.4}$$

To transfer to the interpretation of the solution from the FBT point of view, we introduce new variables. M(t) = x(t)/No - the avalanche gain coefficient at the moment of time t, Mo=N/No - maximal avalanche gain in GAPD.

We write (6.1.4) in the form

$$M(t) = \frac{M_0}{(1 + \beta(t) \cdot M_0)} \tag{6.1.5}$$

where $\beta(t) = (1 - \frac{1}{M_0}) \cdot \exp(-\frac{t}{\tau_0})$

The model given above describes the exponential change of the FB factor from the maximal value $\beta_{max} = (1 - \frac{1}{M_0})$ to zero that corresponds to the maximal current through GAPD.

In Fig. 6.1.2 we show the results of calculations of the avalanche front line for different values of N_o. The lower are the threshold values the faster augment the avalanche front. It corresponds to the idea that higher probability of the charge carrier ionization in APD corresponds to the fast front of the avalanche process.

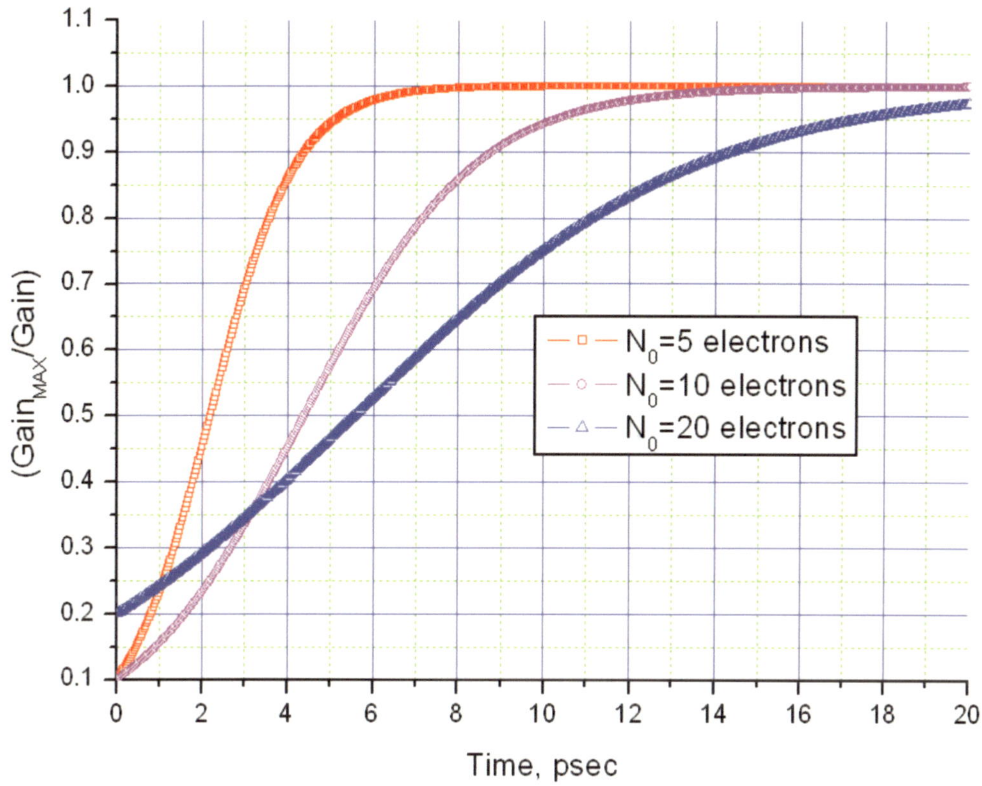

Fig. 6.1.2. Simulation of raising current in avalanche versus time by using Logistic model

6.2 GAPD Analysis. FBT interpretation

Paper [11] suggests a model where the feedback factor was given in the form $\beta = \frac{V_{fb}}{V_{br}}$.

Proceeding from the supposition that in GAPD the voltage should change by the value V-V_{br} with the avalanche suppressing due to the voltage drop on the amplifying cell to suppress the avalanche, we think it is the very similar FB voltage

V-V_{br}=V_{fb}. Let the FB factor be equal to β=V_{fb}/V_{br}. We consider the charge gain in the GAPD cell whose capacity is C_o. The cell discharge charge is Q_o=C_o*V_{fb}. If the threshold number of electrons N_o, the avalanche gain coefficient of the charge M_x~Q_o/e*N_o = C_o*V_{fb}/e*N_o. On the other hand, we can define the voltage of a separate gain at $V=V_{br}$ when $\beta\sim1$, so the choice $\beta=\frac{V_{fb}}{V_{br}}$ will be justified if M_o is chosen in the corresponding way. From the expression (6.2.1) we obtain the value M_o.

$$M=\frac{C_0\cdot V_{fb}}{e\cdot N}=\frac{M_0}{1+\frac{V_{fb}}{V_{br}}\cdot M_0} \tag{6.2.1}$$

The result of the solution is given by way of the charge gain coefficient.

$$M_0=\frac{C_0\cdot V_{fb}}{e\cdot N_0-C_0\cdot V_{fb}\cdot\frac{V_{fb}}{V_{br}}}=\frac{Q_0}{Q_i+\frac{V_{fb}}{V_{br}}\cdot Q_0} \tag{6.2.2}$$

In this expression Q_o is an intensified charge from the cell and Q_i is the initial charge that switches the avalanche in GAPD. Thus, we have managed to present the FB factor in the desired form and, at the same time, keep the general kind of the expression for the avalanche gain in the form $M=\frac{M_o}{1+\beta\cdot M_o}$.

Let us discuss now the feedback model when the avalanche restrictions happen due to limitation of the charge available for the avalanche multiplication in the cell. As the main parameter is the type and concentration of GAPD material doping, we will try to connect our abstract model with silicon parameters. The maximal charge available for multiplication in the cell is defined by the cell and SCR dimensions and equals to $Q_{max}=C_o*V_{fb}\sim N_a^{1/2}$ (we suppose that the basic material is silicon of the N type with donor concentration N_d). The possible charge of the avalanche gain Q_{av} is defined by the electric field intensity in the gain area, $Q_{av}=e*N_o*M\sim N_a^m$ where m>1.2 than the length of the avalanche region.

As the change of the field intensity per length unit in low-ohm material is higher than in high-ohm material, the rate of the avalanche gain change in the low-ohm material in the SCR avalanche part is higher and the avalanche generates in smaller space. It leads to its greater instability and imposes strict requirements to the time of FB establishment, as it was indicated above (formula (5.7)). In this case the FB mechanism due to the limitation of the charge available for gain in the cell is more preferable.

There should exist conditions when FB due to charge quantity limitation in the cell is possible and this condition will be $Q_{av}\geq Q_{max}$. In carrying out this inequation, the maximal charge is defined by the cell geometry and not by electric field intensity in it. Meeting this requirement will allow one to develop GAPD with high field intensity without fear of destroying the cell with high currents due to local instability of the avalanche process.

Let us study a simple model shown in Fig. 6.1 the abrupt P-N junction on the N-type silicon. If N_d>>N_a the dimension W_{depl} of SCR at the bias voltage equal to breakdown voltage is known and the maximal charge accumulated in the cell at the excess of breakdown voltage by $\Delta V=V-V_{br}$ is defined from equation (6.2.3). Let us suppose that the region of the avalanche multiplication is W_{av}~$0.1*W_{depl}(V_{br})$, and R_o – is the dimension of the region of avalanche current filament. The maximal cell charge is:

$$Q_{max} = C_0(N_d, V_{br}) \cdot \Delta V \approx \frac{\varepsilon_0 \cdot \varepsilon_{Si} \cdot S}{W_{depl}(V_{br})} \cdot \Delta V = \frac{\varepsilon_0 \cdot \varepsilon_{Si} \cdot \pi \cdot R_0^2}{\sqrt{\frac{2 \cdot \varepsilon_0 \cdot \varepsilon_{Si} \cdot V_{br}}{e \cdot N_d}}} \cdot \Delta V \tag{6.2.3}$$

To calculate the maximally possible charge of the avalanche gain let us suppose that the electric field in the region of the avalanche gain W_{av} is constant E=const and the ionization coefficient α(E) is known. We think that an avalanche appears as a micro-plasma current filament of a constant and independent of material thickness and $S_0/S_1=1$ (see chapter 6). Then, the maximally achievable charge of the avalanche multiplication can be calculated from equation (6.2.4).

$$Q_{av} \approx e \cdot N_0 \cdot \alpha(E) \cdot S_1 \cdot W_{av} = e \cdot N_0 \cdot \alpha \cdot (-\frac{E_c}{E}) \cdot S_1 \cdot W_{av} \tag{6.2.4}$$

In the expression α_o, E_c the model parameters and e- is the electron charge. Let us suppose that the field intensity E is equal to the maximal for the given PN junction at voltage $V=V_{br}$.

$$E = E_{max} = \frac{e \cdot N_d \cdot W_d(N_d, V_{br})}{\varepsilon_0 \cdot \varepsilon_{Si}}$$

As a result, we obtain expression (6.2.5)

$$Q_{av} = e \cdot N_0 \cdot \alpha \cdot (-\frac{\varepsilon_0 \cdot \varepsilon_{Si} \cdot E_c}{e \cdot N_d \cdot W_d(N_d, V_{br})}) \cdot S_1 \cdot W_{av}(N_d, V_{br}) \tag{6.2.5}$$

where $W_{av}(N_d, V_{br}) \approx 0.1 \cdot \sqrt{\frac{2 \cdot \varepsilon_0 \cdot \varepsilon_{Si} \cdot V_{br}}{e \cdot N_d}}$

In Fig. (6.2.1) and (6.2.2) calculation results are presented in the given formulas (6.2.3) and (6.2.5) for N_o=1, R_o=2,3,5,7,10,15 um, S_1=1мкм^2 and ΔV=0.1,0.25,0.5, 1.0, 2.0, 3.0, 5.0 V. Although the dimension of the current filament is not taken into account in calculations of the dependence that shows the maximal charge change of the avalanche gain $Q_{av}(N_d,W_{av}(N_d,V_{br}))$ achievable in multiplying, while it is easy to make the corresponding correction, the main thing is to connect it rightly with real GAPD topology and technology. The observed decrease of the charge at high N_d concentrations is explained by the effect of decrease $W_{av}(N_d,V_{br})$.

In Fig. (6.2.1) the red line shows how the maximal charge $Q_{av}(N_d,W_{av}(N_d,V_{br}))$ achievable in multiplying changes if the donor concentration in the base silicon changes in the limits from 10^{12} cm^{-3} to 10^{16} cm^{-3}, and the straight lines show how the available charge in the cell changes $Q_{max}(N_d,W_{av}(N_d,V_{br},R_o))$ at various values of the avalanche region radius R_o and ΔV=1. The region where the feedback mechanism operates to restrict the avalanche at the expense of the limitation of the charge available for gain in the cell is defined by the inequation $Q_{av} \geq Q_{max}$.

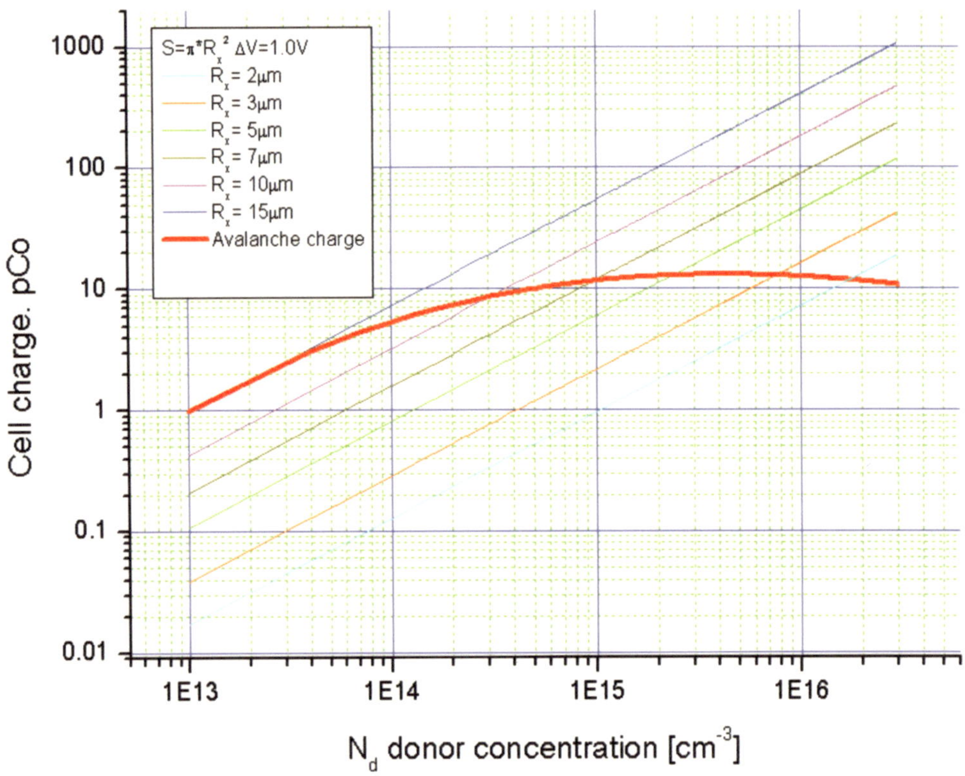

Fig. 6.2.1. The selection optimal size of cell for GAPD. The concentration of donors in cell must be selected near cross point of red line of maximum avalanche charge and the lines of different radius of cell.

In Fig. (6.2.2) the red line shows how the maximal charge $Q_{av}(N_d,W_{av}(N_d,V_{br}))$ changes if it is attained in multiplying and the donor concentration in the base silicon changes in the limits from 10^{12} cm^{-3} to 10^{16} cm^{-3}, while the straight lines show how the charge available in the cell changes $Q_{max}(N_d,W_{av}(N_d,V_{br},R_o))$ at various voltages ΔV if $R_o=5$. The region of application of the feedback mechanism to restrict the avalanche at the expense of the limitation of the charge available for gain in the cell is defined by the inequation $Q_{av} \geq Q_{max}$.

Let us discuss an example of application of the obtained dependences. If we choose silicon with $N_d=10^{16}$ cm^{-3} at the expected $\Delta V=1$ the size R_o of the charge region available for multiplication should be less than 3 μm (for example, 2 μm) and the mode is implemented when the maximal avalanche gain is restricted by the cell geometry and not by the intensity of the cell electric field. At $\Delta V=2$ the size R_o of the charge region available for multiplication should be less than 2 μm. In another Figure we see that if we choose silicon with $N_d=10^{16}$ cm^{-3} at the chosen $R_o=5$ the excess voltage ΔV for the charge multiplication should be less than 0.5 (for example, 0.25); if $R_o=2$ we can choose $\Delta V=2$ but not larger. The discussed examples show that the implementation of the restriction mode of the charge in the cell is quite complicated due to the complex interconnections of many parameters. It is necessary to calculate the GAPD construction from the point that the emerged avalanche will discharge the cell to the full extent and switch off due to the carriers' deficit, with no unstable unlimited by anything avalanche process. In practice, the discussed above mechanisms of avalanche quenching are implemented simultaneously, both in the form of decreasing the voltage in the cell and achieving depletion of free multiplication carriers in the avalanche region. To answer this question it is necessary to conduct more elaborate studies.

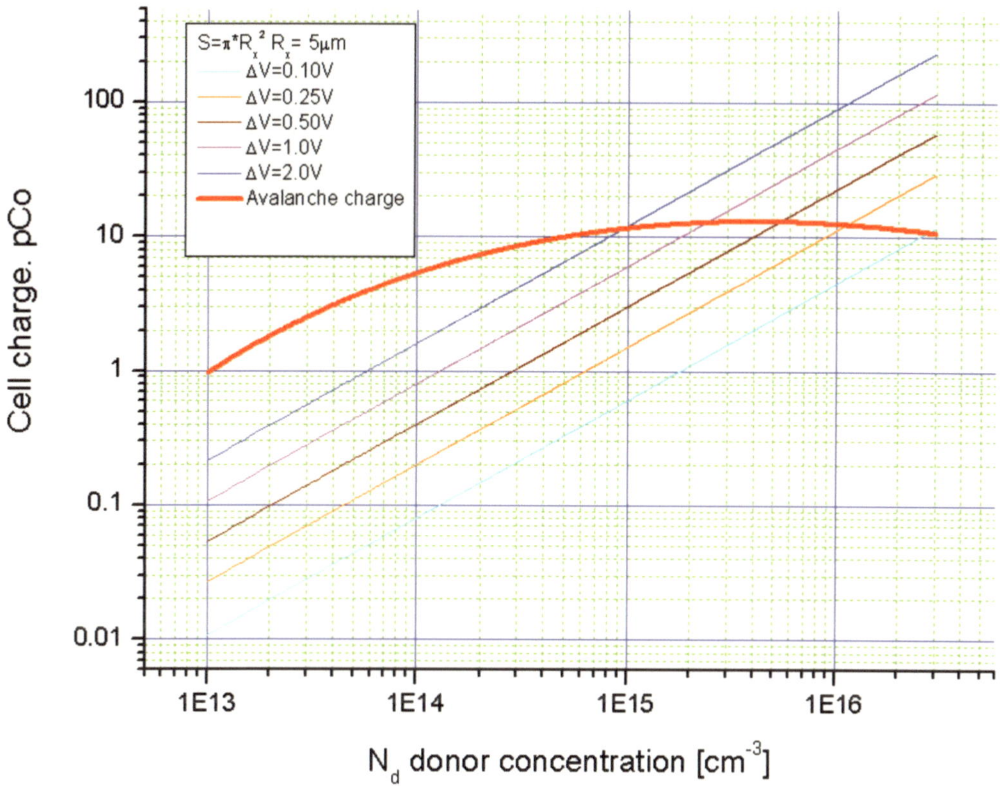

Fig. 6.2.2. The selection optimal over voltage $\Delta V=(V_{bias} - V_{br})$ for GAPD. The red line is the maximal avalanche charge can be the limit for over voltage for given GAPD with fixed size and concentration donor in cell.

6.3 APD equivalent scheme – SPICE model

A model to shape an avalanche front in APD on the basis of n-SiC/p-Si P-N junction is suggested for application in paper [11]. The form of the FB factor (its time dependence during the front shaping) was obtained on the basis of experimental data, with an initial supposition that the n-SiC layer properties determine the characteristic time of FB establishment.

The full equivalent schemes of MRS APD and Micro Channel Avalanche Photo Diode (MCAPD) are shown in Fig. 6.3.1. We are interested in one of its elements – a nonlinear source of current I(t) (APD) controlled by current (R_{fb}, D_{fb}, C_{fb}). The source amplifies the photocurrent $I_o(t)$ M times and by a certain time dependence of the FB factor β(t) it correctly describes the shaping of the APD front.

Let us consider the method of definition of time dependence of the FB factor to be applied in the further suggested model. The method can be applied to linear APD, but not to GAPD.

Let the APD under study be under the effect of a rectangular light pulse that generates current $I_o(t)$ in APD. An intensified current pulse I(t) can be given in the form

$$I(t) = I_0(t) \cdot \frac{M_0}{1 + \beta(t) \cdot M_0} \tag{6.3.1}$$

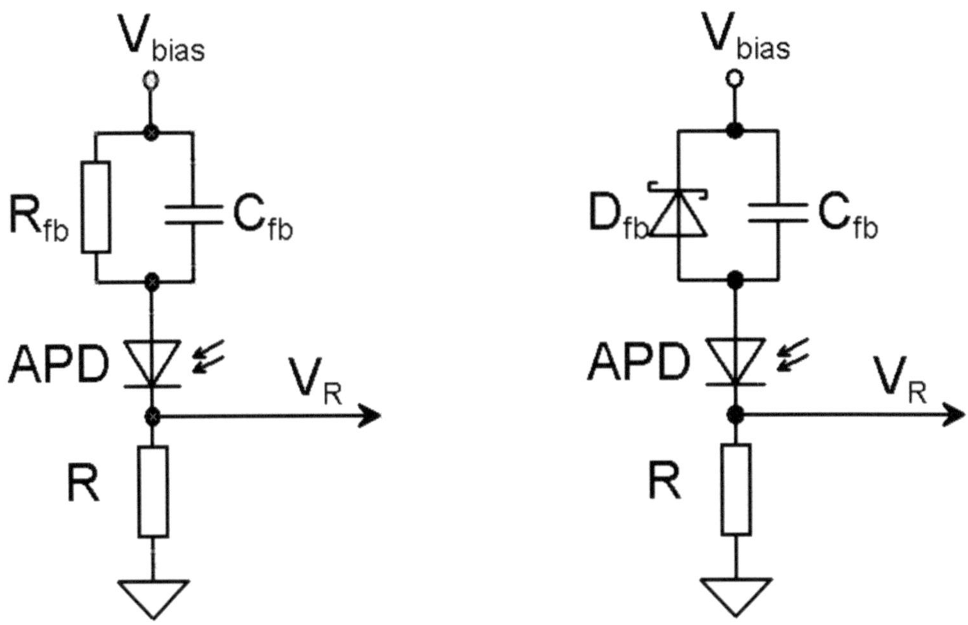

Fig. 6.3.1. Two different model were investigated for analysis by SPICE. APD - is source of current that controlled by voltage. R_{fb} - linear element of FB can be used for MRS APD [14], the D_{fb} is not linear element of FB and can be used for MC APD [15].

We suppose that M_o=const does not depend on time $M_o \neq M_o(t)$ and only β=β(t).

Then it is easy to obtain the dependence for the FB factor from (5.3.1)

$$\frac{I_0(t)}{I(t)} = \beta(t) + \frac{1}{M_0} \tag{6.3.2}$$

As we suppose that M_o^{-1}=const any time dependence of the FB factor is obviously reflected by equation (6.3.2).

For test of method were defined of FB factors for APD of two types described in papers [14] and [15].

The dependences that correspond to the definition are presented in Table 6.1.1 as fragments of the SPICE code and the simulation result and its comparison with experimental data is given in Fig. 6.3.2 and 6.3.3.

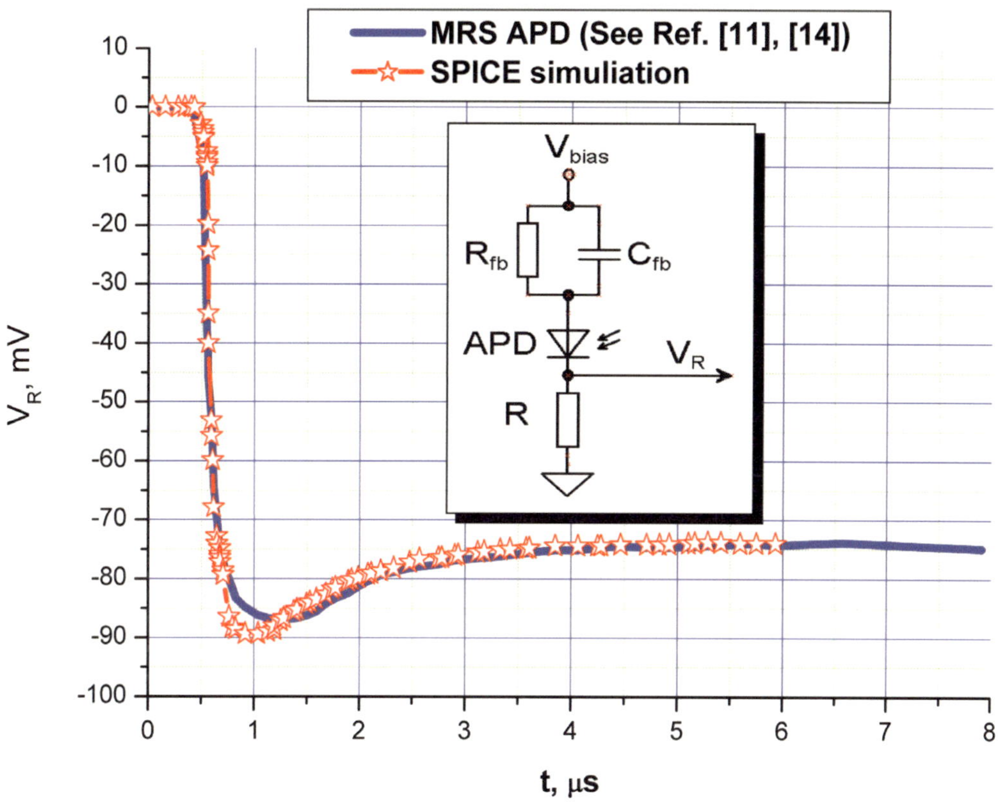

Fig. 6.3.2. Simulation result for MRS APD [11], [14].

Model MRS	R_{fb}=1K	C_{fb}=1200pF	R=1K	
Model MP APD	C_{fb}=5pF	D_{fb}(CJO)=10pF	D_{fb}(BV)=100mV	D_{fb}(IBV)=20nA

Table 6.1. (Parameters for simuliation)

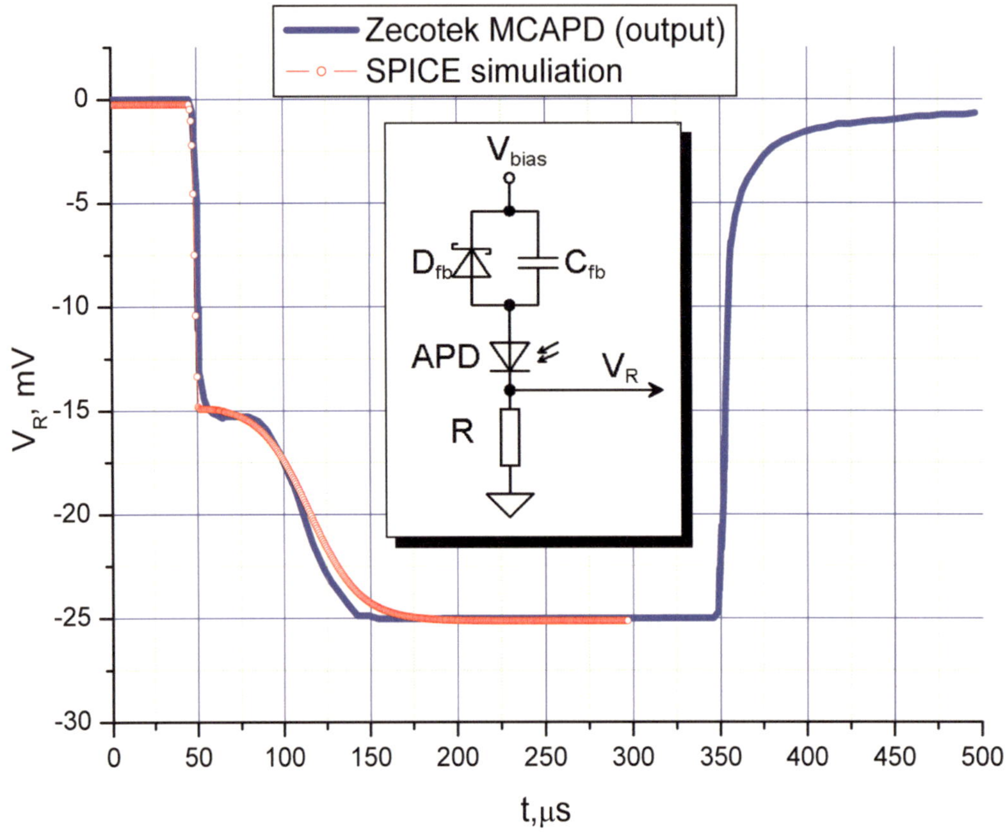

Fig. 6.3.3. Simulation result for MCAPD [15].

APD SPICE model:

*.FUNC DPWR(D) {I(D)*V(D)}*
*.FUNC BPWR(Q) {IC(Q)*VCE(Q)+IB(Q)*VBE(Q)}*
*.FUNC FPWR(M) {ID(M)*VDS(M)}*
*.FUNC HOTD(D,MAX) {IF((V(D)*I(D)>MAX),1,0)}*
*.FUNC HOTB(Q,MAX) {IF((VCE(Q)*IC(Q)+IB(Q)*VBE(Q)>MAX),1,0)}*
*.FUNC HOTF(M,MAX) {IF((VDS(M)*ID(M)>MAX),1,0)}*
.PARAM LOW3MIN={IMPORT(LOW3MIN.OUT,LOW3THRES)}
.PARAM HIGH3MAX={IMPORT(HIGH3MAX.OUT,HIGH3THRES)}
.PARAM LOWLVDS={IMPORT(LOWLVDS.OUT,LOWLIMIT)}
.PARAM HILVDS={IMPORT(HILVDS.OUT,HILIMIT)}
.PARAM LIMTLVDS={IMPORT(LIMTLVDS.OUT,LVDSLIMITS)}
*.FUNC SKINAC(DCRES,RESISTIVITY,RELPERM,RADIUS) {((PI*RADIUS*RADIUS)/ ((PI*RADIUS*RADIUS)-PI*(RADIUS-SKINDEPTHAC(RESISTIVITY,RELPERM))**2))*DCRES}*
.FUNC SKINDEPTHAC(RESISTIVITY,RELPERM) {503.3(SQRT(RESISTIVITY/(RELPERM*F)))}*
*.FUNC SKINTR(DCRES,RESISTIVITY,RELPERM,RADIUS,FREQ) {((PI*RADIUS*RADIUS)/*

*((PI*RADIUS*RADIUS)-PI*(RADIUS-SKINDEPTHTR(RESISTIVITY,RELPERM,FREQ))**2))*DCRES}*
.FUNC SKINDEPTHTR(RESISTIVITY,RELPERM,FREQ)
{503.3(SQRT(RESISTIVITY/(RELPERM*FREQ)))}*
*BG1 0 PINA I = {V(IN)*POWER((40/(40-(V(0)-V(PINA)))),1)*

7. Conclusion

The FBT approach allows to hide information about real physics processes in different types of APD. We can exchange a physical model describing a complex behavior of the system by a simple universal model with predictable results. The basic steps are in definition of main statistical parameters and their connections with external conditions of APD. Practically in all cases it is possible to implement a formal transfer from the physical model of an avalanche generation to the description in FBT. The formal separation of processes into those of avalanche multiplication and avalanche suppression or restriction allows a simpler understanding of the physical origin of the phenomena. Based on FBT notions, it is possible to construct simple enough model to describe processes of avalanche gain. This model can be connected with important parameters of APD under study and allows to determine the link of some technological parameters with APD properties.

Application of the FBT approach to construct the SPICE model allows one to facilitate the procedure of definition of the equivalent APD circuit main parameters and correctly describe the peculiarities of the front signal shaping.

8. References

[1] F.L. Lewis, (1993). Applied Optimal Control and Estimation, Prentice-Hall.
[2] John Doyle, Bruce Francis, Allen Tannenbaum, (1990). Feedback Control Theory, Macmillan Publishing Co.
[3] William McC. Siebert, (1986). Circuits, signals, and systems, Published by MIT Press, McGraw-Hill in Cambridge, Mass, New York .
[4] K. Jradi, et.al., (2011). Computer-aided design (CAD) model for silicon avalanche Geiger mode systems design: Application to high sensitivity imaging systems, Nuclear Instruments and Methods in Physics Research Section A. Vol. 626-627, pp. 77-81
[5] S. M. Sze, (1981). Physics of Semiconductor Devices, New York: Wiley.
[6] S. L. Miller, (1955). Avalanche breakdown in germanium, Phys. Rev. 99, pp. 1234-1241
[7] D. Bisello, et al., (1995). Metal-Resistive layer-Silicon (MRS) avalanche detectors with negative feedback, Nuclear Instruments and Methods in Physics Research Section A, Vol. 360, Issue 1-2, pp.83-86
[8] F. Zappa, A. Lacaita, C. Samori, (1997). Characterization and Modeling of Metal-Resistance-Semiconductor Photodetectors, IEEE Trans. On Nucl. Sc., Vol. 44, No. 3
[9] McKay, K. (1954). Avalanche Breakdown in Silicon, Physical Review 94:pp. 877-880
[10] Z. Sadygov et al., (1991). Proc. SPIE. Optical Memory and Neural Networks, pp. 158-168
[11] V. V. Kushpil, (2009). Application of simple negative feedback model for avalanche photodetectors investigation, Nuclear Instruments and Methods in Physics Research Section A, Vol. 610, Issue 1, pp.204-206
[12] (2003). Encyclopedia of Optical Engineering, pp. 128-130

[13] Gershenfeld, Neil A. (1999).The Nature of Mathematical Modeling. Cambridge, UK: Cambridge University Press.

[14] A.V Akindinov, et al., (1997). New results on MRS APD, Nuclear Instruments and Methods in Physics Research Section A, Vol. 387, pp.231-234

[15] Z. Sadygov et al., (2010). Microchannel avalanche photodiode with wide linearity range, arXiv.org>physics>ArXiv:1001.3050

11

Multiscale Computer Aided Design of Microwave-Band P-I-N Photodetectors

Mikhail E. Belkin
Moscow State Technical University of Radio-Engineering, Electronics and Automation, Faculty of Electronics, Joint Research Laboratory "Microwave and Optoelectronic Devices", Moscow Russian Federation

1. Introduction

Long wavelength InP-based p-i-n photodetectors (PD) are ubiquitous in modern optoelectronic circuits due to their inherent combination of ultra-high speed, high sensitivity in the most popular for modern telecom systems spectral range of 1.3-1.6 μm, and low bias voltages features that are impossible in principle for Si, GaAs or Ge counterparts. Typical material systems for the telecom spectral range are GaInAsP and GaInAs on InP substrate (Capasso et al., 1985). Now in the number of classical and present-day works (see, e.g. Bowers & Burrus, 1987; Beling & Campbell, 2009) is well-proved that compound semiconductor p-i-n photodetectors have the valid merits such as: high responsivity (up to 1 A/W), lowest dark current (below 10 fA), ultra-high bandwidth (up to 100 GHz and above), possibility of monolithic optoelectronic receiver module creation on common InP substrate.

Through a wide time period this type of PD found the most intensive application inside a receiver of the long-haul digital fiber-optic systems. But recently the p-i-n PDs for the wavelength of 1.3 and 1.55 μm are advancing in various RF and microwave photonic apparatus, particularly, in microwave-band optoelectronic oscillators (X.S.Yao, 2002), frequency converters (J. Yao, 2009), as well in so called photonic antennas for the base stations of radio over fiber (RoF) systems (Sauer et al., 2007). While designing a receiving module for the above devices and systems the stage of circuit schematic development referred to active and passive, electronic and photonic components' modeling would be most complicated and labour-intensive. Based on microwave IC design practice in this case the modeling is realized as a description of electrical and optical features of the integrated functional elements by means of mathematical equations, equivalent circuits or tables.

At present time computer-aided design (CAD) exploitation for innovative high-tech production R&D acceleration is a common way (Minixhofer, 2006). This is especially important for microwave semiconductor component base with measurement equipment and experimental work's cost being considerably more expensive compare to that in lower frequency bands. Using modern CAD software the next two approaches may be relevant for this case: by means of so-called physical models and by means of equivalent circuit models. A physical modeling is generally the most accurate but at the same time the most complicated. It is executed through a computer simulation of dedicated physical processes running in the semiconductor chip's epitaxial layers. On the other hand, equivalent circuit model is simpler and includes current and voltage sources as well as passive elements that subject to frequency band are built on a linear circuitry with lumped (for RF band) or distributed (for microwave band) parameters. Correct simulation of the real device features is ensured by the account of nonlinearity and lag effects in the sources and passive elements. Their parameters and characteristics are specified by the DC or low-frequency measurements also by the small- or large-signal mode measurements in operating band and adjusted due to parametric optimization process. The main advantage of the approach is in clarity of device operation scripts and in versatility for the same class components when the parameter spread could be simply taken in account by proper changing the numerical value of the separate element. Therefore, a simulation issue optimal decision based on criteria of accuracy and computation time would rely on multiscale combination of the physical and equivalent circuit models. For computations the means of one or a number of CAD tools could be applied. Such approach might be called as multiscale or end-to-end design.

Generally a transparent way for microwave photonic devices' computer design lies in simulating with a help of so called optoelectronic CAD (OE-CAD) tool (Lowery, 1997). Nevertheless, our review of modern commercial OE-CAD software has shown up their lack of development to simulate a microwave circuitry. An alternative approach is to model by two classes of CAD tools: so-called technology CAD or TCAD software (Blakey, 2008) and microwave-band electronic CAD or MW-ECAD software (Kielmeyer, 2008). By means of the first one a designer could realize the analysis procedures for a number of semiconductor chip physical, electrical and optical features and its DC characteristics; also simulate semiconductor laser, photodiode, transistor, or diode structure's small-signal frequency response. But for designing a number of relevant microwave photonics devices, e. g. optically controlled microwave amplifiers and switches, optoelectronic oscillators, transmitter and receiver optoelectronic modules, and so on one needs to know quite a number of the active semiconductor device parameters specifying its functioning in the operating bandwidth and in large-signal mode: output intercept point, harmonic and intermodulation distortions, large-signal output reflection coefficient. It is impossible to simulate the above features by the means of a TCAD tool. One more significant drawback of the tool is that the account of device package and assembling parasitic elements impacts is also impractical. This matter results in the solution error up to 30% even at the lower part of the microwave band that is growing along with a modulation frequency is being increased (Belkin & Dzichkovski, 2009).

The both issues are overcome by the application of ECAD tool optimizing for microwave band operation. The MW-ECAD tool special preference is in powerful electromagnetic (3D) analysis and optimization resources of microwave electronic circuitry and layouts design, for instance low-noise and power amplifiers, highly complex receiver optoelectronic modules. So introducing a convenient interface between these CAD tools one could realize multiscale end-to-end design of the device or module with microwave bandwidth. Usually for this goal well-known HSPICE interface is feasible. In this case, input data are multilayer semiconductor heterostructure with the certain cross section, and output data are a whole number of the outer parameters that totally and correctly characterize its features in microwave band. Also it might be a layout drawing of the module in development.

Thus, there exists a need for a reliable circuit-level photodetector model suitable for a designer of receiving modules of modern optical fiber-based systems and microwave photonics devices. In this chapter our recent results of the combined TCAD&ECAD and solitary MW-ECAD simulations for designing a long wavelength microwave-band p-i-n photodetector as a basic component of a multichannel analog system are presented.

2. General review

Generically, a photodiode as any electronic diode is a nonlinear device with a saturation effect and when it works in large-signal mode the fundamental signal is usually accompanied by a remarkable level of nonlinear distortions (Yu & Wu, 2002). The main distortion sources of the PD's conversion characteristic include absorption saturation, transit-time effects, and electric field screening arising from the space charge in the depletion layer (Williams et al., 1996). The latter effect also arises due to series resistance outside a junction region of a PD.

Nevertheless, in digital optical fiber systems handling with typical receiving power levels lower than 1 mW a photodiode is able to represent as linear device. On the other hand, in order to secure the needed spurious free dynamic range (SFDR) in the above microwave photonics devices and RoF systems the power levels may be up to 100 mW. For example a simple analysis of an opto-electronic oscillator model using quasi-linear interpretation results in the next expression for self-oscillating regime (X.S.Yao, 2002):

$$I_{ph}R \geq \frac{V_\pi}{\pi}, \tag{1}$$

where I_{ph} is the PD photocurrent; R is the PD load impedance; V_π is so called half-wave voltage of Mach-Zehnder modulator. Relationship between the photocurrent I_{ph} and the PD optical power P_0 is addressed to the well-known formula:

$$I_{ph} = P_0 S_0 , \tag{2}$$

where S_0 is the PD current responsivity. From (1) and (2) one can clearly conclude that self-oscillating condition could be simplified through increasing optical power at PD input. Namely, for typical values of R=50 Ω, V_π=5 V, S_0=0.5 A/W it would be more than 64 mW. Naturally, it is unpractical to represent photodiode in such a regime as a linear element.

Large-signal operation of PD in the above-mentioned devices and system called out an additional issue that a pre-amplifier following the PD must concurrently have low noise figure and enough high linearity. A good chance to circumvent this issue is in utilizing a special type high-linear PD and a scheme without pre-amplifier (Joshi & Datta, 2009). Note that operating the photodiodes at larger photocurrent has added benefit of reducing overall noise figure of the device or link. But an employment of special type PDs that are structurally more complicate than standard ones might degrade the economical features of optical receivers. Thus, a direct linearity comparison of the widespread top-illuminated PD and middle-power microwave transistor amplifier is a subject of practical interest.

As known from RF and microwave amplifier design experience (Kenington, 2000), device linearity in large-signal mode is regularly characterized by so-called harmonic and inter-modulation distortions (IMD). Among them a two-tone third-order IMD (IMD3) is the most severe issue in design practice. The reason is in their closer positioning to carrier signals to allow rejecting by filters. Also for super wideband microwave photonic devices and systems when the bandwidth is more than an octave the two-tone second-order IMD (IMD2) is in great importance too. Generally, two-tone IMD is a product of passing two unmodulated carriers at the frequencies of f_1 and f_2 through a nonlinear device. When viewed in the frequency domain, output spectrum of the circuit includes besides f_1 and f_2 a lot of overtones. In the general case of distortions created by any order nonlinear circuit new frequencies will be generated in the form:

$$f_i = mf_1 \pm nf_2 \tag{3}$$

where m and n are positive integers (including zero), $m+n$ is equal to the order of the distortion. Then in the case of second-order nonlinearity each of the two tones will have a second harmonic and additional sum and difference beatings will occur at frequencies of $f_1 \pm f_2$. Also in the case of third-order nonlinearity each of the two tones will have a third harmonic and additional sum and difference beatings will occur at frequencies of $2f_1 \pm f_2$ and $2f_2 \pm f_1$. These overtones are known as second-order and third-order intermodulation products correspondingly. One can calculate the above distortions through carrier transport equations (Williams et al., 1996) but the most simple and accurate way lies in measuring them in microwave band.

An obvious methodological drawback of IMD is referred to the dependence from powers of input signals that make impossible to compare devices at different input powers. To cope with it another well-known figure of merit for microwave amplifier linearity so called output intercept point (OIP) (Kenington, 2000) is widespread for microwave-band optoelectronic devices acting in large-signal mode. Following RF and microwave amplifiers, a rough estimate of the photodiode's 2-order OIP (OIP2) and the 3-order OIP (OIP3) can be produced by the following simple equations.

$$OIP2 = P_{out} - IMD2, \tag{4}$$

$$OIP3 = P_{out} - IMD3/2, \tag{5}$$

where P_{out} is output AC power of the fundamental tone in the photodiode load. Each parameter is usually expressed as a power in dBm. However, a common practice to define

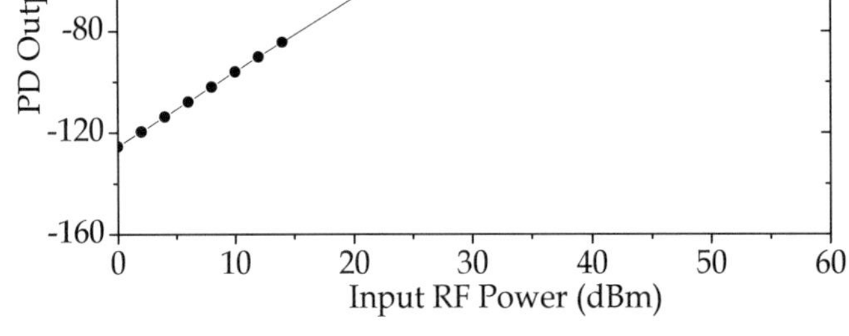

Fig. 1. A typical example of OIP3 determination for a photodiode

No	Description of the PD under test	Small-signal bandwidth	Output Intercept Point	Reference
1	Large optical cavity p-i-n waveguide InGaAs photodiode	20 GHz	OIP3 (-9 V, 1 GHz)=15.3 dBm OIP3 (-9 V, 18 GHz)=2.4 dBm	Jiang et al., 2000
2	Uni-traveling carrier waveguide integrated InGaAs photodiode	20 GHz	OIP3 (-5 V, 1 GHz)=24 dBm OIP3 (-5 V, 18 GHz)=8 dBm	Liao et al., 2003
3	Backside illuminated uni-traveling carrier InGaAs photodiode	10 GHz	OIP3 (-3 V, 10 GHz)=32 dBm OIP3 (-5 V, 10 GHz)=40 dBm	Gustavsson et al., 2005
4	GRIN lens-coupled top-illuminated InGaAs photodiode	6.5 GHz	OIP3 (-5 V, 0.83 GHz)=43 dBm OIP3 (-9 V, 0.83 GHz)=49 dBm	Joshi et al., 2008
5	P-i-n waveguide InGaAs photodiode	20 GHz	OIP3 (-4 V, 1.1 GHz)=32.6 dBm	Draa et al., 2008
6	Charge compensated modified uni-traveling carrier InGaAs photodiode	23 GHz	OIP3 (-7 V, 0.31 GHz)=52 dBm	Beling et al., 2008
7	Partially-depleted absorber back-illuminated mesa structured photodiode	10 GHz	OIP3 (-1 V, 1 GHz)=30 dBm OIP3 (-3 V, 1 GHz)=38 dBm OIP3 (-5 V, 1 GHz)=43 dBm	Tulchinsky et al., 2008
8	Top-illuminated InGaAs p-i-n photodiode	8 GHz	OIP3 (-2 V, 1 GHz)=17 dBm OIP3 (-8 V, 1 GHz)=30 dBm OIP2 (-2 V, 1 GHz)=33 dBm OIP2 (-8 V, 1 GHz)=46 dBm	Godinez et al., 2008
9	Top-illuminated InGaAs planar p-i-n photodiode	3.2 GHz	OIP3 (-16 V, 0.83 GHz)=49 dBm	Joshi & Datta, 2009
10	Ge n-i-p on silicon-on-insulator substrate photodiode	4.5 GHz	OIP3 (-5 V, 1 GHz)=20 dBm OIP2 (-5 V, 1 GHz)=40 dBm	Ramaswamy et al., 2009
11	Backside illuminated InGaAs p-i-n photodiode array	8 GHz	OIP3 (-2 V, 5 GHz)=32.5 dBm	Itakura et al., 2010
12	Modified uni-traveling carrier InGaAs photodiode with cliff layer	24 GHz	OIP3 (-5 V, 0.32 GHz)=45 dBm OIP3 (-9 V, 0.32 GHz)=50 dBm	Li et al., 2010

Table 1. Results of photodetector's output intercept point measurements

OIP relies on a graphical representation of measured amplitude curves for fundamental and second- or third-order IMD tones. Following it Figure 1 illustrates a typical example of OIP3 determination for a photodiode (Jiang et al., 2000) extrapolated from the measured data of fundamental (squares) and IMD3 (circles) plots. In the Figure's logarithmic scale on the both axes the graphs represent straight lines where the fundamental tone has a slope of 1:1 and IMD3 tone has a slope of 1:3.

Up-to-date there is a series of papers referred to OIP measurements. Table 1 lists the results of some of them.

At a glance the values of photodiode OIP were eventually increased and for today the highest OIP of more than 50 dBm possesses charge-compensated modified uni-traveling carrier InGaAs photodiode (Beling et al., 2008). Standard vertical-illuminated p-i-n PDs have some lower OIP values up to 46-49 dBm. The general tendencies for all types of PDs consist in decreasing OIP as the test frequency and the negative voltage bias are increased.

Another known property of nonlinear circuits is that vector addition of the output fundamental and harmonic components also determines a phase variation of the resultant output signal when the input level varies. This effect is a requisite of dynamic systems and as the figure of merit for microwave amplifier linearity is called amplitude-to-phase (AM-to-PM) conversion for single-tone mode that is usually expressed as a certain phase deviation, in degrees or radians at predetermined input power (Pedro & Carvalho, 2003). In the case of photodetectors the similar parameter in terms of power-to-phase conversion (PPC) is being investigate by a joint research team (Tailor et al., 2011) for emerging microwave photonics-based timing applications in the fields where it is important to precisely know the arrival time of light pulse such as RoF, phased-array radars, optical fiber coherent communications and so on. Our results of MW-ECAD's photodiode nonlinear modeling will be highlighted in section 4.

3. Combined TCAD&ECAD simulation

To illustrate effectiveness of multiscale approach, the end-to-end design of the photodiode modules with bandwidth near thirty gigahertz was fulfilled. At the first step following the modern CAD tools capabilities (see the Introduction), the simulation of the super high-speed pin-photodiode was carried out with the aid of Synopsys Sentaurus TCAD software tool[1]. Sentaurus package allows for correct modeling of the wide spectrum of semiconductor devices due to transport equations and various physical models combining. There is fundamental capacity of photodetector simulation too but up to date the examples of it are not frequently (see e. g. Zakhleniuk, 2007).

3.1 P-i-n photodiode heterostructure TCAD simulation

For the simulation a widespread mesa-construction with semi-insulator-buried and vertically-illuminated p-i-n photodiode heterostructure similar to partially-depleted absorber type (Tulchinsky et al., 2008) was selected (Figure 2). In the Figure: region 1 is p^+-type contact layer, region 2 – p-type buffer layer, region 3 – i-type absorption layer, region 4

[1] http://synopsys.com/home

- n-type buffer layer, region 5 - n^+-type contact layer. The mesa's material composition, layers initial doping levels and thickness selected for simulation are: region 1 - $Ga_{0.47}In_{0.53}As$, $5 \cdot 10^{18}$ cm^{-3}, 0.5 μm; region 2 –InP, $3 \cdot 10^{17}$ cm^{-3}, 3 μm; region 3 - $Ga_{0.47}In_{0.53}As$, $3 \cdot 10^{16}$ cm^{-3}, 1 μm; region 4 – InP, $3 \cdot 10^{18}$ cm^{-3}, 0.3 μm; region 5 - InP, $8 \cdot 10^{18}$ cm^{-3}, 2 μm. The substrate's doping level and thickness are $2 \cdot 10^{18}$ cm^{-3} and 300 μm correspondingly. The chip dimensions are 200x200 μm^2, the mesa's top diameter is 50 μm.

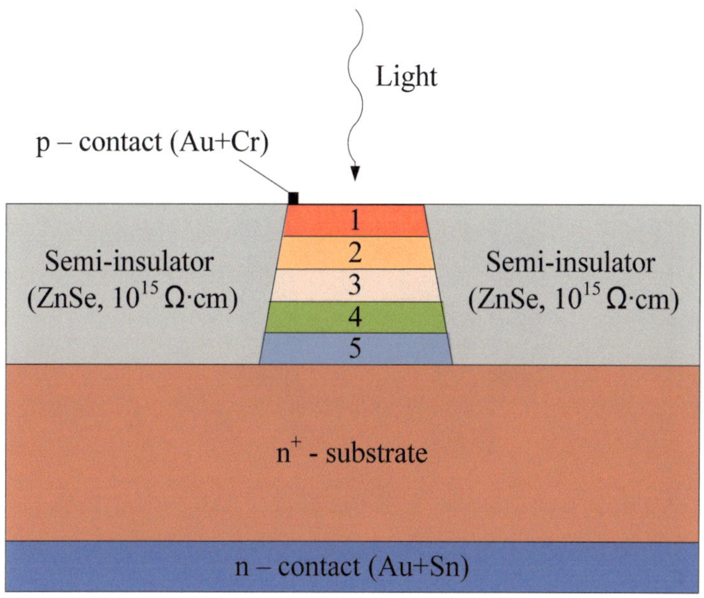

Fig. 2. The cross section of the photodetector in research

Such double-heterojunction p-i-n photodiode structure as will be shown below is lightly slower than more popular single-heterojunction one (Bowers & Burrus, 1987) but has a number of physics and technological advantages with minor deterioration of operating speed (Belkin et al., 2008):

- the possibility of the buffer and absorption layers precise doping;
- decrease of growth defects (growth pits, packing defects, absorption layer surface erosion);
- volatile components (such as As and P, and Zn doping agent) evaporation exclusion;
- the possibility of using InP "wide-gap window" effect that permits to collect photons on absorption layer without their surface recombination in the air-absorption layer interface;
- the possibility of desired form photodiode mesa-structure creation, using indium phosphide and solid solution etching selectivity, and small-area photodiodes forming selectivity;
- the possibility of compensating in part space-charge effect due to different thickness of the layers 2 and 4.

All the advantages above allow minimizing photodiode dark current ensuring maximum current responsivity.

The SYNOPSYS Sentaurus TCAD platform was used to model the Fig. 2's photodiode structure. The simulating procedure based on the following methodology. As the first step optical field distribution profile was inspected (Figure 3). In the result we determined that optical field is absorbed completely in the depletion layer 3 of Fig. 2 (dark horizontal stripe in Fig. 3). Then, by optical power-dependence reverse current-voltage characteristics computation we found that optimal photodiode operating voltage was -10 V (Belkin et al., 2008). The final step was AC small-signal current responsivity-frequency characteristics analysis and optimization in dependence of the absorption layer doping density, N_D, thickness, s and the photosensitive window width, w (in 2D-model).

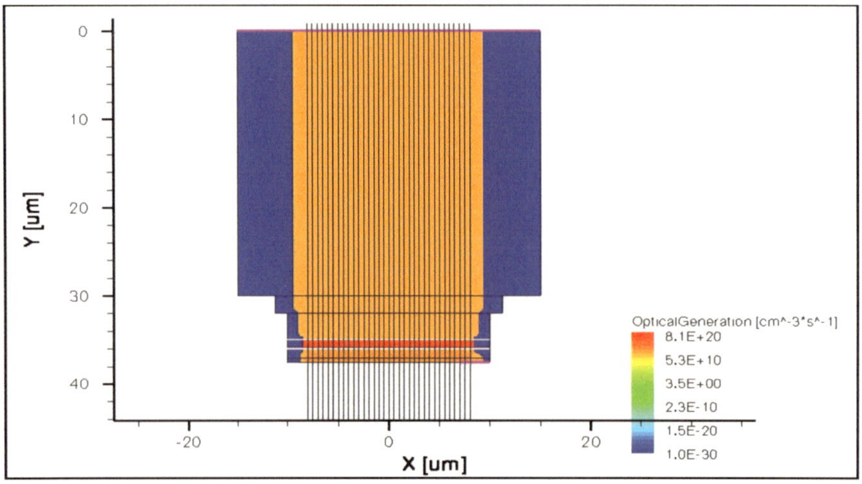

Fig. 3. Optical field distribution profile (The PD structure is upside-down to that of Fig. 2)

As known (Bowers & Burrus, 1987), the general limitations on the speed of p-i-n photodetectors are: 1) carrier drift time across depletion layer, 2) diode inherent capacitance charge/discharge time, 3) charge trapping at heterojunctions, 4) diffusing time out of undepleted regions, 5) package parasitic effects. Among them, in photodetector heterostructure with target bandwidth of 30 GHz the major effect is due to the first two factors because the hole trapping time even for abrupt p-InP/i-GaInAs heterojunction was estimated to be about 5 ps (Wey et al., 1995).

In the course of simulations by means of the program developed, some data for technology research and photodiode structure design, that are optimized for all the three above parameters, are received. First, optimal absorption layer doping density with s=1 μm and w=25 μm was investigated (Figure 4). Absorption region doping level analysis showed: 1) photodiode bandwidth was increased with the level diminishing, 2) there was no sense to provide lower than $N_D= 10^{16}$ cm^{-3} as the bandwidth became almost invariable (The latter is important from the technology point), 3) to meet the above project goal of 30 GHz the simply realized by well-known liquid-phase epitaxy $N_D=1{\cdot}10^{-16}$ cm^{-3} is quite enough. That's why the above value was used in the future simulations.

Furthermore, optimal absorption layer thickness with $N_D=1{\cdot}10^{-16}$ cm^{-3} and w=25 μm was investigated (Figure 5). Absorption layer thickness analysis showed: 1) as for super high-speed p-i-n photodiode it is necessary not to diminish, according to known results (see e.g.

Agethen, et al., 2002), but to increase the thickness for bandwidth enhancement because of the need to consider not only time of carrier drift through depletion region, but diode's self-capacitance charge/discharge time too; 2) optimal thickness of absorption region for photodiode modeled is near s=1 μm as for the larger ones the bandwidth became almost invariable. That's why the above value was determined as optimal for the project goal achievement.

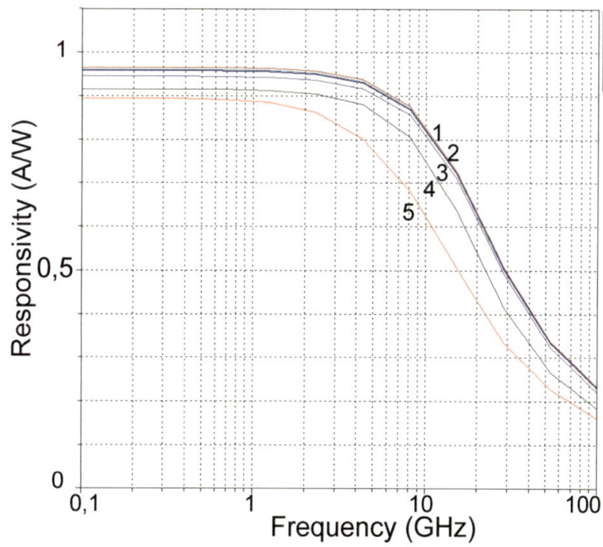

Fig. 4. The bandwidth vs. absorption region doping density simulation (1: $N_D=8 \cdot 10^{15}$ cm^{-3}; 2: $N_D=1 \cdot 10^{16}$ cm^{-3}; 3: $N_D=2 \cdot 10^{16}$ cm^{-3}; 4: $N_D=3 \cdot 10^{16}$ cm^{-3}; 5: $N_D=4 \cdot 10^{16}$ cm^{-3})

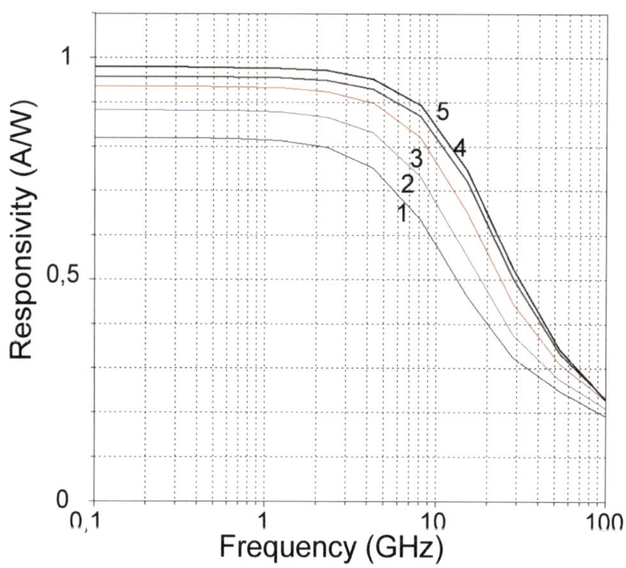

Fig. 5. The bandwidth vs. absorption region thickness simulation (1: s=0.3 μm; 2: s=0.55 μm; 3: s=0.8 μm; 4: s=1 μm; 5: s=1.3 μm)

Finally, the optimal photosensitive window width with N_D=1·10^{-16} cm^{-3} and s=1 μm was investigated (Figure 6). Photosensitive window width analysis showed: 1) the photodiode bandwidth was increased with the width diminishing; 2) the project goal value of 30 GHz is arrived when w≤22 μm that relatively similar with the other results (Bowers & Burrus, 1987; Agethen, et al., 2002).

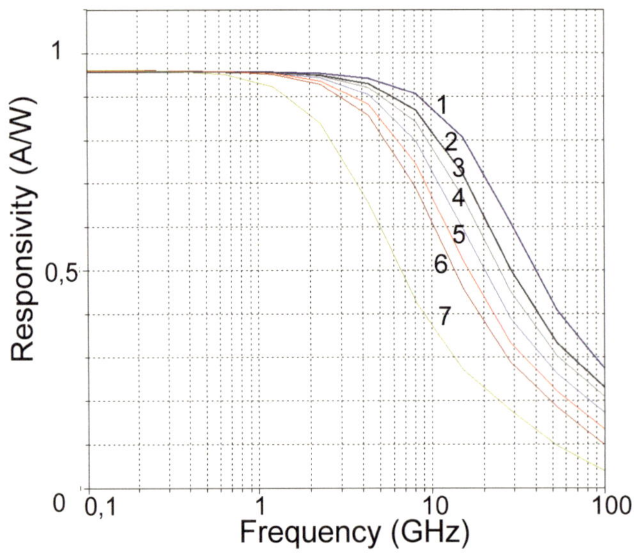

Fig. 6. The bandwidth vs. photosensitive window width simulation (1: w=18 μm; 2: w=20 μm; 3: w=22 μm; 4: w=24 μm; 5: w=26 μm; 6: w=28 μm; 7: w=30 μm)

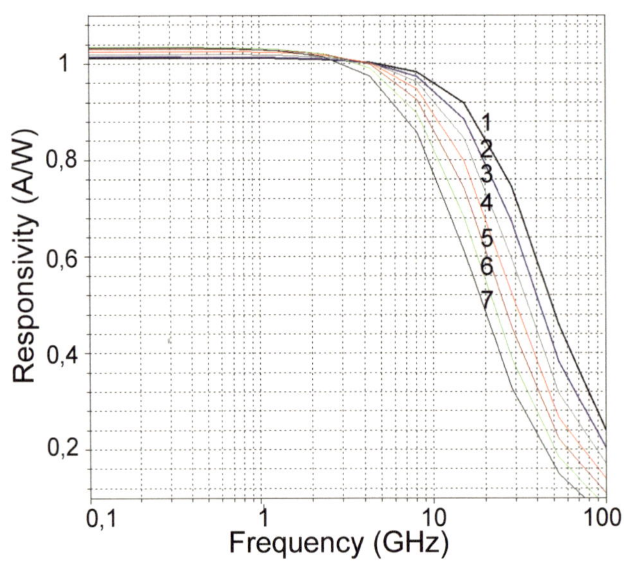

Fig. 7. The bandwidth vs. photosensitive window width simulation for a usual top-illuminated p-i-n photodiode structure with single heterojunction (1: w=18 μm; 2: w=20 μm; 3: w=22 μm; 4: w=24 μm; 5: w=26 μm; 6: w=28 μm; 7: w=30 μm)

Our last TCAD modeling was referred to the photodetector bandwidth estimation due to additional epitaxial layer introduction (Fig. 2, region 2). For this purpose the same photodiode structure response with closed region 2 was simulated (Figure 7). From the Figure one may conclude that the photosensitive window width needed for the project goal of near 30 GHz was increased slightly (up to 24 μm). So its effect has not to simplify substantively the mesa etching process.

3.2 P-i-n photodiode MW-ECAD simulation

As mention above for end-to-end design of a microwave-bandwidth photodiode an appropriate ECAD tool is required. The most convenient way for this microwave network research is the approximation by a physical equivalent circuit (Figure 8). In the Figure, I_{ph} is the complex photocurrent source (frequency F dependent), R_d, C_d, R_s are the p-i-n photodiode heterostructure differential resistance, intrinsic capacitance and serial resistance respectively. Furthermore, L_w and C_c are the connecting wire inductance and case capacitance approximation.

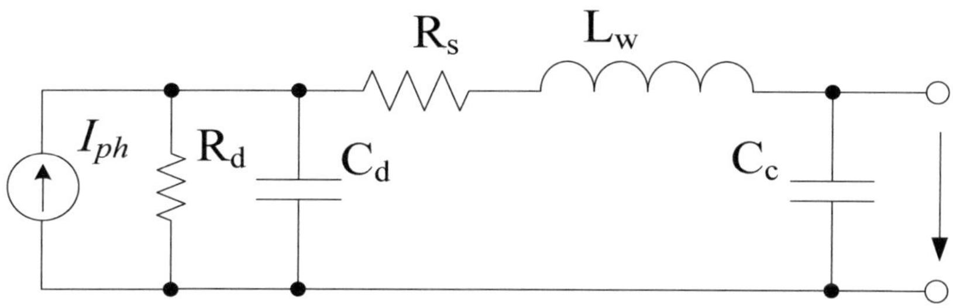

Fig. 8. The p-i-n photodiode electrical equivalent circuit

Based on the usual RF circuit simulation technique the photodiode transducer gain k_{pd} can be determined from this scheme as:

$$k_{pd} = I_L / I_{ph} \tag{6}$$

where I_L is complex load current. Then from the Figure 8:

$$\dot{k}_{ph} = \frac{(1 + j\omega C_c z_L)^{-1}(1 + j\omega\tau)}{1 - \omega^2 L_w C_d + j\omega C_d R_s + z_L(1 + j\omega C_d z_L)^{-1}} \tag{7}$$

where $\omega = 2\pi F$, z_L is load impedance, τ is transit time.

The values of $I_{ph}(F)$, τ, R_d=3.5 MΩ, C_d=80 fF, R_s=21 Ω were found analytically from the Sentaurus program. Also the values of L_w and C_c=0.11 pF were determined experimentally by the microwave vector network analyzer. Figure 9 illustrates some results of the AWRDE [2] ECAD environment simulation. It follows from the graphs that the designer can realized twofold expansion of the photodiode 3 dB bandwidth (20 to 40 GHz) owing to the appropriate fitting of the connecting wire inductance L_w.

[2] http://web.awrcorp.com

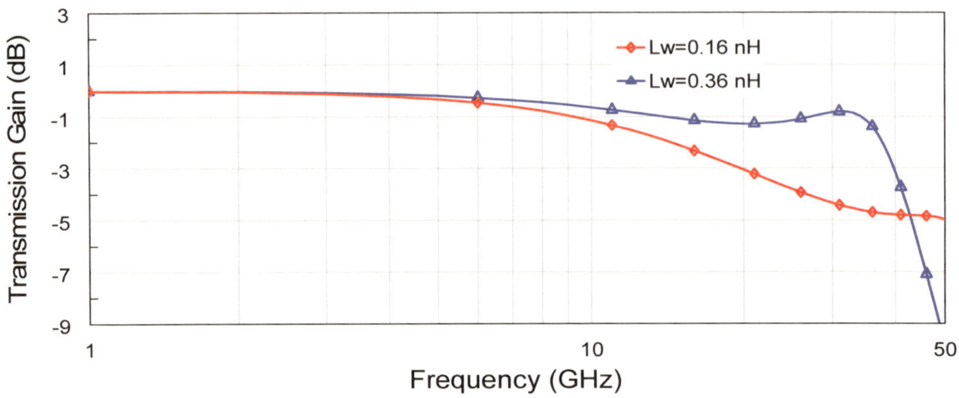

Fig. 9. The photodiode packaging effect simulation

3.3 Experimental verification

Based on the above simulation results some p-i-n photodiode samples were produced and mounted into the microwave microstrip test fixture. The typical time response of the photodetector sample measured by the two-section mode-locked laser and high-speed sampling oscilloscope is shown in the Figure 10. The pulse duration from the Figure is near 10 ps, so the accuracy of the above simulation results is verified.

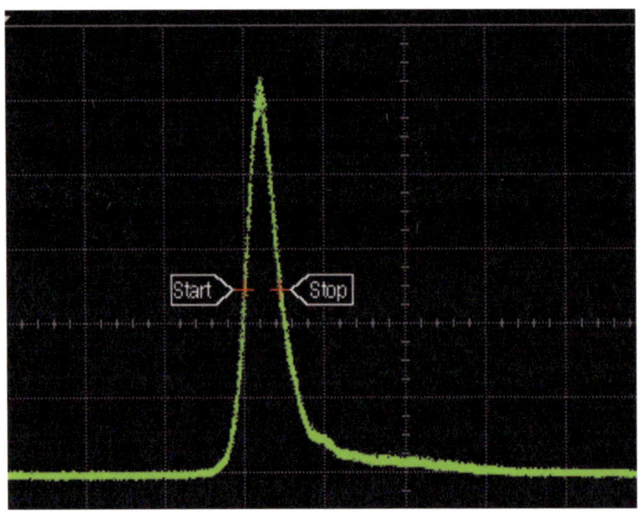

Fig. 10. Experimental photodiode impulse response (horizontal scale 25 ps/div.)

Thus, the methodology of the end-to-end computer-aided design for a p-i-n photodiode with microwave bandwidth is addressed. First, using the technology CAD tool a designer can define and optimize the dependence of the major photodiode structure parameters (absorption layer doping density and thickness, photosensitive window diameter, so on) on small-signal bandwidth and get the data for the following calculations. Then the electronic MW-CAD simulation based on the photodiode electrical equivalent circuits is performed that allow to take into account various effects in real receiver module with microwave bandwidth: LC package parasitics, matching networks, front-end amplifier and so on. To confirm the actuality and accuracy of the proposed methodology the specific p-i-n photodiode with bandwidth of near 30 GHz was simulated. The simulation results were experimentally verified with fine accuracy.

4. Solitary MW-ECAD simulation

The main concept of this approach can be formulated as follows: a designer of analog receiving modules usually does not know a lot about the PD heterostructure's specific parameters, usually has not an access to a specific cost-expensive TCAD tool but he has facilities suitable for measuring transmittance and reflectance (or S-parameters) of the photodiode chip under test in the optical system's modulation band. Then equipped with experimental results, and equivalent circuit interpretation as well as AWRDE or another MW-ECAD tool facilities he builds up a non-linear "optoelectronic" equivalent circuit and if necessary some linear or nonlinear electronic circuits behind it.

4.1 P-i-n photodiode nonlinear model

Figure 11 represents cascaded equivalent circuit of PD that does it quite effective for simulation by means of a highly-developed MW-ECAD tool. There are two sub-circuits inside it: linear (right) and nonlinear (left). The linear sub-circuit simulating the frequency distortions due to parasitic elements impact agrees with small-signal PD model of Fig. 8. The nonlinear sub-circuit includes a diode D1 and a capacitance of p-n junction that is changed in according with applied reverse voltage (element PNCAP). To supply the nonlinear elements a DC source V2 is introduced. Inductance L1 is assigned for DC/AC isolation; resistance R2 models PD's DC termination. Non-ideality, phase shift, and delay of the optical-to-electrical conversion process are modeled by voltage-controlled current source U1.

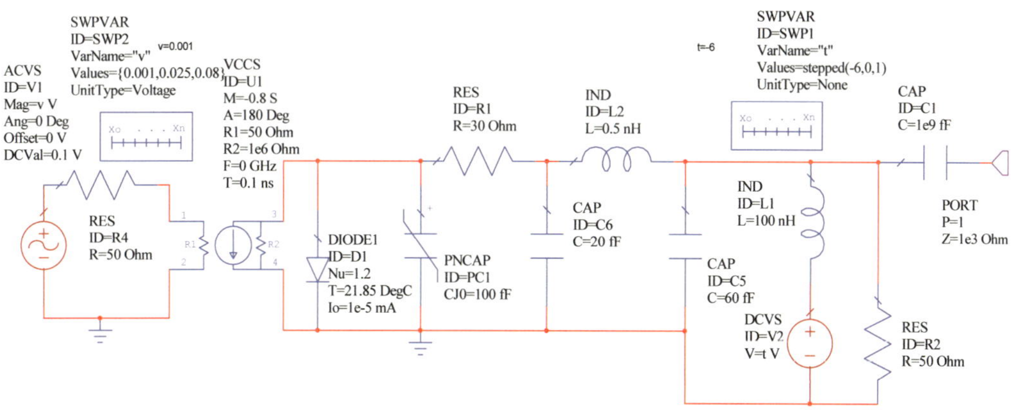

Fig. 11. MW-ECAD nonlinear p-i-n photodiode model based on electrical equivalent circuit

For the simulation an equivalent electrical AC source V1 is placed at the circuit input. A frequency of this source is tuned through the microwave-band modulation bandwidth, and its amplitude is in exact numerical agreement with a power of intensity-modulated optical signal. As it follows from Figure 11 for the given model experiment the next values are assumed: an equivalent power of optical carrier is 100 mW (DC voltage 0.1 V), equivalent modulated optical power is changed in the range of 10-100 mW (0.01-0.1 V) that meets the optical modulation rate between 10 and 100 percentage. A reasonableness of the optoelectronic circuit's simulation by ECAD tool is founded on the proposed method of equivalent voltage which uses identical linear dependences of a current vs. voltage in an electrical circuit and a photocurrent vs. optical power in an optoelectronic circuit of a photoreceiver.

Thus, the next linear or nonlinear effects are taken into account in the given model: (i) nonlinearity of depletion layer capacity; (ii) frequency distortions due to parasitic elements of the PD package and receiver circuitry; (iii) reflections from photosensitive window; (iv) dark current and its dependence of temperature and bias voltage; (v) DC photocurrent generation due to optical carrier impact; (vi) deviation of the optical signal modulation rate; (vii) electric field screening due to series resistance.

4.2 Simulation results and discussion

At the first step, to calibrate the model the device-under-test small-signal frequency characteristics are simulated. As the result, 3-dB bandwidth is limited in the range of 20-30 GHz depending on reverse bias voltage and agreed well with the Figure 9 at the bias of 10 V. Then, the above-described three nonlinear distortion parameters depending on equivalent input optical power and reverse bias (Ub) are simulated through one-button operations. Figure 12 shows an example of equivalent AM-to-PM distortion characteristics at the fundamental frequency 15 GHz.

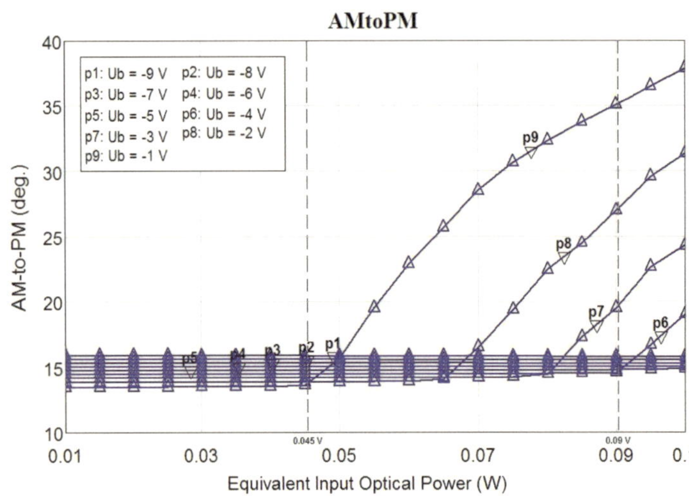

Fig. 12. Photodiode power to phase conversion characteristics

Also Figure 13 depicts an example of OIP2 characteristics for the input tones of 5 GHz and 6 GHz, and distortion tone of 11 GHz. At last Figure 14 shows an example of OIP3 characteristics for the input tones of 5 GHz and 5.1 GHz, and distortion tone of 4.9 GHz.

A combined outcome from Figures 12-14 follows that large-signal conversion features of a p-i-n photodetector are notably degraded as far as negative reverse DC bias is increased that is quite coincided with the experimental data of Table 1. In particular, Figure 12 depicts that the PPC linearity threshold is twofold as the PD bias is decreased from -1 to -4 V. Also from Figure 13 one can see that second-order output intercept points are almost constant vs. optical power and their numerical deviations vs. bias are near 26 dB (from 47 dBm at -8 V to 21 dBm at -1 V). In spite of this, shown in Figure 14 third-order output intercept points begin to degrade with optical power at the bias more than -4 V, and their numerical deviations vs. bias are near 17 dB (from 44 dBm at -8 V to 27 dBm at -1 V). Note that the results of power-to-phase conversion simulation are comparable in principle with recent experimental data (Tailor et al., 2011), also the results of OIP2 and OIP3 simulation are closed to recent experimental data of Table 1.

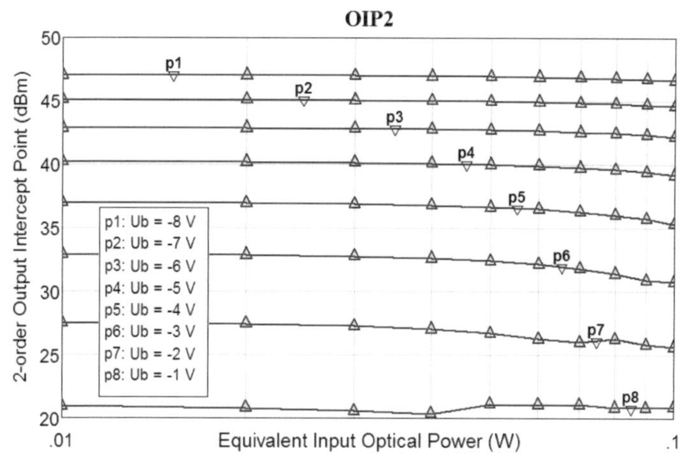

Fig. 13. Output second-order intercept point characteristics

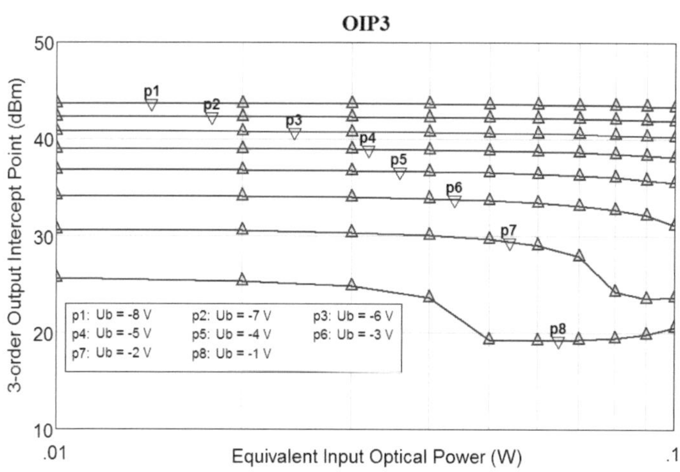

Fig. 14. Output third-order intercept point characteristics

Generally, based on our model experiments we are able to state that simple in design and producing cost-effective top-illuminated photodetector has been inspected holds somewhat (3-5 dB) lower linearity as compared to a special high-linear charge-compensated modified uni-traveling carrier photodiode (see Table 1). But the latter has at the same time significantly higher bandwidth running up to THz band (Nagatsuma et al., 2007). Nevertheless the researched type of photodetector thanks to inherent cost-effectiveness also has a potential for RoF systems and microwave photonics applications with moderate frequency and power requirements.

For example, the typical values of the allowable AM-to-PM distortion and OIP3 for middle-power reasonable-noise microwave amplifiers are not more than 10° as well near 30-35 dBm correspondently. Following it we are able to conclude: (i) even a standard p-i-n photodetector is more linear circuit element than pre-amplifier and can be used without it in the most photoreceiving schemes of microwave photonics devices and RoF system's photonic antennas; (ii) to secure a comparable p-i-n photodetector linearity the reverse bias voltage must be less than -4 V at optical power up to 100 mW.

5. Comparison of OE-CAD and MW-ECAD photodiode simulations

As we communicated in Introduction, the available optoelectronic CAD tools are unpractical for designing nonlinear device operating in microwave band. That's why it is in essential importance to state the numerical value of error while simulating for instance noise features of a photodetector in microwave band using MW-ECAD and OE-CAD software. Our direct comparison was fulfilled by AWRDE and advanced OE-CAD tool VPItransmission Maker™ version 8.6.

In the case of VPI software a library p-i-n photodiode model considering thermal and shot noise sources and dark current of 0.15 nA was selected. On the other hand, for photodiode MW-ECAD simulation the equivalent circuit model of Fig. 11 was used. For the both simulating experiments a broadband Gaussian-distributed white noise generator model with identical power levels was employed as a source. Varying its levels output noise power was measured in the band from 1 GHz to 50 GHz by the noise analyser model in the bandwidth of 1 Hz. Figure 15 depicts the simulation results.

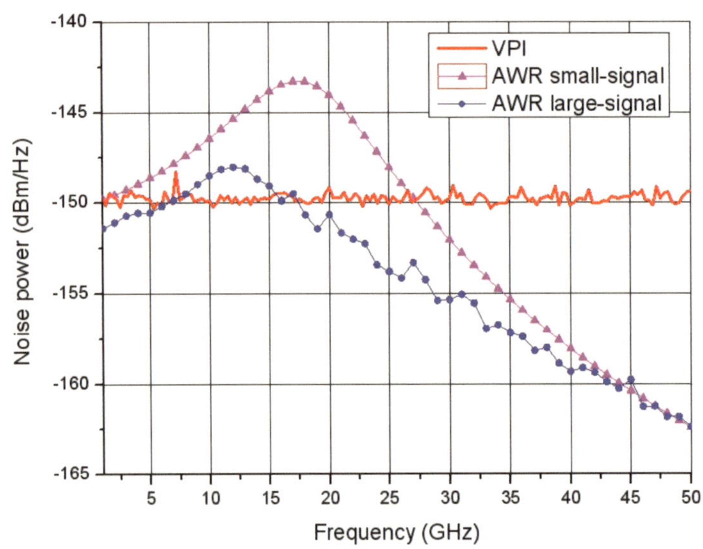

Fig. 15. OE-CAD tool vs. MW-ECAD tool simulation

As shown from Figure, simulated by VPI (solid curve) spectral density of average output noise power represents frequency and noise source power independent horizontal line that is unrealistic. On the other hand, simulated by AWRDE results follow the real frequency characteristic of the photodiode under test and differ subject to small-signal (triangles) or large-signal (circles) power regime of the noise source. The numerical value of error is up to 10 dB.

6. Conclusion

This chapter presented a multiscale approach to the combined TCAD&ECAD and solitary MW-ECAD simulations for designing a microwave-band p-i-n photodetector as a basic component of the receiving modules for radio-over-fiber systems and microwave photonics devices. Within the limits of the first approach, primarily, using a TCAD tool a designer can define and optimize the dependence of the major photodiode structure parameters on small-signal bandwidth. Then an electronic MW-CAD simulation based on the photodiode electrical equivalent circuits is performed that allow to take into account various effects in real receiver module with microwave bandwidth. To confirm the actuality and accuracy of the proposed methodology the specific p-i-n photodiode with bandwidth of near 30 GHz was simulated. The main concept of the second approach can be formulated as follows: a designer of analog receiving modules usually does not know a lot about the PD heterostructure's specific parameters, usually has not an access to a specific cost-expensive TCAD tool but he has facilities suitable for measuring transmittance and reflectance of the photodiode chip under test in the optical system's modulation band. Then equipped with experimental results, and equivalent circuit interpretation as well as AWRDE or another MW-ECAD tool facilities he builds up a non-linear "optoelectronic" equivalent circuit and if necessary some linear or nonlinear electronic circuits behind it. Following the simulation results we have concluded: (i) even a standard p-i-n photodetector is more linear circuit element than pre-amplifier and can be used without it in the most photoreceiving schemes of microwave photonics devices and RoF system's photonic antennas; (ii) to secure a comparable p-i-n photodetector linearity the reverse bias voltage must be less than -4 V at optical power up to 100 mW.

The modeling results are validated by the experimental data for TCAD&ECAD similation and by close agreement with the known measured data for MW-ECAD large-signal simulation.

7. Acknowledgment

This work was supported by the Russian Ministry of Education and Science program "Progress of scientific potential of the higher school (2009-2011)". Author wishes to thank Dr. E. Portnoi from Ioffe Physico-Technical Institute RAS for the mode-locked laser's photodiode measurement.

8. References

Agethen, M., et.al. (2002). InGaAs pin Detectors for Frequencies above 100 GHz, *IEEE Indium Phosphide and Relative Materials Conference, IPRM'2002*, A8-2, pp. 673-676

Beling, A., Pan, H., Chen, H. & Campbell, J.C. (2008). Measurement and modeling of high-linearity modified uni-traveling carrier photodiode. *IEEE Photonics Technology Letters*, Vol.20, No.14, (July 2008), pp. 1219-1221, ISSN 1041-1135

Beling, A. & Campbell, J.C. (2009). InP-Based High Speed Photodetectors, *IEEE Journal of Lightwave Technology*, Vol. 27, No 3, (February 2009), pp. 343-355

Belkin, M.E., Dzichkovski, N.A. & Indrishenok.V.I. (2008). Super high-speed p-i-n-photodiode heterostructures computer simulation. *Journal of Nano and Microsystem Techniques*, No 10 (99), (October 2008) pp. 23 - 28, (in Russian)

Belkin, M.E. & Dzichkovski, N.A. (2009). Research of microwave-bandwidth p-i-n photodetectors, *Proceedings of EUROCON* 2009, p.193-196, St. Peterburg, Russia, May 2009

Blakey, P.A. (2008). Technology CAD. In: *RF and Microwave Circuits, Measurements, and Modeling*, M. Golio, J. Golio, (Ed.), Boca Raton, 34-1 - 34-11, ISBN 13-978-0-8493-7218-6, London

Bowers , J. E. & Burrus C.A. Ultrawide-Band Long-Wavelength p-i-n Photodetectors, *IEEE Journal of Lightwave Technology*, vol. 5, No 10,(October 1987), pp. 1339-1350

Capasso, F., et. al. (1985). Photodetectors, In: *Lightwave Communications Technology*, Vol. 22, Part D, W.T. Tsang, (Volume Ed.), Academic Press Inc.

Draa, M.N., et al. (2008). Frequency behaviors of third order intercept point for a waveguide photodiode using three laser two-tone setup. *Proceedings of 21st Annual Meeting of the IEEE Lasers and Electro-Optics Society. LEOS 2008*, pp. 284-285, Acapulco, Mexico, November 9-13, 2008

Godinez, M.E., et al. (2008). RF Characterization of Zero-Biased Photodiodes. *IEEE Journal of Lightwave Technology*, Vol. 26, No 24, (December 2008), pp. 3829-3834

Gustavsson, M., Hedekvist, P.O. & Andrekson, P.A. (2005). Uni-Travelling-Carrier Photodiode Performance with X-band Modulation at High Optical Power. *IEEE Microwave and Wireless Components Letters*, v.15, No.5, (May 2005), pp. 297-299, ISSN 1531-1309

Itakura, S., et al. (2010). High-Current Backside-Illuminated Photodiode Array Module for Optical Analog Links. *IEEE Journal of Lightwave Technology*, Vol. 28, No 6, (March 2010), pp. 965-971, ISSN 0733-8724

Jiang, H. et al. (2000). The Frequency Behavior of the Third-Order Intercept Point in a Waveguide Photodiode. *IEEE Photonic Technology Letters*, Vol.12, No.5, (May 2000), pp. 540-542

Joshi, A., Datta, S. & Becker, D. (2008). GRIN Lens Coupled Top-Illuminated Highly Linear InGaAs Photodiodes. *Photonics Technology Letters*, Vol.20, No.17, (September 2008), pp. 1500-1502, ISSN 1041-1135

Joshi, A. & Datta, S. (2009). Highly Linear, High Power Handling Photodiode for RF Photonic Links, *Proceedings of ECOC 2009*, p. 9.2.4, Vienna, Austria , 20-24 September 2009

Kenington, P.B. (2000). *High-Linearity RF Amplifier Design*. Artech House, ISBN 1-58053-143-1, Norwood, MA

Kielmeyer, R. (2008). Computer Aided Design (CAD) of Microwave Circuitry. In: *RF and Microwave Circuits, Measurements, and Modeling,* M. Golio, J. Golio,, (Ed.), Boca Raton, 31-1 - 31-8, ISBN 13-978-0-8493-7218-6, London

Li, Z., et al. (2010). High-Saturation-Current Modified Uni-Traveling-Carrier Photodiode With Cliff Layer. *IEEE Journal of Quantum Elecronics,* Vol.46, No.5, (May 2010),pp. 626-632, ISSN 0018-9197

Liao, T.S. et al. (2003). Investigation of the High Power Integrated Uni-Traveling Carrier and Waveguide Integrated Photodiode, *IEEE 2003 MTT-S International Microwave Symposium Digest*, pp. 155-158, vol.1, Philadelphia, PA, 8-13 June, 2003

Lowery, A.J. (1997). Computer-aided photonics design. *IEEE Spectrum,* Vol. 34, No 4 (April 1997), pp. 26-31, ISSN: 0018-9235

Minixhofer, R. TCAD as an integral part of the semiconductor manufacturing environment. (2006), *Proceedings of International Conference on Simulation of Semiconductor Processes and Devices, SISPAD,* pp. 9 - 16. Monterey, California, September 6-8, 2006

Nagatsuma, T., Ito, H. & Ishibashi, T. (2007). Photonic THz Sources Using Uni-Traveling-Carrier Photodiode Technologies. *Proceedings of LEOS 2007, International Topical Meeting of Laser and Electro-Optic Society,* pp. 792-793.

Pedro, J. C. & Carvalho, N. B. (2003). *Intermodulation Distortion in Microwave and Wireless Circuits,* Artech House, ISBN 1580533566, Boston, London

Ramaswamy, A. et al. (2009). Experimental analysis of two measurement techniques to characterize photodiode linearity. *Proceedings of MWP '09, International Topical Meeting on Microwave Photonics,* pp. 1-4, ISBN 978-1-4244-4788-6. Valencia, Spain, October 14-16, 2009

Sauer, M., Kobyakov, A. & George, J. (2007). Radio over Fiber for Picocellular Network Architectures, *IEEE Journal of Lightwave Technology,* vol. 25, no. 11, (November 2007), pp. 3301-3320

Taylor, J.A., et al. (2011). Characterization of Power-to-Phase Conversion in High-Speed P-I-N Photodiodes. *IEEE Photonics Journal,* Vol. 3, No.1, (February 2011), pp. 140 - 151, ISSN 1943-0655

Tulchinsky, D.A. et al. (2008). High Current Photodetectors as Efficient, Linear and High-Power RF Ouput Stages. *IEEE Journal of Lightwave Technology,* Vol. 26, No.4, (February 2008), pp. 408–416

Wey, Y.-G. et al. (1995). 110-GHz GaInAs/InP Double Heterostructure p-i-n Photodetectors, *IEEE Journal of Lightwave Technology,* vol. 13, No 7,(July 1995), pp. 1490-1499

Williams, K. J., Esman, R. D. & Dagenais, M. (1996). Nonlinearities in p-i-n Microwave Photodetectors., *IEEE Journal of Lightwave Technology,* vol. 14, No 1,(January 1996), pp. 84-96

Yao, J.P. (2009). Microwave Photonics, *IEEE Journal of Lightwave Technology,* vol. 27, No 3, (February 2009), pp. 314-335

Yao, X.S. (2002). Opto-electronic Oscillators, In: *RF Photonic Technology in Optical Fiber Links,* W. S. Chang(Ed.), 255-292, Cambridge University Press, Cambridge, UK

Yu, P. K. L. & Wu, M. C. (2002). Photodiodes for high performance analog links, In: *RF photonic technology in optical fiber links,* W. S. Chang, (Ed), 231-254, Cambridge university press, Cambridge, UK

Zakhleniuk, N.A. (2007). Theory and Modelling of High-Field Carrier Transport in High-Speed Photodetectors. *Proceedings of NUSOD'07 Numerical Simulation of Optoelectronic Devices*, pp. 77-78, ISBN 978-1-4244-1431-4, Newark, DE, September 24-28, 2007

12

Mathematical Modeling of Multi-Element Infrared Focal Plane Arrays Based on the System 'Photodiode – Direct-Injection Readout Circuit'

I.I. Lee and V.G. Polovinkin
A.V. Rzhanov Institute of Semiconductor Physics Siberian Branch of Russian Academy of Sciences
Russia

1. Introduction

The direct-injection readout circuit was proposed in 1973 by A. Steckl and T. Koehler (Steckl & Koehler, 1973). With such readout circuits, first hybrid multi-element InfraRed Focal Plane Arrays (IR FPAs) were implemented (Steckl, 1976; Iwasa, 1977). The direct-injection readout circuits are being used in the majority of hybrid multi-element Long-Wave InfraRed (LWIR) FPAs, which hold more-than-70% a share in the world market of thermography systems (Rogalski, 2000).

A diagram of FET-based readout circuit is shown in Fig.1a, the most popular design involving a charge-coupled device, Fig.1b (Longo, 1978; Felix,1980; Takigawa, 1980; Rogalski, 2000). On application of a dc voltage U_G to the input gate, a certain voltage, defined by the surface potential under the input gate of the photodetector channel, sets across the photodiode, and the current generated in the IR detector is integrated at the storage capacitor C_{int}. On applying of a transfer pulse Ft (Fig.1a), the charge accumulated in the storage capacitor transfer to the column read bus.

The equivalent circuit of the direct injection readout circuit is shown in Fig. 2.

Normally, the current through the photodiode is in the range of $10^{-7}\div10^{-10}$A, the input FET being therefore operated in subthreshold mode. The transconductance of the input FET channel is:

$$g_{in} = \frac{\partial I_{in}}{\partial V_G} = \frac{qI_{in}}{N^* kT} \quad (1)$$

Here, q is the electron charge, k is the Boltzmann constant, $N^*= (C_{OX}+C_D^*+C_{SS}^*)/C_{OX}$ (C_{OX} is the specific capacitance of gate dielectric, $C_{SS}^*=qN_{ss}^*$ is the specific capacitance of fast surface states, C_D^* is the specific capacitance of the depletion region, T is temperature, and the asterisk * indicates that the parameter value is taken for the conditions with surface Fermi level equal to $3/2\ \varphi_{FB}$.

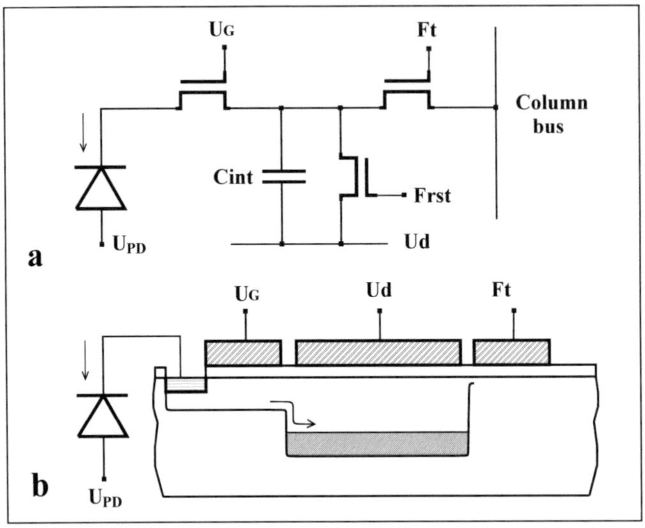

Fig. 1. Readout circuit with charge direct injection.

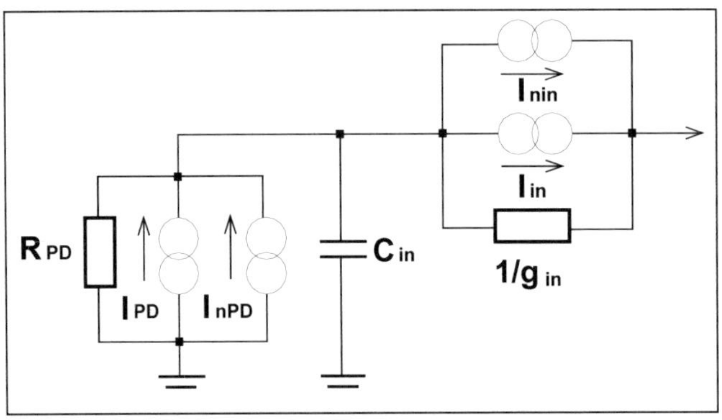

Fig. 2. The equivalent circuit of the photodetector channel of a hybrid IR FPA based on the system 'photodiode - direct-injection readout circuits'. Here, I_{PD} is the current that flows through the photodiode, I_{nPD} is the noise current that flows through the photodiode, R_{PD} is the dynamic resistance of the photodiode, I_{in} is the current integrated under the storage gate of the readout circuit, I_{nin} is the noise current integrated under the storage gate, and C_{in} is the capacitance of the input direct injection.

From the equivalent circuit, all basic relations for the readout circuit of interest can be inferred. In particular, for injection efficiency η_I we obtain (Longo, 1978; Felix,1980):

$$\eta_I = \frac{g_{in}}{g_{PD}+g_{in}+j\omega C_{in}} \tag{2}$$

where $g_{PD}=1/R_{PD}$, R_{PD} is the dynamic resistance of the photodiode.

Analysis of transfer characteristics of the system 'photodiode - direct-injection readout circuit' has revealed several drawbacks inherent to such readout circuits (Longo, 1978; Felix,1980; Takigawa, 1980; Rogalski, 2000):

- an IR FPA responsivity value comparable with that of IR photodiodes operated in photovoltaic regime can be reached only on the condition that the injection efficiency approaches unity. This requirement entails rigid requirements on the dynamic resistance of photodiodes, these requirements often being hard to meet, especially in LWIR FPAs;
- since the bias voltage across the photodiode is defined by the surface potential of the input-gate modulated channel, for multi-element IR FPAs the variations of transfer characteristics (fixed-pattern noise) and the excess noise level proved to be largely defined by the spread of input-FET threshold voltages ΔV_{th} (Longo, 1978; Gopal, 1996).

Several circuit designs were proposed to eliminate the above drawbacks. For instance, R. Bluzer and R. Stehlik (Bluzer & Stehlik,1978) proposed to use, with the aim of reducing the input impedance of readouts, the so-called buffered direct injection, implemented through introduction of an inverting amplifier in between the input direct injection and input FET gate.

J. Longo *et al.* (Longo, 1978) proposed a design of direct injection readout circuit capable of diminishing the spread of photodiode bias voltages to the typical difference of threshold voltages of two closely spaced FETs. Nonetheless, being capable of improving certain characteristics of direct injection readouts, the additional components (the amplifier normally involves 4 to 6 FETs) take off a substantial fraction of surface area from the photodetector cell, doing simultaneously the direct injection readout circuits out of their main advantageous features, low energy consumption and a large storage capacity, high in comparison with readout circuits of other types.

The advances in silicon technology and in the technology of IR photodiodes based on narrow bandgap semiconductors have allowed a substantial increase of photodiode dynamic resistance and a reduced variation of photodiode bias voltages across FPA from the values ~ 20-50 mV, typical of the technology level of the 1970s, to ~ 1-10 mV. For this reason, as a rule, the simplest design of direct injection readout circuit shown in Fig.1 is mainly used.

An analysis of the system 'photodiode – direct injection readout circuit' given in (Steckl & Koehler 1973; Steckl, 1976; Iwasa, 1977; Rogalski, 2000; Longo, 1978; Felix,1980; Takigawa, 1980; Gopal, 1996) was carried out versus photodiode biasing, although the bias voltage across the photodiode is defined by the surface potential in the input-FET channel and can be adjusted through the proper choice of the potential U_G applied to the input-FET gate. That is why such estimates do not allow the prediction of responsivity of multi-element IR FPAs and evaluation of fixed-pattern noise level and Noise-Equivalent Temperature Difference (NETD) figures of thermography systems based on such IR FPAs. In other words, such estimates do not permit numerical modeling of multi-element IR FPAs.

In the present article, we propose a mathematical model of the system 'photodiode – direct injection readout circuit', a computer program, and an analysis procedure for thermography systems based on multi-element IR FPAs; these model and analysis procedure were developed to solve the above-indicated problems. We describe a procedure for revealing the effects due to the spread of electrophysical parameters of photodiodes and Si readouts based on plotting calculated fixed-pattern noise, detectivity, and NETD histograms of thermography systems based on multi-element IR FPAs. It is significant that all dependences were obtained as dependent on the input-gate voltage of readouts; this enables a comparison with the experiment and formulation of requirements to electrophysical and design parameters of photodiodes and Si readouts necessary for implementation of design characteristics.

2. A mathematical model for the system 'IR photodiode - direct-injection readout circuit'

Basic assumptions adopted in the proposed mathematical model of the system 'IR photodiode – direct injection readout circuit' were first formulated in (Kunakbaeva, Lee & Cherepov, 1993) and further developed in (Kunakbaeva & Lee, 1996; Karnaushenko; Lee et al., 2010; Gumenjuk-Sichevska, Karnaushenko, Lee & Polovinkin, 2011). In calculating the transfer characteristics of the system, for the input FET we employ the model of long-channel transistor in weak inversion (Overstraeten, 1975). The noise charge Q(t) integrated under the storage gate of the direct injection readout circuit is calculated as a McDonalds function expressed in terms of spectral current density $S_i(\omega)$ (Buckingham, 1983):

$$\overline{Q^2}(t) = \frac{1}{\pi}\int_0^{\infty} \frac{S_i(\omega)}{\omega^2}(1-\cos\omega t)d\omega \tag{3}$$

The spectral density of the noise current in the input-gate modulated channel is given by

$$S_i(\omega) = 4kTg_{in}\alpha_1\left|1-\eta_I\right|^2 + \frac{2\pi B_{in} I^2_{in}}{\omega}\left|1-\eta_I\right|^2 + \left(2qI_{PD} + \frac{4kT\alpha_2}{R_{PD}}\right)\left|\eta_I\right|^2 + \frac{2\pi B_{PD}}{\omega}\left|\eta_I\right|^2 \tag{4}$$

The first and second terms in (4) stand for the thermal noise of the input FET and for the 1/f noise induced by this FET (here, α_1 and α_2 are coefficients) and the third and forth terms stand for the thermal noise and 1/f noise of the photodiode, here, $B_{in}= 2\pi K I_{in}^2/WL(C_{OX}+C_D^*)^2$, K is a coefficient (D'Souza, 2002), and $B_{PD} = \alpha_3(I_{PD} - \eta_K I_P)^2$, α_3 is a coefficient (Tobin, 1980). Substituting the first term of (4) into (3), after integration we obtain the following analytical expression for Q_1 - the thermal noise of the gate-modulated channel expressed as the amount of noise electrons:

$$\begin{aligned} Q_1^2 = {} & \frac{2kTC_{in}g_{in}R_{PD}\alpha_1}{q^2(1+g_{in}R_{PD})}\left[1-\exp\left(-T_{in}\frac{1+g_{in}R_{PD}}{R_{PD}C_{in}}\right)\right] + \\ & + \frac{2kTg_{in}\alpha_1}{q^2(1+g_{in}R_{PD})^2}\left[T_{in} - \frac{R_{PD}C_{in}}{1+g_{in}R_{PD}}\left(1-\exp\left(-T_{in}\frac{1+g_{in}R_{PD}}{R_{PD}C_{in}}\right)\right)\right] \end{aligned} \tag{5}$$

where T_{in} is the integration time.

In estimating the 1/f noise of the input FET Q_2, also expressed as the amount of noise electrons, while performing integration, we have to take the fact into account that the readout regime involves a high-pass filter with transfer characteristic $\omega^2/(\omega_0^2 + \omega^2)$, where $\omega_0 = \pi/T_{in}$.

$$Q_2^2 = \frac{2KI_{in}^2}{qWL(C_{OX}+C_D)^2}\int_0^{\infty}\frac{1+(C_{in}R_{PD}\omega)^2}{(1+g_{in}R_{PD})^2+(C_{in}R_{PD}\omega)^2}\frac{1}{\omega(\omega_0^2+\omega^2)}\sin^2\frac{\omega T_{in}}{2}d\omega \tag{6}$$

We substitute the third term of (4) into (3) and obtain, after integration, the following analytical expression for Q_3 - the noise component due to photodiode current noise expressed as the amount of noise electrons:

$$Q_3^2 = \frac{(g_{in}R_{PD})^2}{q^2(1+g_{in}R_{PD})^2}\left(qI_{PD}+\frac{2kT\alpha_2}{R_{PD}}\right)\left(T_{in}-\frac{R_{PD}C_{in}}{1+g_{in}R_{PD}}\left(1-\exp\left(-T_{in}\frac{1+g_{in}R_{PD}}{R_{PD}C_{in}}\right)\right)\right) \tag{7}$$

The forth term in (4) stands to allow for the $1/f$ noise of the photodiode:

$$Q_4^2 = 2\alpha_3^2(I_{PD}-\eta_K I_P)^2\frac{(g_{in}R_{PD})^2}{q^2}\cdot \int_0^{\infty}\frac{1}{(1+g_{in}R_{PD})^2+(C_{in}R_{PD}\omega)^2}\frac{1}{\omega(\omega_0^2+\omega^2)}\sin^2\frac{\omega T_{in}}{2}d\omega \tag{8}$$

The detectivity D* of the photodetector channel in which photosignals from the IR photodiode are read into the direct injection input is given by (Rogalski, 2000):

$$D^* = \frac{\lambda}{\hbar c}\frac{(A_{PD}T_{in}/2)^{1/2}\eta_I\eta_K}{\left(Q_1^2+Q_2^2+Q_3^2+Q_4^2+Q_{other}^2\right)^{1/2}} \tag{9}$$

where A_{PD} is the photodiode area and Q_{other} stands for all other noise components such as, for instance, the measuring channel induced noise.

3. Examples of model calculations

3.1 Modeling procedure for multi-element IR FPAs with direct injection readout circuits

The modeling procedure for the system 'IR photodiode - direct injection readout circuits' can be described using the simple IR photodiode model:

$$I_{PD} = \eta_K I_P + I_0\left[1-\exp(-\beta V_{PD})\right]+\frac{V_{PD}}{R_S} \tag{10}$$

Here, η_K is photodiode quantum efficiency, I_0 is the photodiode saturation current, R_S is the photodiode shunt resistance, and I_P is the current induced by background radiation.

The calculation starts with specifying the values of electrophysical and design parameters of readout circuits, photoelectrical parameters of photodiodes, and operating conditions of IR FPA, which all are listed in Table. 1.

At a given input gate voltage V_G, the voltage across the photodiode can be determined from the condition $I_{in} = I_{PD}$. We identify the points of intersection of the current-voltage characteristic of the photodiode and the transfer characteristics of the input FET to determine the photodiode bias voltage.

Designation	Parameter	Numerical value
μ	Mobility of minorities in the inversion channel of FET	500 $cm^2\ V^{-1}\ c^{-1}$
N_D	Donor concentration in the substrate	$7 \cdot 10^{14}\ cm^{-3}$
C_{OX}	Specific capacitance of gate dielectric	$1.24 \cdot 10^{-7}\ F \cdot cm^{-2}$
V_{FB}	Flat-band voltage	0 V
N_{ss}^{*-}	Surface-state density	$1 \cdot 10^{9}\ cm^{-2} \cdot eV^{-1}$
W, L	Length and width of input-gate modulated channel	30 μm, 3 μm
C_{in}	Input capacitance of FPA cell	0.5 pF
α_1	Numerical coefficients	2
α_2	Numerical coefficients	2
α_3	Numerical coefficients	10^{-3}
K	Numerical coefficients	$1.5 \cdot 10^{-24}\ F^2 \cdot cm^{-2}$
A_{PD}	Photodiode area	$9 \cdot 10^{-5}\ cm^2$
η_K	Photodiode quantum efficiency	0.8

Table 1. Design and electrophysical parameters of readout circuits

Figure 3 shows the curves of currents I_{in} and I_{PD} for the photodiode model (10). With the parameters of photodiode, input FET, and radiation environment adopted in Fig. 3, the voltage drop across the photodiode is zero at V_{G0} = 1.198 V. Given the voltage V_G, the quantities η_I, I_{in}, and D^* can be calculated. In the next cycle, a new value is assigned to V_G, and all the characteristics are to be recalculated. In this way, we obtain the main performance characteristics of the system 'photodiode-direct injection readout' as a function of the gate voltage at the input gate. The dependences in Fig. 3 can be used in a joint analysis of the effect due to noise characteristics of photodiodes, measured versus photodiode bias voltage, and the responsivity of photodetector channels based on the system 'photodiode – direct injection readout circuits'.

Figures 4a, b, c, and d shows the calculated curves of photodiode bias voltage $V_{PD}(V_G)$, injection efficiency $\eta_I(V_G)$, input current $I_{in}(V_G)$, and detectivity $D^*(V_G)$, respectively.

At V_{PD} = 0 V, the product $g_{in}R_{PD}$ for curves 1-5 in Fig. 4 equals respectively 794, 11.1, 11.1, 3.4, and 1.4. As it is seen from Figs. 4a and 4b, the dependences $V_{PD}(V_G)$ and $\eta_I(V_G)$ small informative in evaluating the uniformity level of transfer characteristics and responsivities of photodetector channels, more helpful here being the dependences $I_{in}(V_G)$ and $D^*(V_G)$ (see Figs 4c and 4d). As it is seen from Fig. 4, curves 1, at $g_{in}R_{PD}$ >100 the photodiode-direct injection readout system can be considered "ideal". With increasing the gate voltage V_G, as the photodiode bias voltage approaches zero, the injection efficiency η_I tends to unity, the current integrated in the readout circuit becomes roughly equal to I_{PD}, $I_{in} \approx I_{PD}$, and the detectivity reaches $D^* \approx 2.87 \cdot 10^{11}$ $cm \cdot Hz^{1/2} \cdot Wt^{-1}$, this value being comparable with the theoretical limit of D^* for a photodiode with η_K=0.8 operating in photovoltaic regime in BLIP mode. With further increase of V_G, for the "ideal" system the performance characteristics of the photodetector channel based on the system 'photodiode – direct injection readout circuit' (namely, η_I, I_{in}, and D^*) become almost independent of both V_G and electrophysical and design parameters of readout circuits. In the latter case, performance characteristics of IR FPAs based on direct injection readout circuits can be evaluated using standard simplifications (Longo, 1978; Felix,1980). In "non-ideal" systems, in which the relation $g_{in}R_{PD}$ >>1 is fulfilled not too strict (see curves 2-5 in Figs. 4c and 4d), curves I_{in} (V_G) and D*(V_G) show a different behavior. The current I_{in} integrated in the readout circuit increases with increasing V_G. The dependences D*(V_G) exhibit a pronounced maximum. Being considered as a function of photodiode electrophysical parameters and background illumination current, the maximum detectivity is attained at a voltage V_G at which the photodiode gets driven by 5-30 mV in reverse direction (curve 4 in Fig. 4); this detectivity rather weakly depends on the value of $g_{in}R_{PD}$. For instance, as the value of $g_{in}R_{PD}$ decreases from 794 to 3.4 (curves 1 and 4 in Fig. 4) the detectivity D* falls in value from $2.87 \cdot 10^{11}$ cm $\cdot Hz^{1/2} \cdot Wt^{-1}$ to $1.96 \cdot 10^{11}$ cm $\cdot Hz^{1/2} \cdot Wt^{-1}$, i.e. within a factor of 1.5. Note that the values of $g_{in}R_{PD}$ for curves 2 and 3 are identical; nonetheless, the dependences I_{in} (V_G) (Fig. 4c) and D* (V_G) (Fig.4d) for those cases differ substantially.

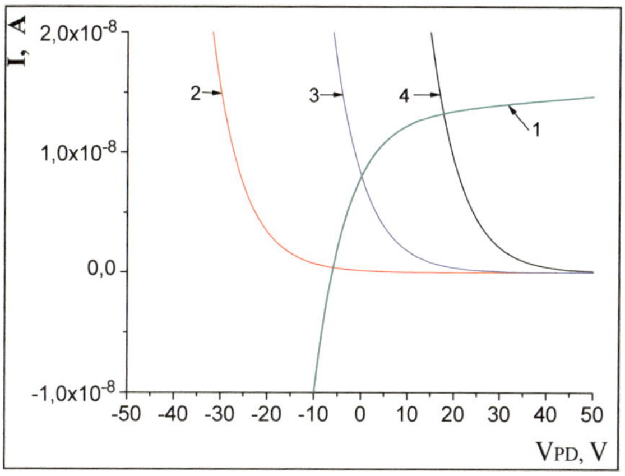

Fig. 3. Current-voltage characteristics of photodiode (curve 1) and input FET in weak inversion (curves 2-4) for three values of V_G, 1.17, 1.198, and 1.22 V. $I_0=5 \cdot 10^{-9}$ A, $R_S=3 \cdot 10^7$ Ohm, $I_P = 1 \cdot 10^{-8}$ A, and η_K =0.8.

The factors causing these differences can be clarified if one considers the calculated dependences of Fig. 5; in this figure, in addition to curves D* (V_G), also dependences $Q_1(V_G)$, $Q_2(V_G)$, $Q_3(V_G)$, and $Q_4(V_G)$ are shown. In the "ideal" system (see Fig.5 a), with increasing the voltage V_G, when the injection efficiency approaches unity, the total noise charge Q_5 becomes defined just by the photodiode noise current Q_3, having almost the same magnitude as the noise induced by background radiation fluctuations (9) since $I_{in} \approx I_P$ and $\eta_I \approx 1$. In "non-ideal" systems (see calculated dependences in Fig. 5 b-e), the charge Q_5 grows in value with increasing V_G. The increase of the noise and the related reduction of D* is primarily defined by the growth of Q_3 and Q_4. For the dependences shown in Fig. 5 b-e the difference between the curves D*(V_G) is primarily defined by the growth of the 1/f noise of photodiode (component Q_4). The difference of the dependences $Q_4(V_G)$ is due to the higher current I_{in} and a lower value of η_I for curves 4. With identical values of $g_{in}R_{PD}$ at zero voltage drop across the photodiode (compare Fig. 5b and Fig. 5c), a better IR FPA responsivity can be reached with diodes exhibiting a larger dynamic resistance on their biasing in reverse direction by 10-30 mV. For state-of-the-art level of silicon technology (C_{ox} ~ (0.5-1.2)$\cdot 10^{-7}$ F$\cdot$cm^{-2}, N*≈ 1), the noise induced by the input FET (cp. curves 1 and 2) is only substantial at voltages $V_G < V_{G0}$ even if the product $g_{in}R_{PD}$ has a value close to unity, and this noise does not limit the IR FPA detectivity. Knowing of the values of individual noise components enables goal-directed optimization of electrophysical and design parameters of photodiodes and readout circuits.

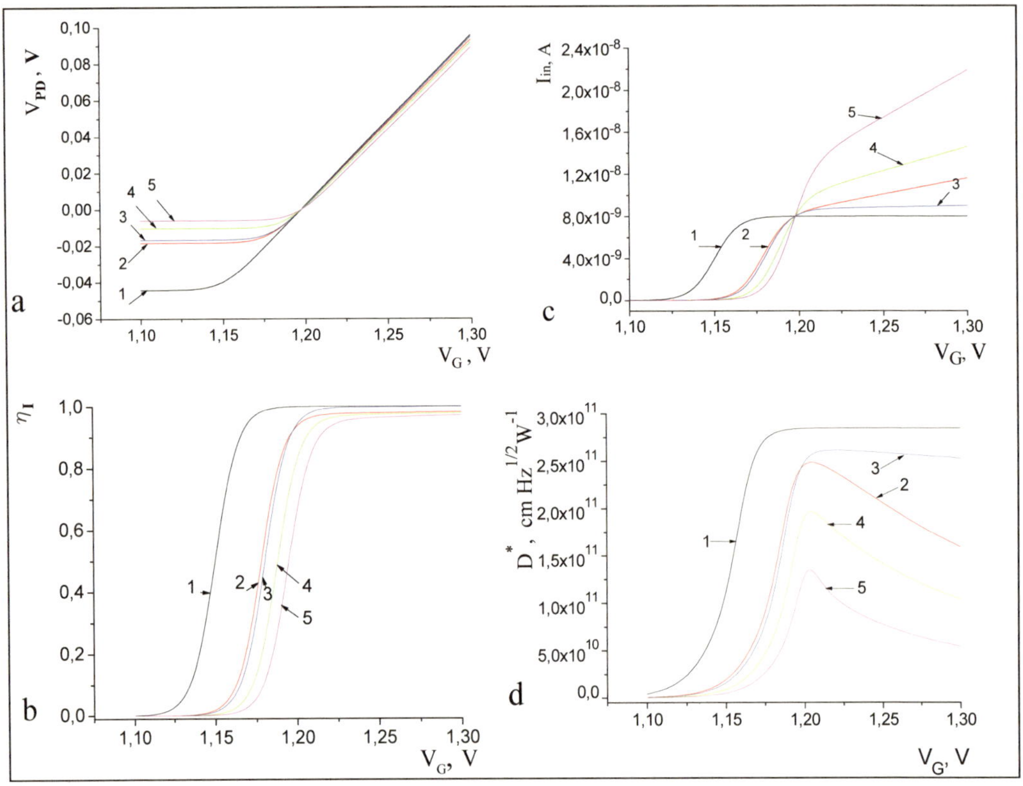

Fig. 4. Calculated dependences of performance characteristics of a photodiode-direct injection system on input-gate voltage. a – photodiode bias voltage, b – injection efficiency η_I, c – current I_{in} integrated in the readout circuit, d - detectivity D*. Integration time T_{in} $=5\cdot10^{-4}$ s, $I_P=1\cdot10^{-8}$ A, $\eta_K=0.8$, for curve 1 - $I_0=1\cdot10^{-11}$ A, $R_P=1\cdot10^{11}$ Ohm; for curve 2 - I_0 $=5\cdot10^{-10}$ A, $R_P=3\cdot10^{7}$ Ohm; for curve 3 - $I_0=7\cdot10^{-10}$ A, $R_P=3\cdot10^{8}$ Ohm; for curve 4 - $I_0=2\cdot10^{-9}$ A, $R_P=2\cdot10^{7}$ Ohm; for curve 5 - $I_0=5\cdot10^{-9}$ A, $R_P=1\cdot10^{7}$ Ohm.

Important figures of merit of multi-element IR FPAs are the fixed-pattern noise and the responsivity uniformity of photodetector channels. The spread of photoelectric parameters is defined by the variation of electrophysical parameters of readouts and by the variation of photoelectrical parameters of photodiodes. For direct injection readouts operated in weak inversion regime, the main effect is due to input-FET threshold variations rather than due to non-uniformity of geometric dimensions of input-FET gates.

Histograms of photodiode bias voltages and injection efficiencies calculated with allowance for FET threshold variations are shown in Fig. 6. In performing the calculations, it was assumed that the spread of threshold voltages obeys a normal distribution law, and a total of 1000 realizations were considered. As it is seen from Fig. 6, the spread of FET threshold voltages leads to a spread of photodiode bias voltages (Fig. 6 a), and also to a spread of injection efficiencies η_I and, as a consequence, to variation of transfer characteristics of the 'photodiode-direct injection readout' system.

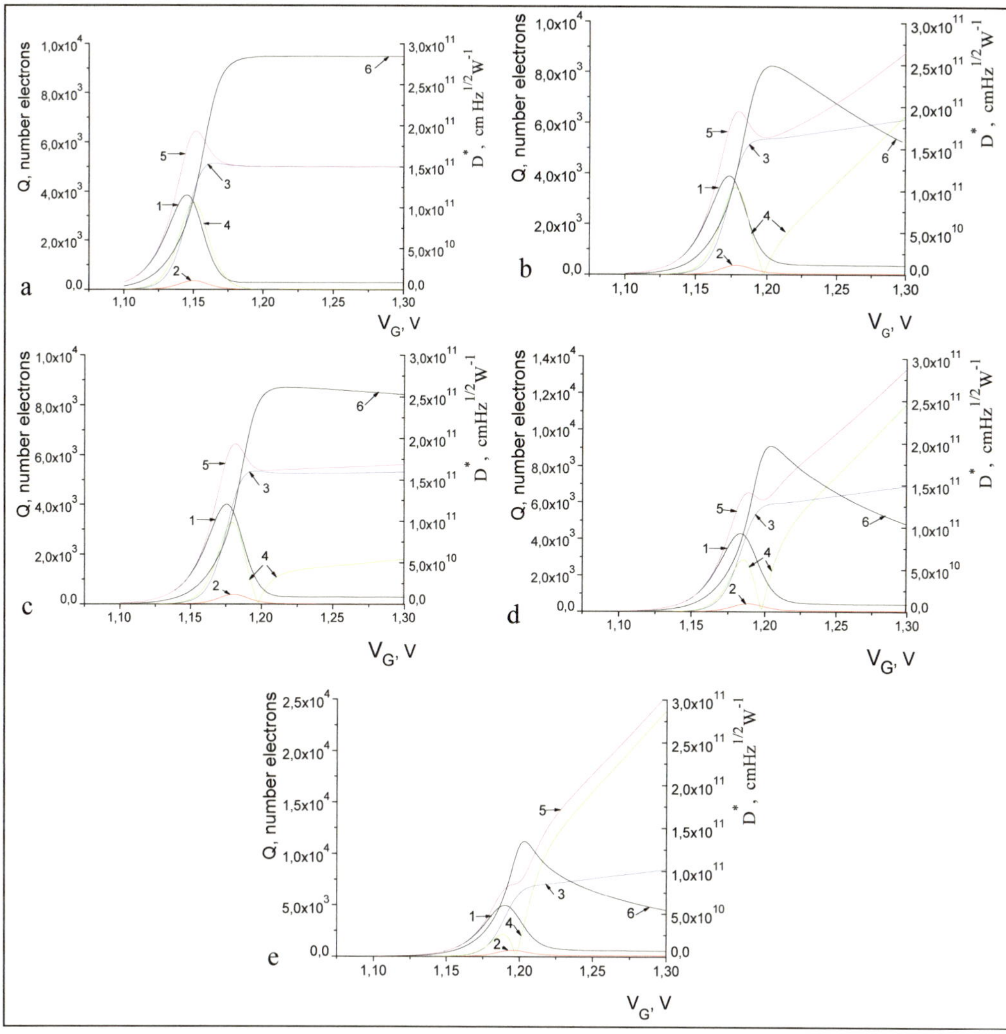

Fig. 5. Calculated dependences of $D^*(V_G)$ (curve 6, right axis) and $Q_1(V_G)$, $Q_2(V_G)$, $Q_3(V_G)$, $Q_4(V_G)$, and $Q_5(V_G)$ (curves 1-5, left axis). $Q_5=(Q_1^2+Q_2^2+Q_3^2+Q_4^2)^{1/2}$. The values of photodiode electrophysical parameters adopted in the calculations are the same as in Fig. 4 for curves 1-5, respectively.

Figure 7 shows the calculated curve I_{in} (V_G) that illustrate the influence of the dispersion of threshold voltages with $\sigma(V_{th})$ = 10 mV on the characteristics of a multi-element IR FPA (the values of photoelectrical parameters for IR photodiodes are the same as those adopted in Fig. 4 for curves 2). It is seen from Fig. 7 a-b that for a "non-ideal" system the dispersion of threshold voltages brings about a fixed-pattern noise and a spread of currents integrated in the readouts, I_{in} (V_G). For multi-element IR FPAs, with allowance for the spread of threshold voltages, at a voltage V_G =1.205 V (Fig.5b), for which the detectivity of a single channel D* attains a maximum, the spread of currents I_{in} integrated in the readouts falls in the range from $4 \cdot 10^{-9}$ to $1.15 \cdot 10^{-8}$ A (Fig. 7a), whereas the current integrated in the "ideal" system is $I_{in} = 0.8 \cdot 10^{-8}$ A.

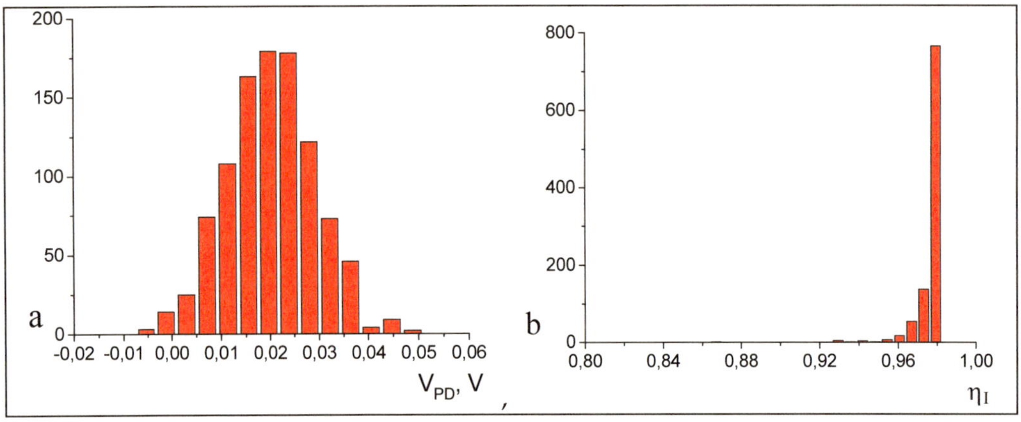

Fig. 6. Effect due to the spread of FET threshold voltages on the characteristics of multi-element IR FPAs; a – histogram of photodiode voltages, b – histogram of injection efficiencies η_I for I_0=5 $\cdot 10^{-10}$ A, R_P =3 $\cdot 10^7$ Ohm, I_P =1 $\cdot 10^{-8}$A, η_K =0.8, and V_G=1.225 V. The standard deviation of input-FET threshold voltages is $\sigma(V_{th})$ = 10 mV.

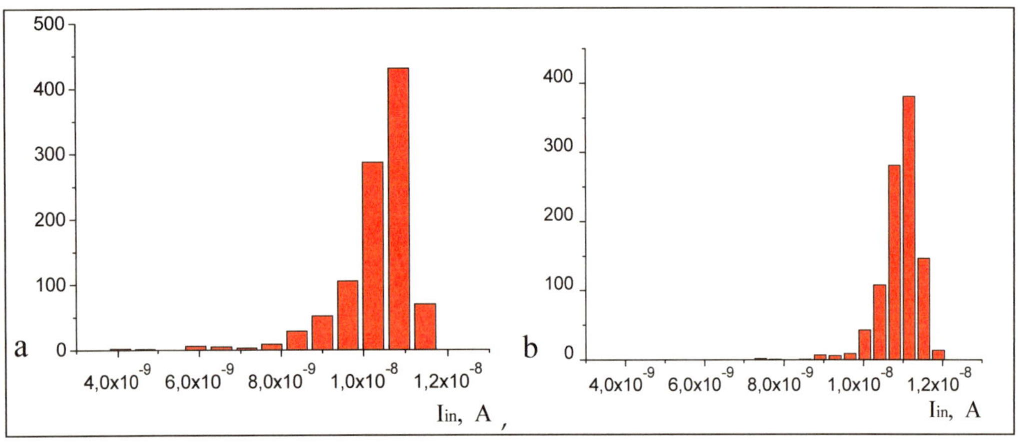

Fig. 7. Calculated histograms $I_{in}(V_G)$ for a normal distribution of input-FET threshold voltages with $\sigma(V_{th})$ = 10 mV. For Figs. 7a and 7b the voltages V_G are respectively 1.205 Vand 1.225 V. The calculations were performed for photodiodes with parameter values the same as those for curves 2 in Fig. 4.

Figure 8 show histograms $D^*(V_G)$ calculated with allowance for the spread of threshold voltages. In the calculations of detectivity, the accumulation time was defined by the charge capacity of the readout circuit and by the maximal (over 1000 realizations) current $I_{in}(V_G)$.

In multi-element IR FPAs, in view of non-uniformity of threshold voltages, at V_G =1.205 V about 1% of photodetector channels have a detectivity D* lower than $1 \cdot 10^{11}$ cm $\cdot$Hz$^{1/2}$ $\cdot$Wt^{-1} (see Fig. 8b), whereas for a single channel we have D* = $2.48 \cdot 10^{11}$ cm $\cdot$Hz$^{1/2}$ $\cdot$Wt^{-1}. In the case of multi-element IR FPAs because of the spread of threshold voltages, there arises a necessity to adjust the gate voltage (for the values of electrophysical parameters of photodiodes adopted in the calculations, Fig. 8b); the optimum value here is V_G =1.225 V. The increase of V_G results in a substantial reduction of fixed-pattern noise (see Fig.7b). In the latter situation, the minimal detectivity is not lower than $1.85 \cdot 10^{11}$ cm·Hz$^{1/2}$·Wt^{-1}, and more than 98% pixels have a detectivity D* greater than $2.0 \cdot 10^{11}$ cm·Hz$^{1/2}$·Wt^{-1}. With further increase of V_G, "dark" parasitic photodiode current components grow in value and the detectivity of the majority of photodetector channels decreases (Fig. 8c).

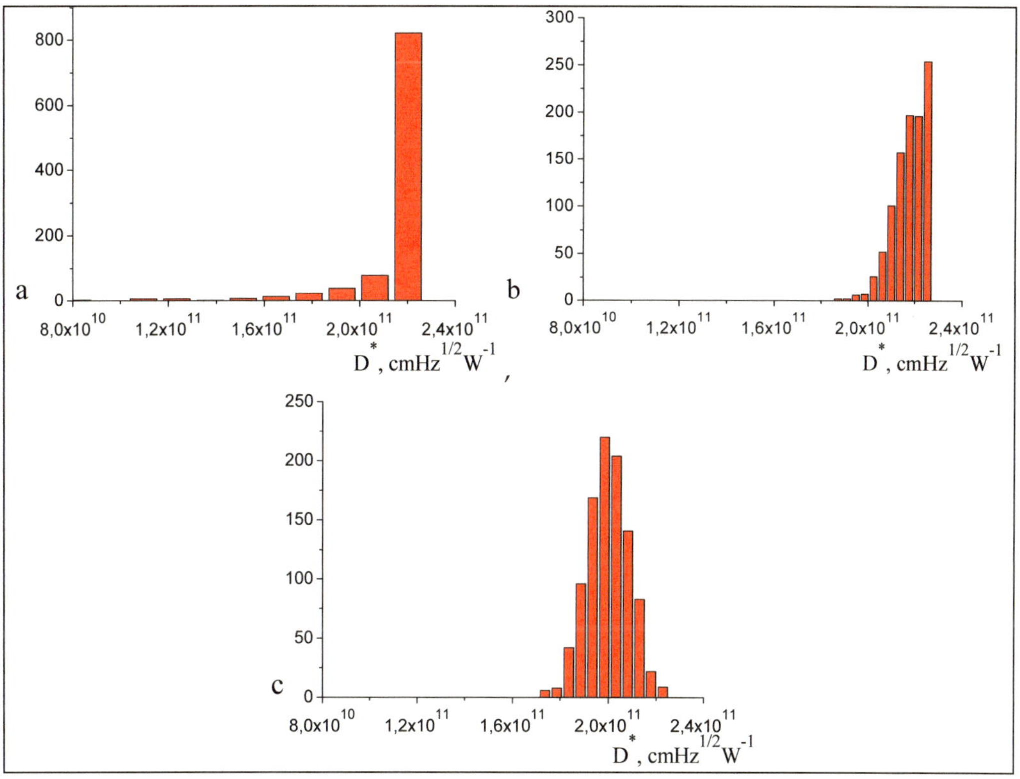

Fig. 8. Histograms D*(V_G) calculated for FET threshold voltages normally distributed with $\sigma(V_{th})$ = 10 mV for V_G = 1.201, 1.225, and 1.249 V (respectively Figs. 8a, 8b, and 8.6c). The calculations were performed for photodiodes with parameter values adopted for curves 2 in Fig. 4.

The effect due to dispersion of threshold voltages in fabricated readout circuits is illustrated by a comparison of calculated histograms $I_{in}(V_G)$ and D*(V_G) in Fig. 8 with histograms calculated for $\sigma(V_{th})$ = 2 mV, Fig. 9. As it is seen from the calculated dependences in Fig. 9, in the latter case the detectivity of the multi-element IR FPA is at the level of about 90% of D* of the single photodetector channel.

Along with non-uniformity of input-FET threshold voltages, the photoelectric parameters of photodiodes, η_K, R_{PD}, and I_0, also inevitably display some scatter of values. Figure 10 shows histograms of I_{in} (Fig. 10a) and D* (Fig.10b) calculated on the assumption that the distributions of R_{PD} and I_0 obey normal distribution laws.

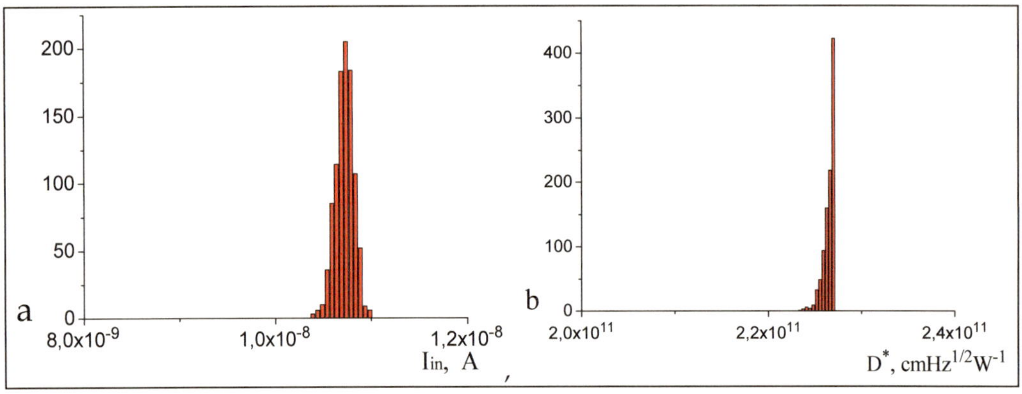

Fig. 9. Influence of the threshold voltage spread on the characteristics of multi-element IR FPAs; a – histogram of I_{in}, b – histogram of D*. The standard deviation of input-FET threshold voltages is $\sigma(V_{th})$ = 2 mV, I_0=5 ·10-10 A, R_{PD} =3 ·107 Ohm, I_P =1 ·10-8 A, η_K =0.8, V_G =1.213 V.

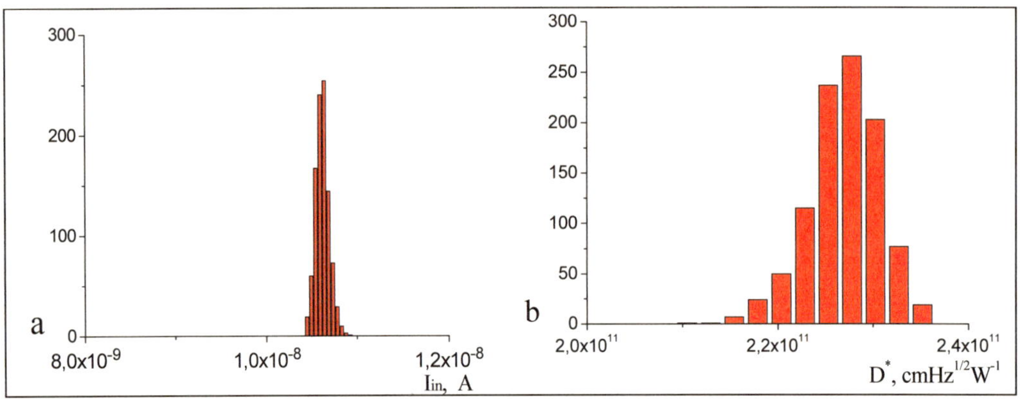

Fig. 10. Influence of the spread of photoelectric parameters of photodiodes on the performance characteristics of multi-element IR FPAs; a – histogram of I_{in}, b – histogram of D*. Parameter values adopted in the calculations: $\sigma(R_{PD})$ = 3 ·106 Ohm, $\sigma(I_0)$ = 1 ·10-10 A, I_0=5 ·10-10 A, R_{PD} =3 ·107 Ohm, I_P =1 ·10-8 A, η_K =0.8, V_G =1.225 V.

Plotting histograms of performance characteristics of multi-element IR FPAs, namely, currents integrated in the readout circuits and detectivity) versus the input-gate voltage V_G allows one to carry out numerical experiments and compare their results with experimental data. Such histograms can be considered generalized characteristics of multi-element IR FPAs indicative of their quality.

3.2 Modeling procedure for multi-element IR FPAs with direct injection readout circuits using the experimental-current voltage characteristics of photodiodes

The developed procedure enables performing an analysis of the system 'photodiode –direct-injection readout' using the experimental current-voltage characteristics of photodiodes. Figure 11 shows a family of "dark" and "light" current-voltage characteristics of one hundred photodiodes measured under room-temperature background conditions. The photodiodes were fabricated on the basis of a $Hg_{1-x}Cd_xTe$ variband heteroepitaxial structure. The stoichiometric composition of the photosensitive layer was x=0.225 (Vasilyev, 2010).

In performing the calculations, the experimental current-voltage characteristics were approximated with the expressions:

$$I_{PD}(V_{PD}) = C_0 + C_1 \exp(C_2 V_{PD}) + C_3 \exp(C_4 V_{PD}) + C_5 V_{PD} \tag{11}$$

where the values of coefficients C_0-C_5 were chosen individually for each photodiode.

The characteristics of multi-element IR FPA were calculated by the procedure, described in section 3.1, that allows calculation of all photodetector-channel characteristics, $\eta_I(V_G)$, $I_{in}(V_G)$, and $D^*(V_G)$.

Figure 12 shows the histograms $D^*(V_G)$ and $I_{in}(V_G)$ calculated for the standard deviation value of threshold voltages of the readouts $\sigma(V_{th})$ =10 mV. An analysis of dependences calculated with lower values of $\sigma(V_{th})$ suggests that the use of silicon technology permitting values $\sigma(V_{th}) < 2$–3 mV will allow a substantial (by 20–30%) reduction of the fixed-pattern noise level. The changes in the calculated detectivity histograms proved to be less substantial since the spread of detectivity values is primarily defined by the variations of photodiode quantum efficiency (Vasilyev, 2010).

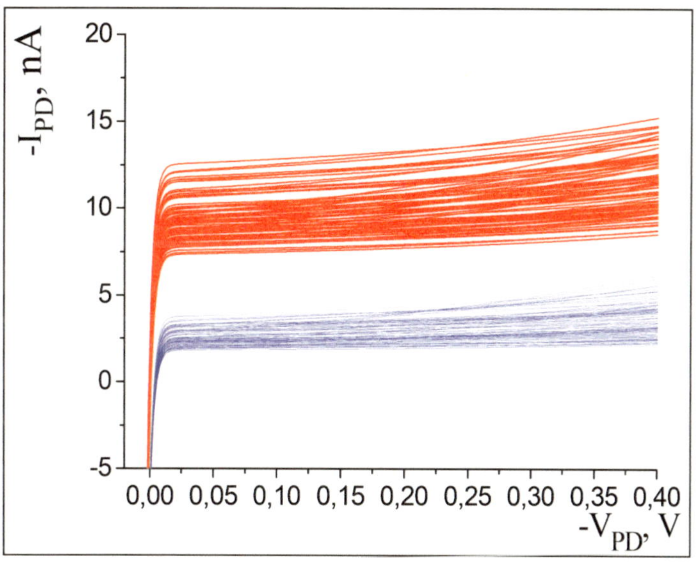

Fig. 11. A family of experimental current-voltage characteristics of $Hg_{1-x}Cd_xTe$ photodiodes; "dark" characteristics are shown with blue lines, and characteristics measured under room-temperature radiation background with red lines are shown with red lines.

It should be noted that the numerical values of coefficients α_1, α_2, α_3, and K were borrowed from literature sources. Experimentally measured values of these coefficients or more elaborated model for photodiode noises can easily be incorporated into the computer problem, allowing an improved accuracy in predicting the characteristics of IR FPAs.

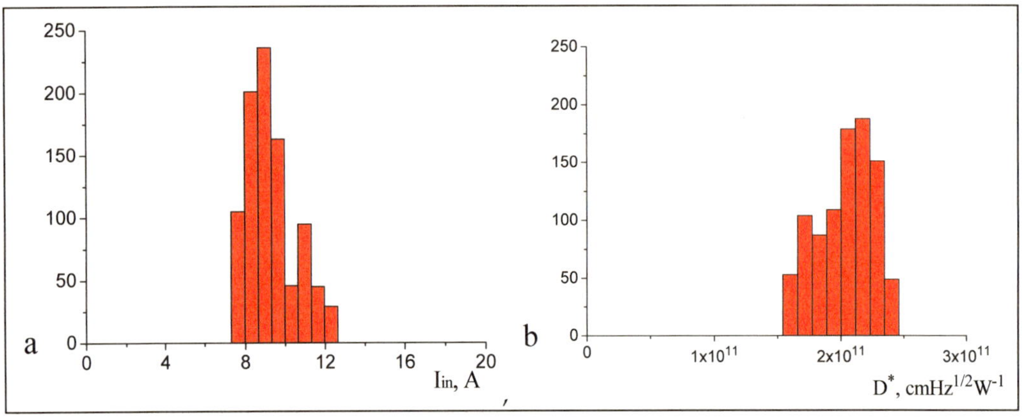

Fig. 12. Influence of the spread of photoelectric parameters of photodiodes on the performance characteristics of multi-element IR FPAs; a - histogram of I_{in}, b - histogram of D*. Parameter values adopted in the calculations: $\sigma(R_{PD}) = 3 \cdot 10^6$ Ohm, $\sigma(I_0) = 1 \cdot 10^{-10}$ A, $I_0 = 5 \cdot 10^{-10}$ A, $R_{PD} = 3 \cdot 10^7$ Ohm, $I_P = 1 \cdot 10^{-8}$ A, $\eta_K = 0.8$, $V_G = 1.225$ V.

3.3 Analysis of LWIR thermography systems based on $Hg_{1-x}Cd_xTe$ photodiodes

The simple photodiode model (10) proved to be insufficient for adequately predicting the current-voltage characteristics of $Hg_{1-x}Cd_xTe$ photodiodes in the spectral region 8 to 14 µm. In modeling current-voltage characteristics of $Hg_{1-x}Cd_xTe$ photodiodes, all main mechanisms of charge transport in *p-n* junctions must be taken into account, including the background radiation current $\eta_K I_P$, the diffusion current, the thermal generation/recombination current, the tunneling current through localized levels within the bandgap of the depleted region of the *p-n* junction, the interband tunneling current (Anderson,1981), the Shockley-Reed-Hall current through traps in quasi-neutral *n* and *p* regions, the Auger current, and the radiative generation/recombination current in the *p-n* junction and in the quasi-neutral regions (Anderson, 1981; Rogalski, 2000). All the currents listed above are independent currents to be taken into account additively, except for the thermal generation-recombination and trap-assisted tunneling through traps of the Shockley-Reed-Hall type in the depleted region of the *p-n* junction, since the rates of the latter processes is determined by the trap occupation. The two latter mechanisms were modeled in the approximation of balance equations for charge carriers at trap levels (Anderson, 1981; Gumenjuk-Sichevska, 1999; Sizov, 2006; Yoshino,1999). We assumed the presence of localized donor-type centers in the bandgap with energy E_t=0.6-0.7E_g over the valence-band edge (Krishnamurthy, 2006). In calculating the band-gap energy of the material as a function of stoichiometric composition and temperature, we used the expression (Rogalski, 2000).

$$E(x,T) = \left[-0.302 + x\left[1.93 + x(-0.81 + 0.832x)\right] + 5.32 \cdot 10^{-4} \cdot (1-2x)\frac{-1822.0 + T^3}{255.2 + T^2} \right] \quad (12)$$

The main electrophysical parameters of $Hg_{1-x}Cd_xTe$ photodiodes adopted in the calculation of their current-voltage characteristics are listed in Table 2.

By way of example, Figure 13 shows the current-voltage characteristics of $Hg_{1-x}Cd_xTe$ photodiodes for different long-wave spectral-response cutoffs, and Figure 14 shows the calculated dependences of R_{PD} A_{PD} on the cutoff wavelength of the spectral response of the photodiodes.

As it was shown in (Gumenjuk-Sichevska, 1999; Sizov, 2006), the photodiode model that was used in the present calculations permits calculation of the current-voltage characteristics as a function of the stoichiometric composition of $Hg_{1-x}Cd_xTe$, temperature, and electrophysical parameters of substrate material, the obtained dependences being consistent with those reported in the literature (Rogalski, 2000; Yoshino,1999).

Figure 15 shows the calculated dependences of the maximum values of $D^*(V_G)$ of a photodetector channel based on the system 'HgCdTe photodiode – direct injection readout circuit' on the cutoff wavelength λ_2. In calculating the background radiation flux reaching the photodiodes in the spectral region from λ_1 to λ_2 (λ_1 =4 μm and λ_2 is the cutoff wavelength), for the blackbody temperature a value 300 K was adopted; the aperture angle was assumed defined by the relative aperture of the optical system F/f = 0.5, and for the transmission of the optical system, a value 0.9 was adopted.

Designation	Parameter	Numerical value	
		Case 1	Case 2
N_a, N_d	Acceptor and donor concentrations	$1.0 \cdot 10^{22}$ m^{-3}, $2.0 \cdot 10^{21}$ m^{-3}	$2.0 \cdot 10^{22}$ m^{-3}, $2.0 \cdot 10^{21}$ m^{-3}
N_t, N_{tV}	Concentration of traps in the p-n junction and in the quasi-neutral regions	$6.0 \cdot 10^{21}$ m^{-3}, $2.0 \cdot 10^{21}$ m^{-3}	$3.0 \cdot 10^{21}$ m^{-3}, $1.0 \cdot 10^{21}$ m^{-3}
τ_n, τ_p, τ_{nV}, τ_{pV}	Electron and hole lifetimes in the p-n junction and in the quasi-neutral regions	$0.2 \cdot 10^{-6}$ s, $0.2 \cdot 10^{-6}$ s, $6.0 \cdot 10^{-6}$ s, $6.0 \cdot 10^{-6}$ s	$0.2 \cdot 10^{-6}$ s, $0.2 \cdot 10^{-6}$ s, $1.0 \cdot 10^{-7}$ s, $1.0 \cdot 10^{-7}$ s
E_t	Energy position of trap level	E_t=0.7 E_g eV	
P	Interband matrix element	$8.3 \cdot 10^{-10}$ eV·m	
W_c^2	Squared matrix element for the tunneling of charge carriers from the trap level into the band	$3 \cdot 10^{-67}$ J^2·m^3	
Δ	Spin – orbital interaction constant	0.96 eV	
ε_r	Static dielectric permittivity	17.5	
A_{PD}	Photodiode area	$9 \cdot 10^{-5}$ cm^2	

Table 2. Electrophysical parameters of $Hg_{1-x}Cd_xTe$ photodiodes.

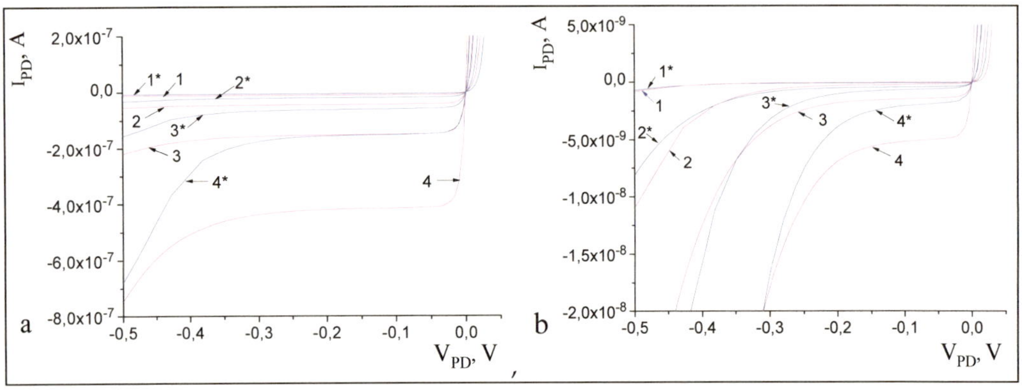

Fig. 13. Calculated current-voltage characteristics of $Hg_{1-x}Cd_xTe$ photodiodes at temperatures 77 K (a) and 60 K (b). For curves 1 - 4 (1* - 4*) the cutoff wavelength is respectively λ_2 = 11, 12, 13, and 14 μm. The electrophysical parameters of photodiodes are listed in Table 2; case 1 - curves 1-4, case 2 - curves 1* - 4*.

The detectivity of the photodetector channel at temperature T=77 K for photodiodes with electrophysical-parameter values indicated as case 1 in Table 2 (curve 1 in Fig. 15) at λ_2 > 10 μm becomes lower than Background Limited Performance (BLIP) detectivity (curve 3). On cooling the photodiodes to temperature 60 K, the detectivity of photodetector channel approaches BLIP detectivity at wavelengths below ~ 13 μm (curve 3). For photodiodes with electrophysical-parameter values indicated as case 2 in Table 2.2, due to a larger dynamic resistance, the long-wave spectral-response cutoff at which the FPA detectivity attains its maximum value shifts towards longer wavelengths (see curves 2 and 4 in Fig. 15).

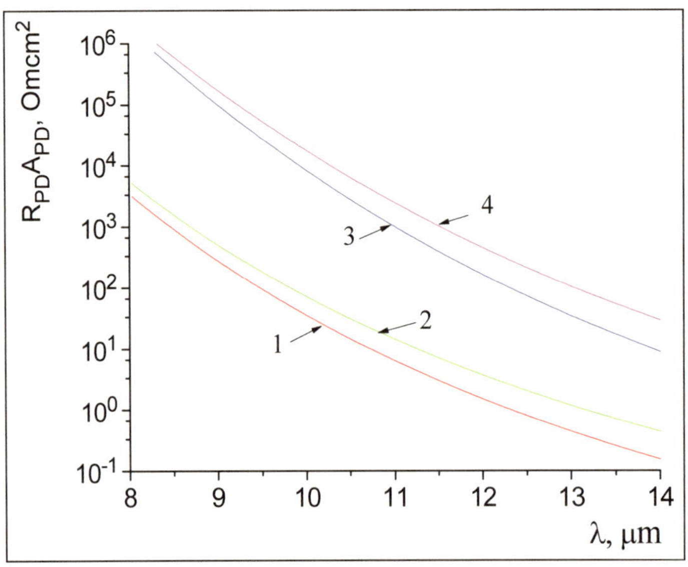

Fig. 14. Calculated dependences of R_{PD} A_{PD} on the long-wave cutoff of photodiodes; curves 1 and 2 refer to temperature T=77 K, and curves 3 and 4, to temperature T=60 K. Electrophysical-parameter values: curves 1 and 3 - case 1, curves 2 and 4 - case 2.

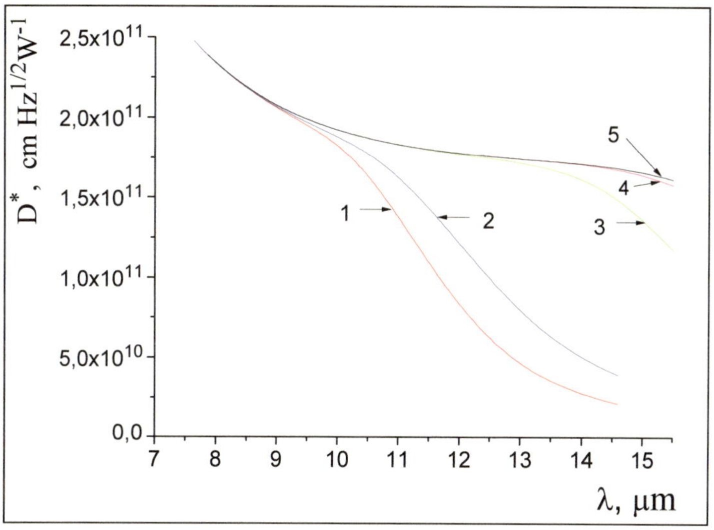

Fig. 15. Calculated curves $D^*(\lambda_2)$ for a photodetector channel based on the system '$Hg_{1-x}Cd_xTe$ photodiode-direct injection readout'; 1, 2 - photodiode temperature 77 K, 3, 4 - photodiode temperature 60 K. Curves 1, 3 and 2, 4 were calculated for parameter values indicated in Table 1 as cases 1 and 2, respectively, for direct injection charge capacity $Q_{in}=5 \cdot 10^7$ electrons. Curve 5 - calculated dependences for BLIP detectivity.

Figure 16 shows the curves $D^*(V_G)$ and the dependences $Q_1(V_G)$, $Q_2(V_G)$, $Q_3(V_G)$, $Q_4(V_G)$ and $Q_5(V_G)$ calculated for λ_2 = 10, 12, and 13 μm. The calculations were performed for photodiodes with parameter values indicated in Table 2 as case 1.

The detectivity of the photodetector channel (Fig. 16a) at λ_2 = 10 μm is close to BLIP detectivity D^*; this detectivity is limited by the photodiode current noise Q_3 and at voltages V_G >1.2 V it is almost independent of V_G. For λ_2 = 12 μm (Fig. 16b) the maximum detectivity of the photodetector channel is $D^* \approx 8.34 \cdot 10^{10}$ cm·Hz$^{1/2}$·Wt^{-1} (BLIP detectivity is D^*= 1.77·10^{11} cm·Hz$^{1/2}$·Wt^{-1}). For λ_2 = 13 μm (Fig. 16c) the maximum detectivity of the photodetector channel is $D^* \approx 4.7 \cdot 10^{10}$ cm·Hz$^{1/2}$·Wt^{-1} (BLIP detectivity is D^*= 1.74·10^{11} cm·Hz$^{1/2}$·Wt^{-1}). The degradation of detectivity at voltages V_G > 1.225 V is primarily defined by the increase in the photodiode 1/f-noise level (Q_4 in Fig. 16b, c).

An increase of the direct injection storage capacitance and the related increase of the accumulation time both lead to a greater contribution made by the 1/f-noise. Figure 17 shows the curves $D^*(V_G)$ and the dependences $Q_1(V_G)$, $Q_2(V_G)$, $Q_3(V_G)$, and $Q_4(V_G)$ similar to those shown in Fig. 16b yet calculated for direct injection storage capacity Q_{in} = 2·10^8 electrons.

A comparison between the dependences in Figs. 16 and 17 shows that the maximal detectivity rather weakly depends on the storage capacity of the readout circuit. Yet, because of the increased contribution due to the 1/f noise of photodiodes, with increasing the input-gate voltage the dependence of D^* on V_G becomes more clearly manifested. That is why with increasing the storage capacity of direct injection readouts and, hence, with increasing the accumulation time, the requirements imposed on the standard deviation of threshold voltages become more stringent.

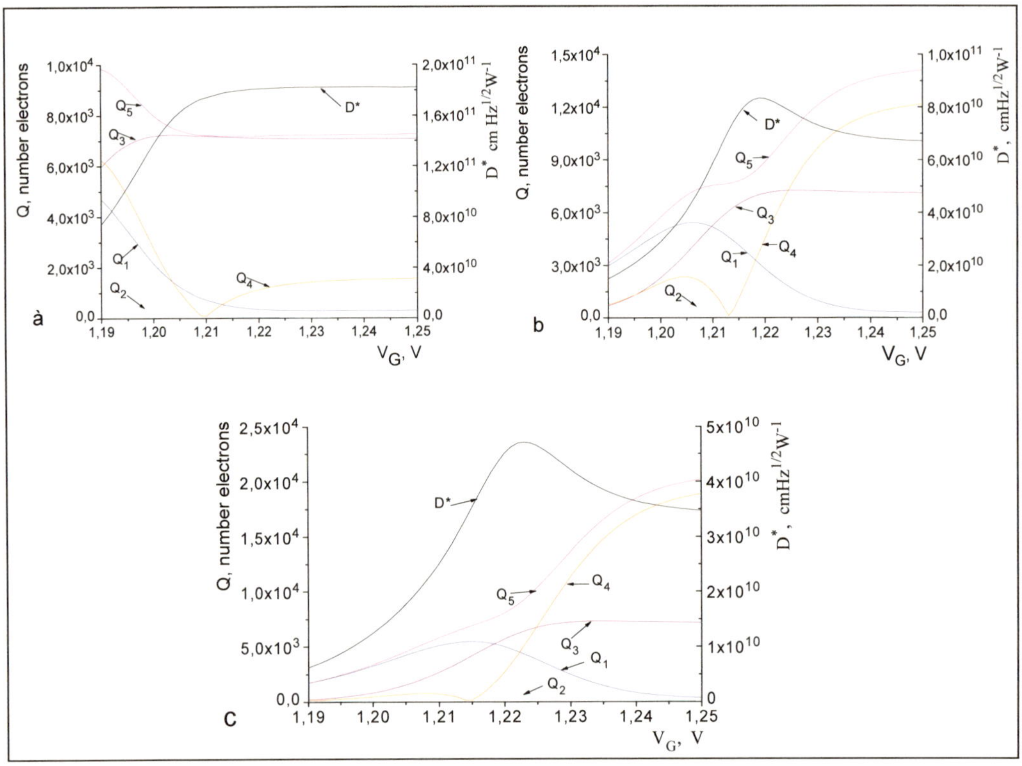

Fig. 16. Curves $D^*(V_G)$ (right axis) and dependences $Q_1(V_G)$, $Q_2(V_G)$, $Q_3(V_G)$, $Q_4(V_G)$, and $Q_5(V_G)$ (left axis). Temperature 77 K, direct injection charge capacity $Q_{in}=5\cdot 10^7$ electrons; a – $\lambda_2=10$ μm, b – $\lambda_2 = 12$ μm, c – $\lambda_2 = 13$ μm.

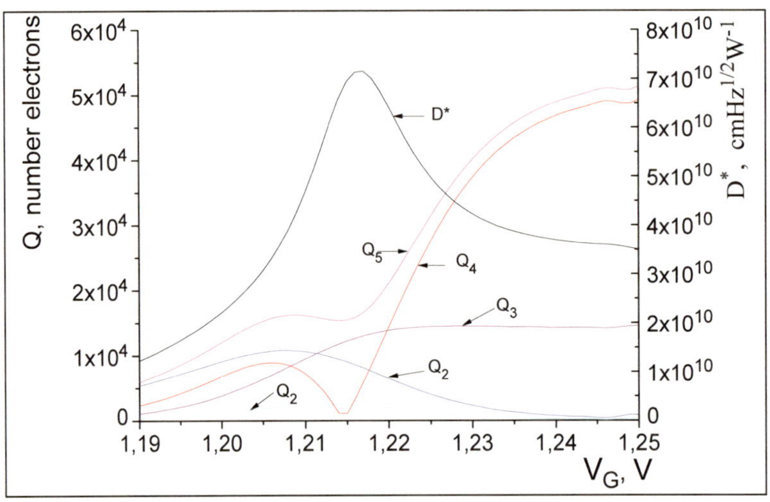

Fig. 17. Calculated curves $D^*(V_G)$ (right axis) and dependences $Q_1(V_G)$, $Q_2(V_G)$, $Q_3(V_G)$, $Q_4(V_G)$, and $Q_5(V_G)$ (left axis). The photodiode temperature is 77 K; the storage capacity of direct injection readout is $Q_{in}=2\cdot 10^8$ electrons; $\lambda_2 = 12$ μm. The values of electrophysical parameters of photodiodes are given in Table 2.2, case 1.

4. Temperature resolution of thermography systems based on multi-element IR FPAs with direct injection readouts

For thermography systems based on multi-element photodetectors, an important figure of merit is the noise-equivalent temperature difference (NETD) (Rogalski, 2000). The NEDT value can be calculated by the formula (Taubkin, 1993).

$$\mathrm{NETD} = \frac{\sqrt{1/2T_{in}}}{\frac{F^2}{4f^2}(A_{PD})^{1/2} G_{Op} \int_{\lambda 1}^{\lambda 2} \frac{d(dR/d\lambda)}{dT} S(\lambda) D_\lambda^*(\lambda_1, \lambda_2, T)\, d\lambda} \quad (13)$$

Here, $dR/d\lambda$ is the blackbody spectral luminous exitance, F/f is the relative aperture of the optical system, f is the focal distance, G_{Op} is the optical transmission of the system, and $S(\lambda)$ is the atmospheric transmission.

An analysis of NETD starts with calculating the current-voltage characteristics of photodiodes and the background radiation level. Then, currents integrated in the readout circuit and detectivity are calculated.

Figure 18 shows the calculated curves of NETD versus the cutoff wavelength λ_2.

For an "ideal" thermography system (curve 1), in which the integration time is equal to the frame time, T_{in} =20 ms in the present calculations, and the detectivity of photodetector channels is close to BLIP detectivity, we obtain the well-known dependences NETD(λ) that show that the temperature resolution improves with increasing the wavelength λ_2. The dependencies in Fig. 18 reveal the causes of the losses in temperature resolution owing to insufficiently high dynamic resistance of photodiodes or limited storage capacitance of silicon direct injection readout circuits of thermography systems in comparison with the theoretical limit. It can be inferred from the dependences $D_\lambda^*(\lambda_2)$ (Fig. 15) and NETD(λ_2) (Fig. 18) that the temperature resolution of thermography systems based on the photodiode - direct injection readout system attains a maximum value at some wavelength λ_2 and, then, decreases (see curves 2 and 3 in Fig. 18). The position of NEDT maximum is defined by electrophysical parameters of HgCdTe photodiodes, by the optical transmission of the system, and by the photodiode temperature. Yet, as it follows from the calculated curves of NETD in Fig. 18 (curves 4-7) the main factor limiting the temperature resolution of the thermography systems is the storage capacitance of silicon readout circuits.

With increasing the cutoff wavelength λ_2, the intensity of background radiation coming to photodetector also increases and, as a result, the integration time T_H decreases, with simultaneous shift of the maximum temperature resolution of the thermography system towards short wavelengths in comparison with curves 2 and 3 in Fig. 18. The charge capacity of readout circuits $Q_{in} = 5 \cdot 10^7$ electrons presents a maximal value reported in literature for multi-element matrix IR FPAs (at 30-μm photocell pitch) (Rogalski, 2000). At such values of integration capacitance of readout circuits the temperature resolution of thermography systems (curve 5) already with $\lambda_2 = 8$ μm is more than one order of magnitude and with $\lambda_2 = 14$ μm, two orders of magnitude lower than that of the "ideal" thermography system (curve 1). The storage capacitance can be increased through implementing a new architecture of readout circuits (curves 6 and 7 in Fig. 18) (Lee, 2010). A cardinal solution here is implementation of ADC in each pixel of the readout circuit (Zhou, 1996; Martijn, 2000; Fowler, 2000). In (Bisotto, 2010), each pixel was provided with a 15-bit ADC, and the effective storage capacitance was in excess of $3 \cdot 10^9$ electrons. This has enabled an increase in integration time up to the frame time and, in this way, allowed reaching a NETD of LWIR thermography systems based on multi-element IR FPAs amounting to 2 mK, the latter value being close to the maximum theoretically possible NETD.

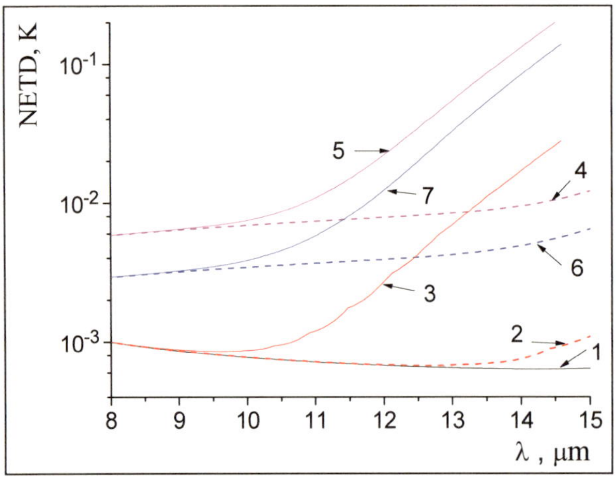

Fig. 18. Calculated curves NETD(λ_2). In the calculations, the values $F_p/f = 0.5$ and $\lambda_1 = 5$ μm were adopted. Curve 1 was calculated for $D^*_\lambda = D^*_{\lambda BLIP}$ and T_H=20 ms; curves 2 and 3 were calculated with allowance for the dependence of D^*_λ on the long-wave spectral-response cutoff, integration time T_{in}=20 ms; curves 4-7 were calculated for conditions with limited storage capacitance of direct injection readouts (curves 4 and 5 - $Q_{in} = 5 \cdot 10^7$ electrons, curves 6 and 7 - $Q_{in} = 2 \cdot 10^8$ electrons); curves 2, 4, and 6 – photodiode temperature 60 K, curves 3, 5, and 7 – photodiode temperature 77 K. The electrophysical parameters of photodiodes are given in Table 2, case 1.

The dependences $D^*(V_G)$ and NETD(λ_2) calculated for different values of the storage capacity of readout circuits on the input-gate voltage V_G and shown in Figs. 15-18 define the ultimate performance characteristics of thermography systems in the spectral range 8-14 μm. In fact, these dependences are will also be observed for a single-element FPA based on the system '$Hg_{1-x}Cd_xTe$ photodiode - direct injection readout circuit' since they were plotted using just the maximum values of D^*. For multi-element IR FPAs, the non-uniformity of input-FET threshold voltages leads to an increased fixed-pattern noise level and, for some part of photodetector channels, to a considerable reduction of D* and NETD in comparison with the dependences shown in Figs. 15 and 18. An additional factor causing an increase in the fixed-pattern noise, a decrease of detectivity, and worsened temperature resolution of multi-element IR FPAs is non-uniformity of the stoichiometric composition of the $Hg_{1-x}Cd_xTe$ substrate.

For multi-element IR FPAs with $\lambda_2 \leq 10$ μm, detectivity rather weakly depends on the gate voltage V_G and, hence, on the dispersion of threshold voltages (see Fig.16a). The fixed-pattern noise level and the spread of NETD values of multi-channel IR FPAs are primarily defined by the scatter of the long-wave photosensitivity cutoff owing to non-uniform stoichiometric composition of the $Hg_{1-x}Cd_xTe$ substrate.

Figure 19 shows calculated histograms of D*, I_{in} and NETD values of thermography systems for $Hg_{1-x}Cd_xTe$ photodiodes with photoelectric parameters given in Table 2, case 1. In the calculations, it was assumed that the dispersion of threshold voltages and the non-uniformity of the stoichiometric composition of substrate material obey normal distribution laws with parameter values indicated in the caption to the figure (the total number of realizations in the calculations was 200).

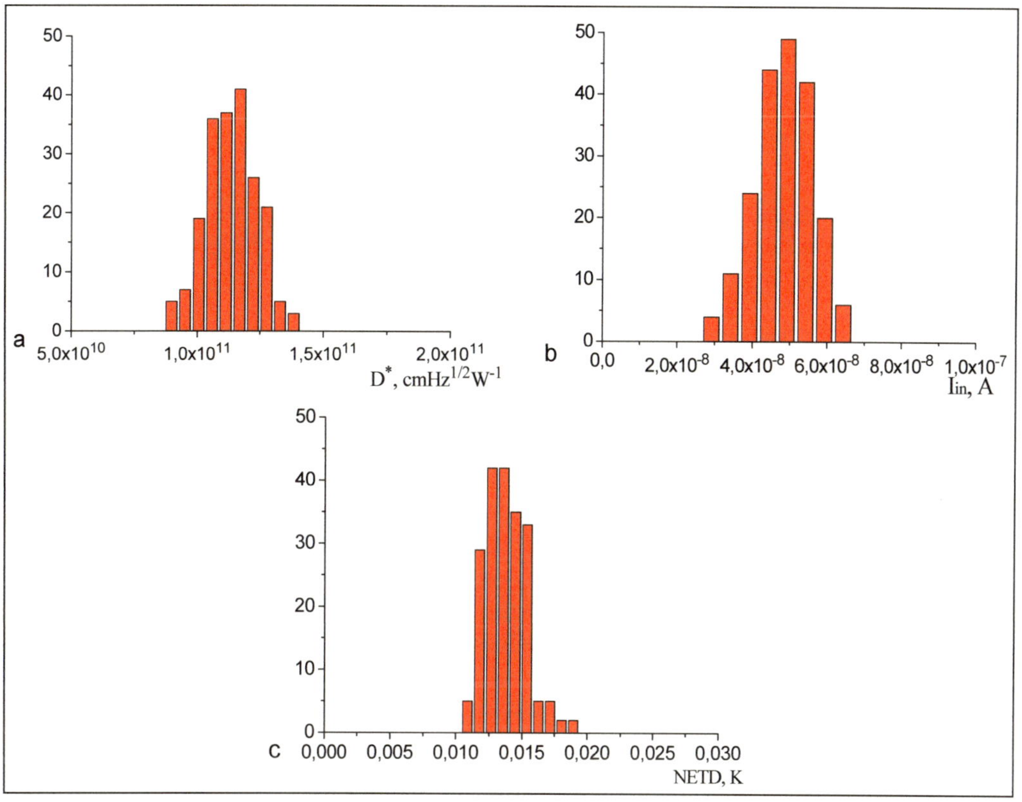

Fig. 19. Calculated histograms of performance characteristics of multi-element IR FPAs; a - detectivity D*, b – currents I_{in} integrated in the readout circuits, c - NETD. Standard deviation of input-FET threshold voltages - $\sigma(V_{th})$=5 mV. Standard deviation of the stoichiometric composition of $Hg_{1-x}Cd_xTe$ substrate - $\sigma(x)$ = 0.1%. Average stoichiometric composition - x=0.2167, T=77 K, storage capacitance of readout circuit $Q_{in}=5 \cdot 10^7$ electrons.

The photocurrent level was calculated with allowance for the long-wave cutoff of photosensitivity individually for each photodiode. The integration time was defined by the magnitude of the storage capacitance of the readout circuit and by the maximum level of photodiode current over 200 realizations. At T=77 K, the mean stoichiometric composition x=0.2167 ensures a long-wave photosensitivity cutoff λ_2=11 μm at standard dispersion of stoichiometric composition $\sigma(x)$ = 0.1%, typical of the present-day state-of-the-art technology level (Rogalski, 2000).

For a single channel with c λ_2=11 μm, at the same values of electrophysical and design parameters of photodiodes and readout circuits the maximum sensitivity D* is $1.4 \cdot 10^{11}$

cm·Hz$^{1/2}$·W^{-1}, D_{BLIP}* =1.8·10^{11} cm·Hz$^{1/2}$·W^{-1}, and NETD=10.8 mK (see curve 5 in Fig. 18). The dispersion of the stoichiometric composition of substrate $\sigma(x)$ = 0.1% results in a spread of long-wave cutoff wavelengths in the interval from 11 to 11.8 μm, this being the main factor causing NETD degradation.

With increasing the long-wave cutoff wavelength λ_2, requirements to the uniformity of threshold voltages under the input gates of direct injection readout circuits and requirements to the uniformity of the stoichiometric composition of substrate both become more stringent. Figure 20 shows calculated histograms of currents I_{in}, detectivities D* and NETD values of thermography systems for $Hg_{1-x}Cd_xTe$ photodiodes with photoelectric parameters indicated in Table 2 as case 1. At mean stoichiometric composition x=0.21055, $\sigma(x)$ = 0.1% and T=77 K the non-uniformity of the stoichiometric composition of substrate material results in a spread of long-wave photosensitivity cutoffs in the interval from 12 to 13.2 μm. The input-gate voltage value V_G=1.23 V appears to be optimal for the radiation environment conditions and values of electrophysical and design parameters of photodiodes and readout circuits adopted in the calculations.

Note that for a single channel at λ_2=12 μm the maximum detectivity D* is 8.4·10^{10} cm·Hz$^{1/2}$·W^{-1} and NETD = 0.022 K (Fig. 18, curve 5). Taking the dispersion of threshold voltages with $\sigma(V_{th})$ = 5 mV and substrate stoichiometric composition with $\sigma(x)$ = 0.1% into account results to two-three-fold degradation of NETD in a considerable fraction of photodetector channels.

Figure 21 shows calculated histograms of currents I_{in}, detectivities D* and NETD values of thermography systems for $Hg_{1-x}Cd_xTe$ photodiodes with photoelectric-parameter values adopted in Fig. 20 yet under more stringent conditions in terms of the dispersion of FET threshold voltages and stoichiometric composition of $Hg_{1-x}Cd_xTe$ substrate, $\sigma(V_{th})$ = 2 mV and $\sigma(x)$ = 0.03%. A comparison between the histograms in Figs. 20 and 21 shows that more stringent requirements imposed on the uniformity of threshold voltages and stoichiometric composition of substrate allow a substantial reduction of the fixed-pattern noise and a considerable improvement of NEDT values of thermography systems.

With the adopted values of electrophysical parameters of Si readout circuits and $Hg_{1-x}Cd_xTe$ photodiodes (Tables 1 and 2), thermography systems based on multi-element IR FPAs intended for operation at liquid-nitrogen temperature in the spectral range up to 13-14 μm with maximum possible NETD values (Fig. 18, curve 5) can be implemented at an acceptable level of fixed-pattern noise by:

- increasing the dynamic resistance of $Hg_{1-x}Cd_xTe$ photodiodes;
- adhering to more stringent requirements in terms of uniformity of input-FET threshold voltages and stoichiometric-composition uniformity of the substrate.

However, requirements to $\sigma(V_{th})$ and $\sigma(x)$ more stringent than the requirements that were adopted in calculating data in Fig. 21 presently cannot be met by silicon CMOS technology and synthesis processes of epitaxial $Hg_{1-x}Cd_xTe$ layers (Phillips, 2002).

Achieving photosensitivity in the spectral region 12-14 μm necessitates cooling the hybrid IR FPA assembly to a temperature below liquid-nitrogen temperature, see curves 4 and 6 in Fig. 19.

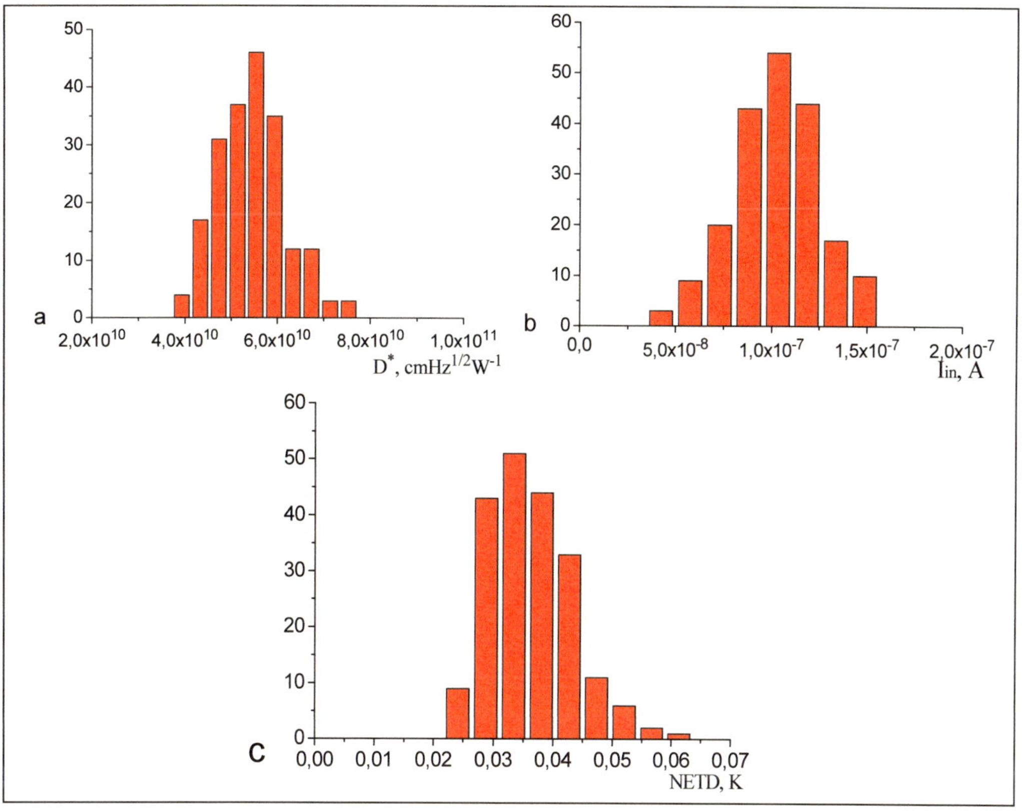

Fig. 20. Calculated histograms of performance characteristics of multi-element IR FPAs; a - detectivity D*, b – currents I_{in} integrated in the readout circuits, c - NETD. Standard deviation of input-FET threshold voltages - $\sigma(V_{th})$ = 5 mV. Standard deviation of the stoichiometric composition of $Hg_{1-x}Cd_xTe$ substrate $\sigma(x)$ = 0.1%. Average stoichiometric composition x=0.2105, T=77 K. Storage capacitance of readout circuit - $Q_{in}=5 \cdot 10^7$ electrons.

Figure 22 shows calculated histograms of currents detectivities D*, I_{in} and NETD values of thermography systems for $Hg_{1-x}Cd_xTe$ photodiodes with photoelectric parameters indicated in Table 2 as case 1. The parameter values used in the calculations were the same as those in Figs. 19 and 20, and the photodiode temperature was assumed to be 60 K.

At mean stoichiometric composition x=0.2105, 0.1% dispersion of stoichiometric composition and temperature T=60 K, the long cutoff wavelength will fall into the wavelength interval from 12.6 to 14 μm.

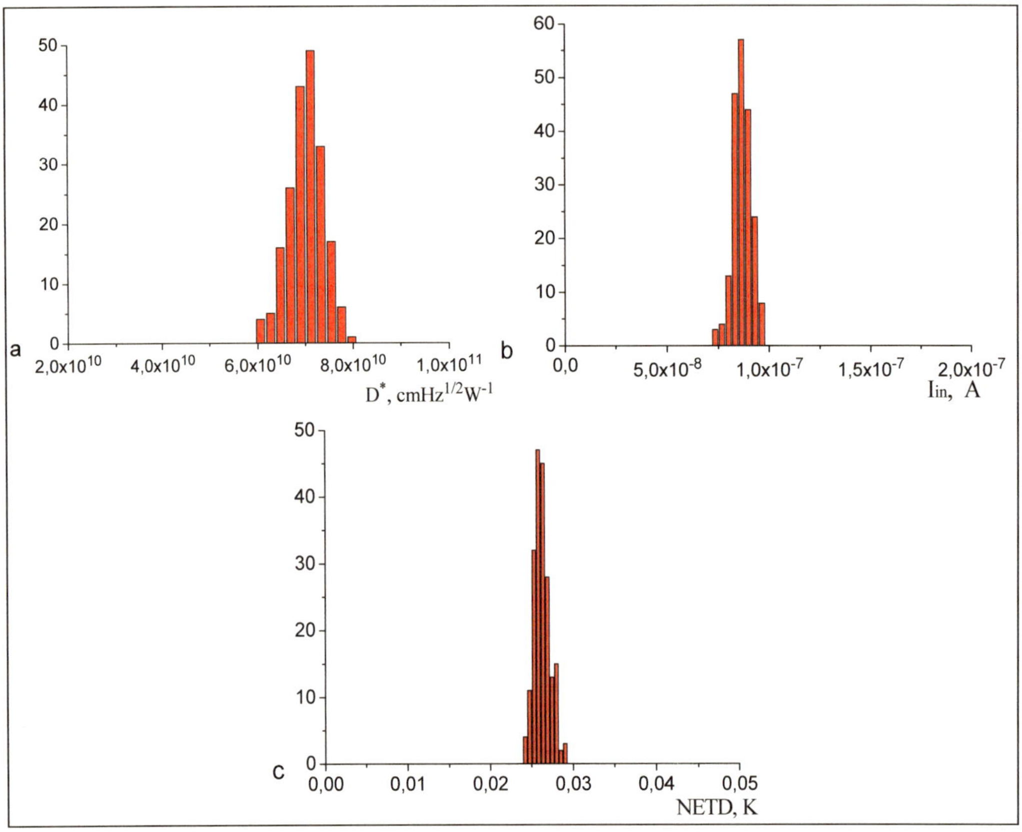

Fig. 21. Calculated histograms of performance characteristics of multi-element IR FPAs; a - detectivity D*, b – currents I_{in} integrated in the readout circuits, c - NETD. Standard deviation of input-FET threshold voltages - $\sigma(V_{th})$ = 2 mV. Standard deviation of the stoichiometric composition of $Hg_{1-x}Cd_xTe$ substrate $\sigma(x)$ = 0.03%. Average stoichiometric composition x=0.2126, T=77 K.

As it follows from the calculated dependencies shown in Fig. 22, the cooling of the hybrid assembly down to temperature 60 K will allow implementation of thermography systems with NEDT values close to maximum possible figures, limited only by the value of the storage capacitance of silicon readout circuits at an acceptable level of fixed-pattern noise. The possibility of variation of the cooling temperature of hybrid assembly in the calculations allows formulation of requirements to required accuracy in maintaining the temperature of cooled hybrid assembly, i.e. requirements to be imposed on the cryostat and cooling system.

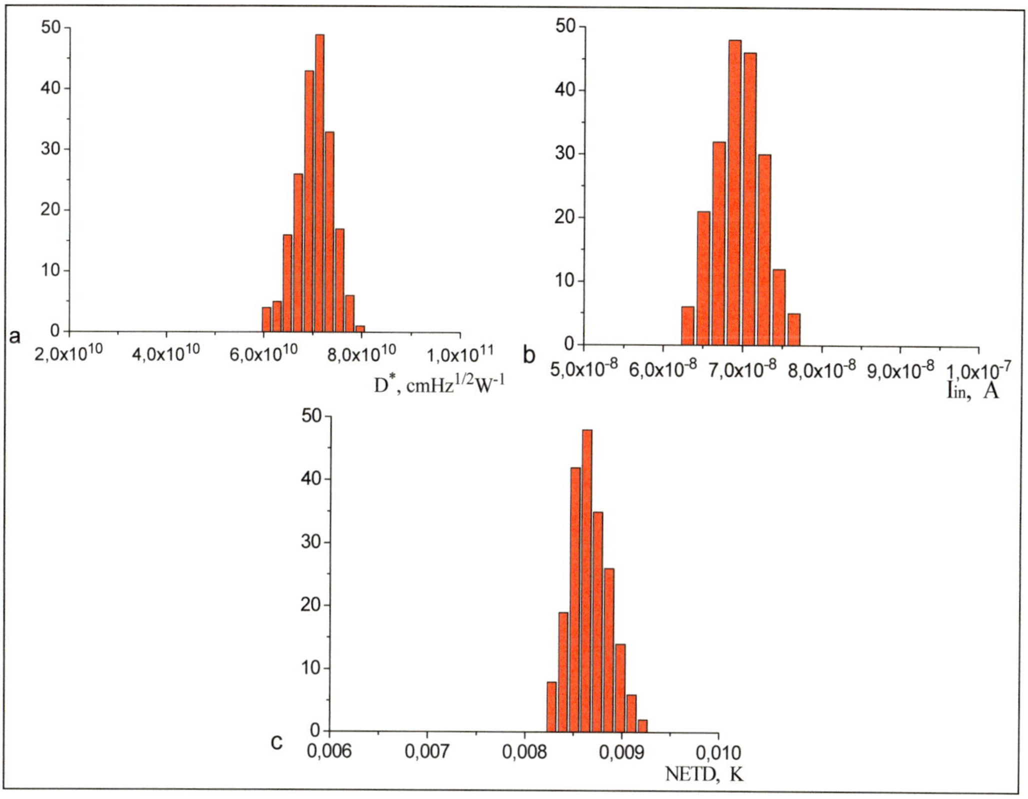

Fig. 22. Calculated histograms of performance characteristics of multi-element IR FPAs; a - detectivity D*, b – currents I_{in} integrated in the readout circuits, c - NETD. Standard deviation of input-FET threshold voltages - $\sigma(V_{th})$ = 5 mV. Standard deviation of the stoichiometric composition of $Hg_{1-x}Cd_xTe$ substrate $\sigma(x)$ = 0.1%. V_G = 1.26 V. Average stoichiometric composition x=0.2105, T=60 K.

5. Conclusion

A mathematical model was developed to analyze the performance characteristics of IR FPAs based on the system 'IR photodiode – direct injection readout circuit'. The proposed mathematical model is based on the solution of the self-consistent problem for the current through photodiode and for the current integrated in the readout circuit, and also on the calculation of the noise charge Q(t) in terms of noise-current spectral density $S_i(\omega)$. Such an approach allows one to determine the main performance characteristics of thermography systems based on multi-element IR FPAs versus the gate voltage of photodetector channels, to calculate histograms $D^*(V_G)$, $I_{in}(V_G)$, and NETD as functions of non-uniformity of photoelectric parameters of photodiodes and silicon readout circuits, and to compare predicted performance characteristics of IR FPAs with experimental results.

The analysis of performance characteristic of multi-element FPAs can be performed with setting photodiode current-voltage characteristics either in analytical form ("classical" photodiode model, analytical model of $Hg_{1-x}Cd_xTe$ photodiodes) or as an approximation of experimental current-voltage characteristics of photodiodes. The proposed approach will be helpful in identifying the main factors that limit the performance characteristics of multi-element IR FPAs, and in formulating requirements to photoelectric characteristics of IR photodiodes, design and electrophysical parameters of readout circuits, and requirements to silicon technology used to fabricate direct injection readout circuits enabling the achievement of required performance characteristics of multi-element IR FPAs and thermography system based on such FPAs.

6. References

Anderson W. (1981). Tunnel contribution to $Hg_{1-x}Cd_xTe$ and $Pb_{1-x}Sn_xTe$ p-n junction diode characteristics. *Infrared Physics*, Vol. 20, (1981), pp. 353-361

Anderson W. & Hoffman H. (1982). Field ionization of deep levels in semiconductors with applications to $Hg_{1-x}Cd_xTe$ p-n junctions. *J. Appl. Physics*, Vol. 53, N.12, (1982), pp.9130-9145

Bisotto S. et al. (2010). A 25µm pitch LWIR staring focal plane array with pixel-level 15-bit ADC ROIC achieving 2mK NETD. *Proc. of SPIE*, (2010) vol.7884, 7884oJ-1

Bluzer N. & Stehlik R. (1978). Buffered direct injection of photocurrents into charge-coupled devices. *IEEE Transaction on Electron Devices*, Vol.ED-25, No.2, (1978), pp.160-167

Buckingham M. (1983). Noise in Electronic Devices and Systems. *Ellis Horwood Ltd*, (1983)

D'Souza F. et al. (2002). 1/f noise in HgCdTe detectors. *Proc. SPIE "Infrared detectors and Focal plane arrays VII"*, No.4721, (2002), pp. 227-233

Felix P., et al. (1980). CCD readout of infrared hybrid focal plane arrays. *IEEE Transaction on Electron Devices*, Vol. ED-27, No.1, (1980), pp.175-188

Fowler B.l, Gamal A. & Yang D. (2000). Technical for pixel level analog to digital conversion. *Proc. SPIE*, Vol.3360, (2000), pp.124-138

Gopal V. (1996). Spatial noise limited NETD performance of a HgCdTe hybrid focal plane array. *Infrared physics and technology*, Vol.37, (1996), pp.313-320

Gumenjuk-Sichevska J. & Sizov F. (1999). Currents in narrow-gap photodiodes. *Semicond. Sci. Techn*. Vol. 14, (1999), pp.1124-1133

Gumenjuk-Sichevska J.; Karnaushenko D.; Lee I. & Polovinkin V. (2011). Infrared photodetectors based on the system $Hg_{1-X}Cd_XTe$ photodiode - direct-injection readout circuit. *Opto-electronics review* Vol.19, No.2, (2011), pp.63-70

Iwasa S. (1977). Direct coupling of five-micrometer (HgCd)Te photovoltaic detector and CCD multiplexer. *Optical engineering*. Vol. 16, N.3, (1977), pp. 233-236

Karnaushenko D.; Lee I. et al. (2010). Infrared focal plane arrays based on systems photodiode-direct injection readout circuits. *Opticheskii Zhurnal*, Vol. 77, No.9, (2010), pp.30-36

Krishnamurthy S. et al. (2006). Tunneling in long-wavelength infrared HgCdTe phodiodes. *Journal of Electronic Materials*, Vol. 35, No. 6, (2006), pp.1399-1402

Kunakbaeva G. & Lee I. (1996). Selection of the spectral range for infrared vision systems based on $Cd_xHg_{1-x}Te$ multielement photodiodes. *Optoelectronics, Instrumentation and Date Processing*, Vol.5, (1996), pp.19-26

Kunakbaeva G., Lee I. & Cherepov E. (1993). The system photodiode - direct-injection input CCD for multi-element FPAs. *Radiotekhnika i Electronika*, Vol.5, (1993), pp.922-930

Lee I. (2010). A new readout integrated circuit for long-wavelength IR FPA. *Infrared Physics and Technology*, Vol. 53, Issue 2, (2010), pp. 140-145

Longo J.T., et al.. (1978). Infrared focal plane in intrinsic semiconductors. *IEEE J. Solid State Circuits*, Vol. SC-13, N.1, (1978), pp.139-157

Martijn H.& Andersson J. (2000). On-chip analog to digital conversion for cooled infrared detector arrays. *Proc. SPIE "Infrared detectors and Focal plane arrays VI"*, Vol.4028, (2000), pp. 183-191

Mikoshima H. (1982). 1/f noise in n-channel silicon -gate MOS transistors. *IEEE Transaction on Electron Devices*, ED-29 - (1982), pp. 965-970

Overstraeten R., Declerck G. & Muls P. (1975). Theory of the MOS transistor in weak inversion - new method to determine the number of surface states. *IEEE Transaction on Electron Devices*, ED-22, (1975), pp. 282-288

Phillips J., Edwall D. & Lee D. (2002). Control of Very-Long-Wavelength Infrared HgCdTe Detector-Cutoff Wavelength. *Journal of Electronic Materials*, Vol. 31, No. 7, (2002), pp.664-668

Reimbold G. (1984). Modified 1/f trapping noise theory and experiments in MOS transistors biased from weak to strong inversion - influence of interface states. *IEEE Transaction on Electron Devices*, ED-31, (1984), pp. 1190-1198

Reimbold G. (1985). Noise associated with charge injection into a CCD by current integration through a MOS transistor. *IEEE Transaction on Electron Devices*, ED-32, (1985), pp. 871-873

Rogalski A. (2000). Infrared detectors. *Gordon and breach science publishers*, Canada, 2000

Sizov F. et al. (2006). Gamma radiation exposure of MCT diode arrays. *Semicond. Sci. Technol.*, Vol. 21, (2006), pp. 356-363

Steckl A.& Koehler T. (1973). Theoretical analysis of directly coupled 8-12 µm hybrid IR CCD serial scanning. *Proc. Int. Conf. Application of CCD's*, (1973), pp.247-258

Steckl A. (1976). Infrared charge coupled devices. *Infrared Physics.* Vol. 16, (1976), pp. 65.

Takigawa H., Dohi M. & Ueda R. (1980). Hybrid IR CCD Imaging Arrays. *IEEE Transaction on Electron Devices*, Vol. ED-27, No.1, (1980), pp.146-150

Taubkin I. & Trishenkov M. (1993). Minimum temperature difference resolvable with the IR imaging method. *Opticheskii Zhurnal*, Vol. 5, (1993), pp. 20-23

Tobin S., Iwasa S. & Tredwell T. (1980). 1/f noise in (Hg, Cd)Te photodiodes. *IEEE Transaction on Electron Devices.* ED-27 (1980), pp. 43-48

Vasilyev V. et al. (2010). 320.256 HgCdTe IR FPA with a built-in shortwave cut-off filter. *Opto–Electron. Rev.*, Vol.18, No. 3, (2010), pp.236-240

Yoshino J. et al. (1999). Studies of relationship between deep levels and RA product in mesa type HgCdTe devices. *Opto-Electronics Review*, Vol. 7, (1999), pp.361-367

Zhou Z., Pain B., et al. (1996). On-focal-plane ADC: Recent progress at JPL. *Proc. SPIE, "Infrared readout electronics III"*, Vol. 2745, (1996), pp.111-122

Part 3

Photodetection Systems

13

Ultrafast Imaging in Standard (Bi)CMOS Technology

Wilfried Uhring[1] and Martin Zlatanski[2]
[1]*University of Strasbourg and CNRS*
[2]*ABB Switzerland Ltd.*
[1]*France*
[2]*Switzerland*

1. Introduction

Around 1822, the French inventor Niépce made the first photographic image by the use of a *camera obscura*. He formed the image passing through the hole on a metal plate with a bitumen coating. After 8 hours of exposure, the bitumen on the illuminated sections of the plate was hardened. By washing the unhardened regions, a print of the observed scene appeared. After Niépce's death in 1833, Daguerre worked on the improvement of the chemical process involving interaction of the plate with light. In 1839 he announced the invention of a new process using silver on a copper plate. This invention reduced the exposure time to 30 minutes and denotes the birth of modern photography. During the following years, improvement on the photographic processes led to increased sensitivity and allowed shorter exposure times. In 1878, Muybridge gave an answer to a popular question at this time: whether all four hooves of a horse are off the ground at the same time during a gallop. By taking the first high-speed sequence of 12 pictures, each picture spaced about 400 ms from the neighbouring one with an exposure time of less than 500 μs. In 1882, George Eastman patented the roll film, which led to the acquisition of the first motion pictures. Four years later, a student of Daguerre, Le Prince, patented a *Method of, and apparatus for, producing animated pictures*. Through its *16 lens receiver*, as he called his camera, and by the use of an Eastman Kodak paper film, Le Prince filmed the first moving picture sequences known as the *Roundhay Garden Scene*, which was shot at 12 frames per second (fps) and lasted less than 2 seconds. Two years later, Edison presented the *Kinetoscope*, a motion picture device capable of acquiring sequences at up to 40 fps. It creates the illusion of movement by conveying a strip of perforated film filled with sequential images over a light source through a mechanical shutter. In 1904, the Austrian physicist Musger patented the *Kinematograph mit Optischem Ausgleich der Bildwanderung* which is capable of recording fast transients and projecting them in slow motion. In acquisition mode, the light is turned off and a rotating mirror, projecting them in slow motion. In acquisition mode, a rotating mirror mechanically coupled to the film shifting mechanism, reflects images onto the film. During the projection, the light is turned on and the same operation is carried out, but at a much slower rate. This high-speed photographing principle was used during the First World War by the German company Ernemann Werke AG to develop the *Zeitlupe*, a 500 fps camera used mainly for ballistic purposes. In 1926, Heape and Grylls constructed *Heape and*

Grylls Machine for High Speed Photography, in which the film shifting mechanism was replaced by a film lining the inside of a rotating drum, driven by an 8 horse power engine (Connel, 1926). The relaxed constraints on the film strength allowed to the 4 tons instrument to reach a frame rate of 5000 fps. In later realizations the drum was powered by an electric motor, leading to much more compact instruments (Lambert, 1937). Nowadays drum cameras are still in use and their fundamental principle is still the same. They employ a rotating mirror mechanically coupled to the drum and use the Miller's principle to perform the shuttering between frames. Miller's principle states that if an image is formed on the face of a mirror, then it will be almost static when relayed by lens to a film. The fastest drum cameras produce frame records at up to 200 000 fps which is limited by the mechanical constraints imposed by the need of a synchronous rotation of the mirror and the drum at very high speeds. Rotating mirror cameras have been developed toward the end of World War II to answer the need of photographing atomic explosions. The image formed by the objective lens is relayed back to a rotating mirror which sweeps the reflected image focused through an arc of relay lenses and shuttering stops on a static film (Fig. 1). The first camera of this type was built by Miller in 1939 and reached a speed of 500 000 fps. The concept was patented in 1946 (Miller, 1946) and in 1955, Berlin Brixner achieved a speed of 1 million fps using the same principle (Brixner, 1955). Cordin's Model 510 rotating mirror camera reaches 25 million fps, by driving the mirror up to 1.2 million rpm in a helium environment using a gas turbine (Cordin 2011). The speed of the drum and the rotating mirror cameras has not been improved for more than 40 years, the limits of this technology being set by the maximum rotation speed of the mirror and the materials (Frank & Bartolick, 2007).

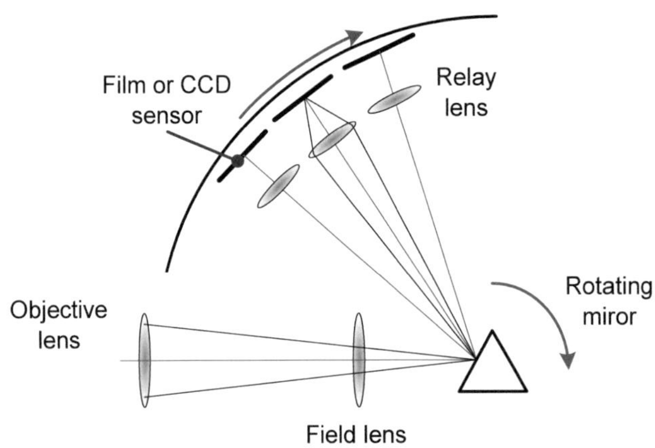

Fig. 1. Principle of the rotating mirror high speed video camera

Shortly after the first CMOS image sensors, the first Charge-Coupled Device (CCD) imagers appeared (Amelio et al., 1970). The CCD, initially developed for semiconductor bubble memory purposes, was invented in 1969 by Boyle and Smith at the Bell Laboratories. After promising initial results, this technology was rapidly adapted to be used for image sensor applications and little-by-little succeeded the photographic film (Smith, 2001). The first high-speed CCD cameras appeared during the 80's and were capable of acquiring up to 185 fps at a resolution of 512 × 512 pixels with a dynamic range of 8 bits. Their CMOS counterparts were able to go even further by integrating the signal conditioning electronics directly on-chip (Krymski et al., 1999; Kleinfelder et al. 2001). In both technologies, the shuttering is operated electronically without the use of a mechanical part and can be potentially operated in few nanoseconds. Nevertheless, in both CCD and CMOS architectures, the maximum frame rate is limited by the time interval required for an image to be read out. Actual CMOS high-speed video cameras are capable of producing about 1 Mega pixel images at a rate of approximately 1000 fps or every other combination of image resolution and frame rate, which maintains the read-out rate to several GS/s. Indeed, by reducing the image size, the frame rate can be increased. A summary of the fastest high-speed video cameras in 2011 is given in Table 1. The maximum amount of data rate at the output (about 100 Gb/s for the fastest available camera) is generally limited by the bandwidth of the output bus. Parallelization of the channels seems to be a unpromising solution, since it leads to an excessive power consumption and unreasonable chip area (Meghelli, 2004; Swahn et al. 2009).

Manufacturer	Model	Resolution	Frame rate @ full resolution (total sampling rate)
Framos	MT9S402 (Sensor)	512×512	2500 ips (655MS/s)
Aptina (ex Micron)	MT9M413 (Sensor)	1280×1024 (10 bit)	500 ips (655 MS/s)
Micron	(Krymski et al. 2003)	2352×1728 (10 bits)	240 ips (975 MS/s)
Optronis	CR5000x2 (Camera)	512×512 (8 bits)	5000 ips (1.3GS/s)
Cypress	LUPA 3000 (Sensor)	1696×1710 8 bits	485 ips (1.4 GS/s)
Vision research	Phantom v640 (Camera)	2560×1600 (8 or 12 bits)	1500 ips (6.1 GS/s)
Vision research	Phantom v710 (Camera)	1280×800 (8 or 12 bits)	7530 ips (7.7 GS/s)
Photron	Fastcam SA5 (Camera)	1024×1000 (12 bits)	7500 ips (7.7 GS/s)
Photron	Fastcam SA2 (Camera)	2048×2048 (12 bits)	1000 ips (4.2 GS/s)
IDT	Y4-S3 (Caméra)	1016×1016 (10 bits)	9800 ips (10 GS/s)

Table 1. State-of-the-art of the fastest electronic high-speed video cameras in 2011

As seen from Table 1, today's high speed video cameras present a temporal resolution in the range of 1 ms in full resolution down to 1 μs in a reduced image size of about 1 thousand pixels. These performances are still far away from those of the rotating mirror cameras. To break the GS/s-order limit, a radically different approach than trying to operate the acquisition and the read-out progressively was required.

2. Ultrafast imaging concepts

2.1 *In situ* storage concept

The conventional high-speed video cameras face the input/output bus bandwidth constraint, which limits their data rate and temporal resolution to about 10 GS/s and 100 μs in full frame format, respectively. To overcome this bottleneck, the solution is simply not to extract the data from the sensor. Indeed, the fastest video sensors reported in the literature employ the *in-situ* concept, which consists in storing the acquired data on-chip rather than continuously extracting it and operate the readout afterwards. The use of this method in

CCD and CMOS optical sensors allowed to push the total sampling rate to about 1 TS/s, i.e. 100 times faster than the fastest high speed video cameras. The architecture of a sensor using the *in-situ* concept is shown in Fig. 2. Each pixel integrates a photodetector, generally a photodiode, and its own *in-situ* memory. Work on this concept started during the 90's and 1 million fps cameras have been reported (Elloumi et al. 1994; Lowrance & Kosonocky, 1997). These first sensors with *in-situ* frame memory, in which each pixel is composed by a photodetector and a linear CCD storage sensor were able to store 30 frames. In 1999, Professor Etoh proposed a sensor with 103-deep CCD storage (Etoh et al., 1999, 2005).

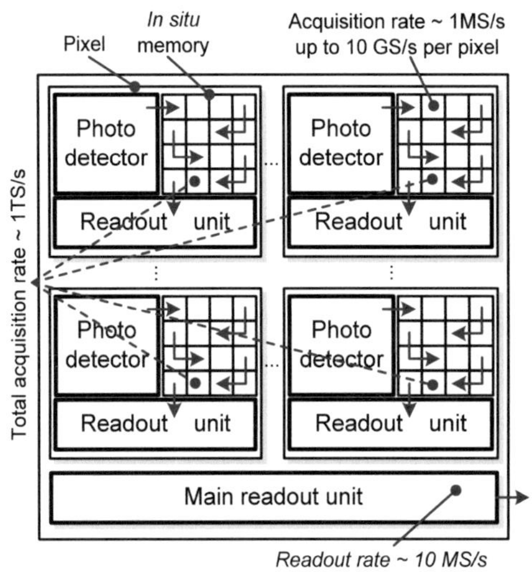

Fig. 2. Architecture of a high speed image sensor with the in-situ storage concept

A comparison between Cordin's Model 121 rotating mirror camera and Shimadzu's Model HPV-1 high-speed CCD imager with *in-situ* storage is made in (Franck and Bartolick, 2007). The test clearly shows the superiority of rotating mirror cameras in terms of spectral range, temporal and spatial resolutions. The authors conclude that at the moment a solid-state replacement for the rotating mirror cameras does not exist, but the *in-situ* based sensor architectures have the potential to reach higher performance through specifically optimized designs. Some studies forecast theoretical rates as high as 100 million fps for *in-situ* CCD detectors (Son et al. 2010), at the cost of reduced image quality.

The development of CMOS technologies allowed the integration of signal processing electronics and photodetecting site on the same substrate, which resulted in an enlargement of the spectrum of applications of solid-state imagers (Bigas et al., 2005). Since the bandwidth of a field-effect transistor in a submicron silicon technology easily exceeds the GHz, a number of these applications were naturally oriented towards high-speed imaging. In 2004, a 12 × 12 pixels demonstrator processed in standard 0.35 μm CMOS technology employing the *in-situ* storage concept achieved more than 10 million fps (Kleinfelder et al., 2004). The memory depth was of 64 and the pixel size of 200 μm × 200 μm. Recently, a 32 × 32 pixel prototype with an *in-situ* memory of 8 frames has been presented (Desouki et al. 2009). Processed in a 0.13 μm standard CMOS, it achieved a rate of 1.25 billion fps. The main drawback of this method is the limited memory depth to a maximum of about several hundred frames. Indeed, the silicon area required increases with the number of stored frames and in 2D imaging, the in-pixel embedded memory decreases drastically the fill factor, e.g. less than 10% for only 8 frames in (Desouki et al., 2009). Nevertheless this is not a real issue, since in practice at these rates, generally only a few frames are analyzed. 3D technologies or die stacking may lead to fill factors as high as 100 %. However, ultrafast optical sensors employing the in-situ concept and featuring high fill factor can be processed in a standard (Bi)CMOS technology if streak-mode images are taken.

2.2 Streak-mode imaging concept

A feature proper to all the devices presented until now is that they produce 2-dimensional *(x,y)* images I_f at equally spaced in time intervals Δt. Consequently, I_f is a function of the two spatial dimensions x,y with a constant time $t_0+n\cdot\Delta t$, where n is an integer:

$$I_f = f\left(x,y,t_0+n\cdot\Delta t\right) \quad (1)$$

This is referred as framing mode photography. The advantage of a frame record is that information in two spatial dimensions is recorded, so the recorded image is an easily recognized version of the subject. However, if finer temporal information is required, another imaging method must be used. Speed and spatial information size of a camera being strongly related, a considerable gain in the frame rate can be achieved if the spatial information is reduced. If a single spatial dimension is selected, a picture containing the continuous temporal evolution of the spatial information is obtained. The recorded image I_s contains the spatial information x which crosses the slit, observed at different times t:

$$I_s = f(x,t) \quad (2)$$

This is denoted as streak-mode imaging. A streak record is made by placing a narrow mechanical slit between the event and the camera. Next, the temporal evolution of the one-dimensional spatial information crossing the slit is swept along a taping material. Thus, a continuous record containing position, time and intensity information is obtained. The recording rate, called sweep speed, translates the distance covered on the record surface for a unity of time and is measured in mm/µs or ps/pixel.

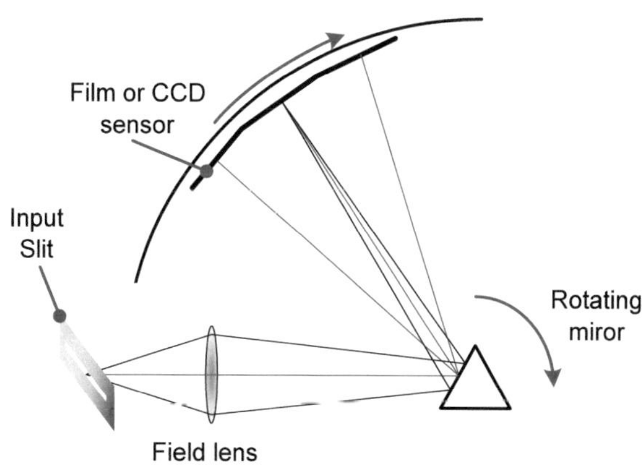

Fig. 3. Rotating mirror camera operating in streak-mode photography

To operate a rotating mirror camera in streak-mode, the relay lenses have to be removed and a slit positioned on the input optical path, (Fig. 3). With such a device, a maximal sweep speed of 490 ps/pixel has been reported (Cordin, 2011). Fig. 4 shows a frame (top) and a streak record (bottom) of a projectile fired through a near-side cut plastic chamber to impact an explosive sample. It clearly shows how a streak record provides continuous inter-frame temporal information of the slit image. The improvement in temporal resolution reaches about 2 orders of magnitude. Streak-mode imaging is ideal for studying events of uniform growth, e.g., an expanding sphere, where the rate of dimensional change is to be measured. To avoid misinterpretation and provide the maximum amount of information, often a combination of streak and framing records are carried out. Thus, framing and streak-mode cameras are not comparable, but rather complementary (Fuller, 2005).

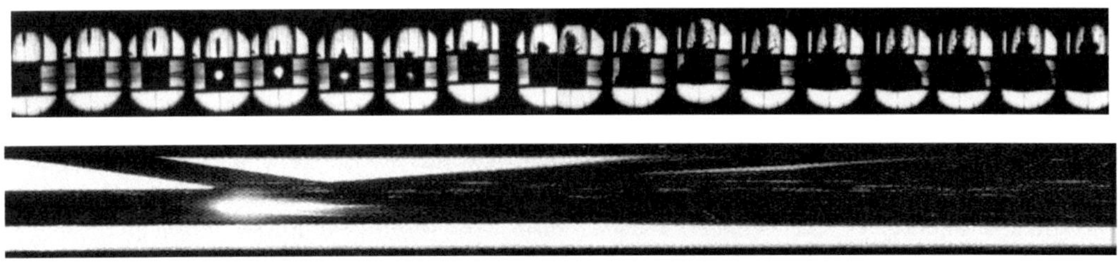

Fig. 4. Illustration of the streak imaging concept: frame record at 100 000 fps (top) and streak record at 0.53 mm/µs (bottom) of a ballistic event acquired by a Cordin Model 330 camera (Cordin, 2011)

The rotating mirror technology is limited by the centrifugal force applied to the mirror by the very high rotation speed. Even the use of light metals as beryllium does not prevent the mirror from explosion when the rotation speed exceeds 2 millions rpm (Igel & Kristiansen, 1997). To overcome this limitation a completely different technology based on a vacuum tube has to be used. The fastest device for direct light measurement available, known as the conventional streak camera allows picosecond-order temporal resolution to be reached. The operation principle of a conventional streak camera is depicted in Fig. 5. The very core of the streak camera is a modified first-generation sealed vacuum image converter tube, known as a streak tube, comprising four main sections: a photon-to-electron converter, an electron bunch focusing stage, an electrostatic sweep unit, and an electron-to-photon conversion stage. On some streak tubes, an internal Micro Channel Plate (MCP) is added in front of the phosphor screen for signal amplification. A mechanical slit is illuminated by the time varying luminous event to be measured and is focalized on the photocathode of the streak tube. The incident photons on the photocathode are converted to photoelectrons with a quantum efficiency depending on the type of the photocathode. A mesh is placed in the proximity of the photocathode and a high static voltage is applied between these two components in order to generate a high electrical field, which extracts the photoelectrons from the photocathode, makes their velocities uniform, and accelerates the pulse of photoelectrons along the tube. At this stage, the photogenerated electrons represent a direct image of the optical pulse, which reached the photocathode surface. When the photoelectrons approach the sweep electrodes, a very fast voltage ramp $V(t)$ of several hundred Volts per nanosecond is applied at a timing synchronized with the incident light. During the voltage sweep, the electrons which arrive at slightly different times, are deflected in different angles in the vertical direction. As far, first a photo-electrical conversion is carried out by the photocathode, and then a translation from time to space is operated through the sweep electrodes. After being deflected, the electrons enter the MCP, where they are multiplied several thousands of times by bouncing on the internal channel walls acting as continuous dynodes and are extracted by the high electrical field applied on both sides of the MCP. At the end of the streak tube the photoelectrons impact against a screen which emits a number of photons proportional to the incident electron density. To prevent dispersion, an objective or a tapper made of a fine optical fiber grid is positioned in-between the phosphor screen and the read-out CCD or CMOS camera to guide the emitted photons.

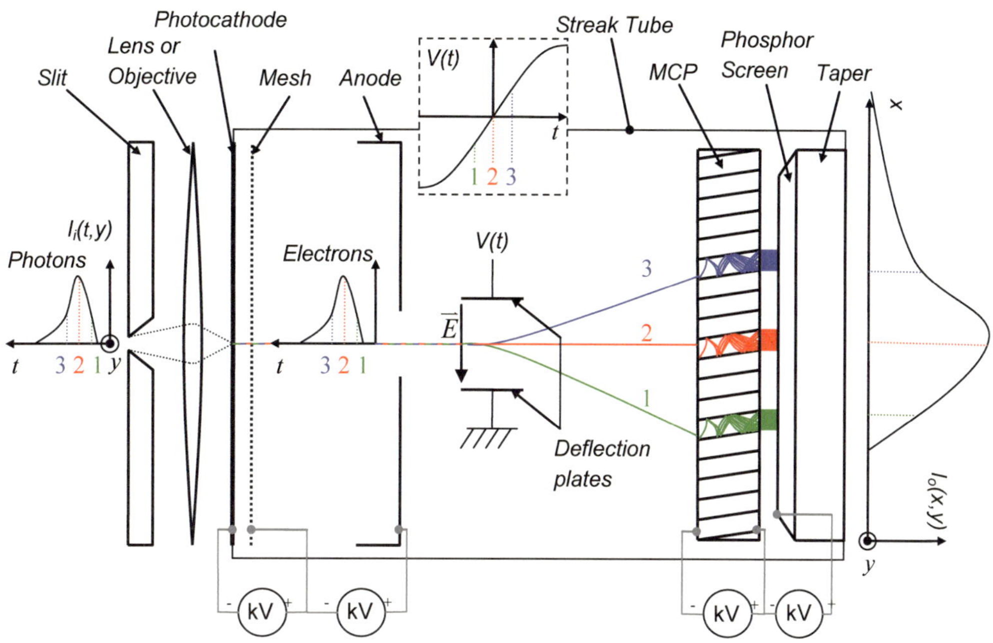

Fig. 5. Operation principle of a conventional streak camera

As the time-varying electric field caused by the voltage ramp between the electrodes is assumed to be spatially uniform, the spatial distribution of the light is directly obtained on the phosphorus screen without temporal modification. Finally, the temporal axis is found along the vertical direction of the screen, the position in the horizontal direction corresponds to the location of the incident light, and the brightness is proportional to the intensity of te respective optical pulses. The relationship between a given vertical position x and the time t depends on the slew rate S_R [V/s] of the sweep voltage $V(t)$ and the deflection sensitivity D_S [V/mm] of the streak tube is:

$$t = t_0 + \frac{D_s}{S_R} \cdot (x - x_0), \tag{3}$$

where x_0 is the position in mm of the slot image on the phosphorus screen when $V(t_0)=0$. As for the rotating mirror camera operating in streak-mode, the sweep speed of the camera is found as the ratio D_S/S_R, generally expressed in ps/mm.

Spectral range	[41 pm, 10 μm] depending on the photocathode type
Sensitivity	Single photon detection capability (with MCP or image intensifier)
Repetition rate	From single shot up to 250 MHz (synchroscan)
Spatial resolution	From 25 μm (ns resolution) to 100 μm (ps resolution)
Temporal resolution	From 200 fs (single shot) or 2 ps (synchroscan) to 300 μs
Sweep speed	From 10 ps/mm (fast sweep unit) up to 5 ms/mm (slow sweep unit)
Observation time	From 60 ps up to 175 ms (phosphorus screen size ∈ [9–35 mm])

Table 2. Performance summary of a conventional streak camera

The conventional streak cameras are very versatile devices with high-end performances (Table 2). The wavelength detection spectrum ranges from the X ray (Scheidt & Naylor, 1999) up to the far infrared (Jiang et al. 1999). Their very high sensitivity permits the detection of a single photon event and the repetition rates extend from single shot operation up to several hundred of MHz. The typical spatial resolution of a conventional streak camera is between 25 μm and 100 μm. Specific tube designs using magnetic solenoid lens can reach a spatial resolution of 10 μm (Feng et al., 2007). Finally their temporal resolutions are very close to the physical theoretic limitation, about 100 fs (Zavoisky & Fanchenko, 1965) but with a poor signal to noise ratio whereas 1 ps can be reached with a high signal to noise ratio (Uhring et al., 2004a). Besides their extreme performances, conventional streak cameras and rotating mirror cameras have drawbacks: they are bulky, fragile, delicate to manufacture, and cost around 100 k€. Moreover there are many applications in which a temporal resolution of about 1 ns is sufficient. For these applications, solid-state technologies can offer interesting alternatives.

2.3 Streak camera alternatives

A high-speed video camera can be used as a streak-mode device by activating several rows on the sensor array and focusing the spatial information of interest on them. This technique is demonstrated in (Parker et al. 2010), where a temporal resolution of 1 μs and a sweep speed of 7.5 mm/μs have been reached, obviously limited by the output data rate.

In 1992, Lai from the Lawrence Livermore National Laboratory (USA) proposed a streak-mode camera in which the sweep is carried out by an externally triggerable galvanometer (Lai et al., 1992, 2003). The signal beam reaches the surface of the moving wedge-gap shaped deflector. It is subjected to multiple reflections between the surfaces of the stationary and deflecting mirror which results in an effective increase of the optical sweep speed. N reflections on the moving mirror result in an output beam angular speed equal to $2N$ times the rotating speed of the deflector. The deflected beam is focused and projected on a CCD sensor. At least 2 ms of trigger delay are required to allow the mirror to reach the targeted deflection speed. A temporal resolution of 30 ns has been reported.

Nowadays there isn't any solid-state streak-mode imaging device with performances situated in-between high-speed video cameras and conventional streak and rotating mirror cameras and in the same time there is a highly potential market for such a device. Indeed, many customers cannot afford the acquisition of a streak camera because of its price and, in the same time, they do not always need picosecond temporal resolution. In many streak-mode photography applications a temporal resolution in the order of several hundreds of picoseconds is often sufficient. The following section presents a solid-state alternative to the conventional streak camera, which should give birth to new commercial and scientific activities, answers existing but unresponded demands and opens new fields of application. The prototypes are designed in a standard (Bi)CMOS technology which ensures quick prototyping and very low production cost estimated to about 10 % of the price of a traditional streak-mode imaging devices.

3. Integrated streak camera

Following the example of high-speed video cameras, the integration possibilities offered by CMOS technologies allow the use of the *in-situ* storage approach and its application to the first streak-mode CMOS imagers. In 1987, Professor Kleinfelder from the University of California proposed a 16 channel electrical transient waveform recording integrated circuit with 128 Sample and Hold (S/H) cells per channel operating in parallel at 100 MHz sampling frequency (Kleinfelder, 1987, 1990). In 2003, he demonstrated a 4 electrical channel 128-deep sampling circuit with up to 3 GHz operation using sub-micron silicon technology (Kleinfelder, 2003). One year later he presented a multi-channel sampling circuit in which the inputs of every channel are connected to an optical front-end (Kleinfelder et al. 2003, 2004). This was the first single-column streak-mode optical sensor. However, the very first integrated streak camera was developed in 2001 by the University of Strasbourg and was based on a pixel-array sensor architecture driven by an electronic temporal sweep unit (Casadei et al. 2003). In the following pages, both pixel array Integrated Streak Camera (MISC) and single-column Integrated Streak Camera (VISC) architectures are discussed.

3.1 Pixel array based integrated streak camera (MISC)

The optical setup required for a correct operation of a MISC is schematized in Fig. 6. The image of a light source illuminating a mechanical slit is uniformly spread along the temporal axis (the rows) of the 2D CMOS sensor through a cylindrical lens. Thus, each pixel of the same row is subjected to the same optical event. The distinctiveness of the integrated streak camera sensor compared to a conventional framing-mode sensors, consists in the way the pixels are operated. Indeed, a sweep unit is used to operate the time-to-space conversion carried out by the deflection plates in conventional streak cameras. The pixels of the sensor operate in photon flux integration mode. The circuit temporally sweeps the columns by shifting the beginning of their integrating phase every ΔT. Consequently, each pixel of a given row is subjected to the same illumination, which is measured at different moments.

The first generation MISC uses a 3T pixel that does not allow the integration to stop (Casadei et al. 2003). In practice, the latter ends with the read-out of the pixels. Consequently, attention must be paid to ensure that no light reaches the sensor's surface until the end of the read-out. The output ΔV_{ij} of the pixel p_{ij} at row i, column j is given by:

$$\Delta V_{ij} = \frac{1}{m} G_c \int_{j \cdot \Delta T}^{\infty} E_i(t) \cdot dt, \tag{4}$$

where G_c [V/n_{ph}] is the global conversion gain including the fill factor and the quantum efficiency, E_i [n_{ph}/s] is the optical power received by row i and m the number of pixels along the temporal axis, i.e., the number of columns. It is assumed that the time origin ($t=0$) corresponds to the first column ($j=0$). Processing the differential of the pixels along the temporal axis, i.e., $S_{ij} = (\Delta V_{i(j+1)} - \Delta V_{ij})$ allows the reconstruction of the luminous event. Nevertheless, during this process, the high frequency noise is amplified and the signal obtained has a poor signal to noise ratio. Moreover, the dynamic range of the sensor is reduced since the first pixel integrates the entire signal and, at the same time, saturation must be avoided. The repetition rate of the acquisitions is limited by the read-out time of the array which is about 20 ms.

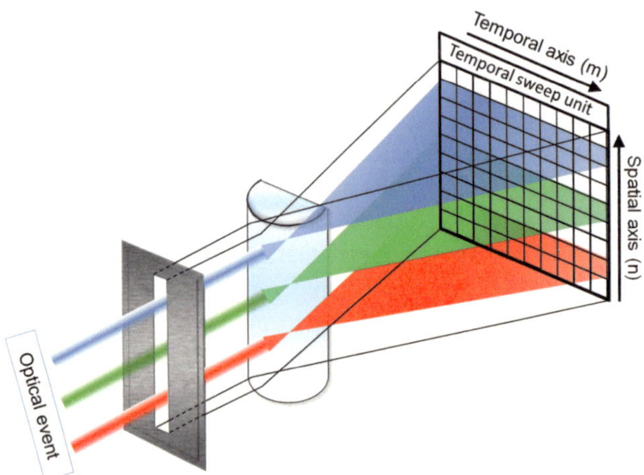

Fig. 6. Optical setup of a MISC

The second generation MISC employs a 6T pixel architecture in order to eliminate the faults of the first generation MISC by featuring a shutter capability. The detailed block diagram of a second generation pixel array-based sensor is shown in Fig. 7 (Morel et al. 2006). The circuit features an array of $n \times m$ active pixels, an m-stage temporal axis generator for the sweep unit, and two selection circuits and an output amplifier for serial data read-out. It temporally sweeps the columns by shifting the beginning and the end of their integrating phase, i.e., by delaying the arrival of the *RPD* and *SH* signals from one column to another. Thus, the integrating procedure starts with the turn off of transistor M_{RPD} and ends with the turn off of the transistor M_{SH} for all the pixels of column i and is delayed from column i-1 by ΔT.

The 6T active pixel is also featuring an analog accumulation capability (Morel et al. 2006). Before acquisition, the potential on the cathode of the photodiode is initialized to V_{RPD} and the potential on the read-out node *RN* to *VRR*. After the integrating phase, i.e., the switching off of transistor $M_{SH,}$ the signal charges are trapped in node *IN*. Next, they are conveyed to the read-out node through M_{TX} acting as a transfer gate, allowing the on-chip accumulation of several repetitive low-light events through successive charge transfers from node *IN* to node *RN*. Anti-blooming is performed by keeping M_{RPD} in sub-threshold conduction after reset. In this circuit, the temporal sweep unit is composed of two identical delay lines with a single cell delay value of ΔT. The first delay line is used to control the reset transistor M_{RPD} whereas the second controls the shutter transistor M_{SH}. The beginning of the acquisition procedure, which consists of feeding the Trigger signal in the temporal sweep unit, is synchronized with the arrival of the optical event, through an external photodiode and a delay box. An externally tunable delay of T_{int} is added between the delay lines and allows the tuning of the integrating time. For this architecture, the output ΔV_{ij} of the pixel p_{ij} is given by:

$$\Delta V_{ij} = \sum_{k=1}^{N} \left(\frac{1}{m} G_c \int_{j \cdot \Delta T}^{j \cdot \Delta T + T_{\text{int}}} E_{ki}(t) \cdot dt \right), \tag{5}$$

where N is the number of accumulations, G_c is the global conversion gain which includes the charge transfer efficiency from the photodiode to node IN and from node IN to node RN and E_{ki} is the k_{nth} optical signal received by row i. The second generation MISC features much higher dynamic range and tunable sweep speed from 140 ps up to 1 ns per pixel. The on-chip accumulation allows the observation of low intensity and repetitive optical signals, comparably to the synchroscan mode of a conventional streak camera. High repetition rates can be reached, since the duration of the charge transfer can be reduced down to 1 ns by appropriately adjusting the V_{RPD} and V_{RR} voltages (Uhring et al. 2011).

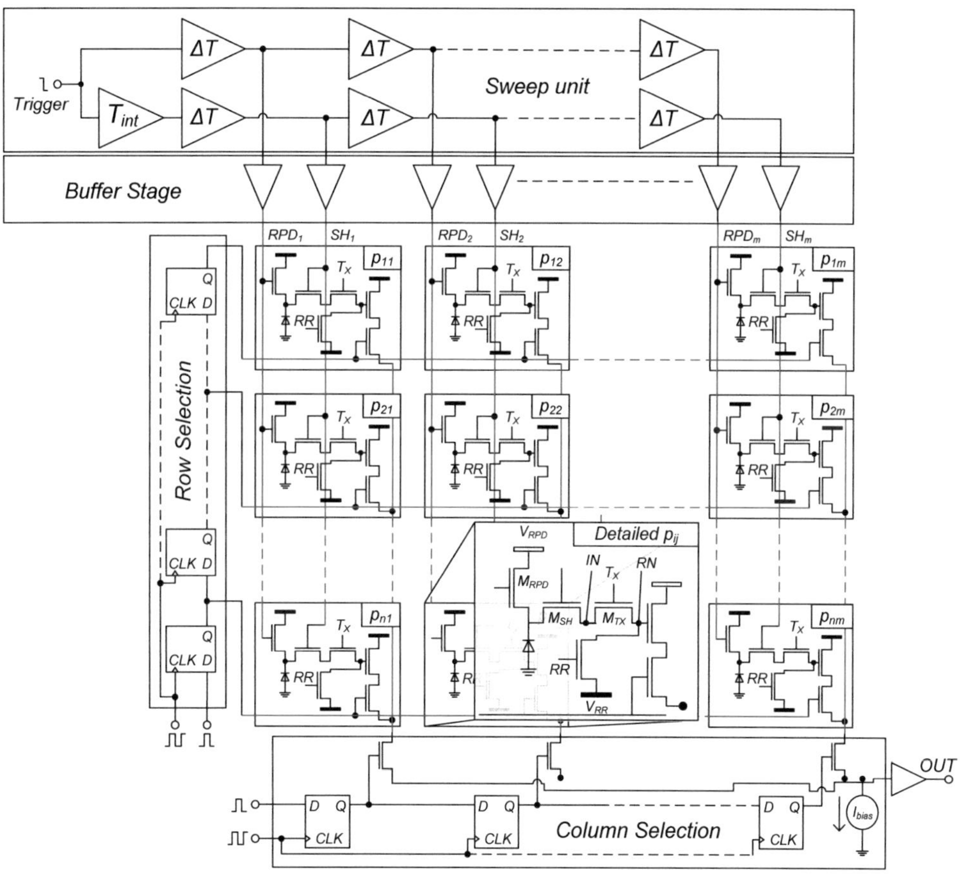

Fig. 7. Block diagram of the second generation MISC and detailed view of a pixel

Through the *in situ* concept and the parallelized operation, the MISC offers a very high speed as the only electronic limitation is the M_{RPD} and M_{SH} transistors bandwidth and turn off transient. The latest prototype with 93 rows channel exhibits a total sampling rate in excess of 650 GS/s (Zlatanski et al. 2010a). The acquisition of a femtosecond laser pulse at 400 nm with an integration duration T_{int} of 400 ps indicates that the temporal resolution is 1.1 ns FWHM (Uhring et al., 2011). Preliminary results obtained with the picoseconds laser diode generator described in (Uhring et al. 2004b) show a temporal resolution very close to the nanosecond at 650 nm whereas the FWHM of the output pulse is increased to more than 3 ns at 808 nm. In 0.35 µm (Bi)CMOS technologies the bandwidth of a MOST switch can easily reach several GHz. The overall temporal response of a MISC in standard CMOS technology is then limited by the photodetector bandwidth. Unfortunately, the MISC suffers from a lack of sensitivity. Indeed, equations 4 and 5 clearly state that the output signal is divided by m, i.e. the memory depth (Zlatanski et al., 2010c). To avoid this, instead of spreading uniformly the light over the whole surface of the sensor, it should be focused to only one column. This is the basis of the vector based integrated streak cameras.

3.2 Vector based integrated streak camera (VISC)

The operation principle of a VISC is schematized in Fig. 8. In this case, the image of the slit is focused on the vector of photodetectors of the sensor through a spherical lens. Each one of the n photodetectors is connected to an amplifier and an m-deep sampling and storage unit. During illumination, the incident light is transformed to an electrical signal, which is amplified by the wideband front-end electronics. The outputs of the front-ends are sampled on the *in-situ* frame storage.

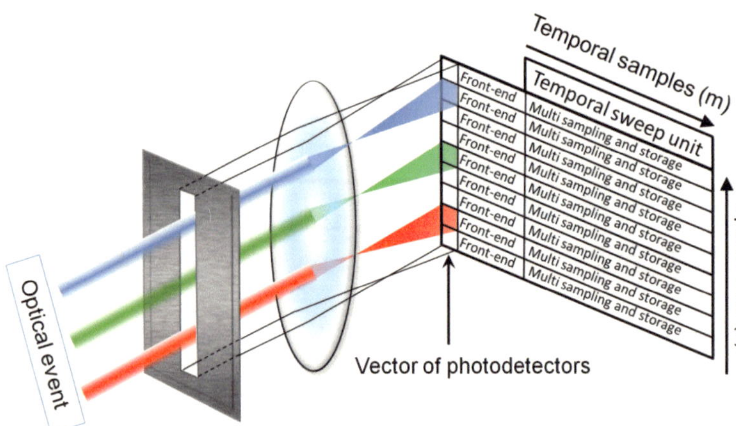

Fig. 8. Optical setup and architecture of a VISC

The different configurations of such a sensor are shown in Fig. 9. One approach is to use a transimpedance amplifier (TIA) coupled to a photodiode as a VISC front-end (Fig. 9 (a)). In this case, the photocurrent provided by the photodetector is converted to voltage with a given transimpedance gain, and is thus a direct image of the light event. At this level the circuit can be seen as a multichannel optical receiver circuit with an analog output. The sampling and storage unit is subjected to less design constraints as the signal to be sampled is in the voltage domain. The output ΔV of a pixel p_{ij} is simply given by:

$$\Delta V_{ij} = G_c \cdot E_i\left(j \cdot \Delta T\right), \tag{6}$$

where G_c is the global conversion gain of the sensor in [V·s/n_{ph}]. The frequency response of the photodiode is only slightly influenced, since the input impedance of a TIA is very low, typically ~100 Ω, and the bias voltage across the photodiode terminals is kept constant. However, a major drawback is the high power consumption as each front-end stage requires about 10 mW to reach a GHz-order bandwidth (Razavi, 2003).

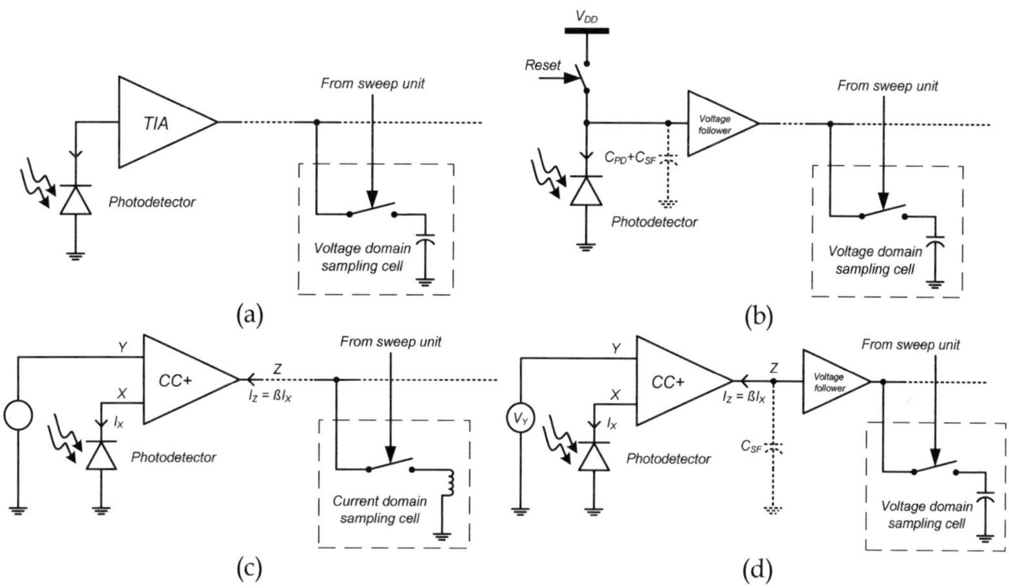

Fig. 9. Different front-end configurations: transimpedance approach (a), integrating approach (b), current to current approach (c) and hybrid integrating approach (d)

Another direct conversion could be obtained by using a Current Conveyor (CC) coupled to a line of *m* current-mode sampling cells (Toumazou et al. 1993; Mehr & Sculley, 1998; Rajaee & Bakthiar, 2005), as shown in Fig. 9 (c). The CC ensures that the voltage applied at *Y* is copied to *X*, keeping the photodetector under constant reverse bias. Next the current is conveyed to the output *Z* with a gain of ß, set by the design. Thus, the CC acts as current buffer between points *X* and *Z*. The value of pixel p_{ij} is given by equation (6), in which G_c, now given in [A·s/n_{ph}], includes the current conveyor and sampling cell gain.

The voltage follower front-end configuration (Fig. 9 (b)) has been used in the first VISC (Kleinfelder, 2004). The photocurrent is integrated on the photodetector and the voltage follower input capacitances and a standard voltage-domain sampling and storage block. The output voltage ΔV of pixel p_{ij} is given by:

$$\Delta V_{ij} = G_c \int_{j\cdot\Delta T}^{(j+\alpha)\cdot\Delta T} E_i(t)\cdot dt \tag{7}$$

where G_c is the global conversion gain and $\alpha = T_{int}/\Delta T$ is the ratio of the integrating time to the sample period with $\alpha\in[0–1[$ in the structure proposed in (Kleinfelder, 2004). The hybrid integrating approach (Fig. 9 (d)) uses a CC to isolate the photodiode from the voltage follower. Thus, the photodiode biasing is maintained constant and the conversion gain is increased as the integrating capacitance is reduced to the one at the input of the voltage follower.

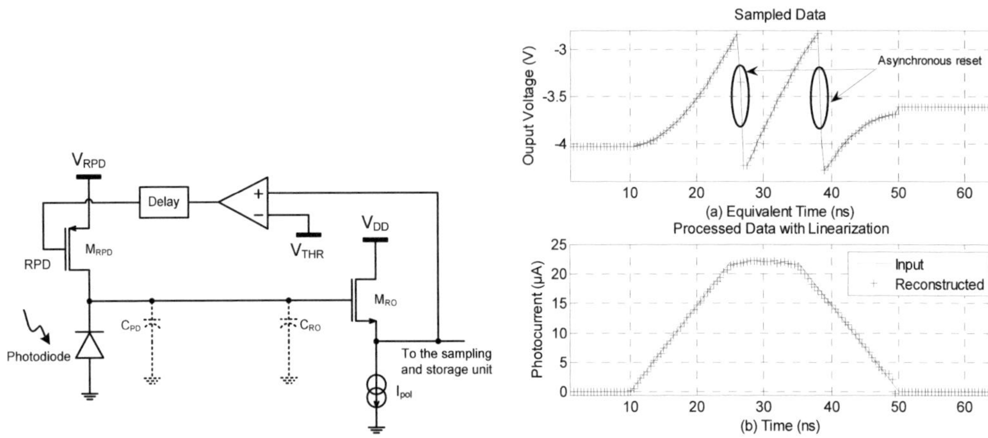

Fig. 10. Architecture of the asynchronous photodiode reset integrating VISC front-end (left) and a simulation result showing the output response to a pulse stimulus (right)

In the latter designs, the synchronous operation of the start and stop integrating phase severely limits the sampling rate to about 100 MS/s. In order to overcome this limitation, a novel asynchronous reset front-end architecture has been developed (Fig. 10-left). It includes a reset transistor M_{RPD}, a read-out transistor M_{RO}, and a feedback circuit which continuously senses the output of the front-end and resets the photodiode when a threshold voltage V_{THR} is attained indicating a soon run out of dynamic range. At circuit power-on, the *RPD* signal is externally triggered in order to provide a correct initial condition. For the fastest operation, the used comparator is an unmatched CMOS inverter. The simulated with Spectre® extracted view of this circuit showed a reset time as short as 600 ps. The photodiode reset phase being performed only when necessary together with its asynchronous with respect to the delay generator operation enable GS/s-order sampling performance associated to the high conversion gain of the integration. Data lost during the reset phase are detected and reconstructed (Zlatanski et al. 2010a).

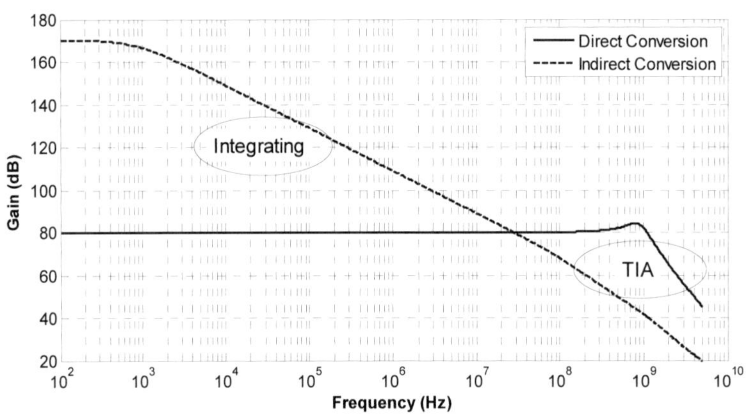

Fig. 11. Frequency responses of direct and indirect conversion architecture

The frequency response of the integrated streak camera depends on the front-end (Fig. 11). The response of the direct conversion front-end is given for a TIA with a gain of 80 dBΩ and a bandwidth of 1.3 GHz. The response of the indirect conversion front-end is given for a total capacitance of 100 fF including the photodiode and the front-end input parasitic capacitance. The cross point of these two responses is in the range of 10 MHz to 100 MHz depending on the integrating capacitance and the TIA. It is seen that the indirect conversion is appropriate for high gain measurements at low frequencies whereas the direct conversion seems to be the right choice when high bandwidth is required.

3.3 Temporal sweep unit

A common and important building block of the ISCs is the temporal sweep unit. The temporal sweep unit depicted in Fig. 12 allows the generation of sweep speeds in the range from 125 ps/pixel up to DC. It consists of a fast and a slow sweep unit. The Fast Sweep Unit (FSU) is useful for sub-nanosecond sampling period operation. It is composed of a Delay-Locked Loop (DLL) and two Voltage-Controlled Delay Lines (VCDLs). The observation time of the camera is set by the DLL reference clock period. The mirror VCDLs are driven by the same control voltages V_{hl} and V_{lh} as the master DLL allowing for asynchronous delay generation with respect to the reference clock. This makes it possible for the proposed generator to be launched from zero-tap position upon external triggering signal, ensuring the synchronization between the beginning of the acquisition procedure and the luminous phenomenon to be captured. The output taps of the VCDLs are delayed by an adjustable delay ΔT given by the ratio of the DLL Clock to the number of column m. The voltage controlled cell is a current starved double inverter with NMOS control transistors. It presents a large delay range and has been optimized for the shortest delay achievable in the CMOS technology employed (Mahapatra et al., 2002). The master DLL ensures the stability over temperature and the absolute precision of the delay. It is important that the VCDLs are

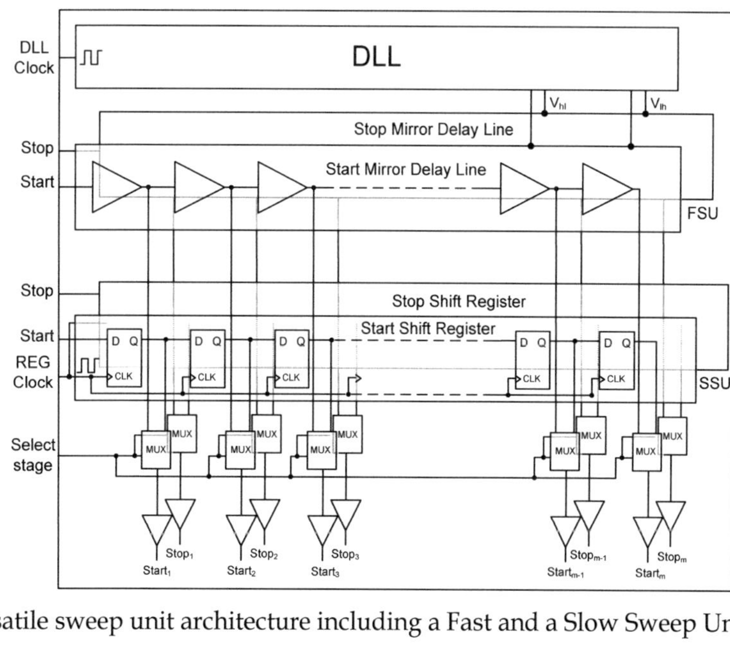

Fig. 12. Versatile sweep unit architecture including a Fast and a Slow Sweep Units

matched to ensure the same temporal shifting. For the MISC architecture, the *Start* and *Stop* signals, which control the M_{RPD} and M_{SH} transistors, respectively, are generated externally. Using two independent control signals allows the reduction the integration time down to 1 ns or less. The VISC architecture requires only one sampling signal, thus the mirror line generating the *Stop* signal is removed. Results from a similar FSU have been reported in (Zlatanski et al., 2011). Sweep speed of 125 ps/pixel up 1 ns/pixel with a drift of less than 0.05%/°C and temporal axis linearity better than 1% have been reported. The measured timing jitter in a completely asynchronous operation mode is less than 70 ps p-p@150ps/pixel and less than 700 ps p-p@770ps/pixel.

The Slow Sweep Unit (SSU) is useful for sweep speed of 1 ns/pixel up to DC. It uses a fast D flip-flop shift register. The sample period ΔT is equal to the period of the register clock *REG Clock*. The linearity of such a delay generator is very high as it is linked to the clock stability. Nevertheless, in an asynchronous operation, the timing jitter of the generated $start_j$ and $stop_j$ signals with respect to the trigger signal *Start* and *Stop* is equal to the sampling period peak-to-peak, i.e. ΔT p-p.

4. Direct optical conversion VISC

4.1 Front-end and sampling cell

A direct optical conversion VISC have been designed with the Austria Micro Systems 0.35µm BiCMOS process. One acquisition channel of the sensor is composed of a transimpedance amplifier as a front-end and a 128-deep memory (Fig. 13, left). The prototype features 64 channels. The temporal resolution of an integrated streak camera is determined by the overall bandwidth of the photodetector and the subsequent electronics. The target is to extend the bandwidth of the front-end beyond 1 GHz, which corresponds to a temporal resolution of less than 1 ns. The challenge consists in keeping the overall sensor bandwidth equal to that of the photodetector alone, while keeping a high gain. Moreover, the power consumption and silicon area should be kept low because of the multichannel architecture. The VISC makes use of a compact and fast voltage-domain sampling scheme shown in Fig. 13-left. An NMOS voltage sample and hold unit is appropriate for the operation as it offers a dynamic range of more than 2 volts and a bandwidth of more than 6 GHz.

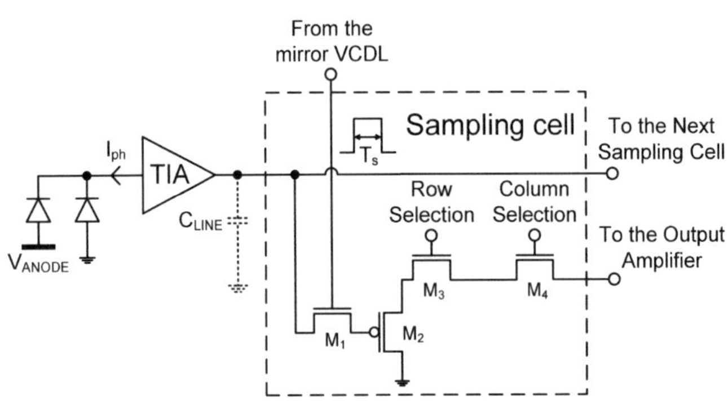

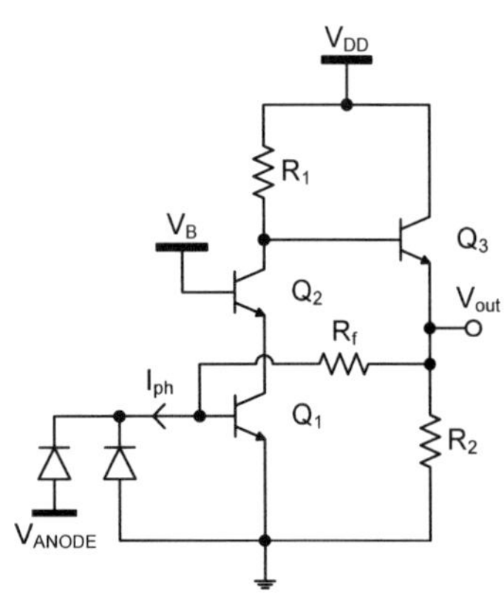

Fig. 13. A channel of the VISC with a double photodiode, a high speed TIA and the detail of a sampling cell with its read-out electronics (left). The transimpedance amplifier schematic (right).

The choice of the TIA topology is strongly restricted to low power and low area consumption architectures as some hundreds of channels are targeted in the final VISC. The designed shunt-shunt feedback common-emitter TIA, shown in Fig. 13-right, features a Gain-Bandwidth product of 17 THzΩ while consuming less than 7 mW. The simulated with extracted parasitics closed-loop gain and bandwidth are 80 dBΩ and 1.75 GHz, respectively. The circuit occupies only 26 × 116 μm² and shows a good trade-off between gain, bandwidth, and power consumption (Table 3).

Reference	Technology	Gain	Bandwidth	GBZ	Power consumption	Notes
(Oh & Lee 2004)	0.35 μm CMOS	68 dBΩ	1.73 GHz	4.35 THzΩ	50 mW	CE with cascode, shunt peaking
(Huang, & Chen, 2009)	0.18 μm CMOS	66 dBΩ	7 GHz	14 THzΩ	-	Differential, inductive peaking
(Kao et al. 2010)	0.18 μm CMOS	74.6 dBΩ	2.9 GHz	15.6 THzΩ	31.7 mW	Differential, two stage, negative Miller capacitance
(Aflatouni et al. 2009)	0.13 μm CMOS	57 dBΩ	6 GHz	4.25 THzΩ	1.8 mW	RGC-input, series and shunt peaking
This work	0.35 μm BiCMOS	80 dBΩ	1.75 GHz	17.5 THzΩ	7 mW	SiGe Bipolar transistor

Table 3. Broadband transimpedance amplifiers in standard (Bi)CMOS 2011

4.2 Photodiode

The most versatile photodiode in bulk CMOS processes is the N_{WELL}–P_{SUB} photodiode since it features a wide and well-positioned depletion region, allowing the detection of long wavelengths with good responsivity. Because of the low overall speed of this diode, it is often used in a spatially modulated light detector (SMLD) configuration which consists in a finger topology of alternatively illuminated and shielded photodiodes (Genoe, 1998, Huang & Chen, 2009). However, the differential diode arrangement has several important drawbacks. The first is the large occupation area that arises from the number of alternating N_{WELL} fingers. Another drawback, directly originating from the finger topology, is the increased parasitic capacitance. To limit its impact on the electrical bandwidth of the subsequent electronics, the input impedance of the amplifier must be very low. The sensitivity is also severely decreased and more than a factor of 5 in output signal is lost at 800 nm, because of the shielded regions (Tavernier et al., 2006). Finally, a SMLD requires a differential processing to carry out the signal of the illuminated and the shielded detector. This configuration is thus not adapted for use within a VISC, where the spatial resolution must be kept around 20 μm and the number of channels must exceed 100.

It has been shown in (Radovanovic et al. 2005; Hermans & Steyaert 2006) that a P_{DIFF}–N_{WELL} photodiode with a screening N_{WELL}–P_{SUB} photodiode features the fastest overall bandwidth among all basic photodiode structures in bulk CMOS technologies, which can exceed 1 GHz at a wavelength of 850 nm. Unfortunately, the responsivity of the screened photodiodes is 10 times lower than that of P_{DIFF}–N_{WELL}–P_{SUB} double photodiodes at 850 nm and 6 times lower at 650 nm. The practical operation range of this topology is restricted to the short wavelengths, where the photogenerated carriers are efficiently collected by the surface photodiode. Employing a screened P_{DIFF}-N_{WELL} photodiode would lead to an improvement of the temporal resolution, but at the cost of a severe loss in sensitivity.

Since the SMLD arrangement and the screened P_{DIFF}-N_{WELL} photodiode are not adapted for use in an integrated streak camera design, a single-ended photodiode coupled to an equalizer could be employed. However, the roll-off of the N_{WELL}-P_{SUB} photodiode at long wavelengths starts at about 1 MHz, which requires a 3rd order equalizer to achieve an effective compensation up to 1 GHz (Radovanovic et al. 2005). A single-stage, 3rd order equalizer requires not less than 2 transistors, 3 capacitors and 5 resistors. Thus, its implementation may become problematic with respect to the small implementation area available and the high power consumption constraint. Moreover, equalization should better be adaptive in order to cover multiple wavelengths and lower the impact of eventual component parameters spread. Adaptive equalization requires adjustable passive component values and thus additional control signals (Chen & Huang, 2007).

In an integrated streak camera, the raw data at the output of the sensor can be processed off-chip, allowing various post-processing such as filtering, FPN reduction through dark image subtraction, and recurrent light event accumulation. In the same vein, equalization can be carried out off-chip through software processing, provided the transfer function of the photodetector is known. If the roll-off in the frequency response can be compensated, it is better to use a photodiode with a higher responsivity in order to widen the spectral range and increase the signal to noise ratio. For this reason the photodiode with the widest spectral range and the highest responsivity in bulk CMOS technologies, i.e., the double P_{DIFF}-N_{WELL}-P_{SUB} photodiode has been chosen (Radovanovic et al., 2006). The equalization process in explained is Section 6.

5. Characterization of the temporal resolution of the VISC

The temporal resolution of the VISC has been measured by using a femtosecond laser source at 400 nm and 800 nm. The test bench, shown in Fig. 14, operates as follow: a femtosecond laser generates a stream of 100 fs FWHM pulses at 800 nm and 80 MHz. One of the pulses is selected and amplified by a oscillator/amplifier at a repetition rate of 5 kHz. This amplified pulse can then be doubled by an optional second harmonic generator (SHG) in order to generate a wavelength of 400 nm. This laser pulse is fed into an optical fiber which is focused on to the slit of the VISC. The measurement of the laser pulse at the output of the optical fiber with a synchroscan conventional streak camera shows that the pulse width is less than 7 ps FWHM, showing that it can be considered as a Dirac impulse by the system. A set of fixed attenuators have been placed in the optical path to avoid saturation of the sensor. A trigger signal generated by the amplifier ensures the synchronization of the streak camera with the laser pulse. A delay and pulse generator allows the precise adjustment of the trigger with respect to the laser pulse.

The response of the camera at a wavelength of 800 nm is shown in Fig. 15. The laser pulses followed by a secondary trace due to reflection in the optical fiber are clearly observed. As the laser pulse is very short, the obtained response represents the impulse response of the system, which in turn stands for the temporal resolution of the ISC. Thus, the temporal resolution is 710 ps FWHM at 800 nm and 490 ps FWHM at 400 nm. To compensate for the poor performance of photodiodes in standard technologies, the frequency response of camera has been equalized.

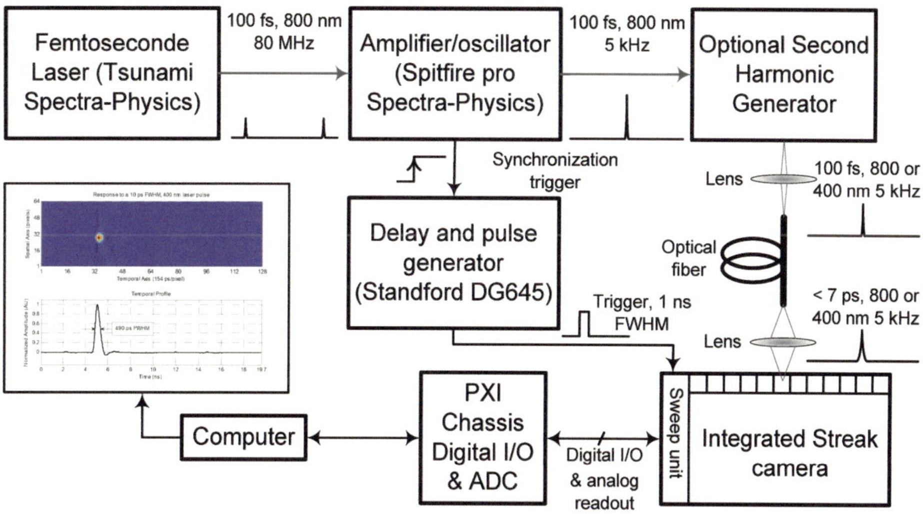

Fig. 14. Experimental setup for impulse response measurement of the VISC

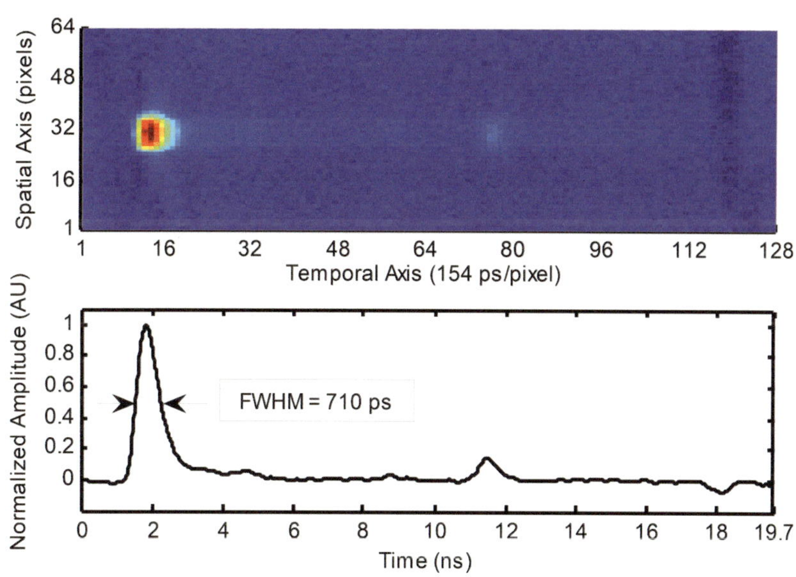

Fig. 15. Impulse response of the VISC at 800 nm

6. Improvement of the temporal resolution

The frequency response of the integrated photodiode has been corrected by using software equalization. The measured frequency response $H_M(z)$ of the TIA VISC, (Fig. 16) is obtained by computing the Fourier transform of the measured impulse response. It is seen that even if the -3 dB bandwidth of the photodiode is as low as 130 MHz at 800 nm, the high frequency attenuation are limited to only 15 dB at 1 GHz. The weak attenuation at high frequency allows the compensation to be efficiently carried out. Assuming that no aliasing occurs in the baseband, if the inverse of the measured transfer function is employed as equalizing filter, a flat frequency response should be obtained. In practice, doing so, the high frequency noise is amplified and becomes dominant over frequencies in which meaningful information is contained. Finally, in the time domain, the event ends completely shaded by noise. To avoid this scenario, the frequency band in which the equalization is operated has to be limited. To do so, we synthesize a filter $H_I(z)$, which exhibits flat response up to a frequency f_{max} beyond which the SNR of the signal becomes unacceptably low, and next cuts off sharply any frequency situated above this bound. Indeed, this filter represents the desired global transfer function of the integrated streak camera. By computing the ratio between $H_I(z)$ and the measured transfer functions $H_M(z)$, we obtain the correcting function $H_C(z)$. Pushing the cut-off frequency of $H_I(z)$ close to f_{max} requires a high-order filter to efficiently attenuate noise. However, an abrupt cut-off approaches the shape of a brick-wall filter, which causes the time response to exhibit non-causal behavior, i.e. a *sinc* function impulse response. To prevent this scenario, the order of the filter must be lowered, which implies a lower cut-off frequency to efficiently attenuate noise. In the time domain, the compromise between filter order and cut-off frequency corresponds to a trade-off between pulse shape and signal to noise ratio. Thus, for given pulse shape and signal to noise ratio, a limit in the maximal achievable temporal resolution exists.

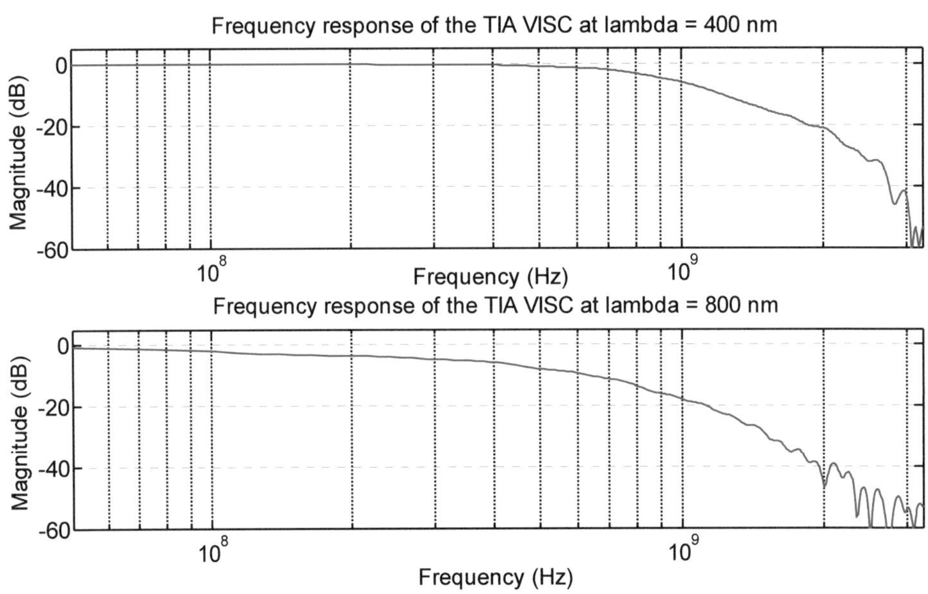

Fig. 16. Frequency response of the TIA VISC at 400 nm and 800 nm

Gaussian-type window has been chosen for the filter, since it matches best the expected and desired pulse shape at the output of a streak camera, excited by a delta function. Application of this global equalization at a wavelength of 800 nm using a 10th-order Gaussian filter $H_I(z)$ with a cut-off frequency of 1 GHz as desired frequency response yields the spectrum of the temporal response $H_E(z)$ shown on Fig. 17. Comparing the measured $H_M(z)$ and equalized $H_E(z)$ responses we notice a significant improvement up to ~2 GHz. Indeed, the low-frequency roll-off is removed, resulting in the suppression of any slow variations in the observation window. The high frequencies have been amplified with a gain of about 15 dB, boosting the high-speed signal variations. Finally, the attenuation of $H_C(z)$ above 2.5 GHz improves the signal to noise ratio of the equalized signal.

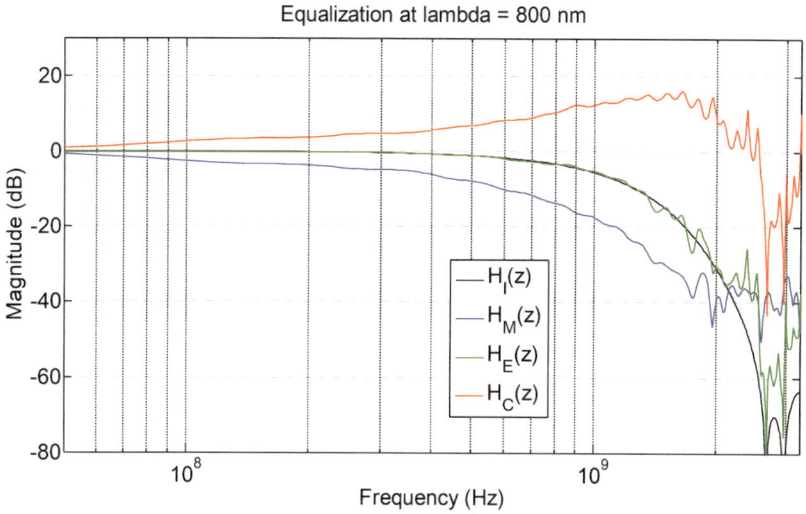

Fig. 17. Spectrum of the measured impulse response H_M (blue), targeted frequency response H_I (black) equalization filter H_C (red) and obtained frequency response after equalization H_E (green)

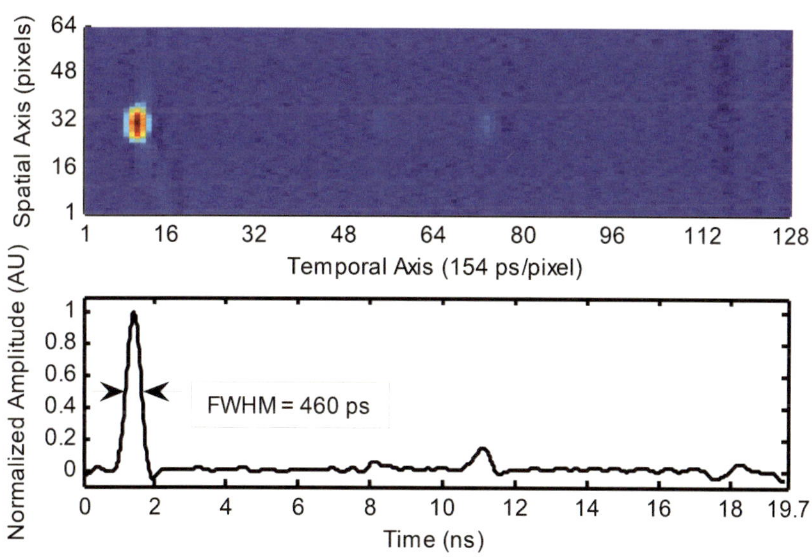

Fig. 18. Equalized impulse response of the TIA VISC at a wavelength of 800 nm

Result of the equalized impulse response of the TIA VISC at a wavelength of 800 nm is shown on Fig. 18. The FWHM is of 460 ps, which is more than 35 % lower than the measured value shown in Fig. 15. Interestingly, after equalization, the camera exhibits similar temporal resolution and pulse shape at both wavelengths of interest. With such a temporal resolution, sub-nanosecond FWHM events can be detected and nanosecond-order phenomena measured with good accuracy. Equalization is a powerful technique, which allows the frequency response of an integrated streak camera to be made uniform over its entire spectral range. This is an important feature, since in streak-mode imaging the incident wavelength is often monochromatic or well known. For example, in one of the most common streak camera applications, the time-resolved spectroscopy, the streak image can be dynamically equalized, such as no distortion is introduced by the degraded behavior of the photodetectors with increasing wavelength.

Sensor	Sampling rate Temporal resolution	Number of channel	Memory depth per channel	Total sampling rate	Notes
UHFR – (Lowrance & Kosonocky, 1997)	1 MHz 1 µs	360 × 360	30	130 GS/s	First ultrafast CCD
ISIS-V4 (Etoh et al. 2006)	1 MHz 1 µs	420 × 720	144	300 GS/s	CCD technology. Colour version exists
3T MISC (UDS/CNRS (Casadei et al. 2003)	1,25 GHz <6 ns	64	64	80 GS/s	First integrated streak camera
6T $MISC_a$ (UDS/CNRS) (Morel et al. 2006)	1,5 GHz <6 ns	64	64	96 GS/s	On chip analog accumulation feature
6T $MISC_c$ (UDS/CNRS) (Zlatanski et al. 2010a)	7 GHz <1.1 ns@400 nm	93	64	650 GS/s	Adjustable sweep speed
$VISC_b$ TIA (UDS/CNRS) (Zlatanski & Uhring 2011)	7 GHz 500 ps@400~800 nm	64	128	450 GS/s	TIA Gain 10000 V/A. First sub 500 ps VISC
VISC Source Follower (Zlatanski et al. 2010a)]	7 GHz < 1 ns	64	128	450 GS/s	Asynchronous reset
Kleinfelder SC (Kleinfelder et al, 2009)	100 MHz 10 ns	150	150	15 GS/s	Synchronous reset
J. Deen (Desouki et al., 2009)	1,25 GHz <1 ns	32 × 32	8	1,28 TS/s	0,130 µm, 1st video camera > 1 Gfps
Near future of integrated streak camera	10 GHz 100 ps	1024	1024	10 TS/s	Do not exist for the moment
Phantom v12.1 in streak-mode (Parker et al. 2010)	1 MHz 1 µs	128 × 8	2 s, i.e. 2 GS	1 GS/s	High speed video camera in streak-mode
Optronis SC10 streak camera	~5 THz 2 ps	~1024	~1024	~5 PS/s	Sweep speed 200 fs/pixel

Table 4. State of the art of ultrafast imaging solid-state sensors, the grey cells refer to other technologies than solid-state sensor with *in-situ* feature

7. Conclusion and perspectives

The concepts of ultrafast imaging and their application in ultrafast BiCMOS optical sensors have been presented. The TIA VISC prototype processed in standard BiCMOS technology reached an overall bandwidth of 1 GHz and a global sampling rate of more than 400 GS/s.

This work demonstrates that sub-nanosecond imaging with an equivalent frame rate of several billion fps is possible using standard (Bi)CMOS technologies. Table 4 summarizes the performances of the state of the art ultrahigh speed imaging solid-state sensors.

Phototransistors or avalanche photodiodes reported in standard technologies could be an interesting alternative to the standard photodiode structures as they can reach a photoelectron amplification factor of 100. The difference in sensitivity between the integrated and the conventional streak camera can thus be reduced down to one order of magnitude. Consequently, the sensitivity of the integrated streak camera has a high potential of evolution. In the near future, integrated streak cameras could be a competitive alternative to conventional streak-mode imaging instruments for applications in which a temporal resolution of about several hundreds of picoseconds is required.

8. References

Aflatouni F., & Hashemi H. (2009). A 1.8mW Wideband 57dBΩ Transimpedance Amplifier in 0.13µm *CMOS, IEEE Radio Frequency Integrated Circuits Symposium*, 2009, pp. 57

Amelio G. F., Tompsett M. F., & Smith G. E. (1970). Experimental Verification of the Charge-Coupled Semiconductor Device Concept, *The Bell System Technical Journal*, vol. 49, 1970, pp. 593-600

Bigas M., Cabruja E., Forest J. & Salvi J. (2006). Review of CMOS Image Sensors, *Microelectronics Journal*, vol. 37, 2006, pp. 433-451

Brixner B. (1955). One Million Frame per Second Camera, *Journal of the Optical Society of America*, vol. 45, no. 10, 1955, pp. 876-880

Casadei B., Le Normand J. - P, Hu Y., & Cunin B. (2003). Design and Characterization of a Fast CMOS Multiple Linear Array Imager for Nanosecond Light Pulse Detections, *IEEE Trans. on Instrumentation and Measurement.*, vol. 52, no. 6, 2003, pp. 1892-1897

Chen W. -Z, & Huang S. -Hao. (2007). A 2.5 Gbps CMOS Fully Integrated Optical Receiver with Lateral PIN Detector, *IEEE Custom Integrated Circuits Conference*, 2007

Connell H. W. (1926). The Heape and Grylls Machine for High-speed Photography, *Journal of Scientific Instruments*, Vol. 4, 1926, pp. 82-87

Cordin. (2011). *Rotating mirror camera specifications*, 10/05/2011, Available from: http://www.cordin.com/

Desouki M., Deen M. J., Fang Q., Liu L., Tse F., & Armstrong D. (2009). CMOS Image Sensors for High Speed Applications, *Sensors*, vol. 9, 2009, pp. 430-444

Elloumi M., Fauvet E., Goujou E., & Gorria P. (2004) The Study of a Photosite for Snapshot Video, *SPIE High Speed Imaging and Photonics*, vol. 2513, 2004, pp. 259-267

Etoh T. G., & Mutoh H. (2005). An Image Sensor of 1 Mfps with Photon Counting Sensitivity, *Proceeings of SPIE*, vol. 5580, 2005, pp. 301-307

Etoh T. G., Mutoh H., Takehara K., & Okinaka T. (1999). An Improved Design of an ISIS for a Video Camera of 1000000 fps, *Proc. of SPIE*, vol. 3642, 1999, pp. 127-132

Feng J. et al. (2007). An X-ray Streak Camera with High Spatio-temporal Resolution, *Applied Physics Letters*, vol. 91, 2007, no. 13

Frank A. M., & Bartolick J. M. (2007). Solid-state Replacement of Rotating Mirror Cameras, *Proc of SPIE.* vol. 6279, 2007, 62791U

Fuller P. (2005), Some Highlights in the History of High-Speed Photography and Photonics as Applied to Ballistics, *High-Pressure Shock Compression of Solids VIII Shock Wave and High Pressure Phenomena*, 2005, pp. 251-298

Genoe J., Coppe D., Stiens J.H., Vonekx R.A., Kuijk M. (2001).Calculation of the current response of the spatially modulated light CMOS detector, *IEEE Transactions on Electron Devices*, Vol. 48, 2001, pp. 1892

Hermans C., & Steyaert M. S. J. (2006). A High-speed 850-nm Optical Receiver Front-end in 0.18-μm CMOS, *IEEE Solid-State Circuits*, vol. 41, no. 7, 2006, pp. 1606–1614

Huang S. H., & Chen W. Z. (2009). A 10 Gb/s CMOS Single Chip Optical Receiver with 2-D Meshed Spatially-Modulated Light Detector, *IEEE Custom Integrated Circuit Conference*, 2009, pp. 129

Igel, E. A. & Kristiansen M. (1997). *Rotating mirror streak and framing cameras*, SPIE, ISBN 0-8194-2461-7

Jiang Z., Sun G. G., & Zhang X. C. (1999). Terahertz Pulse Measurement with an Optical Streak Camera, *Optics Letters*, vol. 24, no. 17, 1999, pp. 1245–1247

Kao T. S.-C., Musa F. A., Carusone A. C. (2010). A 5-Gbit/s CMOS Optical Receiver With Integrated Spatially Modulated Light Detector and Equalization, *IEEE Transactions on Circuits and Systems – I: Regular Papers*, vol. 57, no. 11, 2010, pp. 2844–2857

Kleinfelder S. (1987). Development of a Switched Capacitor based Multi-channel Transient Waveform Recording Integrated Circuit, *IEEE Transactions on Nuclear Science*, vol. 35, no. 1, 1987, pp. 151–154

Kleinfelder S. (1990). A 4096 Cell Switched Capacitor Analog Waveform Storage Integrated Circuit, *IEEE Transactions on Nuclear Science*, vol. 37, no. 3, 1990, pp. 1230–1236

Kleinfelder S. (2003). Gigahertz Waveform Sampling and Digitization Circuit Design and Implementation, *IEEE Trans. on Nuclear Science*, vol. 50, no. 4, 2003, pp. 955–962

Kleinfelder S., & Kwiatowski K. (2003). Multi-Million Frames/s Sensor Circuits for Pulsed-Source Imaging, *IEEE Nuclear Science Symposium*, vol. 3, 2003, pp. 1504–1508

Kleinfelder S., Chen Y., Kwiatkowski K., & Shah A. (2004). High-speed CMOS Image Sensor Circuits with In Situ Frame Storage, *IEEE Transactions on Nuclear Science*, vol. 51, no. 4, 2004, pp. 1648–1656

Kleinfelder S., Chen Y., Kwiatowski K., & A. Shah (2004). Four Million Frames/s CMOS Image Sensor Prototype with on focal Plane 64 Frame Storage. *Proc. of SPIE*, vol. 5210, 2004, pp. 76-83

Kleinfelder S., Sukhwan L., Xianqiao L., & El Gamal E. A. (2001). A 10 000 Frames/s CMOS Digital Pixel Sensor, *IEEE Journal of Solid-state Circuits*, vol. 36, 2001, pp. 2049–2059

Kleinfelder S., Wood Chiang S. -H., Huang W., Shah A., & Kwiatkowski K. (2009). High-Speed, High Dynamic-Range Optical Sensor Arrays, *IEEE Transactions on Nuclear Science*, vol. 56, no. 3, 2009, pp. 1069–1075

Krymski A. et al. (1999). A High-speed, 500 Frames/s, 1024 × 1024 CMOS Active Pixel Sensor, *Symposium on VLSI Circuits*,1999, pp. 137–138

Krymski A.I., Bock N.E., Nianrong T., Van Blerkom D., & Fossum E.R. (2003). A High-speed, 240-frames/s, 4.1-Mpixel CMOS Sensor, *IEEE Transactions on Electron Devices*, vol. 50, no. 1, 2003, pp. 130–135

Lai C. C. (1992). A New Tubeless Nanosecond Streak Camera Based on Optical Deflection and Direct CCD Imaging, *Proceedings of SPIE*, vol. 1801, 1992, pp. 454–468

Lai C. C., Goosman D. R., Wade J. T., & Avara R. (2003). Design and Field Test of a Galvanometer Deflected Streak Camera, *Proc. of SPIE*, vol. 4948, 2003, pp. 330–335

Lambert R. (1937). A Rotating Drum Camera for Photographing Transient Phenomena, *Review of Scientific Instruments*, vol. 8, 1937, pp. 13–15

Le Normand J-P, Zint V., & Uhring W. (2011). High Repetition Rate Integrated Streak Camera in Standard CMOS Technology, *Sensorcomm 2011*, 2011, pp. 322-327

Lowrance J. L., & Kosonocky W. F. (1997). Million Frame per Second CCD Camera System, *SPIE*, vol. 2869, 1997, pp. 405–408

Mahapatra N. R., Garimella S. V., Tareen A. (2000). An Empirical and Analysis Comparison of Delay Elements and a New Delay Element Design, *IEEE Computer Society Workshop on VLSI*, 2000, pp. 81–86

Meghelli M. (2004). A 132-Gb/s 4:1 Multiplexer in 0.13-Mm SiGe Bipolar Technology, *IEEE Journal of Solid-state Circuits*, vol. 39, no. 12, 2004, pp. 2403–2407

Mehr I. & Sculley T. L. (1998). Oversampling Current Sample/Hold Structures for Digital CMOS Process Implementation, *IEEE Trans. on Circuits and Systems–II: Analog and Digital Signal Processing*, vol. 45, no. 2, 1998, pp. 196–203

Miller C. D. (1946). U. S. Patent 2400887, 1946.

Morel F., Le Normand J.-P., Zint C.-V., Uhring W., Hu Y., & Mathiot D. (2006). A New Spatiotemporal CMOS Imager With Analog Accumulation for Nanosecond Low-Power Pulse Detections, *IEEE Sensors Journal*, vol. 6, 2006, pp. 11-20

Oh Y. - H., & Lee S. - G. (2004). An Inductance Enhancement Technique and its Application to a Shunt-peaked 2.5 Gb/s Transimpedance Amplifier Design, *IEEE Transactions on Circuits and Systems–II: Express Briefs*, vol. 51, no. 11, 2004, pp. 624–628

Parker G. R., Asay B. W., & Dickson P. M. (2010). A Technique to Capture and Compose Streak Images of Explosive Events with Unpredictable Timing, *Review of Scientific Instruments*, vol. 81, 2010, no. 016109

Radovanović S., Annema A. – J., & Nauta B. (2006). High-Speed Photodiodes in Standard CMOS Technology, *Springer*, 2006

Radovanovic S., Annema A.-J., & Nauta B. (2005). A 3-Gb/s Optical Detector in Standard CMOS for 850-nm Optical Communication, *IEEE Journal of Solid-State Circuits*, vol. *40*, no. 8, 2005, pp. 1706–1717

Rajaee O. & Bakhtiar M. S. (2005). A High Speed, High Resolution; Low Voltage Current Mode Sample and Hold, *IEEE Symp. Circuits and Syst.*, vol. 2, 2005, pp. 1417–1420

Razavi B. (2003). Design of Integrated Circuits for Optical Communications, *McGraw-Hill*, 2003.

Scheidt K.,& Naylor G.(1999). 500 fs Streak Camera for UV-hard X-rays in 1 kHz Accumulating Mode with Optical-jitter Free-synchronisation, *4th Workshop on Beam Diagnostics and Instrumentation for Particle Accelerators*, 1999, pp. 54–58

Smith G. E. (2001). The Invention of the CCD, *Nuclear Instruments and Methods in Physics Research, A*, Vol 471, 2001, pp. 1–5

Son D. V. et al. (2010). Toward 100 Mega-Frames per Second: Design of an Ultimate Ultra-High-Speed Image Sensor, *Sensors*, vol. 10, 2010, pp. 16–35

Swahn T., Baeyens Y., & Meghelli M. (2009). ICs for 100-Gb/s Serial Operation, *IEEE Microwave Magazine*, vol. 10, no. 2, 2009, pp. 58–67

Tavernier F., Hermans C., & Steyaert M. (2006). Optimised equaliser for differential CMOS photodiode, Electronic Letters, vol. 42, no. 17, 2006, pp. 1002–1003

Toumazou C., Lidgey F. J., & Haigh D. G. (1993). Analogue IC Design: The Current Mode Approach, *Institution of Engineering and Technology,* 1993

Uhring W. Zint C.V. Bartringer J. (2004b). A low-cost high-repetition-rate picosecond laser diode pulse generator, *Proc. of SPIE,* Vol 5452, 2004, pp. 583-590

Uhring W., Jung M., & Summ P. (2004a). Image Processing Provides Low-frequency Jitter Correction for Synchroscan Streak Camera Temporal Resolution Enhancement, *Proc. of SPIE,* vol. 5457, 2004, pp. 245–252

Zavoisky E. K., & Fanchenko S. D., Image Converter High-speed Photography with 10^{-9}–10^{-14} sec Time Resolution, *Applied Optics*, vol. 4, no. 9, 1965, pp. 1155-1167.

Zlatanski M. Uhring W. Zint C-V. Le Normand J-P., Mathiot D. (2010a). Architectures and Signal Reconstruction Methods for Nanosecond Resolution Integrated Streak Camera in Standard CMOS Technology, *DASIP,* 2010, pp. 1-8

Zlatanski M., & Uhring W. (2011). Streak-mode Optical Sensor in Standard BiCMOS Technology, *IEEE sensor conference,* 2011, In Press

Zlatanski M., Uhring W. Le Normand J-P. Zint C-V. Mathiot D. (2010b). 12 × 7.14 Gs/s rate Time-resolved BiCMOS Imager, *8th IEEE International NEWCAS,* 2010, pp. 97-100

Zlatanski M., Uhring W., Le Normand J.P., & Mathiot D. (2011). A Fully Characterizable Asynchronous Multiphase Delay Generator, *IEEE Transactions on Nuclear Science,* Vol. 58, 2011, pp. 418-425

Zlatanski M., Uhring W. Le Normand J-P. Zint C-V. Mathiot D. (2010c). Streak camera in standard (Bi)CMOS (bipolar complementary metal-oxide-semiconductor) technology, *Measurement science & technology,* Vol. 21, 2010, pp. 115203

14

Photoconductors for Measuring Speckle Motion to Monitor Ultrasonic Vibrations

Jonathan Bessette and Elsa Garmire
Dartmouth College
USA

1. Introduction

This chapter explains a unique miniature photoconductive detector design that has been specifically engineered with a speckle-monitoring application in mind. Small arrays of these detectors have been fabricated with silicon-on-insulator (SOI) technology and in semi-insulating GaAs, and have been put to work measuring surface vibrations in laboratory experiments. The important characteristics of these photodetectors are their small physical dimensions, on the order of 10 μm, with an ability to measure ultrasonic bandwidths extending into the megahertz regime, and with high internal gain. The advantage of photoconductors is their capability of generating multiple electrons of photocurrent for each photon captured.[1]

2. Performance of a general photodetector

At the most fundamental level, electronic photodetection is a method of counting photons by converting them into some form of electronic current. There are many ways to do this, including all the familiar detector technologies such as photodiodes, photomultiplier tubes, and photoconductors, to name a few. For the purposes of this discussion, all these detectors can be abstracted into a simple detector model consisting of three primary parameters: gain, noise, and bandwidth.

Gain can be neatly expressed as the number of electrons per unit time of signal current either generated or modulated by the detector divided by the number of photons per unit time incident upon it. A gain of one, for example, indicates that one electron flows for every photon collected.

Noise is a measure of the electronic current generated or modulated due to random fluctuations in either the arrival or conversion rate of photons, or due to the random generation of the flow of electrons within the detection circuit not caused by photons at all.

Bandwidth is a measure of the maximum frequency response of the detector to a photonic signal. The higher the bandwidth, the faster the detector can respond to changes in the level

---

[1] Much of this chapter comes from the PhD thesis of Jon Bessette: "Silicon-on-Insulator Photoconducting Mesas For High-Speed Laser Speckle Monitoring Applications," Dartmouth College, June, 2010.

of illumination. Bandwidth limits may be imposed by the internal physics of the device or additionally by a combination of factors in the circuit implementation.

Gain, noise, and bandwidth are always intimately connected in any detection system and improved performance in one parameter often comes at the price of decreased performance in another. It is important therefore to understand the connections between them for any given detector technology and the detector application.

A simple model combines these three elements into a description of the signal from a generic photodetector in the useful form of a signal-to-noise ratio (SNR). This ratio indicates a lower bound of sensitivity for which a detector is useful in a given operating regime. The convention adopted here is to express the SNR as the ratio of signal power to noise power:

$$SNR_{gen} = \frac{(e\Gamma A\Delta I)^2}{\sigma^2_{noise}(\Gamma, B)} \tag{1}$$

The numerator is given in units of current squared, where e is the electron charge, Γ is the detector gain, A is the detector area, and ΔI is the change in optical intensity incident on the detector during the time $2/B$. The noise term σ^2 is expressed in units of current squared and is generally a function of the gain Γ, bandwidth B, and the specifics of the detector. There are multiple sources of noise in the signal of any detector that measures changes in incident optical power. The predominant forms of noise in the simple model presented here are photon shot noise, current shot noise, and circuit thermal noise. These noise sources are all well understood and characterized by circuit designers and so will only be summarized here:

- Photon shot noise is caused by the random arrival of discrete photons in a given photon stream. The power in the photon shot noise is proportional to the average incident optical power, and the variance can be given in units of current as $\sigma^2_\varphi = 2e^2\Gamma^2 FAI_o B/h\nu$, where F is an excess noise factor created by probabilistic gain mechanisms and I is the intensity incident on the detector (Liu, 2005).
- Current shot noise is caused by the random arrival of discrete electrons in a given electric current. Its power is proportional to the average background current i and its variance is given as $\sigma_i^2 = 2ei\Gamma^2 FB$.
- Circuit thermal noise is created by random thermal motion of the electrons in a conductor. It is independent of the current flowing through the load resistor R_L and is present even when no electric potential is applied. The thermal noise variance is approximately $\sigma_K^2 = 4k_B TB/R_L$.

In general, the different noise sources are assumed to be independent of each other. This means that the variance of the combined noise amplitude is the sum of the variances of the individual noise sources, i.e. $\sigma_{noise}^2 = \sigma_\varphi^2 + \sigma_i^2 + \sigma_K^2$. To quantify the expected performance of a given detector in a particular operating regime, we must explore the magnitude of the various gain and noise terms for different classes of detectors. Because rather small area detectors are required for our application, we will restrict ourselves to the discussion of semiconductor-based detectors. These can be manufactured on the appropriate scales and easily configured for array operation, which can greatly enhance their effectiveness.

2.1 Gain and noise

The fundamental operating principle of all solid-state photosensitive devices is the same. A photon with sufficient energy interacts with a semiconductor crystal, temporarily changing the distribution of electron energies within the crystal. One electron gains enough energy to attain an energetic conductive state, where it is free to move about within the crystal. The promoted electron leaves behind a vacancy, called a hole, which can also move about within the crystal. Together, these are referred to as an electron-hole pair, and in an ideal photodetector, one such pair is created for every absorbed photon. Eventually, if left in the crystal long enough, the electron-hole pair will recombine, giving up the extra energy in the form of heat. This happens on a characteristic time scale called the recombination lifetime, τ_r.

While excited, the electron and hole will drift in the presence of an electric field, creating an electric current. This current can be detected by connecting the active area of the semiconductor into an electronic circuit. How the electric field is applied and the nature of the electrical contacts defines the class of detector.

2.1.1 Photoconductors

Photoconductors are uniformly doped semiconductors whose conductance changes in proportion to the number of photons absorbed per unit time. When an electrical potential is placed across the photoconductor, the amount of current flowing through the photoconductor indicates the amount of light incident on it. The photoconductor is perhaps the simplest kind of semiconductor photodetector. It is perhaps surprising, then, that such a simple device may exhibit internal gain.

The photoconductor gain is the ratio of electron collection at the contacts to electron-hole pair generation by photons within the channel. The gain is the ratio of recombination time to the average transit time of charge carriers across the photoconductive channel. This ratio is, in turn, determined (at low applied voltages) by the photoconductor's effective combined carrier mobility $\mu = (\mu_e + \mu_h)$, its carrier recombination time τ_r, its channel length L and the applied voltage V according to

$$\Gamma = V\mu\tau/L^2 \tag{2}$$

(for a square detector, $\Gamma = V\mu\tau/A$ where A is area). This gain process is fundamentally probabilistic for each individual photon absorbed, and so necessarily the mechanism produces an excess noise factor. The excess noise factor for a uniformly illuminated photoconductor with such a gain, governed by Poisson statistics of a uniform random variable, is 2 (Liu, 2005).

It is important to note that the gain does not scale linearly with voltage indefinitely. At least two effects may limit the maximum gain seen inside a photoconductive element. The first of these is drift velocity saturation, which occurs when the electric field is great enough to yield carrier drift velocities on the order of the carrier thermal velocity. Increasing the electric field beyond this saturation value - about 30 kV/cm for silicon - does not cause the carriers to drift any faster (Neamen, 2004). Consequently, the transit time is not reduced with greater field strength and gain saturates.

A second mechanism that limits the internal gain is space-charge current limitation (Liu, 2005; Bube, 1960; Rose 1958), when charges build up on either electrode due to the effective capacitance of the device. When this charge exceeds the total number of free carriers within the photoconductor, the excess charge is injected into the device and carries a current that is limited by the dielectric relaxation of the capacitor (Rose, 1954). These so-called space-charge currents screen the photogenerated carriers from any additional electric field, so that increasing the applied potential difference beyond a certain threshold does not increase the photocurrent, and the gain saturates.

The threshold voltage that limits the gain is called the space-charge voltage. It is equal to the applied potential that causes the charge on the effective capacitor to exceed the number of free carriers within the photoconductive channel. For a photoconductor with ohmic contacts and charge carriers of both sign, this is explicitly

$$V_{SC} = \frac{\oint(\sigma_0 + \sigma_{ph})\mathbf{dv}}{e\mu C} \tag{3}$$

where σ_0 is the dark conductivity, σ_{ph} is the photogenerated conductivity, and the integral is over the volume of the photoconductor. The value given for capacitance depends on the exact geometry and material properties of the photoconductor, but for the simplest model of parallel electrodes and a block-shaped semiconductor, the capacitance is simply $C = WD\varepsilon/L$ and the space-charge voltage is (Liu, 2005)

$$V_{SC} = \frac{(\sigma_0 + \sigma_{ph})L^2}{\epsilon\mu}. \tag{4}$$

The gain limit for a space-charge limited device becomes

$$\Gamma_{max} = \frac{\tau(\sigma_0 + \sigma_{ph})}{\epsilon}. \tag{5}$$

Note that the gain limit is a function of photoconductivity (i.e., depends on the amount of light being absorbed), but that this does not greatly affect the performance unless the photoconductivity is comparable to the dark current.

For the dimensions and material of devices investigated in this chapter, it is not the saturation velocity limit but rather the space-charge current that limits the photoconductive gain.

As far as noise is concerned, the dark current shot noise of a photoconductor can be significant in certain operating regimes. For an ideal photoconductor, this noise figure scales linearly with the applied voltage and the cross-sectional area-to-length ratio of the active region. This can be deduced by plugging in the ideal photoconductor dark current into the definition of shot noise

$$\sigma_{i,d}^2 = 2eV\frac{WD}{\rho_0 L}B, \tag{6}$$

where σ_0 is the dark resistivity of the photoconductor. We will see in a following section that this noise figure is dominated by the circuit thermal noise at high bandwidths.

2.1.2 Photodiodes

Photodiodes are rectifying junctions created by adjoining two regions of a semiconductor crystal that are doped with different impurities. The impurities change the equilibrium concentrations of conductive electrons and holes. When a semiconductor is *p*-doped, it has an excess of mobile positive charges (holes). Likewise when it is *n*-doped, the semiconductor has an excess of negative charges (electrons). Note that the total charge of the crystal remains neutral - this implies there are fixed negative charges in the *p*-doped regions and fixed positive charges in the *n*-doped regions.

When two regions of opposite doping are brought into contact, diffusion results in excess mobile charges from each region spreading into the neighboring region, leaving behind some fixed charge. The fixed charge is of opposite polarity on either side of the junction and together the two zones are called the space-charge region.

The space-charge region supports an electric field which points from the *n*-doped side of the junction to the *p*-doped side of the junction. When a photon is absorbed in this region, the resulting electron and hole are quickly swept out through opposite contacts as a result of the electric field. Reverse-bias is generally applied, which means a relatively positive potential is placed on the *n*-doped side of the junction. This widens the junction and increases the maximum field in the space-charge region.

Once mobilized by the absorption of the photon, the electron drifts toward the (positively charged) *n*-doped side of the space-charge region and the hole toward the (negatively charged) *p*-doped side of contact, and the circuit registers a blip of electric current. Once the carriers reach the end of the space charge region, they eventually recombine.

In contrast to the ohmic photoconductor contacts, the *pn* junction barrier under reverse bias acts as a completely blocking contact for the drift of both charge carriers, so that no additional carriers are injected into the space-charge region. Since the photo-induced charge only transits the active region once, the photodiode produces only one unit of electronic current for every photon absorbed. The basic photodiode thus has a maximum of unity internal gain.

The dark current is generally very low for most photodiodes, on the order of pico-amperes for small diodes, and it only decreases as the detector area is reduced. Consequently, the dark current noise is dominated at higher bandwidths by the circuit thermal noise.

2.1.3 Metal-semiconductor-metal photodiodes

Metal-semiconductor-metal photodiodes are created by making contact with a semiconductor using a metal that forms a Schottky potential barrier. Each metal-semiconductor junction exhibits the rectifying properties of a diode. Charge carriers diffuse across the junction to establish a space-charge layer at each contact, similar to *pn* junction diodes. However, unlike *pn* diodes, the current across the Schottky junction barrier is carried by the drift of semiconductor majority-charge carriers and not minority carrier diffusion.

When a bias is placed across the contacts, one of the junctions is forward-biased and one is reverse-biased. It is the reverse-biased contact that limits the current in the rectifying regime.

Another distinction that separates MSM photodiodes from *pn* junction diodes is that they can exhibit photoconductive gain greater than unity, despite the non-ohmic contacts and rectifying Schottky barriers. The source of this gain is the same mechanism as for photoconductors, namely multiple charge carriers make a transit across the semiconductor region for every photon absorbed, despite the presence of the Schottky barriers at the contacts. This is possible because carriers can be injected via thermionic emission over the Schottky barrier (Neamen, 2003; Soares, 1992; Burm, 1996) to maintain charge neutrality. That is to say, the Schottky barrier is not a completely blocking contact - it merely presents an additional potential barrier that increases the device resistance. This is a key difference between *pn* junction photodiodes and Schottky barrier photodiodes.

The dark current in MSM photodiodes can be many orders of magnitude greater than the dark current in reverse-biased photodiodes. Consequently, the dark current noise in MSM photodiodes is of greater concern. The dark current can be estimated by the following formula for the reverse-bias saturation current density in units of current per cross-sectional area (Neamen, 2003):

$$J_{sT} = A^{*}T^{2}\exp(\frac{-e\Phi_B}{k_B T}) \tag{7}$$

where A^* is called the Richardson constant with units of $AK^{-2}cm^{-2}$ (A is amperes) and is specific to the semiconductor, T is temperature in Kelvin, and Φ_B is the Schottky barrier height in units of eV.

As an example, consider a silicon Schottky diode with chromium contacts, similar to the SOI prototype construction discussed later in this chapter. The Richardson constant for silicon is approximately $A^* = 114\ AK^{-2}cm^{-2}$ and the barrier height is approximately 0.49 V. Thus, at room temperature, the reverse saturation (dark) current is approximately $J_{sT} \approx (114\ AK^{-2}cm^{-2}) \times (300\ K)^2 \times \exp(-0.49\ eV/0.026\ eV) = 67\ mA/cm^2$. For a cross sectional area of 10 μm × 10 μm, this gives about 67 nA of dark current, which has a significant impact on noise at low bandwidths.

It should be noted that the Schottky barrier height can be a significant function of both the applied voltage and the level of illumination. This occurs through several mechanisms, including image force lowering (the Schottky effect), (Neamen, 2003; Soares, 1992) charge tunneling from the metal across the barrier (Soares, 1992), and photogenerated charge accumulation (Soares, 1992; Carrano, 1998). By these processes, charges are pulled to either contact where they establish compensating electric fields. The effect is to lower the potential barrier with additional applied voltage and consequently modify the current-voltage characteristics of the diode - as the reverse bias voltage increases, so does the reverse saturation current.

2.2 Small area detectors in the thermal noise limit

By way of example, let us compare the predicted performance of photoconductors and photodiodes in the face of noise for a megahertz bandwidth and low detector area situation. The parasitic capacitance for both photodiodes and photoconductors for small detectors is usually somewhere in the 10 pF range or greater. If we want to operate with a 1 MHz bandwidth, we require an RC time constant such that R_LC = 1/(2π×1MHz) = 160 ns. This means we need to use a load resistance no greater than RL = 16 Ω, and the thermal noise generated by such a load at room temperature T will be

$$\sigma_K = \sqrt{4k_BTB/R_L} \approx 1\,\text{nA} \tag{8}$$

This noise figure is quite significant and dominates the other primary sources of noise, photon shot noise and dark current noise. Take the photon shot noise for example. A detector with an active area of 100 μm^2 under 1 mW/cm^2 illumination sees a total of 1 nW optical power P_0. At an optical wavelength of 1 μm, this corresponds to a photon shot noise of

$$\sigma_\Phi = (\sqrt{2P_0F/h\nu})e\Gamma B \approx \Gamma \times 10\,\text{pA}.$$

This is less than thermal noise for gains less than 100.

The dark current noise is often negligible as well. For small area photodiodes, the dark current itself is negligible compared to the thermal noise. (Consider that the dark current scales with detector area and 1 mm^2 detectors have dark currents in the pA range.) Photoconductors have considerably more dark current, but in most cases the noise associated with it will not be more significant than the thermal noise. For instance, according to Eq. 6, the dark current noise of an ideal photoconductor with a dark resistivity of 10 kΩ-cm that is 10 μm thick, 10 μm wide and 10 μm long, and operated with 1 V applied at 1 MHz bandwidth is

$$\sigma_{i,d} = \sqrt{2eVWDB/\rho_0 L} \approx 200\,\text{pA}$$

This is still less than the thermal noise, even for the highest allowable resistor load.

If we examine the SNR for photoconductors and photodiodes using the above approximations and Eq. 1, we obtain

$$SNR = \frac{(e\Gamma A\Delta I)^2}{4k_BT/R_L} \tag{9}$$

where the gain Γ_{pd} = 1 for photodiodes and the gain Γ_{pc} is a function of geometry and applied voltage for photoconductors. In this approximation, the SNR scales with the square of the internal gain. This implies that the performance of a photoconductor will exceed the performance of a photodiode whenever the gain is greater than one.

In Fig. 1, a direct comparison between the SNR for photoconductors and photodiodes is plotted as a function of detector area, WD, assuming a square detector, load resistance of RL = 1 kΩ, a bandwidth of B = 1 MHz, a photoconductor recombination time τ_R = 10 ns, a photoconductor mobility of 1850 cm^2=V•s, and an applied photo-conductor voltage of 1 V. The signal is assumed to be a 50% modulation of uniform 1 mW/cm^2 illumination. The conductivity is 10^{-3} $\Omega^{-1}cm^{-1}$, giving a maximum photoconductive gain of Γ_{max} = 9.6. For this example, the threshold for enhanced photoconductor performance is at 43 μm, and gain saturation occurs at widths below 13 μm.

Below a certain threshold of detector area, the photoconductor exhibits a superior SNR to the photodiode. This is a direct consequence of the internal gain mechanism of the photoconductor: the noise is the same for both detectors, but the gain is for a photodiode is limited to one, whereas the gain for the photoconductor can be significantly greater than one.

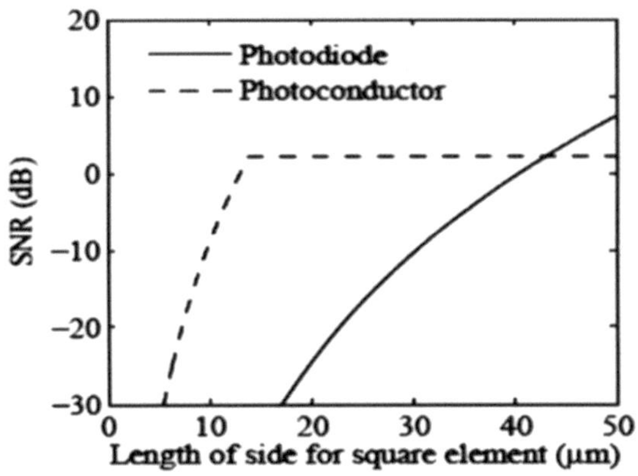

Fig. 1. Plot of SNR for a photoconductor and SNR of a photo-diode of the same square size as a function of the side-length.

2.3 Summary

In the limiting case of the thermal noise regime, where photon shot noise and current shot noise are dominated by the thermal noise of the load resistance, a detector with high internal gain is preferred. This thermal noise limit is approached at low light levels and at higher bandwidths, precisely the operating conditions required by speckle monitoring detectors for ultrasound vibration measurement. Both photoconductors and metal-semiconductor-metal detectors can exhibit internal gain at megahertz bandwidths with detector areas appropriate for direct speckle monitoring, and so merit further investigation.

3. Photoconductor development

This section describes the design and development process for the photoconductor prototypes built and tested at Dartmouth College for use in speckle monitoring applications. It also discusses several of the device characterization experiments performed on the resulting devices.

3.1 Gallium arsenide photoconductors

The first prototypes for speckle monitoring were constructed from GaAs (Heinz, 2004; Heinz, 2007; Heinz, 2003). They consisted of a raised mesa of GaAs 100 μm wide by several hundred μm long divided into four active areas defined and separated by superposed contacts. Each active area was approximately 100 $\mu m \times 40$ μm.

Although quite sensitive to speckle motion, these devices suffered from a lack of bandwidth beyond some tens of kilohertz, and so were unsuitable for systems requiring ultrasonic bandwidths into the megahertz regime. It was determined that the limiting mechanism was most likely space-charge buildup, a result of high contrast in conductivity between the illuminated active regions and the shadowed contact regions.

Two potential solutions were investigated to reduce the effect of space-charge in the GaAs photoconductors and increase the individual detector bandwidth. The first of these

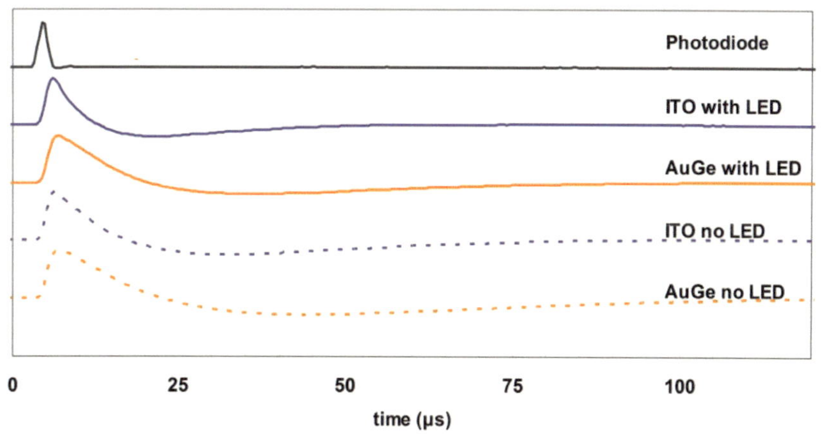

Fig. 2. Plot of GaAs photoconductor response to a 3 µs long laser diode pulse for various devices with and without LED background illumination. The response from a silicon photodiode is shown for comparison. The values on the y-axis are normalized (arbitrary units) (Bessette, 2006).

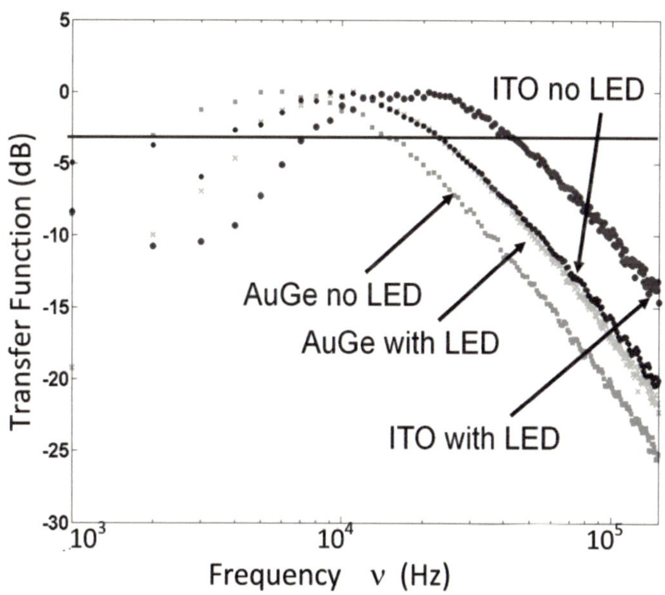

Fig. 3. Fast Fourier transform of AuGe and ITO device impulse response with and without red LED background illumination. The graph shows that the ITO devices are inherently faster than their AuGe counterparts with and without the 32 mW/cm^2 background illumination. The -3 dB rolloff for the un-illuminated AuGe device is around 10 kHz, while the ITO device rolls off at 20 kHz. With background illumination, the ITO device rolls off around 40 kHz (Bessette, 2006).

solutions was to use a transparent electrical contact, thereby lessening contact shadowing and reducing the impact of the resulting space-charge. To that end, a new GaAs prototype design was developed and produced (Bessette, 2006). The major improvement in this detector generation was the use of indium-tin oxide (ITO) for electrical contacts.

The second proposed solution for improving bandwidth was to provide a uniform background illumination with an LED during device operation. This would reduce space-charge buildup by shortening the space-charge relaxation time via increased average conductivity. Both of these measures were evaluated for effectiveness in a single impulse-response style experiment. Figure 2 shows the oscilloscope traces from different versions of the detector in response to an impulse-modulated laser diode. The settle time is seen to improve significantly with the presence of LED background illumination in both ITO and AuGe contacted devices. The settle time is also remarkably shorter for the ITO device, compared to the AuGe device. The photodiode response is limited by the diode laser driver which has a 200 kHz rolloff frequency. There was indeed an improvement in response time for the devices when using transparent contacts and background illumination, but as seen in Fig. 3, the gain in frequency response was still not enough to obtain ultrasonic bandwidth.

3.2 Silicon-on-insulators as a solution

To altogether eliminate the problems associated with persistent space-charge, a photoconductor array made from a relaxation-limited semiconductor was proposed. Silicon is a logical choice, since it is relaxation time-limited, responsive to visible and near-infrared wavelengths, and the most commonly used semiconductor material. The challenge with silicon is that in bulk form, the carrier diffusion length, or the distance on average that an unconstrained, excited electron or hole will diffuse before recombining, can be on the order of centimeters (Neamen, 2003). This rules out the kind of mesa architecture utilized in the GaAs array design, since the photo-generated carriers would escape from each element and diffuse throughout the crystal, destroying the spatial resolution required for speckle monitoring.

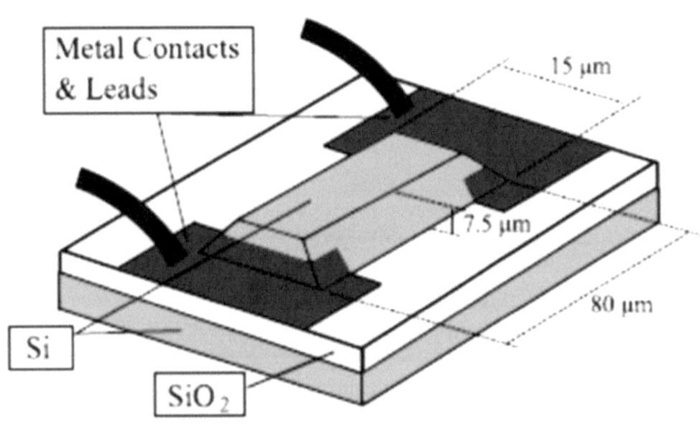

Fig. 4. Each detector element consists of an island of silicon bonded to an insulating layer of silicon dioxide grown on the silicon substrate. These structures are achieved by anisotropic etching through photolithographic masks. Electrical connection to each island is made by metal contacts, also defined by photolithography.

A new design was proposed incorporating silicon-on-insulator (SOI) technology, which consists of a thin layer of silicon crystal, on the order of 10 μm thick, fusion-bonded to a layer of insulating amorphous silicon dioxide grown on a supporting silicon wafer. Photoconductive islands are defined by etching through the thin top layer of silicon on every side, as is shown in Fig. 4. The mutual isolation of the islands in the array ensures that carrier diffusion will not compromise the spatial resolution of each detector element.

3.3 Fabrication details

The fabrication of SOI detector arrays, although they are comprised of only a few component layers, is not a trivial procedure to develop or execute. The difficulty arises as a consequence of both the small feature-size and the high aspect-ratio of individual photoconducting islands. The structures must be considered as fully three-dimensional, which makes processing in a top-down manner, the traditional approach to two-dimensional semiconductor wafer processing, more difficult.

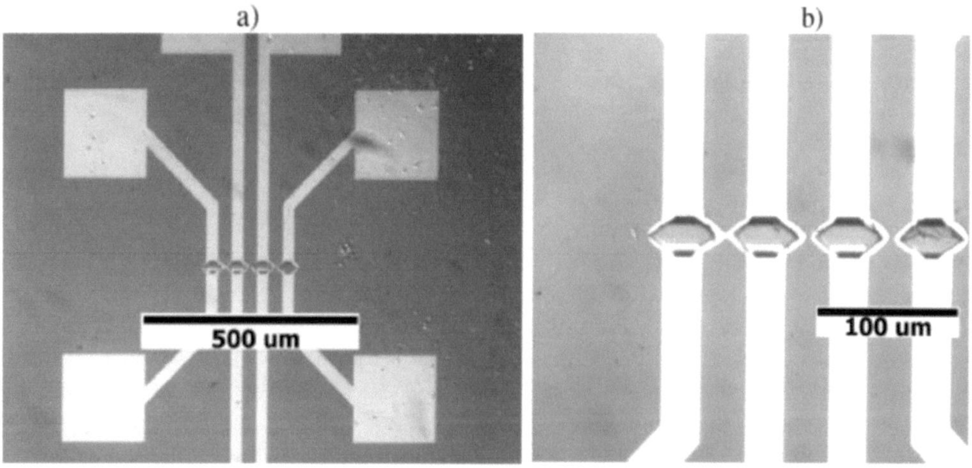

Fig. 5. a) Photograph of a finished four-element SOI photoconductor device array, including metal contact pads. In this design, each element can be contacted individually; b) Close-up of a finished four-element array. This sample was presumably underexposed during the contact layer photolithograpy step, resulting in a short circuit around the perimeter of each element after the metal etch step. This array was not actually used in experiments.

The following is an overview of the most current fabrication protocol for the SOI devices as designed for this project and executed in the available microfabrication facilities. An outline of microfabrication protocol:

1. RCA cleaning routine to remove surface impurities from wafer
2. Oxide layer grown on outer silicon layer
3. Photolithography step to define oxide etch
4. Oxide etch with HF, leaving oxide as a hard mask
5. Silicon etch with KOH
6. Metal contaminant removal and wafer cleanup
7. Optional removal of remaining oxide mask layer and optional oxide under-etch filling with hard-baked photoresist

8. Thermal deposition of chromium and gold layers to provide adhesive conducting layer
9. Photolithography step to define electrical contacts
10. Gold etch
11. Chrome etch
12. Mount wafer dice into electronic DIP package, connect leads to contacts

Optical micrographs of a finished sample that incorporates four individually addressable elements into a small array are shown in Fig. 5.

3.3.1 Anisotropic etching of silicon device layer

In order to facilitate the required electrical connections using top-down deposition of electrical contacts, a potassium hydroxide (KOH) anisotropic silicon etch step was chosen to define the mesa structures from the thin SOI device layer. KOH attacks the different crystal planes of silicon at different rates. Chiefly, there is a high degree of selectivity between the etch rate in the (100) direction of the crystal and the etch rate in the (111) direction. The (100) direction etches hundreds of times faster than the (111) direction. The result of etching with a mask edge aligned parallel with the <110> plane (that is, generally speaking, a mask edge aligned with the wafer flat of a (100) oriented wafer) is a sloped vertical etch exposing the <111> crystal plane, as is seen in Fig. 6(a).

This is in contrast to an isotropic chemical etch such as with nitric acid (HNO_3) based etchants, which would result an under-etched mesa, as in Fig 6(b). Under-etching is bad in this context for two reasons. Firstly, it would be very difficult to connect these devices electrically without removing the top oxide mask. Removing the oxide mask is undesirable since such a step would also partially etch the SOI buried oxide layer, further complicating electrical connection. Secondly, since the devices are almost as thick as they are wide, isotropic etching would eliminate the majority of the actual device. This problem only gets worse as the device dimensions are decreased. For these reasons the anisotropic etch was chosen.

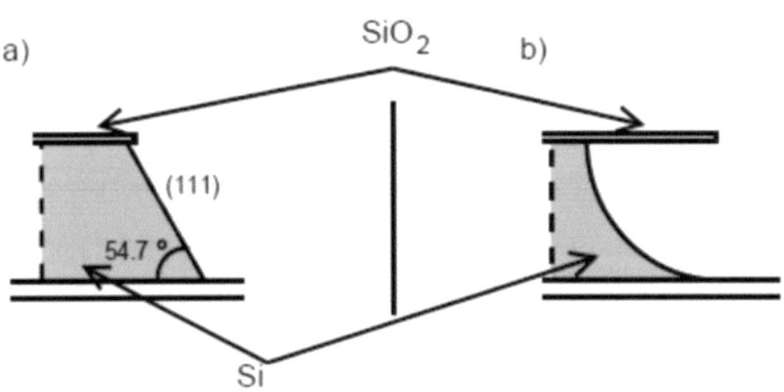

Fig. 6. Sketch of the side-view of silicon facets due to chemical etching a) Sloped vertical facet resulting from anistropic KOH etching of (100) oriented silicon. b) Under-etching resulting from an isotropic etch in which the device is etched laterally as much as it is etched vertically.

One difficulty encountered in utilizing such an anisotropic etch on these structures is the undesired etching of convex corners. There are many other crystal directions besides (110) which etch faster than the (111) direction. When convex corners are exposed to the etchant, all of these other directions get etched, in addition to the desired downward etching (Lee, 1969; Wu, 1989). This results in convex corners that are etched inward, rapidly degenerating the structure from that direction. Fig. 7 shows the result of this effect, and demonstrates how this undesirable effect is exacerbated as device dimensions are reduced.

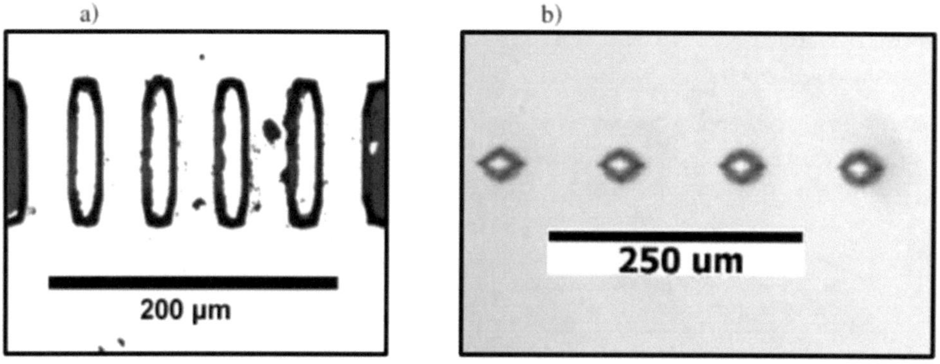

Fig. 7. The devices in prototype (a) are much longer than wide and the device layer thickness of silicon is only 7.5 μm, so the convex corner etching, while apparent, is not as significant. The devices in prototype (b) were meant to be more square, and the device layer thickness is 10 μm, which exaggerates the effect of the convex corner etching.

Convex corner etching is particularly problematic for these three dimensional structures, whose height is on the same order as their width. A common solution to this problem is to extend the etch mask over these corners (Wu, 1989; Biswas, 2006). The idea is to add extra sacrificial material that must be etched in the fast etching directions of the corner, thereby giving enough time for the vertical etch to finish before the protected corner material begins to etch. See Fig. 8 for a diagram of such a mask solution, after (Wu, 1989). Certainly, any new prototype including the anisotropic etch step during fabrication should include some form of convex corner protection in the mask design.

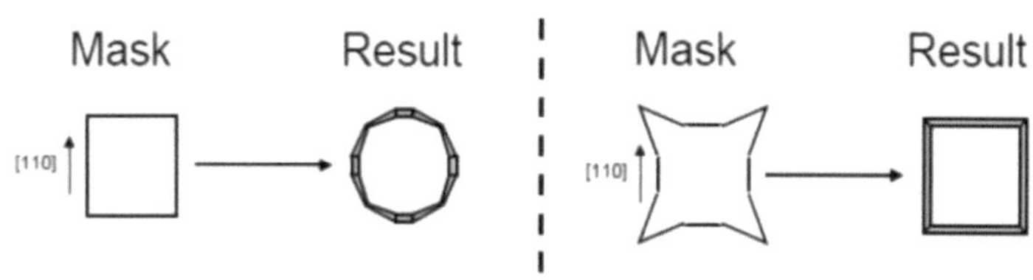

Fig. 8. Sketch of typical results of mask patterns for anisotropically etched structures with convex corners. The triangular overhangs on the mask help protect the desired corner integrity.

3.3.2 Electrical contacts

Early versions of the prototype used only a sputtered gold contact layer, but this was found to be insufficiently adhesive. In later prototypes, the contact layer was deposited via thermal deposition of a thin layer of chromium (to promote adhesion) of approximately 30 Å thickness followed by thermal deposition of a conductive gold layer of approximately 2000 Å. These contacts proved to be robust enough to avoid excessive flaking, but they are not ohmic, as illustrated by the current versus applied voltage curve of a single device shown in Fig 9. The metal-semiconductor junctions formed by the contacts do exhibit rectifying behavior that typifies MSM photodiodes rather than a pure photoconductor. The next section discusses the implications of this result.

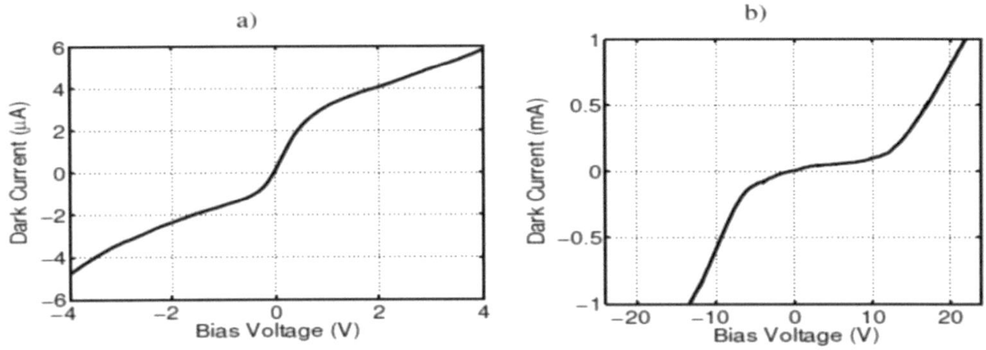

Fig. 9. Current versus bias voltage for an SOI detector with zero incident optical power a) near zero bias and b) extending into diode breakdown regions. The shape of the curve is typical of MSM photodiodes (Soares, 1992).

3.4 Characterization

After the prototypes were constructed, they were evaluated on different aspects of their performance, including the current-voltage relationship, gain, bandwidth, and noise.

3.4.1 Current versus voltage

Figure 9 shows the current versus voltage for one SOI device both near zero bias and over an extended range. The devices behave as metal-semiconductor-metal diodes already introduced in a previous section. Near zero bias (see Fig. 9(a)) there is a greater effective conductivity until one of the metal-semiconductor junctions is sufficiently reverse-biased to partially rectify the current, at which point the effective conductivity decreases. Note that the reverse bias current densities of these Schottky diodes are orders of magnitude greater than the equivalent *pn* junction diodes. Additionally, the barrier-lowering effects of the applied electric field result in a continuously rising reverse current. At even greater biases, the reverse bias current increases at a much faster rate due to ionization multiplication effects and eventually junction breakdown (Neamen, 2003; Tu, 1992).

Even without improving the contacts to be more ohmic, SOI MSM devices can still have the internal gain mechanism that makes them ideal for speckle monitoring applications with higher bandwidths. This has been discussed earlier and is illustrated in the next section on device characterization.

To optimize these devices for an extended range of ohmic behavior, some experimentation with the contact materials and deposition procedure will be required. Ohmic contacts in silicon devices are generally achieved by introduction of a heavily doped silicon region near the contact, or by the use of silicide contacts (a silicon compound with a metal such as titanium, cobalt, tungsten, or platinum) rather than pure metals (Sze, 2007).

Additional doping would require at least one additional photolithographic and dopant implantation step, and is perhaps best to be avoided if possible. The formation of silicide contacts, however, requires only a slight modification of the contact deposition process and no additional photolithography. Silicide formation can be achieved in any one of four ways, including *direct reaction, chemical vapor deposition, co-evaporation,* or *co-sputtering* (Sze, 2007). Direct reaction is the easiest method to implement - it involves a metal deposition step followed by high temperature annealing. The metal component of the silicide is deposited (for instance by thermal evaporation or sputtering) onto the desired contact area. During the annealing step, some of the silicon device is consumed and becomes part of the contact silicide layer. A combination of careful literature review and experimentation should be sufficient to determine the appropriate metal choice and processing protocol for the formation of properly ohmic contacts with less rectifying effect on the current-voltage behavior of the photoconductors.

There is an apparent asymmetry in the biasing direction. This is possibly due to a difference in the metal-silicon junction size on either side of the element, which would make the reverse bias current in one direction greater than in the other. All elements in this array showed approximately the same shift with biasing applied in the same orientation.

3.4.2 Gain

The photo gain of an SOI device was measured as a function of bias voltage and compared to a silicon *pn* junction photodiode. The raw photocurrent measurements were converted to gain by calibration with a silicon photometer. In Fig. 10, we see that the silicon photodiode exhibits the expected flat gain throughout reverse bias. The gain value is approximately 0.85, indicating an internal quantum efficiency of about 85%.

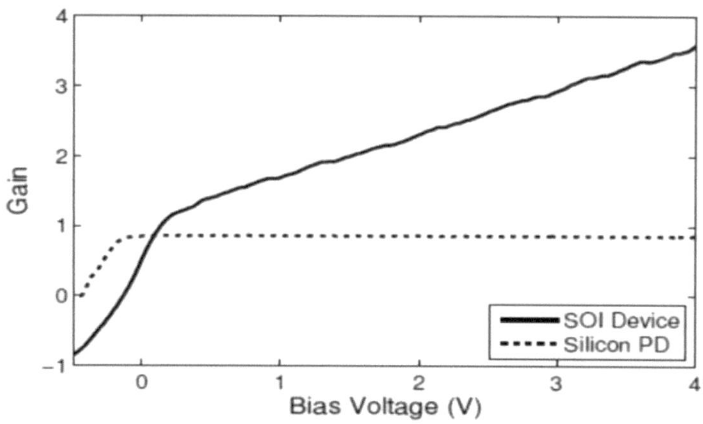

Fig. 10. Measured photo gain for an SOI photodetector (solid) compared to a reverse-biased silicon photodiode (dashed) as a function of bias voltage.

In contrast, the SOI device starts with a small gain near zero bias (due to the Schottky barrier induced fields). The gain increases sharply with applied voltage, quickly surpassing the *pn* junction diode gain, until the bias reaches 0.20 V. At this point, the gain continues to rise linearly, but at a reduced rate.

Two related phenomena were discovered in the behavior of the photodetectors that were not initially predicted. Firstly, the gain increases monotonically with bias voltage up to a maximum and then declines again. It was thought that the gain would saturate, but not decline. Secondly, the gain is not independent of incident illumination, but actually decreases with increased light levels.

Figure 11 shows the block diagram of an experiment performed to demonstrate this behavior. A sweeping DC bias voltage was applied to one SOI element under illumination from two diode lasers. The first laser was a near-IR InGaAs diode that provided a constant background level of illumination. The second laser was an AlGaAs diode operating at 661 nm that provided a low level AC signal (10 kHz). The resulting photocurrent was filtered with a lock-in amplifier and recorded as a function of the bias voltage to obtain the curves shown in Fig. 12.

The peaking phenomenon at some bias voltage and the reduction in gain at high illumination levels can be understood in the context of carrier traps within the photoconductor crystal (Rose, 1958; Macdonald, 2001). The presence of traps in silicon and SOI that affect carrier lifetimes is well noted (MacDonald, 2001; Celler, 2003).

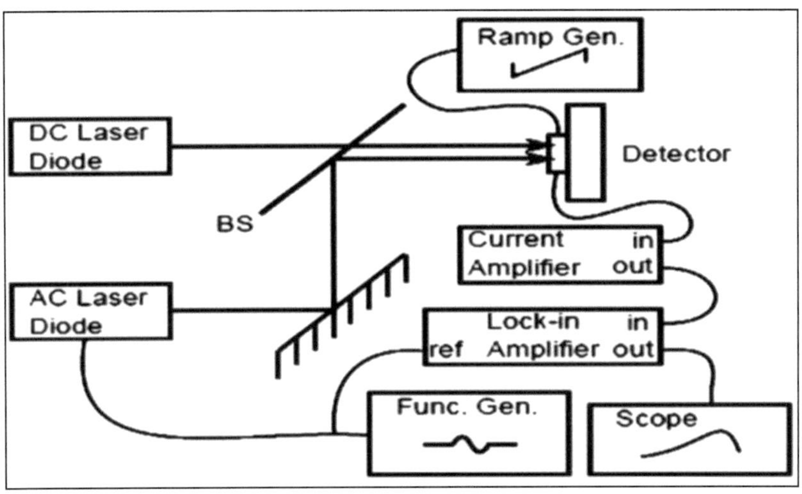

Fig. 11. Block diagram for experimental small-signal gain measurements as a function of bias voltage and background illumination.

Some trapping behavior can increase the effective lifetime of charge carriers; for instance, the presence of shallow traps in effective thermal equilibrium with the conduction band. These states capture some of the photogenerated carriers and prevent them from recombining and returning to the ground state. Since charges in these trapping states communicate with the conduction band, they can carry currents and contribute a photocurrent component with a long-lived response.

Another trap-related phenomenon that can contribute to the apparent long carrier-lifetimes is electron-hole separation due to the trapping of one of the charge carriers (MacDonald,

2001). For example, if holes are trapped in a state where electrons cannot recombine with them, a corresponding excess of free electrons will build up and contribute to the photoconductance (even if the holes do not contribute to the photoconductance).

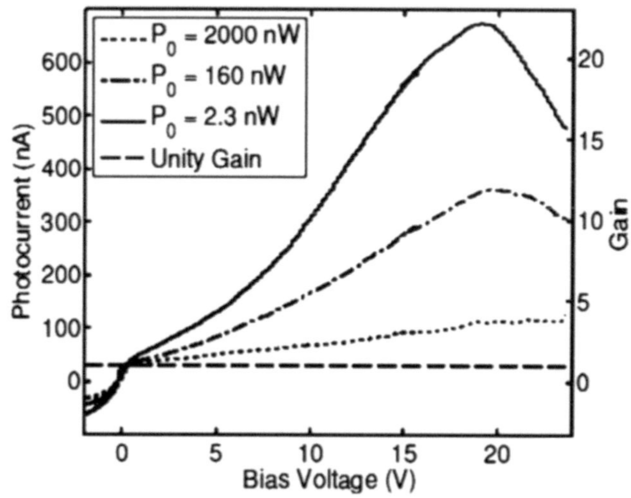

Fig. 12. SOI photodetector gain over a larger interval of bias voltage, for three levels of background illumination. The gain is a strong function of background illumination, presumably due to the sensitivity of the recombination time to trap-filling. Gain includes reflectivity losses.

The trapping states can become saturated by the two methods, as suggested by the curves in Fig. 12, both of which increase the density of excited charge carriers. The first is to raise the bias beyond the threshold to induce space-charge currents. The second is to expose the detectors to increasing levels of background illumination. In either case, the traps will begin to fill up and eventually become saturated.

At low carrier concentrations, these traps can significantly lengthen the effective carrier lifetime. Since the gain is proportional to lifetime (see Eq. 2), the gain is greatest when these traps are relatively empty and available to receive carriers. This phenomenon is also evident in the following impulse response and frequency measurements.

Since we know how much background optical power it takes to reduce the excess gain, we can make an estimate as to the density of these trapping states once we determine their effective lifetime. That will be determined in the next section.

3.4.3 Impulse response measurement

To determine if the prototypes were fast enough to operate at ultrasonic frequencies, an impulse-response measurement was made by exposing a device to a 10 ns square wave optical pulse from a 780 nm wavelength diode laser. Fig. 13 shows the recorded. response.

Beyond a certain bias level, the decay time of the response drops off dramatically, presumably due to the filling of shallow traps by excess dark current. Background illumination from a secondary laser diode also reduces the decay time of the detector response, presumably due to shallow trap filling.

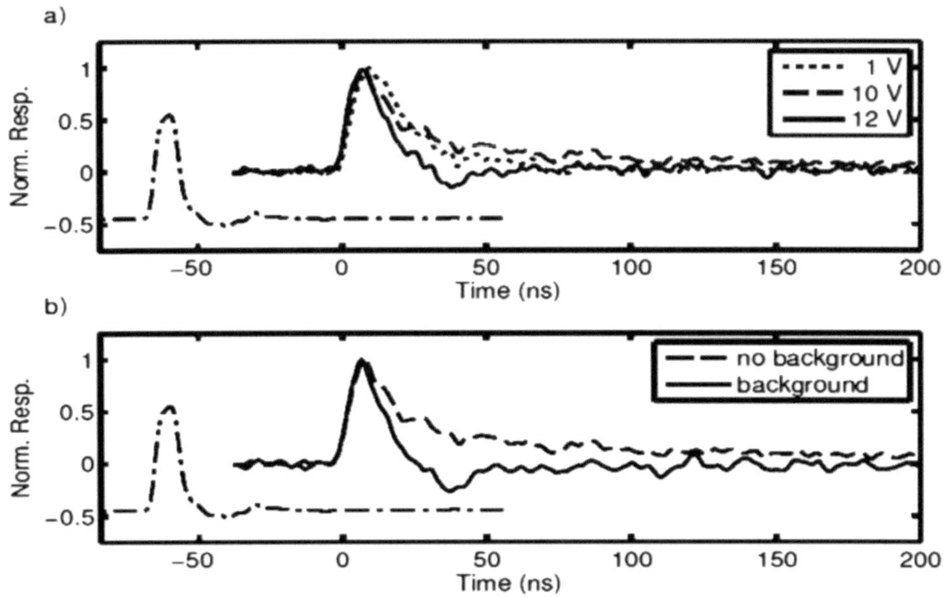

Fig. 13. Photodetector response to a 10 ns square pulse from a 780 nm diode laser under a) variable bias voltage with no background illumination and b) 10 V, bias with and without DC background illumination. Photodiode response for comparison is time-shifted to an earlier time for clarity.

Two things are immediately apparent. First, the bias voltage and background illumination greatly affect the decay time of the response. Second, even at its greatest, the detector response time is on the order of 50 ns. This indicates that the device should adequately measure signals well into the megahertz range with maximum internal gain, already measured greater than 20.

The impulse response appears to be composed of two components, a fast initial response and a slower decay. The slower component can be reduced by adding more bias voltage or by adding more background illumination. Fig. 14(a) shows the actual detector response, under minimal background illumination, modeled well by a double exponential decay with two time-constants, one at τ_1 = 6 ns and one at τ_2 = 90 ns. This secondary response, with the long decay rate, is responsible for the high gain at lower bandwidths up to a few MHz, as can be seen in Fig. 14(b). It is suppressed by either the excessive addition of bias voltage or by the application of strong background illumination, also illustrated in Fig. 14. Carrier lifetimes in this range, tens or hundreds of nanoseconds - much less than in bulk silicon - are often found in SOI structures (Mendicino, 1999; Rong, 2004). The lifetime shortening is due to the proximity of the active area to Si/SiO_2 interfaces.

We can make an order-of-magnitude estimate of the lower limit of trap carrier density by assuming that all of the high-intensity background light is absorbed by the long-lived trapping states, and furthermore that the states are evenly distributed throughout the device.

The power required to saturate those states multiplied by their effective recombination lifetime equals the total number of states in the device. We can see from Figs. 11 and 12 that the traps seem to saturate when 1 μW or more is absorbed by the device. With a response time of 90 ns and a device volume on the order of 10^3 μm^3, this indicates that the trap density is at least 4×10^{12} cm^{-3}. Where they exist, these traps are probably considerably more dense, since a) the background light is not exclusively absorbed by the trapping states and b) the trapping states are most likely associated with the device interfaces and not distributed evenly through the device volume.

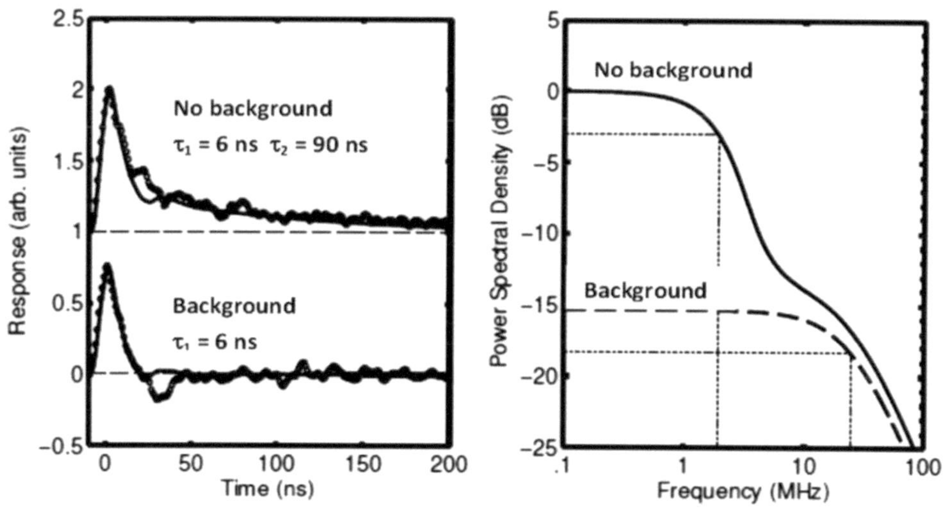

Fig. 14. a) Measured impulse response at 10 V bias (circles) overlaid with a fit of the double exponential impulse response (solid) for the detector with no background illumination and a single exponential response for the detector with significant background illumination. b) The frequency response of the fitted impulse responses. Note that detectors with sufficient (but not too much) bias voltage and low levels of background illumination show a significantly boosted response with bandwidths up to 2 MHz.

3.4.4 Detector noise

As already discussed, the best performance of the detector will be achieved when the SNR is maximized for a given level of illumination. This SNR value depends on both the noise levels and gain of the device. As long as the noise is dominated by the thermal noise term, which is unaffected by the bias voltage, the highest bias voltage (below saturation) and thus the highest possible gain should be used to maximize the SNR. Fig. 15(a) shows both the measured gain for the SOI device together with the combined thermal and current shot noise levels based on the dark current measurements. Assuming a 120 Ω load resistance and a 1 MHz detection bandwidth, the overall noise is dominated by the thermal noise component and is virtually unchanging until the bias voltage exceeds 10 V. All the while, the gain continues to rise with increasing bias, improving the signal-to-noise ratio.

However, as the bias voltage is increased, so is the dark current, which in turn increases the current shot noise. At some point, the dark current shot noise will become a significant portion of the overall noise. This increased noise will cancel the benefit of increased gain. The bias beyond which this happens can be determined by plotting both the gain and the noise on the same semilog plot, as in Fig. 15(b). When the slope of the noise versus voltage exceeds the slope of the gain versus voltage on a logarithmic scale, the noise is increasing faster than the gain, and no benefit is obtained by turning up the bias more. The best SNR is obtained for this device somewhere around 15 V. The benefit of using detectors with enhanced internal gain in detector regimes dominated by thermal noise is nicely illustrated by this particular measurement. If the detector gain was limited to unity, about 1/20 of the actual measured value, the best the detector could achieve in terms of SNR is less than 0.2% of the actual measured value. This is obviously a significant improvement.

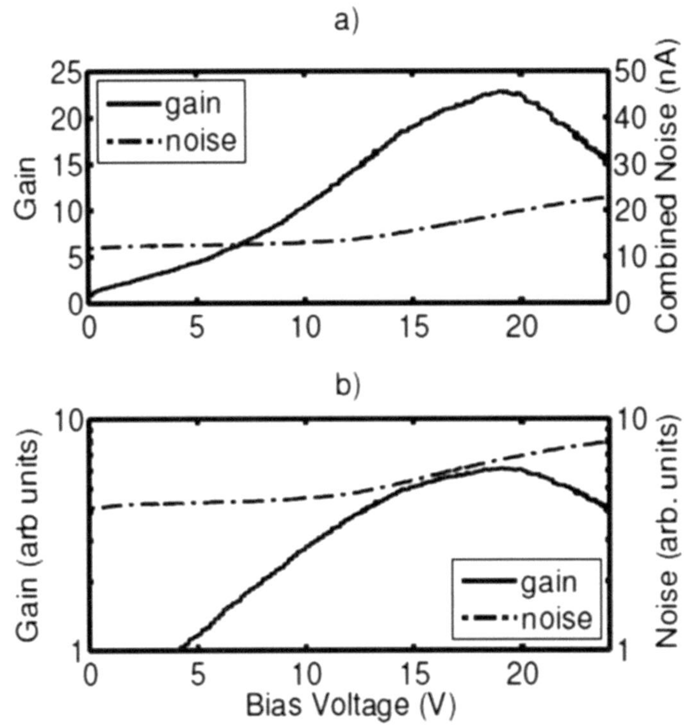

Fig. 15. a) Internal gain and detector noise as a function of bias voltage; b) The same values plotted on a semilog scale. When the slope of the noise exceeds the slope of the gain, the noise increases faster than the gain, and there is no benefit from increasing the bias. Combined detector noise is calculated assuming thermal noise of a 120 Ω load resistance and the shot noise associated with the measured dark current values. A detection bandwidth of 1 MHz is assumed.

3.5 Summary

Gallium aresenide photoconductors of the type discussed in (Heinz, 2004, 2005, 2007, 2008) were determined to be too slow for megahertz detection of speckle motion due to the buildup of space charge, even with the slight improvement of transparent electrical contacts and background illumination. Silicon mesa structures were developed that do not suffer from the same space-charge bandwidth limitations. The mesa architecture provides spatial isolation of each detector, and the prototypes proved to be fast enough to operate at megahertz frequency.

The current-versus-voltage curves for the prototypes show behavior that is more like MSM photodiodes than a true photoconductor, but this is acceptable since MSM devices can have greater than unity gain, which is the important characteristic for high-speed speckle-monitoring. Indeed, the internal gain was measured as a function of bias voltage and found to have a maximum for one device at about 23. The gain is greatest for biases below 20 V and at low levels of illumination due to the trap filling effects of large dark currents and photo-generated charges.

It was also shown that when operating at MHz bandwidths, the noise is indeed dominated by the thermal noise component of the load resistance even at relatively high bias voltages. The SOI device was determined to have a maximum SNR roughly 500 times greater than an equivalently sized silicon *pn* junction diode operating under the same conditions.

4. Optical generation and detection of acoustic waves with an experimental direct speckle monitoring system

This section discusses experimental demonstrations of speckle monitoring-based vibration measurement using the photo-conductor mesa detectors coupled with optical techniques for acoustic excitation. There are many reasons to develop a system that exploits the optical excitation of acoustic waveforms as well as their optical detection. Fully non-contact systems can be less invasive, less disruptive to the device under test, and in some cases more rapidly deployed. Extensive work on the subject is available in the literature (Scruby, 1990; Davies, 1993; Pierce, 1998) that will only be summarized briefly here. Some understanding of the different mechanisms involved will help us better understand the experimental results obtained from tests with the photo-conductive mesa detectors.

4.1 Optical generation of acoustic waves

Generation of ultrasound waves with laser beams in untreated surfaces can be divided into two categories (Davies, 1993):

i. The thermoelastic regime describes acoustic waveforms generated by expansion and compression resulting from the localized heating caused by an optical pulse. Typical peak optical intensities supplied to the surface in this regime are < 10^7 W cm^{-2} (Scruby, 1990).

In the thermoelastic regime, the acoustic source (the bit of solid material that actually moves in direct response to heating, setting the acoustic wave in motion) is roughly a disc with a diameter equal to the width of the impinging laser beam and a depth determined by the thermal conductivity of the material and the rise-time of the laser pulse (Scruby, 1990; Davies, 1993; Pierce, 1998). For metal samples, and for nanosecond pulses (from a Q-switched Nd:YAG pulse, for instance), the source depth is on the order of microns, but can grow to hundreds of microns for millisecond rise times (Scruby, 1990; Pierce, 1998). As a result of heating, the disc exhibits expansion, predominantly in the plane of the surface (Pierce, 1998).

ii. The ablation regime pertains to acoustic waveforms generated by the momentum transfer of evaporated material ejected from a rapidly, locally heated component of the test object's surface. Typical peak optical intensities supplied to the surface in this regime are > 10^7 W cm^{-2} (Scruby, 1990).

In the ablation regime, the acoustic source is a cloud of vaporized material forming a plasma just above the surface and moving rapidly away from it. The result is an acoustic impulse directed predominantly normal to the surface.

In addition to peak powers and the *temporal* profile of the laser beam, the *spatial* profile of the laser beam as it intersects the solid surface is crucial in determining the final acoustic waveform injected on the test object (Monchalin, 2007; Khang, 2006]. The beam shape can be tweaked to encourage or discourage coupling into one or more acoustic modes. For instance, an optical grating can be imposed on the object surface with a spacing corresponding to the wavelength of the desired surface acoustic wave. This can greatly enhance coupling into that mode. Since no effort was made to explore this aspect of optical ultrasonic generation during the course of this project, no further discussion will be given to this topic. However, this remains a rich area for study, in particular learning how to promote the in-plane acoustic modes - we have already seen how photoconductive mesas are well suited to the detection of such waves, through speckle monitoring.

4.2 Nd:YAG pulsed laser generation of ultrasound waves

An experiment was performed in which ultrasonic waveforms were produced in solid metal targets by surface-ablation with a Q-switched Nd:YAG laser pulse. The various samples under test were thin strips made from unpolished aluminum, copper, and stainless steel. The resulting acoustic waves were allowed to propagate through the strips and were detected by a speckle monitoring system that observed a point on the test object located some distance away from the acoustic source.

The interest in these experiments is two-fold. Firstly, they show how well the photoconductive mesas monitor direct speckle motion, providing a sensitive method for remote ultrasound detection, especially for detection of in-plane surface motion to which other detection schemes are not as selectively sensitive (Monchalin, 1986). Secondly, they provide a demonstration of the bandwidth capabilities of the SOI photoconductive detector mesa prototypes made in our laboratories.

4.2.1 A_0 antisymmetric mode Lamb waves

Acoustic waves that propagate in the plane of thin plates and membranes can be decomposed into different Lamb wave modes. The two main classes of Lamb wave are the antisymmetric modes, for which the front and back surfaces of the plate move out of phase with respect to the middle of the plate and the symmetric modes, for which the front and back surfaces move in phase. For low frequencies, below some cut-off value that depends on surface thickness and plate material, we saw that only the lowest order modes, A_0 and S_0 exist. Which of these gets excited depends on the type and geometry of the exciting pulse.

Fig 16 shows the experimental arrangement for the first test, in which the Nd:YAG laser is directed onto the back planar surface of the strip. This directs most energy into a direction perpendicular to the surface (flexing), which is the primary motion of the A_0 mode. The antisymmetric mode should be the dominant acoustic waveform resulting from this excitation (Pierce, 1998). The detector output confirms this hypothesis, as there is clear evidence of antisymmetric Lamb wave propagation in the strips, and very little evidence of symmetric Lamb waves. In Fig. 17, we see a comparison of the recorded signal and the theoretical calculations of the A_0 mode propagating through the same distance of an aluminum sheet. The calculations assume a square impulse excitation of the A_0 mode with the bandwidth of the recorded signal. The dispersion relation is $V_p = 2V_S(1 - V_S^2/V_L^2)\beta b/(2\bullet 3^{1/2})$, where V_p is the phase velocity of the mode, V_S is the shear velocity of the material, V_L is the longitudinal velocity of the material, β is the wavenumber of the mode, and b is the plate thickness, according to Chapter 9 of (Cheeke, 2002).

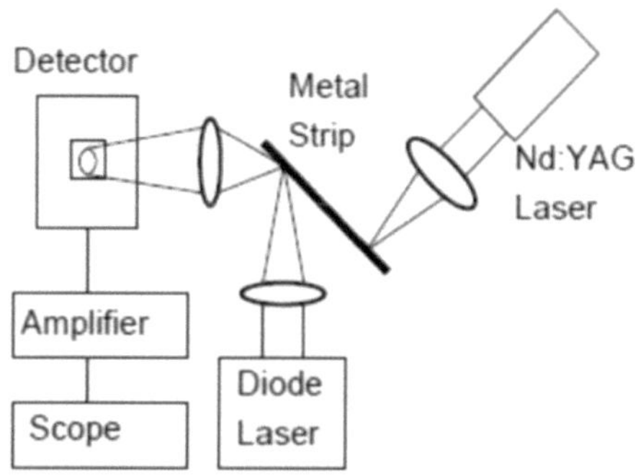

Fig. 16. Experimental arrangement for pulsed Nd:YAG laser excitation of ultrasound in metal strips. Detection of ultrasound was made with a speckle monitoring photoconductor.

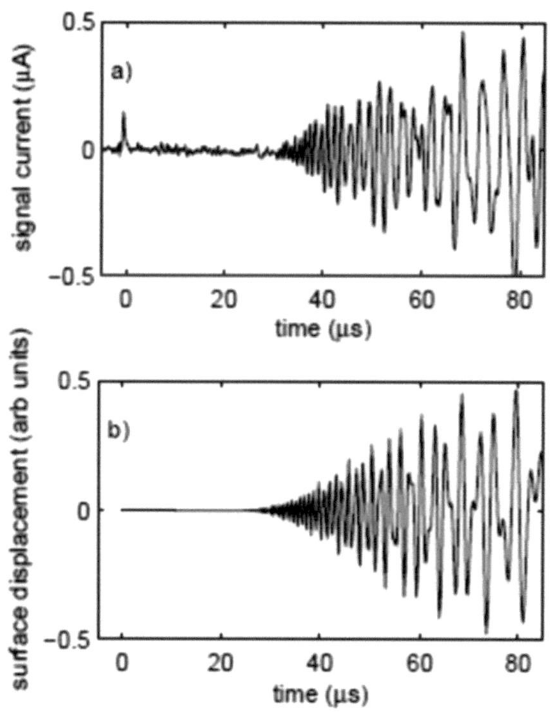

Fig. 17. a) Oscilloscope trace of an acoustic vibration generated with 7 cm of separation between the Nd:YAG pulse and the reference spot. The initial spike in photocurrent comes from the Nd:YAG pulse at time $t = 0$. b) Calculated surface displacement caused by A_0 Lamb waves propagating a distance of 7 cm. One reflection from the far edge of the strip (1 cm further from the detection point) is included to illustrate the effect of interference.

Figure 18a depicts the signal obtained from the copper strip, showing the arrival of multiple distinct waves. By looking the spectrogram in Fig. 18b, it is clear that each of these waves shares the same dispersion relation. Thus each wave is an instance of the same acoustic mode (A_0). These secondary instances are most likely reflections from one of the copper strip edges.

The antisymmetric mode is certainly useful, especially for extracting material properties from samples. For just one example, there is great interest in online monitoring of paper stiffness, consistency, and quality with laser ultrasonics exploiting antisymmetric Lamb waves (Brodeur, 1997). The thickness and shape of thin film structures, especially during fabrication (Pei, 1995), is another parameter of interest to which the propagation of such Lamb waves are sensitive. Any all-optical method for such measurements is certainly worth a consideration.

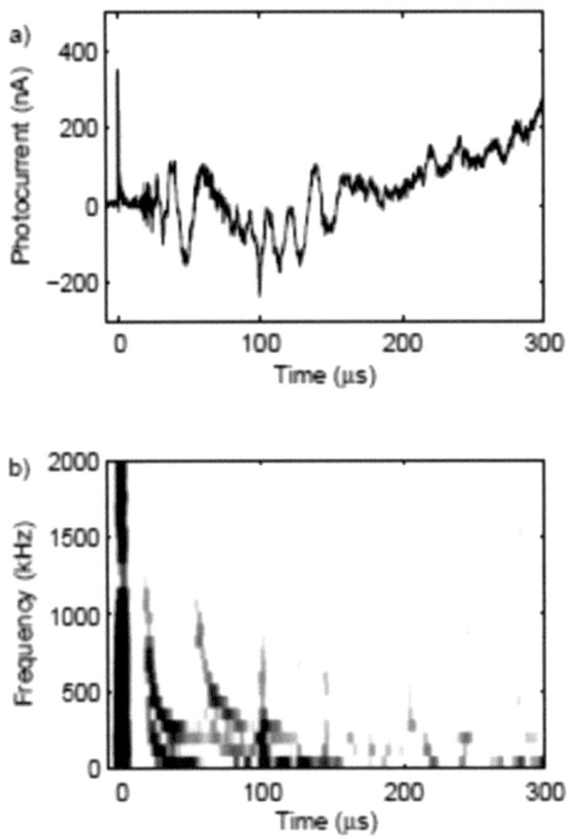

Fig. 18. (a) Oscilloscope trace of ultrasonic waves traveling in a copper strip with (b) corresponding spectrogram showing the spectral content of distinct waves as a function of time. The initial spike at time t = 0 comes from stray Nd:YAG light and serves as a marker for the onset of sonic excitation. At least the first two modes appear to have the same dispersion relation, indicating that they represent the reflection of a wave in the same mode (A_0).

4.2.2 S_0 symmetric mode Lamb waves

The lowest order symmetric mode (S_0) is of particular interest to applications that record the timing of pulse echoes. This would include parts inspections for internal cracks and discontinuities (Lowe, 2002). The reason for this is that the S_0 mode suffers very little dispersion, limiting pulse breakup and allowing for better event timing. Furthermore, at low frequencies, the S_0 mode is dominated by in-plane surface motion (Cheeke, 2002), so that is it is particularly well-suited to observation by a speckle monitoring system.

One problem with using the S_0 mode is that it is more difficult to excite by Nd:YAG laser pulse than the A_0 mode. Striking the flat surface of a plate tends to put most acoustic energy into the A_0 mode. The dominant motion of the S_0 mode is compression in the plane of the plate, so it is easier to excite this mode by striking the thin plate edge, as in the arrangement of Fig. 19. Although edge-excitation has been investigated with traditional contact transducers (Lowe, 2002), this novel arrangement is the first all-optical, end-on excitation technique specifically designed to excite only the symmetric mode. With further improvements, it could be adapted to generate in-plane vibrations for testing in any kind of plate-like structure.

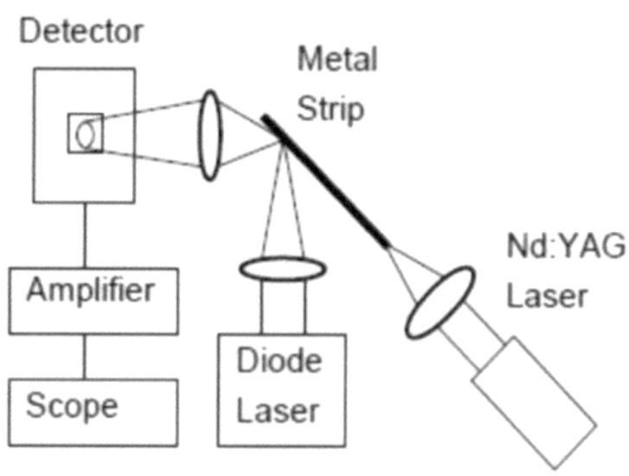

Fig. 19. Edge excitation of metal strip with Nd:YAG laser pulse.

When edge excitation is used, the coupling efficiency into the S_0 mode is dramatically enhanced, as can be seen in the oscilloscope trace in Fig. 20, showing the resulting vibration of such excitation in a stainless steel strip. This arrangement provides better timing information due to the non-dispersive nature of the S_0 waveforms. The waveform in this case is more or less a single pulse, about 10 μs wide, with a slightly exaggerated tail.

The round-trip travel of the wave is clearly discernible from the detector output in the first 300 μs of the recorded signal. The wave first arrives (after traveling just under one length of the strip) after 18 μs, and subsequently every 37 μs or so after that.

Although the Nd:YAG pulses produce ultrasonic pulses in these strips that can be used to investigate the material and geometric properties (for example, sheet thickness and points of internal reflection), this method can hardly be called "nondestructive." The material ablation that causes the acoustic impulse leaves a crater in the object surface. Repeated impulses, especially on delicate structures and thin films, require a gentler approach.

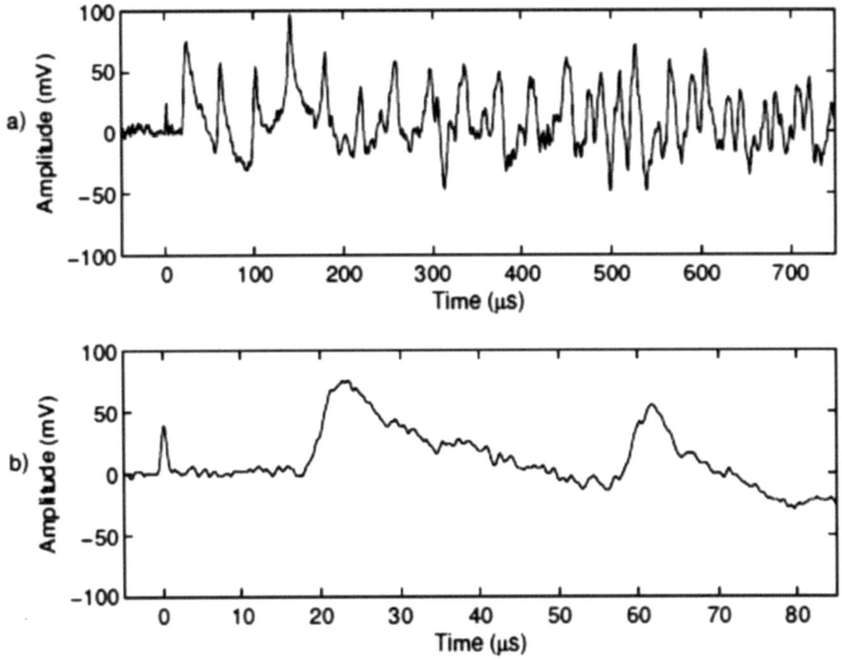

Fig. 20. Vibration signals recovered during edge excitation experiment. The spike at time t = 0 is from scattered Nd:YAG laser light, marking the onset of acoustic excitation. This S_0 waveform is much less dispersive than the A_0 mode, allowing for more precise timing of discrete events, such as reflections. a) For the first 300 μs or so, the waveform is dominated by repeated reflections of the non-dispersive acoustic wave from either end of the strip as it traverses the detection point. b) At a smaller time scale, the time separations between peaks are easier to see. The time (37 μs) between the first pass of the wave and the second pass, or one round trip of the wave, is just slightly more than twice the delay (18 μs) between excitation onset at the far end of the strip and the leading edge of the first pass. This is precisely what is expected from the experimental geometry, depicted in Fig. 19. The mode velocity as measured by the temporal separation of the peaks is 5200 m/s.

4.3 Laser diode generation of ultrasound

To produce ultrasound in a truly nondestructive manner, one must restrict oneself to the thermoelastic regime of optical ultrasound generation. This can be achieved by lowering the peak power below the damage threshold. That approach favors the use of technical finesse over raw power, for which laser diodes are a good alternative to the Q-switched Nd:YAG laser. Consider the experiment in Fig 21, in which the Nd:YAG excitation laser has been replaced by an InGaAs current-modulated diode laser with a peak output power of approximately 50 mW in a surface-normal configuration. Figure 22 shows the detector output for an aluminum foil strip with dimensions 10 cm × 0.5 cm in response to excitation pulses of various widths. A 1 ms excitation pulse produces a large vibration that is easily measured by the speckle-monitoring detector. Even with such low coupling-efficiency, there is some diagnostic capability with this simple configuration, for instance monitoring low-frequency standing waves in small structures.

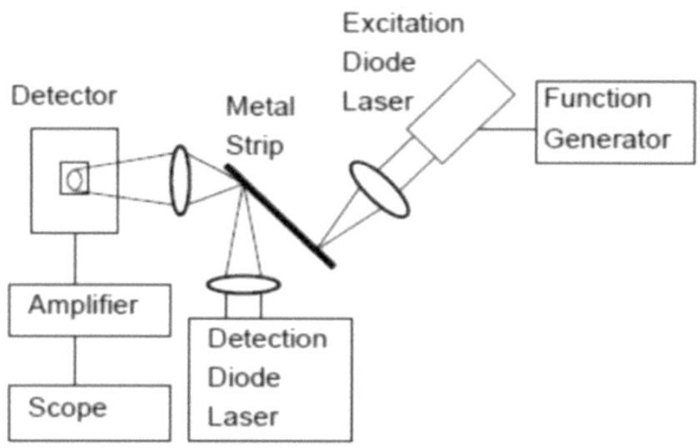

Fig. 21. Experimental setup for measuring the impulse response excited with a current-modulated laser diode. For the data presented in Figures 22 and 23, the metal targets were strips of aluminum foil.

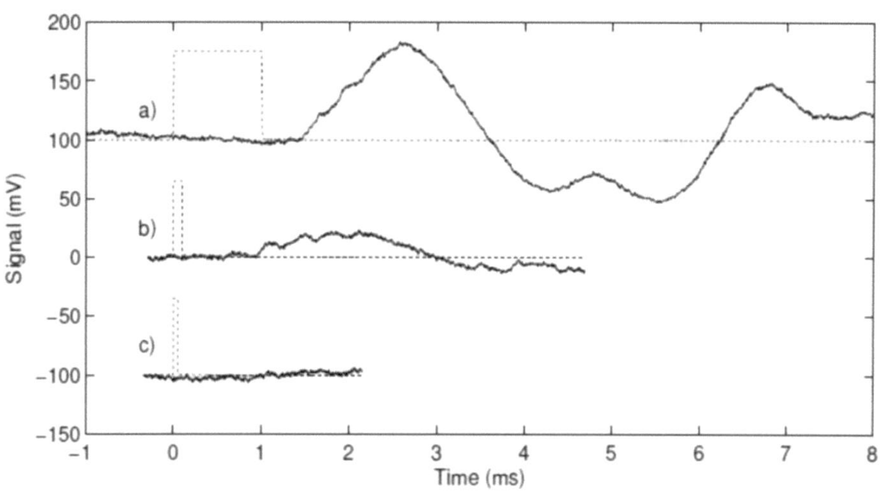

Fig. 22. Waveforms in an aluminum foil strip resulting from thermoelastic generation with an InGaAs laser diode pulse with a peak power of 50 mW and spot size of approximately 1 mm. Dotted lines indicate the period of laser diode excitation, solid lines indicate the detector signal. Three pulse lengths are shown: a) 1000 μs, b) 100 μs, and c) 50 μs. Decreasing the overall optical power delivered to the metal shows a corresponding decrease in the amplitude of the detected signal. For pulse widths less than 50 μs, the signal is no longer discernible from the noise.

Consider the data shown in Fig. 23. Three aluminum foil strips, 10 cm × 0.5 cm, were gently stretched flat and anchored on either end with approximately equal tension. Two of the three strips were left unaltered, but in the third strip a small tear was introduced into the strip's edge halfway between the excitation and detection points. Each strip was excited by a 500 μs square wave impulse from the excitation laser diode. The impulse responses of the two undamaged strips look nearly identical, a fundamental vibration with a period of about 4.7 ms ringing down at approximately the same rate. The damaged strip does not ring as cleanly. Although the fundamental frequency looks approximately the same as for the undamaged strip, the signal is disrupted by some other vibrations. Without analyzing the precise differences in the impulse response signature between the healthy and the damaged strips, it is fair enough to say that there is a clearly discernible difference. Such a distinction could be used to detect the presence of damage.

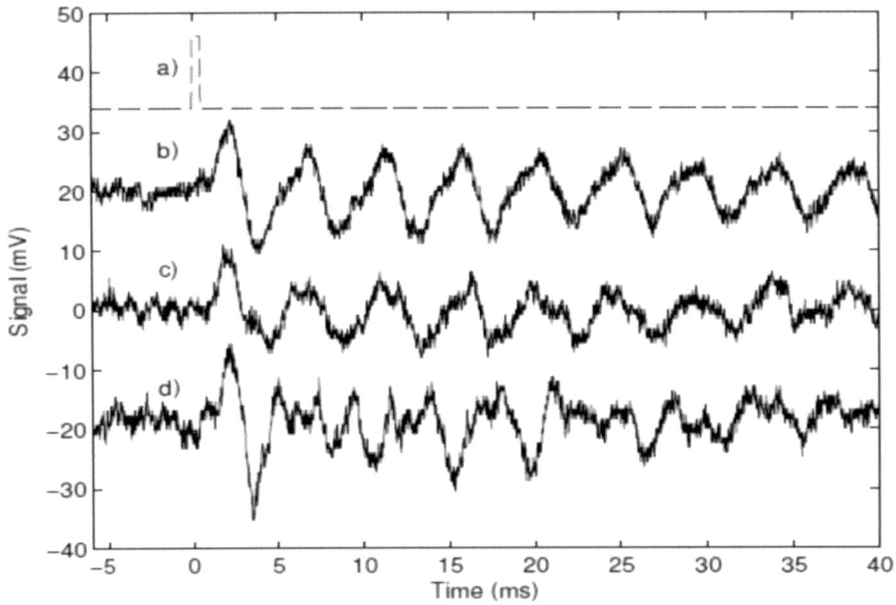

Fig. 23. Response of three identically sized aluminum foil strips under equal tension to a 500 μs rectangular pulse from a diode laser. a) Excitation diode current (arbitrary vertical scale), b) strip 1, no defect, c) strip 2, no defect, and c) strip 3, a tear introduced between points of acoustic generation and detection.

Although useful, this low-frequency, resonance technique is not perfect for every solution. Some applications are best served by a different approach to acoustic imaging that requires temporally and spectrally resolved pulse information such as time-of-flight and dispersion data. This generally precipitates the use of higher-bandwidth acoustic regimes.

The obvious drawback to operating in the thermoelastic regime is the decrease in acoustic power delivered to the object of interest. This problem can be addressed by using selective filtration methods on the detector signal, such as lock-in amplification and input-output signal correlation techniques.

4.4 Summary

In this section it was shown that speckle-monitoring is a viable technique for the remote detection of acoustic waves, in particular ultrasonic waveforms that are useful for material characterization and structural examination. The possibility of coupling speckle-monitoring with optical ultrasound generation is particularly attractive, as it allows for a completely non-contact system. Speckle-monitoring is especially appealing for use in measuring in-plane ultrasonic waves, such as the S_0 Lamb wave mode measured in the edge-excitation experiment, since it is selectively sensitive to that type of motion, unlike most other optical vibration detection schemes. There is an incentive to restrict optical excitation to the lower-power thermoelastic regime, such as by using current modulated CW laser diodes.

5. SOI speckle monitoring: detection limits compared to alternatives

Having laid the framework for characterizing the SOI detector prototype and conducting preliminary experiments, we are in a position to make some statements about the viability of this technique as an alternative to other optical vibration measurements. This section quantifies those comparisons and indicates certain applications to which speckle monitoring with photoconductor arrays is best suited. Numerical values are reported as rough estimates constructed from the data gathered in the previous chapters.

5.1 Pure in-plane-motion

After establishing that measurement with such a system is indeed possible, the most obvious question to ask about a particular measurement scheme for a given application is "how sensitive is it?" To that end, this section will attempt to answer that question for the direct speckle-monitoring method and other comparable techniques under similar circumstances.

Since we have already established that speckle-monitoring is most effective for in-plane measurement, we will focus on that application. As a convenient metric for the lower limits of sensitivity, let us define the quantity δ_{min} to be the minimum detectable displacement that produces a signal with amplitude equal to the noise for a given system output. Since we have seen that sensitivity generally scales in proportion to optical power and inversely to bandwidth, δ_{min} will be given units of m $Hz^{-1/2}$ W, where the power is the total available laser power.

5.2 SOI Detectors, single element detection limit

The minimum detectable photocurrent is equal to the noise current, which from experiments is roughly 20 nA at 1 MHz bandwidth with an overall gain of 20 across a load resistor $RL = 50\Omega$. Since noise scales with the square root of the bandwidth, we can estimate a noise figure of $\sigma_n = 20$ pA•$Hz^{1/2}$. A gain of 20 corresponds to a responsivity of about $\Re = 10$ A/W at 633 nm. This means that the minimum detectable change in collected power per element is $P_{min} = 2$ pW•$Hz^{1/2}$.

The amount of speckle translation required to achieve this change in photocurrent depends on the optical system and how much optical power is collected. For a small scattering spot whose width is S, illuminated with a total power P_S, the speckle size in the detector plane is approximately $\varepsilon = \lambda B/S$, where B is the optical matrix element. Assuming the light is scattered equally in all directions and the scattering spot size $S << B$, the average speckle intensity is approximately $I_0 = P_S\lambda^2/(2\pi\varepsilon^2S^2)$ as shown in (Dainty, 1975, Bessette, 2010). For a small translation δ_s of the speckle pattern, this means the detector will see on average (considering the ensemble of all possible random speckle patterns) a change in collected power given by $\Delta P = I_o \varepsilon^2 (\delta_S/\varepsilon) = P_S\lambda^2\delta_S/(2\pi\varepsilon S^2)$, where we have assumed that the speckle width is matched to the detector width. Setting ΔP = to ΔP_{min} and solving for δ_s, we find that the minimum measurable motion is

$$\delta_{S,min} = 2\pi\varepsilon S^2 \Delta P_{min} / \lambda^2 P_S \quad .$$

Using a typical scattering spot size of 0.5 mm and a detector-matched speckle size of 20 μm, this gives $\delta_{S,min}$ = 160 pm $Hz^{1/2}$ W. This is the minimum detectable translation of the speckle pattern. Typically, there will be optical magnification in this system with a value ~10, so this yields a minimum detectable displacement on the order of δ_{min} = 16 pm $Hz^{1/2}$ W.

For example, with a 1 W laser source and a bandwidth of 1 MHz and using a 50Ω load resistor, the detection limit is approximately δ_{min} = 16 nm for a single element. This value can be brought down even more by using a larger load resistor, as large as the bandwidth limitations will allow.

5.3 Other methods

Classical interferometry can be used to detect in-plane motion under certain conditions. For instance, rather than using one beam to illuminate the object and one reference beam, the differential interferometry approach uses two beams to illuminate the object (Monchalin, 1986). If the input aperture of the system and the illuminating spot are sufficiently small, then what forms in the detector plane is essentially a single speckle. The phase of that one speckle is sensitive to motion parallel to the difference in wave-vectors of the two illuminating beams. If the two beams illuminate the surface at equal but opposite angles with respect to the normal, the setup is sensitive purely to motion in the plane of the object. Using a 0.5 mm spot size and a HeNe laser with a wavelength of 633 nm gives a detection limit δ_{min} = 16 pm $Hz^{-1/2}$ W. With a 1 W laser and 1 MHz bandwidth, a sensor with these parameters has a detection limit on the order of δ_{min} = 16 nm. Again, this can be reduced by increasing the load resistor to the maximum the bandwidth requirements will allow. We see that the detection limit for the direct speckle monitoring method with SOI detectors is roughly equal to the interferometric method. The speckle monitoring method has the greatly added advantage that it does not require two coherent beams to be precisely overlapped at a specific angle in a small spot.

Speckle photography relies on tracking individual speckle features as they move across the detection plane with an electronic detector array. These methods are generally used in low speed applications -- CCD arrays with low read-out rates and long integration times (on the order of milliseconds). A recent report on different algorithms (Cofaru, 2010) finds that resolution is on the order of 10^{-3} pixel widths, generally speaking around δ_{min} = 10 nm. Although this is on the same order as the detection limit of the SOI sensor, the SOI photoconductive mesa sensor described here does so at 1 MHz bandwidth.

Spatial filtering of moving speckles is a subcategory of speckle photography in which image processing of the speckle pattern is done in real-time by spatially filtering the speckle pattern with a periodic mask (for example a simple Ronchi grating) and integrating the filtered light (Komastu, 1976). As the speckle pattern translates across the grating, the integrated signal is periodic with a period equal to the time it takes an individual speckle to translate a distance equal to the grating spacing. The frequency of the integrated signal is proportional to velocity, so this technique is generally used to measure surface velocity. The measurement depends on determining a temporal frequency, so there must be at least one full period observed. This means that the speckle must travel at least the distance of a grating spacing. Since signal-to-noise is best when the speckle is about the same width as the grating spacing (Kamshilin, 2009), the detection limit is approximately 10 μm. This method is not nearly as sensitive to small displacements as the other methods discussed; however it works well for high in-plane velocities when the object is moving a large distance.

5.4 Summary

We have shown that photoconductive mesas directly monitoring speckle motion have an important place among applications requiring higher bandwidth detection of in-plane motion. The detection limit for a single element monitoring in-plane motion is on the order of δ_{min} = 16 pm $Hz^{-1/2}$ W.

6. Arrays of photoconductive mesas

The performance of speckle monitoring systems can be dramatically increased by using multiple detectors to make simultaneous measurements of a single speckle pattern displacement. An array configuration of photodetectors can be used to enhance the signal to noise ratio of displacement measurements. In addition to a lower noise floor, uncertainty in absolute displacement measurements can be mitigated. Additionally, the adverse effects owing to harmonic distortion can be lessened by the use of arrays, all allowing for better overall system calibration.

The signals from multiple array elements can be combined in a way that reduces noise in the same spirit as simple averaging. The contribution of each element to the average, however, must be weighted according to the particular speckle that element is monitoring. The primary challenge is the ambiguity in both the sign and magnitude of the slope in photo-intensity as a function of speckle displacement. If the signals are corrected so that every element has the same phase (i.e., all elements are made to have a positive slope), then the weighted average shows an increase in signal to noise ratio as the number of elements is increased.

A demonstration of the power of arrays for improving signal-to-noise was made by comparing the signals of individual detector elements with a phase-corrected average of 18 combined elements. The 18 independent signals were obtained by using a three-element array to monitor six different random speckle patterns, each undergoing the same displacement. The SNR of the phase-corrected average was superior by approximately 10 dB over the typical single element signal. In this demonstration case, the phases were sorted manually during post processing.

7. Conclusions

This chapter has shown how photoconductive detectors tens of microns in size can provide more sensitivity than photo-diodes. Very small photo-diodes have very small areas, which limits their sensitivity at high bandwidth. They are useful in applications such as single-mode fiber optics telecommunications, because the light is concentrated to a very small area. When signals are as small as a nanowatts, photo-multipliers are often used. However, in applications where spatial resolution at high bandwidth is required, such as monitoring mechanical vibrations, photoconductive elements can have the best useful figure of merit. We have demonstrated a minimum detectable power of P_{min} is 2 pW•$Hz^{1/2}$.

Experiments showed the most sensitivity with photoconductive GaAs mesas, but speed was limited to tens of kHz because of slow space-charge buildup in regions of sharp transition between light and dark. Reducing the shadows with transparent ITO contacts and reducing the background resistivity with continuous illumination with an LED, provided a response time of about 10 μs, three times faster than without these special measures. Nonetheless, GaAs mesas were unable to respond fast enough to measure ultrasonic waves.

The solution was to use photoconductive mesas in silicon, which does not suffer from space charge buildup. Because carriers can diffuse very far in silicon, it is important to use silicon-on-insulator (SOI) technology, to remove any diffusion into the substrate. It is also necessary to electrically isolate mesas by etching all the way through the silicon layer. With careful etching, mesas 15 μm wide were etched and contacted on either side. These photoconducting mesas were demonstrated to respond in 2 MHz with a 10V bias. Analysis of speckle motion measured with these SOI mesas indicated a sensitivity of δ_{min} = 16 pm $Hz^{1/2}$ W.

Preliminary experiments were carried out to demonstrate improvements when arrays of photo-conducting mesas are used, as long as the signals are properly combined. Further work on arrays will enable system calibration as well as measurement. All results, both experimental and theoretical, indicate chapter that this technology is certainly amenable to further development in practical applications of vibration and motion sensing, particularly at high speeds and small distances.

8. References

Bessette, J. & Garmire, E. (2010). Silicon-on-insulator photoconducting mesas for high-speed detection of laser speckle motion. *Journal of Selected Topics in Quantum Electronics*, Vol. 16, No. 1, Jan/Feb 2010, pp. 93-99.

Bessette, J., Gogo, A. & Garmire, E. (2006). Four-point photo-conductance monitoring array for speckle and fringe motion sensing. *Proceedings of SPIE* , Vol. 6379, Oct 2006, pp. 637903(1-10).

Biswas, K., Das, S. & Kal, S. (2006). Analysis and prevention of convex corner undercutting in bulk micromachined silicon microstructures. *Microelectronics Journal*, Vol. 37, No. 8, Aug 2006, pp. 765-769.

Brodeur, P.H., Johnson, M.A., Berthelot, Y.H. & Gerhardstein, J.P. (1997). Noncontact laser generation and detection of Lamb waves in paper. *Journal of Pulp and Paper Science*, Vol. 23, No. 5, May 1997, pp. J238-J243.

Bube, R.H. (Ed) (1960). *Photoconductivity of Solids*. John Wiley & Sons, Library of Congress Control Number lc60010309, New York, 1960.

Burm, J. & Eastman, L.F. (1996). Low-frequency gain in MSM photodiodes due to charge accumulation and image force lowering. *IEEE Photonics Technology Letters*, Vol. 38, No. 1, Jan 1996, pp. 113-115.

Carrano, J.C., Li, T., Grudowski, P.A., Eiting, C.J., Dupuis, R.D. & Campbell, J.C. (1998). Comprehensive characterization of metal-semiconductor-metal ultraviolet photodetectors fabricated on a single-crystal GaN. *Journal of Applied Physics*, Vol. 83, No. 11, June 1998, pp. 6148-6160.

Celler, G.K. & Cristoloveanu, S. (2003). Frontiers of silicon on insulator. *Journal of Applied Physics*, Vol. 93, No. 9, May 2003, pp. 4955-4978.

Cheeke, J.D.N. (2002). *Fundamentals and Applications of Ultrasonic Waves*. CRC Press, ISBN 0-8493-0130-0, Boca Raton, Florida.

Cofaru, C., Philips, W. & Van Paepegem, W. (2010). Evaluation of digital image correlation techniques using realistic ground truth speckle images. *Measurement Science and Technology*, Vol. 21, No. 5, March 2010, 055102 (17pp).

Dainty, J.C. (ed)(1975). *Topics in Applied Physics. Vol. 9, Laser Speckle and Related Phenomena*, Springer-Verlag, ISBN 0387131698, 9780387131696, Berlin and New York.

Davies, S.J., Edwards, C., Taylor, G. S. & Palmer, S. B. (1993). Laser-generated ultrasound: its properties, mechanisms and multifarious applications. *Journal of Physics D: Applied Physics*, Vol. 26, No. 3, Nov 1993, pp. 329-348.

Heinz, P. (2008). *Optical Vibration Detection with a Four-Point Photoconductance-Monitoring Array*. Ph.D. thesis, Thayer School of Engineering at Dartmouth College, 2008.

Heinz, P. & Garmire, E. (2004). Optical vibration detection with a photoconductance monitoring array. *Applied Physical Letters*, Vol. 84, No. 16, April 2004, pp. 3196–3198.

Heinz, P. & Garmire, E. (2005) Low-power optical vibration detection by photoconductance-monitoring with a laser speckle pattern. *Optics Letters*, Vol. 30, No. 22, Nov 2005, pp. 3027–3029.

Heinz, P. & Garmire, E. (2007). Photoconductive arrays for monitoring motion of spatial optical intensity patterns. *Applied Optics*, Vol. 46, No. 35, Dec 2007, pp. 8515–8526.

Jhang, K.Y., Shin, M.J. & Lim, B.O. (2006). Application of the laser generated focused Lamb wave for non-contact imaging of defects in plate. *Ultrasonics*, Vol. 44, No. 1, Dec 2006, pp. e1265-e1268.

Kamshilin, A., Miridonov, S.V., Sidorov, I.S., Semenov, D.V. & Nippolainen, E. (2009). Statistics of dynamic speckles in application to distance measurements. *Optical Review*, Vol. 16, No. 2, March 2009, pp. 160-166.

Komatsu, S., Yamaguchi, I. & Saito, H. (1976). Velocity measurement using structural change of speckle. *Optics Communications*, Vol. 18, No. 3, Aug 1976, pp. 314-316.

Lee, D.B. (1969). Anisotropic etching of silicon. *Journal of Applied Physics*, Vol. 40, No. 11, Oct 1969, pp. 4569-4574.

Liu, J.M. (2005). *Photonic Devices*. Cambridge University Press, ISBN: 9780521551953, UK, 2005-04-21.

Lowe, M.J.S. & Diligent, O. (2002). Low-frequency reflection characteristics of the S_0 Lamb wave from a rectangular notch in a plate. *Journal of the Acoustical Society of America*, Vol. 111, No. 1, Jan 2002, pp. 64-74.

Macdonald, D., Sinton, R.A. & Cuevas, A. (2001). On the use of a bias-light correction for trapping effects in photoconductance-based lifetime measurements of silicon. *Journal of Applied Physics*, Vol. 89, No. 5, March 2001, pp. 2772-2778.

Mendicino, M.A. (1999), in *Properties of Crystalline Silicon*. Hull, R. (ed.) (1999), The Institution of Electrical Engineers, ISBN: 0852969333, London, UK, 1999.

Monchalin, J.P. (2007). Laser-Ultrasonics: Principles and Industrial Applications, chapter 4, in: *Ultrasonic and Advanced Techniques for Nondestructive Testing and Material Characterization*, C.H. Chen (Ed), pp. 79-116, World Scientific Publishing Co., ISBN: 978-981-270-409-2, Singapore.

Neamen, D. (2003). *Semiconductor Physics and Devices*, McGraw-Hill, ISBN: 0072321075, New York.

Pei, J., Degertekin, F.L., Khuri-Yakub, B.T. & Saraswat, K.C. (1995). *In situ* thin film thickness measurement with acoustic Lamb waves. *Applied Physics Letters,* Vol. 66, No. 17, April 1995, pp. 2177-2179.

Pierce, S.G., Culshaw, B. & Shan, Q. (1998). Laser generation of ultrasound using a modulated continuous wave laser diode. *Applied Physics Letters,* Vol. 72, No. 9, March 1998, pp. 1030-1032.

Rong, H., Liu, A., Nicolaescu, R. & Paniccia, M. (2004). Raman gain and nonlinear optical absorption measurements in a low-loss-silicon waveguide. *Applied Physics Letters,* Vol. 85, No. 12, Sept 2004, pp. 2196-2198.

Rose, A. (1955). Space-charge-limited currents in solids. *Physical Review,* Vol. 97, No. 6, March 1955, pp.1538-1544.

Rose, A. & Lampert, M.A. (1959). Photoconductor performance, space-charge currents, and the steady-state Fermi level. *Physical Review,* Vol. 113, No. 5, March 1959, pp. 1227-1235.

Scruby, C.B. & Drain, L.E. (1990). *Laser ultrasonics: techniques and applications.* CRC Press, ISBN 0750300507, New York.

Soares, S.F. (1992). Photoconductive gain in a Schottky barrier photodiode. *Japanese Journal of Applied Physics,* Vol. 31, No. 2A, Feb 1992, pp. 210-216.

Sze, S.M. & Ng, K.K. (2007). *Physics of Semiconductor Devices, 3rd Ed.* John Wiley & Sons, ISBN: 978-0-471-14323-9, New York, 2007.

Tu, S.L. & Baliga, B.J. (1992). On the reverse blocking characteristics of Schottky power diodes. *IEEE Tranactions on Electron Devices,* Vol. 39, No. 12, Dec 1992, pp. 2813-2814.

Wu, X.P. & Ko, W.H. (1989). Compensating corner undercutting in anisotropic etching of (100) silicon. *Sensors and Actuators,* Vol. 18, No. 2, June 1989, pp. 207-215.

15

Remote Optical Diagnostics of Nonstationary Aerosol Media in a Wide Range of Particle Sizes

Olga Kudryashova et al.*
Institute for Problems of Chemical and Energetic Technologies SB RAS
Russia

1. Introduction

Polydisperse gas flows with condensed particles suspended therein are widely spread in nature and play an important role in many branches of the modern engineering and technology. There is a necessity for estimating parameters of disperse media in the manufacturing processes, scientific research, and atmosphere sounding. In industry, the result of this estimation can be employed as the quality measure of a product. When designing and distributing multi-phase systems to study the processes of controlling their parameters, it is also needed to determine values of the quantities characterizing such systems.

Technological processes exploit submicron media alongside with those of large and medium sizes. In such a system the particle sizes vary from tenths of nanometers to tenths of micrometers, and the resultant product may possess significantly different properties, even at small size variations of the condensed phase of a substance used. In this case, it becomes necessary to restore the particle size distribution function taking into account the contribution of all the particles present in the medium. Such functionality may be required from a measuring device when selecting modes of operation or designing sprayer units, controlling the quality of various micro- and nanopowders, monitoring the ecological situation of an area, and studying the dustiness of shop floors. Furthermore, the experimental information on disperse aerosol parameters is needed for evaluating the adequacy degree of a mathematical model accepted to describe an actual process and serves as initial data to calculate working processes in certain devices (Zuyev et al., 1986). The overview of the existing devices to investigate aerosol media has however shown that it cannot be realized using the known instruments.

The available techniques of determining the condensed phase dispersiveness can conditionally be divided into two main groups based on: (i) estimating the particle size in samples selected from a medium under examination and (ii) direct noncontact determination of sizes of the particles present in the medium. The sampling methods have gained a wide distribution, but the essential drawbacks of such methods are both the

---

* Anatoly Pavlenko, Boris Vorozhtsov, Sergey Titov, Vladimir Arkhipov, Sergey Bondarchuk, Eugeny Maksimenko, Igor Akhmadeev and Eugeny Muravlev
Institute for Problems of Chemical and Energetic Technologies SB RAS, Russia

introduction of disturbances into a medium under examination and the complexity of ensuring the representativeness of samples collected. To the noncontact methods of studying aerosols are optical techniques related (Table 1).

Method	Problems to solve	Peculiarities
Spectral transparency	Estimation of the distribution function and concentration. The size range of the measured particles when using the probe radiation of the visible spectrum is 0.01–30 μm.	Requires determining the transmission coefficient within a wavelength interval of $\lambda_{min} \leq \lambda \leq \lambda_{max}$ where about 20-30 measurements need to be conducted at equidistant values of λ. The technical implementation involves broad-spectrum optical radiation sources and selective detectors.
Small-angle scattering	Estimation of the distribution function without preliminary information on physical properties of a substance. When using a laser with a wavelength of 0.63 μm, the size range of the measured particles is 2–100 μm.	The method is very demanding of the accuracy in determining the scattering indicatrix; in this connection, the problem of restoring the particle size distribution function may be solved incorrectly. The technical implementation is relatively simple.
Complete indicatrix	Estimation of the distribution function. When a laser with a wavelength of 0.63 μm is used, the size range of the measured particles is 0.2–100 μm.	Requires measurements in the entire range of scattering angles.
Lidar	Evaluation of the microstructure of an aerosol.	The back scattering is very small. Requires the application of an high-sensitive radiation detector.

Table 1. The optical methods to study aerosols

The methods allow the high-speed characterization of a disperse medium directly during the process of its generation or evolution without introducing any changes into an object under study. Various values that are of interest to a researcher under the given process can be estimated from the change in the scattered, attenuated or reflected radiation passed through an aerosol (Arkhipov, 1987).

The main aerosol parameters are the condensed phase dispersiveness and its concentration. In order to measure simultaneously these characteristics in dynamics and consider the evolution process of an aerosol cloud as applied to media of a broad particle size range, it is necessary to employ several different methods combined into a unified measuring system. Following the analysis of the optical methods for measuring the aerosol dispersiveness, a decision was made to utilize the methods of small-angle scattering and spectral transparency. The chosen methods, mutually supplementing each other, permit (i) determining the dispersiveness of a nonsteady heterogeneous system with its high velocity of travel, (ii) conducting measurements at significant background light, (iii) recording particles of diameters between hundreds of nanometers (the spectral transparency method) and several tenths of micrometers (the small-angle method), and (iv) using less detectors with the possibility of their remote location. When classically implemented, these methods however rely on solving the inverse problem that is incorrect and therefore require a modification in a part of both mathematical result treatment and instrumentation, to increase the informativeness of data obtainable from experiments.

2. Modified method of spectral transparency

The study into the disperse parameters of aerosol media containing 1−100 μm particles has led to a modified method of spectral transparency to measure the mean size and concentration. The method is based on measuring spectral coefficients of particle cloud-induced attenuation of laser radiation with a limited set of probe radiation wavelengths and on calculating averaged attenuation efficiency factors of radiation $\overline{Q}$ (Vorozhtsov et al. 1997). This method does not make it possible to determine the particle size distribution function $f(D)$, but is suitable to measure mean particle sizes, particularly the mean volumetric-surface diameter D_{32} , because it has such merits as simple instrumentation and alignment and can diagnose high-temperature two-phase flows and other aerosol media of a high optical density.

The essence of this method consists in solving the inverse problem for the integral equation:

$$\tau_\lambda = \frac{\pi C_n l}{4} \int_0^\infty D^2 Q(D,\lambda,m) f(D) dD \, , \tag{1}$$

where τ_λ - the optical thickness; C_n - the calculated particle concentration; λ - the probe radiation wavelength; Q - the attenuation efficiency factor for single particles; l - the optical length of the probing; D - the particle diameter; $m = n + i\wp$ - the complex refractive index of the particles material, n - the refractive index; $\wp$ - the absorption coefficient; $f(D)$ - the particle size distribution function.

Most of the unimodal particle distributions occurring in the disperse media physics as well as those characteristic of two-phase flows of various substances are described by the gamma distribution that takes the form:

$$f(D) = aD^\alpha \exp(-bD) \, , \tag{2}$$

where a, α and b - the distribution parameters.

By introducing the notion of the averaged attenuation efficiency factor $\overline{Q}$ in the form:

$$Q(\lambda,m) = \frac{\int_0^\infty Q(\lambda,D,m) D^2 f(D) dD}{\int_0^\infty D^2 f(D) dD} \, , \tag{3}$$

and having replaced the calculated concentration C_n by the mass concentration C_m through

$$C_m = C_n \frac{\pi\rho_p}{6} \int_0^{\infty} D^3 f(D) dD \,, \tag{4}$$

we derive an expression for the optical thickness:

$$\tau_{\lambda i} = \frac{1.5 C_m l \bar{Q}(\lambda, m)}{\rho_p D_{32}} \,, \tag{5}$$

where ρ_p – the material density of the particles; D_{32} – the mean volumetric-surface particle diameter that is calculated by the formula:

$$D_{32} = \frac{\int_0^{\infty} D^3 f(D) dD}{\int_0^{\infty} D^2 f(D) dD} \,, \tag{6}$$

The physical model of the method relies on the interaction between laser radiation and a polydisperse medium through the Mie mechanism and on the conservation of the invariance of the averaged attenuation efficiency factor with respect to a form of the particle size distribution function. The correctness of this assumption is governed by the fact that $\bar{Q}$ is defined by integrals from $f(D)$ and, hence, $\bar{Q}$ is insensitive to the behavior of $f(D)$ in the particle size range under consideration (Prishovalko & Naumenko, 1972).

The averaged attenuation efficiency factor under certain conditions is independent on a form of the particle size distribution function $f(D)$ but is determined by the mean volumetric-surface particle diameter D_{32} at a specified λ, $\bar{Q}_\lambda = f(D_{32})$, and is the most important feature determining the optical properties of polydisperse two-phase media.

The problem of estimating particle sizes by the present method reduces to measuring the optical density of a disperse medium at the two wavelengths λ_1 and λ_2 and to calculating averaged attenuation efficiency factors of laser radiation for the same wavelengths.

The relation of the experimentally measured optical thicknesses at the two wavelengths is equal to the relation of averaged attenuation efficiency factors, both representing the particle size function:

$$\frac{\tau_{\lambda i}}{\tau_{\lambda j}} = \frac{\bar{Q}_2\left(D_{32}, m, \lambda_i\right)}{\bar{Q}_1\left(D_{32}, m, \lambda_j\right)} = F_{ij}\left(D_{32}\right), \tag{7}$$

The measurement range of averaged particle sizes depends on selecting probe radiation wavelengths. Thus, at $\lambda_1 = 0.63$ and $\lambda_2 = 3.39$ μm the D_{32} range is 0.5–4 μm.

The developed method employed the three wavelengths, $\lambda_1 = 0.63$, $\lambda_2 = 1.15$ and $\lambda_3 = 3.39$ μm, and the relations of the experimentally measured optical thicknesses at the three wavelengths, $\frac{\tau_{\lambda_2}}{\tau_{\lambda_1}}$, $\frac{\tau_{\lambda_3}}{\tau_{\lambda_1}}$, and $\frac{\tau_{\lambda_3}}{\tau_{\lambda_2}}$.

The averaged efficiency factors for each wavelength are calculated from precise formulae of the Mie theory which have the form of infinite weakly converging series obtained from the rigorous solution of the diffraction problem of electromagnetic fields on a sphere. The plots of the averaged efficiency factors $\bar{Q}_i(D_{32}, m, \lambda_i)$ versus the mean volumetric-surface diameter D_{32} are shown in Figure 1.

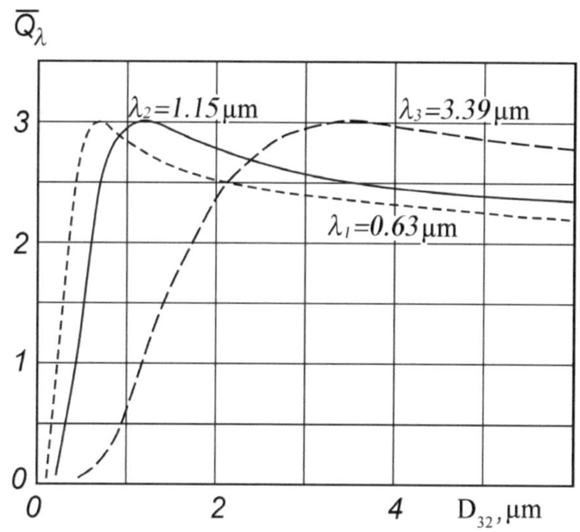

Fig. 1. The averaged attenuation efficiency factors of radiation $\bar{Q}$ plotted against the mean volumetric-surface particle diameter D_{32}

The efficiency factors for single particles were calculated according to the Mie theory using a logarithmic derivative of the Riccati-Bessel function (Deyrmendzhan, 1997). The optical constants for oxide particles were taken from Gyvnak and Burch (1965).

Thus, D_{32} can be directly found from the experimentally measured $\tau_{\lambda i}$ and calculated values of $F(D_{32})$. The particle concentration in the measuring zone is determined by the formula:

$$C_m = \frac{\tau_{\lambda i} \rho_p D_{32}}{1.5 l \bar{Q}(\lambda_i, D_{32})}, \tag{8}$$

where $i = 1;2;3$ on the condition that the optical length of the probing is known or experimentally established, with the values of $\bar{Q}_i$ being determined from the plots (Figure 1).

The practical realization of the multifrequency probing of two-phase media requires solving the problem of selecting radiation wavelengths at which the measurements should be conducted, so that they could be informative with respect to the particle size range (Pavlenko et al., 2005).

The main condition for selecting λ_i during the method implementation is the explicit sensitivity of the dependence of $\bar{Q}$ on the mean volumetric-surface particle diameter D_{32}, with the maximum of the functions $\bar{Q}(D_{32})$ being well-described by the formula:

$$D_{32}^{\circ} = \frac{\lambda}{\pi}\left(1 + \frac{4}{n^2 - 1}\right), \tag{9}$$

where D_{32}° - the mean particle size at which the function $\bar{Q}(D_{32})$ has an extremum; n - the aerosol particle refractive index.

Hence, when the variation range of the mean particles size of a medium under study is assigned *a priori*, the radiation wavelengths must be close to the values corresponding to the following expressions:

$$\lambda_1 = \pi D_{32}^{\min}\left(1 - \frac{4}{n^2 + 3}\right), \tag{10}$$

$$\lambda_2 = \pi D_{32}^{\max}\left(1 - \frac{4}{n^2 + 3}\right), \tag{11}$$

If λ_2 is lower than that determined by formula (11), the experiment will estimate the mean size of a small fraction of particles of a studied medium; and if λ_1 is greater that that determined by formula (10), the measured mean size will characterize a large fraction of the desired distribution.

For the selected wavelengths λ_1 and λ_2, the dependence $\bar{Q}(D_{32})$ has been calculated and the relation $\bar{Q}_{\lambda_2} / \bar{Q}_{\lambda_1} = F(D_{32})$ determined. Table 2 lists data on the relations of the averaged attenuation efficiency factors of radiation $\bar{Q}_{\lambda_2} / \bar{Q}_{\lambda_1} = F(D_{32})$ for various wavelength pairs, depending on D_{32}, from which the limits of measuring the mean particle sizes are established.

In the given case, the maximal measurable particle size D_{32} will be determined by a value at which the function $F(D_{32})$ has a maximum. On the understanding that $\bar{Q}$ is unambiguously dependent on D_{32} only in some interval of particle sizes, generally from zero up to the maximal value (Figure 1), and is an analog of the dependence of the efficiency factor Q on the particle diameter D for an arbitrary particle size distribution, $f(D)$, it follows that particles whose size exceeds $D_{32}^{\max}$ out of the whole range of the disperse medium particle sizes are excluded from consideration. Here, we can introduce a notion of the active particle fraction in the distribution $f(D)$ for which there exists a strongly pronounced dependence of Q on D at different λ.

Thus, the laser measurement of the dispersiveness of two-phase media ascertains the mean size of the active particle fraction starting from zero up to some value that is dependent on the selection of λ_2 and at which the function $F(D_{32})$ reaches the maximum.

D_{32}, μm	F_{21}	F_{31}	F_{32}
1.5	1.117	0.648	0.580
2.0	1.116	0.946	0.847
2.5	1.100	1.139	1.031
3.0	1.125	1.270	1.128
3.5	1.087	1.308	1.203
4.0	1.064	1.296	1.218
4.5	1.089	1.310	1.203

D_{32}, μm	F_{21}	F_{31}	F_{32}
5.0	1.065	1.281	1.202
5.5	1.089	1.303	1.196
6.0	1.063	1.254	1.180
6.5	1.067	1.231	1.152
7.0	1.099	1.226	1.114
7.5	1.056	1.222	1.157
8.0	1.050	1.211	1.151

Table 2. The averaged attenuation efficiency factors for each wavelength: $\lambda_1 = 0.63$ μm, $\lambda_2 = 1.15$ μm, $\lambda_3 = 3.39$ μm

Based on the calculations of the dependences $\bar{Q}(D_{32})$ and $F(D_{32})$ for the corresponding wavelengths, the measurement ranges of D_{32} ($D_{32}^{\min}$, $D_{32}^{\max}$) and the active particle fraction range ($D^{\min}$, $D^{\max}$) of the studied distributions $f(D)$ were determined and are presented in Table 3.

λ_1, μm	λ_2, μm	$D_{32}^{\min}$, μm	$D_{32}^{\max}$, μm	$D^{\min}$, μm	$D^{\max}$, μm
0.63	1.15	0	1.5	0	1.5
0.63	3.39	0	3.5	0	4.0
1.15	3.39	0.1	3.6	0.1	4.0
0.63	5.30	0.8	6.5	0	8.0

Table 3. Ranges of the determinable active fractions of particles

The theory of the modified method of spectral transparency has been well elaborated so far for disperse media that are characterized by the unimodal particle size distribution. However, as it follows from the literature sources, in some cases like heterogeneous combustion, plasma spraying, local man-made aerodisperse systems, there is a possibility for a bimodal particle size distribution.

In order to expand the modified method of spectral transparency for investigating disperse systems of a bimodal particle size distribution, a mathematical model of calculating characteristics of the laser radiation attenuation by such media has been devised (Potapov & Pavlenko, 2000).

From the physical point of view, bimodal disperse media can correctly be described in the form of the following analytical dependence:

$$f(D) = aD^{\alpha}\exp(-bD^{\beta}) + c\exp[-p(D-q)^2], \tag{12}$$

where a, c, b, p, q, α, β – the distribution parameters. Dependence (12) is a sum of the generalized gamma distribution and normal distribution. The position of the first maximum is at point $D_0^{(1)} = \alpha/b$ and that of the second – $D_0^{(2)} = q$.

In the calculation of $\overline{Q}(D_{32})$, a series of D_{32} values and the corresponding distributions of $f(D)$ are preliminary assigned because the analytical dependence of D_{32} on the parameters of distribution (12) cannot be established.

What is important is to assign a form of the bimodal distribution, its determinative characteristics, and the variation mechanism of $f(D)$ when scanning the argument of the function $\overline{Q}(D_{32})$.

The following model of assigning $f(D)$ seems appropriate:

- the form of the bimodal particle size distribution function is determined by such characteristics as:

$$V = D_0^{(2)} - D_0^{(1)},\ W = f\left[D_0^{(1)}\right] / f\left[D_0^{(2)}\right]; \tag{13}$$

- the variation of $f(D)$ upon scanning the argument of the function $\overline{Q}(D_{32})$ occurs on the condition that V and W remain constant.

Let the parameters α, β and q in equation (12) be represented as $\alpha = vk$, $b = 1$, $q = vk/b + V$, where $k = 1, 2, 3 \ldots z$. From the condition of normalization and from formula (13) was a system of equations relative to the parameters a and b in formula (12) derived

$$a(k)E^{11} + b(k)E^{12} = E^{13}, \tag{14}$$

$$a(k)E^{21} + b(k)E^{22} = E^{23}, \tag{15}$$

where $E^{23} = 1$; $E^{13} = 0$, a E^{11}, E^{12}, E^{21} and E^{22} are calculated by the formulae:

$$E^{11} = (vk/b)^{vk} \exp(-vk) - W\left(V + vk/b^{vk} \exp\left[-(bV + vk)\right]\right), \tag{16}$$

$$E^{12} = \exp\left(-pV^2\right) - W, \tag{17}$$

$$E^{21} = \int_0^{\infty} D^{vk} \exp(-bD)\,dD, \tag{18}$$

$$E^{22} = \int_0^{\infty} \exp\left[-p\left(D - (vk/b - V)\right)^2\right] dD, \tag{19}$$

The parameters $a(k)$ and $b(k)$ obtained from solving equations (14) - (15) are used to calculate $f_k(D)$, and $D_{32}(k)$. The found values of $f_k(D)$ are employed in software to compute $\overline{Q}$.

For the bimodality characteristics $V = 1 \div 2$ and $W = 0.25 \div 4$, $\overline{Q}(D_{32})$ was calculated from the suggested model. The character of the dependence $\overline{Q}(D_{32})$ for bimodal and unimodal distributions was shown to be identical. In this case, the maximal value of $\overline{Q}$ is kept at the corresponding value of D_{32} established from the theory:

$$D_{32} = \lambda\left[1 + 4/\left(n^2 - 1\right)\right]/\pi . \tag{20}$$

This indicates that the model of calculating $\overline{Q}(D_{32})$ for bimodal particle size distributions may be regarded as correct.

When the modified method of spectral transparency was applied, the error in determining the particle sizes was established to grow due to the uncertainty of the particle size distribution class. To enhance the accuracy of the particle size determination, the probing of disperse media requires *a priori* information on a class of particle size distribution, either bimodal or unimodal, and requires the established invariance of $\overline{Q}$ as a function of $f(D)$ within the limits of a certain distribution.

In the diagnostics of bimodal disperse media of a strongly pronounced functionality, when particle fractions are at a great distance from each other, it is possible to implement various mechanisms of the interaction between probing radiation and particles, on the basis of which the particle size of each separate fraction can be estimated. The bimodal distribution can be represented in the form:

$$f(D) = f_1(D) + f_2(D) , \tag{21}$$

where $f_1(D)$ and $f_2(D)$ - the first and second functions that describe small and large particle fractions having the modal diameters $D_0^{(1)}$ and $D_0^{(2)}$ and the mean volumetric-surface diameters $D_{32}^{(1)}$ and $D_{32}^{(2)}$, respectively.

From the analysis of the range of the $\overline{Q}$ dependence on D_{32} for various wavelengths, formulae for selecting the probe radiation wavelengths were derived:

$$\lambda_{i1} = D_{32}^{(i)}/\pi , \tag{22}$$

$$\lambda_{i2} = \lambda_{i1}\left(5 - D_{32}^{(i)}/4\right), \tag{23}$$

where $i = 1 , 2$ - the number of particle size distribution functions.

With such a choice of wavelengths, there occurs a fractional interaction between the radiation and the polydisperse particles, and, in contrast to the common techniques, three scattering mechanisms are immediately brought about: Rayleigh scattering, Mie scattering, and scattering by large particles.

The interaction of the radiation of the wavelengths $\lambda = \lambda_{1j}$, where $j = 1,2$ is the number of the radiation wavelengths needed for probing a separate fraction, with the first $f_1(D)$ and the second fractions $f_2(D)$ results in the Mie scattering and the scattering by large particles, respectively. When the radiation having the wavelengths $\lambda = \lambda_{2j}$ interacts with the first and second fractions, there occurs the Rayleigh scattering and Mie scattering, respectively.

Thus, when the radiation passes at the four wavelengths λ_{ij} through a bimodal disperse medium, the optical thicknesses τ are expressed by the following dependences:

$$\tau_{1j} = \frac{1,5l}{\rho_p}\left[M^{(1)}\frac{Q(\lambda_{1j})}{D_{32}^{(1)}} + M^{(2)}\frac{2}{D_{32}^{(2)}}\right], \tag{24}$$

$$\tau_{2j} = \frac{1,5l}{\rho_p}\left[M^{(1)}\frac{J}{\lambda_{2j}} + M^{(2)}\frac{Q(\lambda_{2j})}{D_{32}^{(2)}}\right], \tag{25}$$

where

$$J = \frac{24\pi n \wp}{(n^2 - \wp^2 + 2)^2 + 4n^2\wp^2}. \tag{26}$$

The mean volumetric-surface diameters $D_{32}^{(i)}$ and particle mass concentrations $M^{(i)}$ for each separate fraction are determined from expressions (24) – (26).

To determine disperse characteristics of bimodal heterogeneous media of a strongly pronounced functionality, Ar, He–Ne and CO_2 lasers as well as collimated radiation-based semiconductor lasers can be employed.

To experimentally test the devised method of spectral transparency, the mean particle sizes of the combustion products produced from a model nozzle-free generator were estimated (Arkhipov et al. 2007).

The tests showed that the mean volumetric-surface diameter of the particles is 0.8–2.7 μm. Figure 2 illustrates typical experimental data on the dependences obtained.

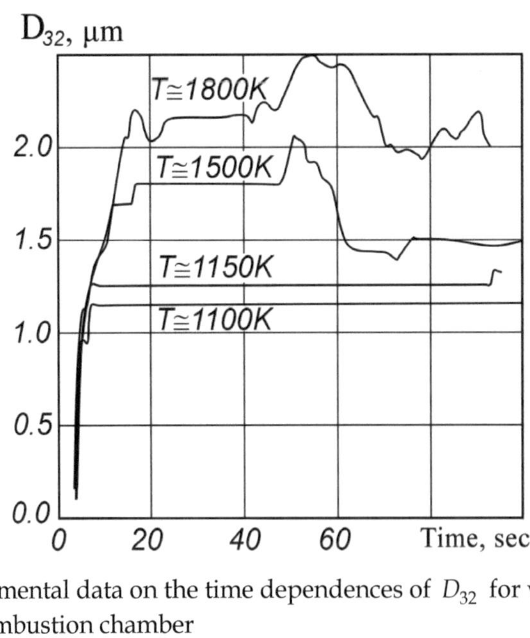

Fig. 2. Typical experimental data on the time dependences of D_{32} for various average temperatures in a combustion chamber

The nonmonotonic behavior of the dependences for higher temperatures is governed by a strong nonstationarity of intraballistic parameters under these conditions.

Data on the particle sizes are qualitatively consistent with the calculation results for particle size distribution spectra at the exit from the nozzle-free fire channel.

3. High-selective optical integral method of estimating the particle size distribution function

A new optical technique has been developed for determining the particle size distribution function in aerosol media with a particle size range of 0.1−5 µm and is based on the spectral transparency method.

The classical method of spectral transparency to estimate the particle dispersiveness relates to inverse problems of aerosol optics (Arkhipov, 1987). The stable solution of the problem of determining aerosol parameters by the spectral transparency method is possible if the transparency is ascertained over the whole wavelength range (Shifrin, 1971). However, nowadays devices can measure the spectral transparency coefficient only at several wavelengths or in some region of wavelengths without selectivity, which undoubtedly reduces accuracy and capabilities of the method (Shaikhatarov et al., 1986).

The method of spectral transparency relies on measuring the spectral transmission factor of optical radiation of a two-phase flow in some range of wavelengths. The initial equation for the method of spectral transparency is as follows (Van de Hulst, 1961):

$$I(\lambda,t)=I_0(\lambda)\exp\left[-\frac{\pi C_n(t)l(t)}{4}\int_0^\infty Q\left(\frac{\pi D}{\lambda},m(\lambda)\right)D^2 f(D)dD\right], \quad (27)$$

where $I(\lambda,t)$ – the intensity of the radiation passed through the aerosol; $I_0(\lambda)$ – the probe radiation intensity; $C_n(t)$ – the calculated concentration of the aerosol condensed phase; $l(t)$ – the optical path length; $Q\left(\frac{\pi D}{\lambda},m(\lambda)\right)$ – the attenuation efficiency factor of the probe radiation; $m(\lambda)$ – the complex refractory index of the aerosol condensed phase.

The problem of finding the aerosol particle diameter distribution $f(D)$ from expression (27) by the known radiation intensity values ($I(\lambda)$ and $I_0(\lambda)$) is incorrect. The said incorrectness is unavoidable for the spectral transparency method. It is suggested that the problem be solved using regularizing algorithms.

The regularizing algorithms are built in such a way as to attract additional *a priory* (with respect to the experiment on the basis of which the inverse problem is set) information on the desired function. With the aid of such information (details on smoothness of the desired solution, its monotoneness, convexity, pertain to the finitely parametric family, etc), such a solution is selected which is close to the true one in some specified sense. The regularizing algorithms allow a stable approximation to the true solution of an incorrect problem, meaning that the approximate solution tends to the true one as the measurement error goes to zero.

One of the regularization ways is the parametrization technique. For the given situation, it will consist in the fact that a form of distribution function (2) is believed to be known *a priori*, in which case the parameters α and b are assigned by the coordinate-wise descent method, and the normalizing coefficient is to be found from the expression:

$$a=\left[\int_{D_{\min}}^{D_{\max}} D^{\alpha}\exp(-bD)dD\right]^{-1}, \tag{28}$$

where $D_{\min}$ and $D_{\max}$ are the minimal and maximal diameters of the particles present in a medium under examination, respectively. Gamma distribution (2) for describing the particle size distribution was chosen in view of its universality as applied to aerosols, with one mechanism of producing a disperse phase (Arkhipov et al., 2006). The attenuation efficiency factor of the probe radiation ($Q\left(\frac{\pi D}{\lambda},m(\lambda)\right)$) is calculated by the formula using specified parameters of the distribution function:

$$Q\left(\frac{\pi D}{\lambda},m(\lambda)\right)=\frac{2}{\left(\frac{\pi D}{\lambda}\right)^{2}}\sum_{n=1}^{\infty}(2n+1)\mathrm{Re}(a_n+b_n), \tag{29}$$

where a_n and b_n are the Mie coefficients and are calculated by the formula:

$$a_n\left(\frac{\pi D}{\lambda},m(\lambda)\right)=\frac{\left[\frac{A_n\left(\frac{\pi D m(\lambda)}{\lambda}\right)}{m(\lambda)}+\frac{n}{\left(\frac{\pi D}{\lambda}\right)}\right]\mathrm{Re}\left[\xi_n\left(\frac{\pi D}{\lambda}\right)\right]-\mathrm{Re}\left[\xi_{n-1}\left(\frac{\pi D}{\lambda}\right)\right]}{\left[\frac{A_n\left(\frac{\pi D m(\lambda)}{\lambda}\right)}{m(\lambda)}+\frac{n}{\left(\frac{\pi D}{\lambda}\right)}\right]\xi_n\left(\frac{\pi D}{\lambda}\right)-\xi_{n-1}\left(\frac{\pi D}{\lambda}\right)}, \tag{30}$$

$$b_n\left(\frac{\pi D}{\lambda},m(\lambda)\right)=\frac{\left[m(\lambda)A_n\left(\frac{\pi D m(\lambda)}{\lambda}\right)+\frac{n}{\left(\frac{\pi D}{\lambda}\right)}\right]\mathrm{Re}\left[\xi_n\left(\frac{\pi D}{\lambda}\right)\right]-\mathrm{Re}\left[\xi_{n-1}\left(\frac{\pi D}{\lambda}\right)\right]}{\left[m(\lambda)A_n\left(\frac{\pi D m(\lambda)}{\lambda}\right)+\frac{n}{\left(\frac{\pi D}{\lambda}\right)}\right]\xi_n\left(\frac{\pi D}{\lambda}\right)-\xi_{n-1}\left(\frac{\pi D}{\lambda}\right)}, \tag{31}$$

where A_n and ξ_n are in recurrence relationships:

$$\xi_n\left(\frac{\pi D}{\lambda}\right)=\frac{2n-1}{\left(\frac{\pi D}{\lambda}\right)}\xi_{n-1}\left(\frac{\pi D}{\lambda}\right)-\xi_{n-2}\left(\frac{\pi D}{\lambda}\right), \qquad (32)$$

$$\xi_0\left(\frac{\pi D}{\lambda}\right)=\sin\left(\frac{\pi D}{\lambda}\right)+i\cos\left(\frac{\pi D}{\lambda}\right), \qquad (33)$$

$$\xi_{-1}\left(\frac{\pi D}{\lambda}\right)=\cos\left(\frac{\pi D}{\lambda}\right)-i\sin\left(\frac{\pi D}{\lambda}\right), \qquad (34)$$

$$A_n\left(\frac{\pi Dm(\lambda)}{\lambda}\right)=-\frac{n}{\left(\frac{\pi Dm(\lambda)}{\lambda}\right)}+\left[\frac{n}{\left(\frac{\pi Dm(\lambda)}{\lambda}\right)}-A_{n-1}\left(\frac{\pi Dm(\lambda)}{\lambda}\right)\right]^{-1}, \qquad (35)$$

$$A_0\left(\frac{\pi Dm(\lambda)}{\lambda}\right)=ctg\left(\frac{\pi Dm(\lambda)}{\lambda}\right). \qquad (36)$$

The limit of summation in expression (29) comes when the condition is fulfilled:

$$\left[\sum_{n=1}^{N}(2n+1)\mathrm{Re}(a_n+b_n)\right]\cdot 10^{-8}>(2(N+1)+1)\mathrm{Re}(a_{N+1}+b_{N+1})\cdot \qquad (37)$$

To solve the problem, the spectral transparency coefficient is used as experimental information and is calculated by the formula:

$$\tau_\lambda^e(t)=\ln\frac{I_0(\lambda)}{I(\lambda,t)}. \qquad (38)$$

Afterwards, there is found a relation of the spectral transparency coefficients obtained from experiments for some wavelengths λ_1 and λ_2:

$$k_e(t)=\frac{\tau_{\lambda_1}^e(t)}{\tau_{\lambda_2}^e(t)}. \qquad (39)$$

Then, a relation of the theoretically obtained coefficients of spectral transparency for various distribution functions (2) is calculated, in accordance with equation (27) for the wavelengths λ_1 and λ_2, by the formula:

$$k_d(t)=\frac{\tau_{\lambda_1}^d(t)}{\tau_{\lambda_2}^d(t)}=\frac{\int_{D_{\min}}^{D_{\max}}Q\left(\frac{\pi D}{\lambda_1},m(\lambda)\right)D^2f(D)dD}{\int_{D_{\min}}^{D_{\max}}Q\left(\frac{\pi D}{\lambda_2},m(\lambda)\right)D^2f(D)dD}. \qquad (40)$$

The cumulative departure of the experimental data $k_e(t)$ from the theoretical $k_d(t)$ is further calculated for all the wavelengths used. In addition, mathematical studies showed that the selection method for pairs of the wavelengths λ_1 and λ_2 has no affect on the accuracy and performance stability of the algorithm suggested and, hence, any selection method can be employed. The comparison of the relations of the spectral transparency coefficients but not their absolute values was undertaken to get rid of the constant factor before the integral sign in expression (27) when calculating theoretical values of the transmission factors of a medium under study (40).

The final step is selecting such a form of the particle size distribution function wherein the departure of the experimental data from the theoretical is minimal.

The direct problem of estimating the spectral coefficient of the probe radiation attenuation for various parameters of the distribution function can thus be solved by the numerical methods.

The optical path length (l) of the probe radiation (say, with high-speed video shooting (Titov & Muravlev, 2008)) in a studied aerosol is then determined, and the mass concentration of the aerosol dispersed phase is calculated by the formula:

$$C_m(t) = \frac{\tau_\lambda^e(t)\rho_p D_{32}}{1.5l(t)\bar{Q}\left(\frac{\pi D}{\lambda}, m(\lambda)\right)}, \tag{41}$$

$$\bar{Q}\left(\frac{\pi D}{\lambda}, m(\lambda)\right) = \frac{\int_{D_{\min}}^{D_{\max}} Q\left(\frac{\pi D}{\lambda}, m(\lambda)\right) D^2 f(D) dD}{\int_{D_{\min}}^{D_{\max}} D^2 f(D) dD}. \tag{42}$$

To verify the operability of the method devised, an experiment to measure the attenuation of the optical radiation by a suspension of chemically pure submicron Al_2O_3 powder in distilled water was performed. Aluminum oxide was chosen because empirical dependences of refractive and absorption indices on probe radiation wavelength for this substance are known (Dombrovskiy, 1982). Neglecting the dependence and using constants as refractive and absorption indices for all the wavelengths was shown by preliminary experiments to result in errors.

The suspended aluminum oxide was exposed to ultrasound in order to grind the resultant agglomerates, following which the whole was placed into a glass cuvette for examination. Prior to measurement, the cuvette was left to stand undisturbed to settle down ungrindable agglomerates. The cuvette that is part of the measuring complex is displayed in Figure 3. The numbers in Figure 3 denote: 1 - glare shield, 2 - collimator outlet, 3 - cuvette, 4 - condenser, 5 - optical waveguide, 6 - direction of optical radiation passage. All the dimensions in Figure 3 are given in millimeters. The optical radiation path length in the medium studied (distance between the inner surfaces of the cuvette glasses) was 5.075 mm.

Before measuring the radiation transmission through the cuvette with suspended aluminum oxide, the radiation transmission through the cuvette filled with pure distilled water was estimated. The transmission spectrum of the cuvette with pure distilled water was used as the reference; it is necessary for all variations of the transmission spectrum of the suspension with respect to the reference spectrum to be caused only by the action of aluminum oxide nanopowder. The reference spectrum and the transmission spectrum of the studied suspension are shown in Figure 4.

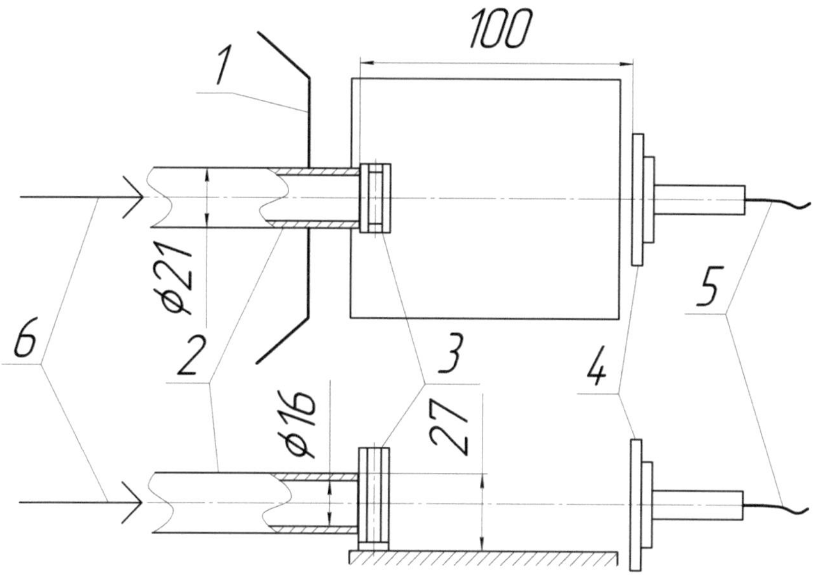

Fig. 3. The cuvette compartment of the measuring complex

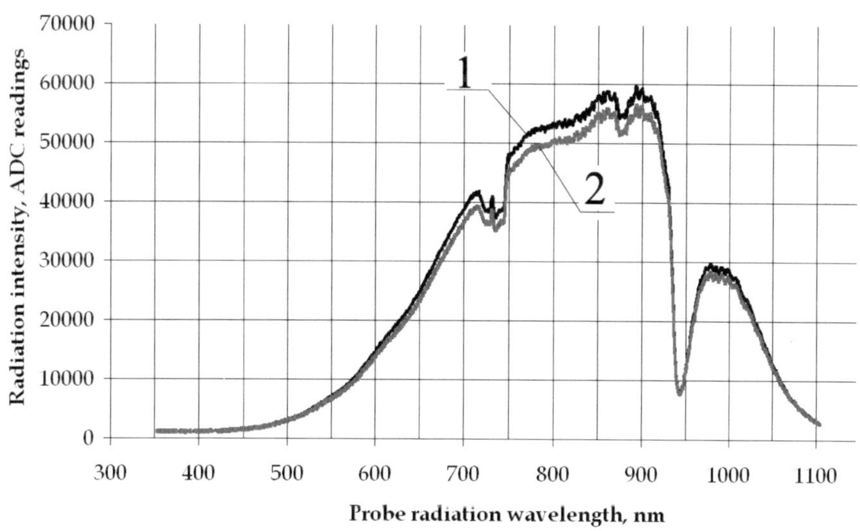

Fig. 4. The optical radiation transmission spectra: 1 - cuvette transmission spectrum with pure distilled water; 2 - cuvette transmission spectrum with Al_2O_3 suspension

From the spectral data presented was the spectral transparency coefficient calculated by formula (38); the resultant dependence of the spectral transparency coefficient on the wavelength is displayed in Figure 5. The moving average method was applied to the dependence shown in Figure 5 in order to eliminate high-frequency noises, resulting in dependence no.1 in Figure 6. The theoretical dependence of the spectral transparency coefficient on the probe radiation wavelength is no.2 in Figure 6, its difference from the experimental data appearing to be minimal.

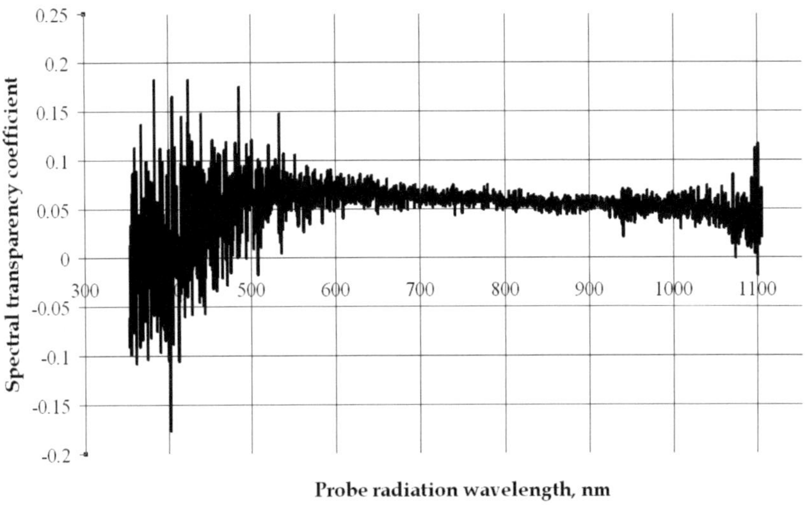

Fig. 5. Experimental spectral transparency coefficient

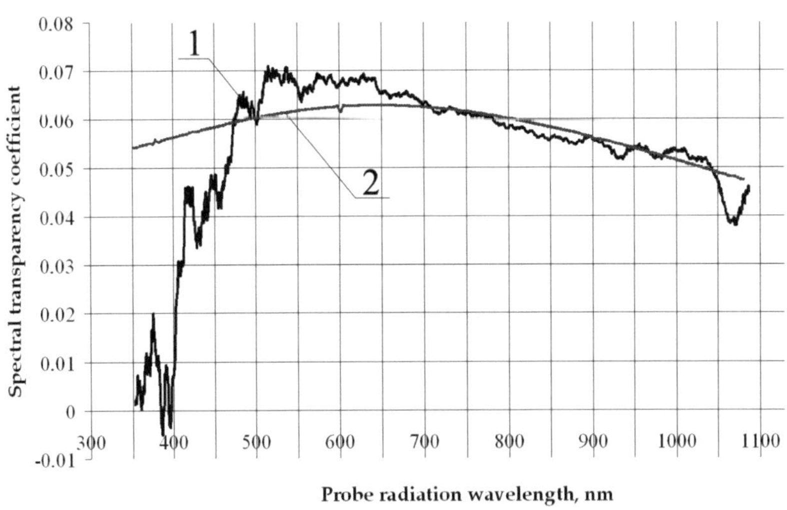

Fig. 6. Experimental and theoretical dependences of the spectral transparency coefficients on the probe radiation wavelength: 1 – experiment; 2 – calculation

The strong difference of the experimental data from the theoretical in the region of wavelengths up to 480 nm is due to the low-level wanted signal therein, as is well illustrated in Figure 4, which nearly coincides with the noise value. This can also be seen from the untreated dependence of the spectral transparency coefficient on the wavelength in Figure 5. This discrepancy can therefore be considered acceptable since instrumental implementation peculiarities are concerned regarding the measurement of optical radiation attenuation, and it does not characterize the algorithm applied.

The found theoretical dependence of the spectral transparency coefficient on the wavelength (curve 2, Figure 6) is consistent with the particle size distribution function shown in Figure 7.

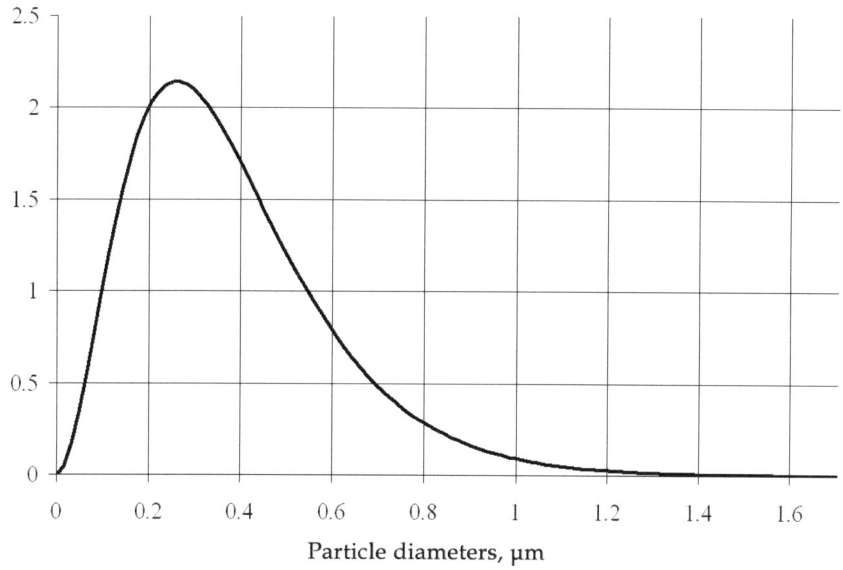

Fig. 7. The restored Al_2O_3 particle size distribution function

The particle size distribution function obtained (Figure 7) is in good agreement with the known data on the aluminum oxide nanopowder studied.

4. Modified method of small-angle scattering

To investigate the aerosol genesis from 2 to 100 µm, a method based on estimating disperse parameters of aerosols from the measured scattered radiation passed through a studied volume has been devised.

The method consists in finding the aerosol particle size range from the measured scattering indicatrix, by means of searching the corresponding parameters of the distribution function. Gamma distribution (2) was accepted as the reference function (Arkhipov, 1987).

Laser beam LS (Figure 8) propagates through a scattering layer with boundaries "1" and "2" to produce some illuminance on plane Y. The distance between the laser and the first boundary of the layer is l_1. The aerosol cloud particles present in the beam scatter radiation. As a consequence, the irradiance of plane Y is determined not only by the direct beam attenuated due to absorption and scattering, but also by the radiation scattered by the particles.

On the assumption of the uniform distributions of concentration and particles sizes in the aerosol cloud, the equation for scattered radiation flux coming on plane Y has the following form:

$$I(y)=\frac{\pi S}{4}\int_{0}^{z}\left[I_0(x)B(x,y)F(x)\right]dx\,. \tag{43}$$

where $F(x)=\int_{0}^{\infty}Q_S(D,\theta(x))D^2 f(D)dD$; $I_0(x)=I_0\exp[-xC_nQ_{ext}]$ - the intensity of the radiation falling on point x; I_0 - the initial radiation intensity; x - the distance between boundary "1" of the scattering layer and point P; Q_{ext} - the attenuation coefficient.

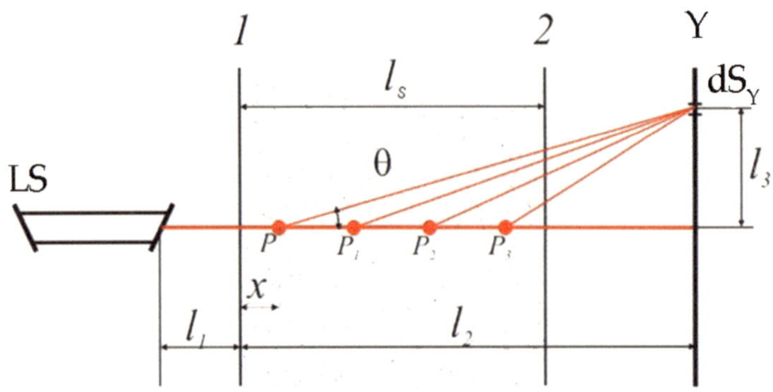

Fig. 8. Diagram of interaction between radiation and aerosol

The radiation scattered from a single particle in the small-angle region under the assumption that the particles are spherical is defined as an analytical dependence in the following form:

$$Q_S(\rho,\theta)=\frac{\rho^2}{4\pi}\cdot\left[\frac{2J_1(\theta\rho)}{\theta\rho}\right]^2, \tag{44}$$

where $\rho=\frac{\pi D}{\lambda}$ - the diffraction parameter (Mie parameter); θ - the radiation scattering angle; $J_1(\theta\rho)$ - the first-order Bessel function of the first kind.

The multiplier $B(x)$ that takes into account the scattered radiation attenuation pursuant to the Buger law is defined by the relation:

$$B(x,y)=\exp\left[-Q_{ext}\frac{z-x}{\cos\theta(x,y)}\right], \tag{45}$$

where $\theta(x,y)=arctg\left(y/(l_2-x)\right)$; $y=l_3$.

The estimation of $f(D)$ from the measured scattering indicatrix $I_e(y)$ reduces to searching the parameters $\{\alpha, b\}$ of distribution (2) and calculating the functional:

$$\Omega = \min_{a,b}\left\{\sum_{i=1}^{n}\left|I_e(y_i) - I(y_i)\right|\right\}, \tag{46}$$

where $I_e(y_i)$ ($i = 1, 2, ..., n$) - the measured values of the scattering indicatrix for discrete values of the detector positions; $I(y_i)$ - the values calculated from (43).

It is possible to restore the droplet size distribution function with a sufficient accuracy if the condition is fulfilled (Gritsenko & Petrov, 1979; Belov et al., 1984):

$$\ln\frac{I_0}{I} \le 1.5, \tag{47}$$

where I_0, I - the illuminance in the central beam before and after its passing through the scattering volume.

To implement the title method, a laser setup has been assembled which comprises (Figure 9): a 1 m³ measuring chamber; a radiator – HeNe laser with 0.632 µm wavelength and 5 mW power; a recording unit consisted of 8 photodiodes placed on the same array; a measuring 8-channel amplifier; ADC and PC; software to record and process measurement information in order to determine calculated and mass functions of particle size distribution, mean volumetric-surface diameter, and aerosol particles concentration.

The laser radiation is 90° angle oriented toward one of the volume faces, 80 Hz frequency modulated, and directed through a scattering medium.

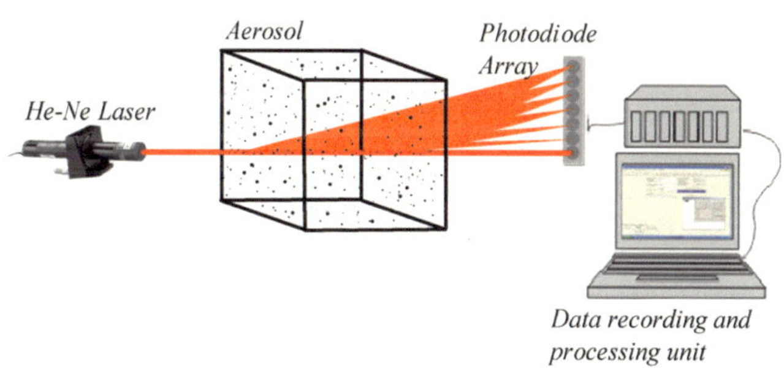

Fig. 9. The laser setup diagram

The optical radiation flux scattered at different angles is recorded by the photodiode array that is located in the plane perpendecular to the laser beam. The photodiode array enables recording the scattered radiation at angles of 0.3–20° relative to the laser beam.

The optical path length is estimated with the aid of a video camera or is set to a fixed value equal to a measuring volume for a steady-state generation process of a two-phase flow.

The aluminum powder measurement results obtained using the setup are collected in Table 4; Figure 10 shows mass functions of the particle size distribution.

Measurement	Distribution coefficient α	Distribution coefficient b	D_{32}, μm	D_{43}, μm
1	1.14	0.7	5.91	7.3
2	0.3533	0.55	6.09	7.91
3	0.53	0.53	6.66	8.54
4	1.24	0.64	6.6	8.2
5	1.1	0.7	5.8	7.2

Table 4. The laser method study results for the fine aluminum powder

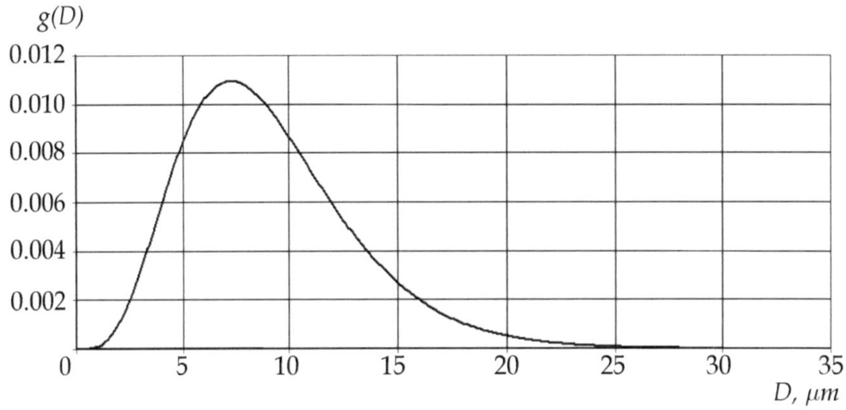

Fig. 10. The mass particle size distribution for the fine aluminum powder

The calculated distribution function can readily be converted into the mass distribution function by the formula (Arkhipov et al., 2006):

$$g(D) = \frac{m}{m_0} f(D), \tag{48}$$

where m - the weight of the particles of size D; $m_{10} = \int_0^\infty mf(D)dD$.

The spread of the aluminum powder dispersiveness estimation results obtained using different methods is not greater than 15%, as demonstrated in Table 5.

An aerosol water cloud produced by pulsed generation was also studied. The results for water and an aqueous glycerol solution are given in Figures 11–13.

	Fine aluminum powder	
	D_{32}, μm	D_{43}, μm
Sieve analysis	7.2	8.3
Microscope	5.7	6.44
Modified method of small-angle scattering	6.2	7.45

Table 5. Comparison of the results of different methods

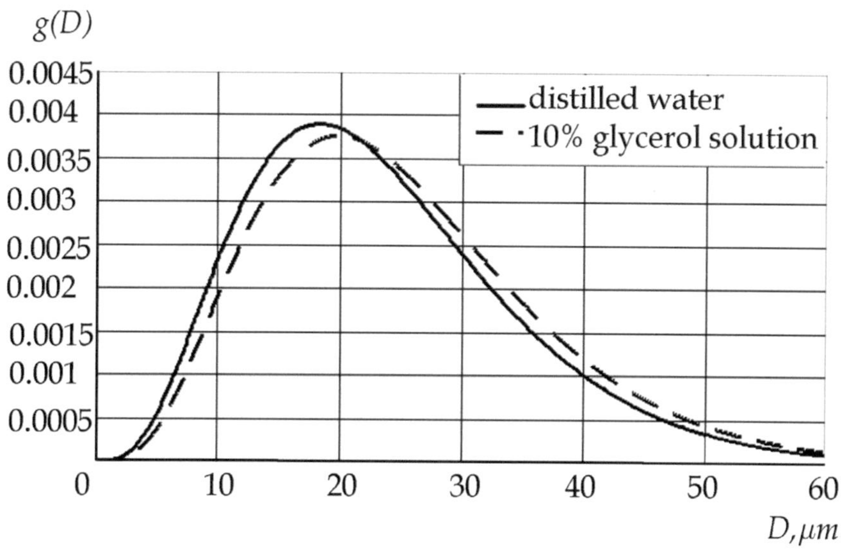

Fig. 11. Mass droplet distribution after spraying (time, 1 s)

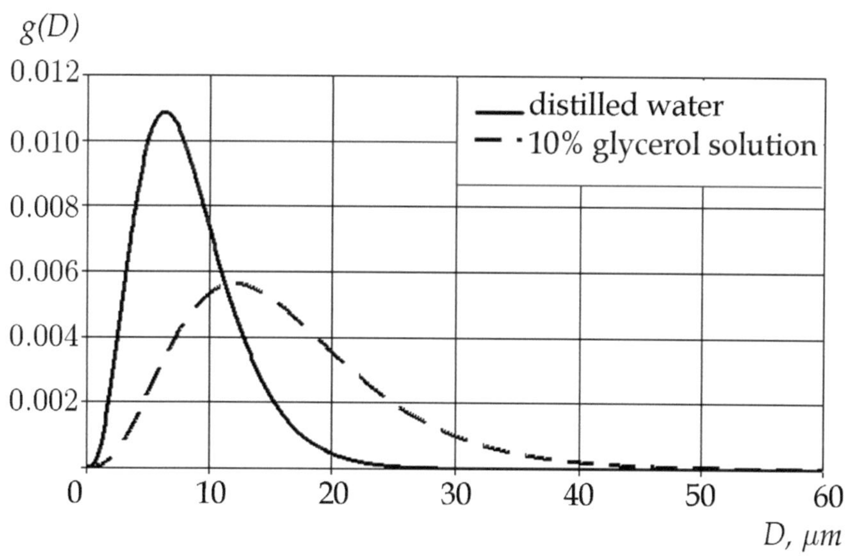

Fig. 12. Mass droplet distribution after spraying (time, 6 s)

The modified small-angle scattering method developed makes it possible to restore the particle size distribution function even from three measured points of the scattering indicatrix.

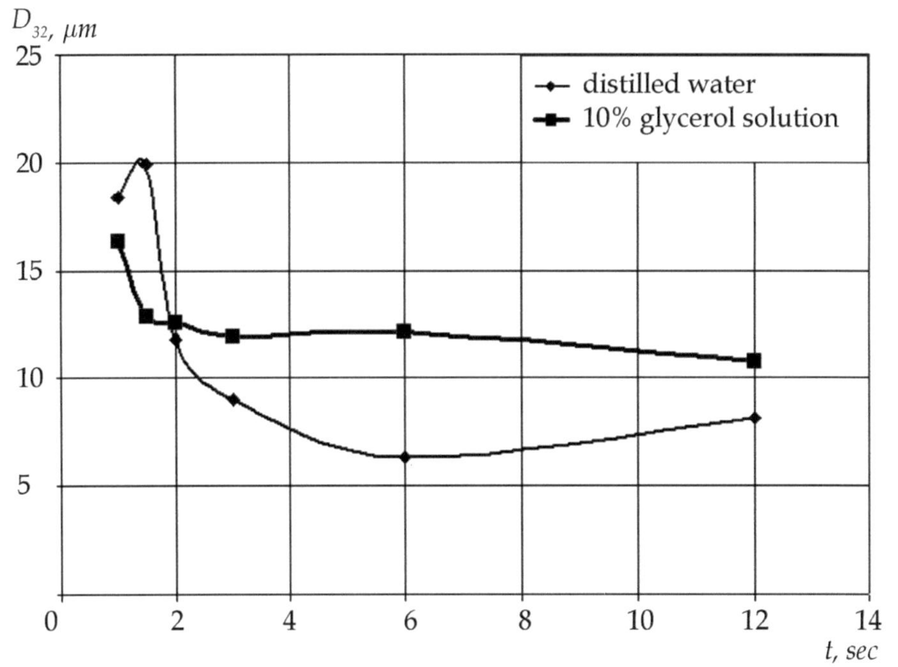

Fig. 13. The change in the mean size D_{32} of water and 10% glycerol solution droplets as a function of the aerosol cloud generation time

5. Conclusion

The optical diagnostics methods for heterogeneous flows have been reviewed and analyzed as the most promising with respect to measuring the dispersiveness and concentration of aerosol particles in dynamics, including the cases of high-velocity flows, significant background illuminance, and similar phenomena accompanying technological processes and affecting measuring instrumentation.

The authors have suggested and implemented a modified method of spectral transparency for estimating the aerosol dispersiveness, which is distinct both in simplicity of instrumental implementation and in possibility of diagnosing two-phase flows of a high optical density.

A new method of determining the particle size distribution function has been developed and is based on the classical method of spectral transparency using information on the probe radiation attenuation in a wide range of wavelengths; the operability of the method has been verified.

The inverse problem of aerosol optics has mathematically been solved for the small-angle scattering method in a part of results processing in order to eliminate incorrect solutions when restoring the particles size distribution function; the possibility of studying the dynamics of generation and propagation of aerosols using the algorithm suggested has been demonstrated.

The developed measuring complex employs data recording and processing algorithms obtained for the first time and combines the modified method of spectral transparency, the high-selective optical integral method of estimating the particle size distribution function, and the modified method of small-angle scattering. The combined application of the methods devised makes it possible to estimate disperse parameters of aerosol media of any nature and of a wide particle size range (0.1–100 µm) under conditions of their high-velocity genesis with a high time resolution. The measuring complex permits studying the genesis of two-phase flows and evaluating the effects on the fractional composition by such processes as particle coagulation, sedimentation, and evaporation.

6. References

Arkhipov, V. (1987). Laser diagnostics methods of heterogeneous flows (in Russian), Tomsk University Press, Tomsk

Arkhipov, V.; Akhmadeev, I.; Bondarchuk, S.; Vorozhtsov, B.; Pavlenko, A. & Potapov, M. (2007). A modified method of spectral transparency to measure aerosol dispersiveness (in Russian), *Atmospheric and Oceanic Optics*, Vol. 20, No. 1, pp. 48–52, ISSN 0869-5695

Arkhipov, V.; Bondarchuk, S.; Korotkikh, A. & Lerner, M. (2006). Production technology and disperse characteristics of aluminum nanopowders (in Russian). *Gornyi Zhurnal. Special Issue. "Nonferrous Metals"*, No.4, pp. 58–64, ISSN 0372-2929

Belov, V.; Borovoi, A. & Volkov, S. (1984). On the small-angle method under single and multiple scattering (in Russian). Izvestia AN SSSR Fizika Atmosfery i Okeana, Vol.20, No.3, pp.323–327

Deyrmendzhan, D. (1997). Electromagnetic radiation scattering by spherical polydisperse particles (in Russian), Mir, Moscow

Dombrovskiy, L. (1982). On a possibility of determining the dispersed composition of a two-phase flow by small-angle light scattering (in Russian), *Teplophysika Vysokikh Temperatur*, No.3, pp.549–557, ISSN 0040-3644

Gritsenko, A. & Petrov, G. (1979). On the role of the multiple scattering in the inverse optics problems of coarse-dispersed aerosols (in Russian). *Optics and Spectroscopy*, Vol.46, No.2, pp.346–349, ISSN 0030–4034

Gyvnak, D. & Burch, D. (1965). Optical and infrared properties of Al2O3 at elevated temperatures. *Journal of Optical Society of America*, Vol. 55, No.6, pp. 625-629

Pavlenko, A.; Arkhipov, V.; Vorozhtsov, B.; Ahmadeev, I. & Potapov, M. (2005). Informative radiation wavelengths under laser diagnostics of aerosol media (in Russian), *XII Joint International Symposium "Atmospheric and Oceanic Optics. Atmospheric physics"*, Tomsk: IAO SB RAS, pp. 148

Potapov, M. & Pavlenko, A. (2000). A mathematical model of calculating characteristics of the laser radiation attenuation for bimodal media (in Russian), *Proceedings of the 1st All-Russian conference "Measurements, automation and simulation in industry and scientific research"*, Biysk, pp. 135–139, ISSN 2223-2656

Prishivalko, A. & Naumenko, E. (1972). The light scattering by spherical particles and polydisperse media (in Russian), IF AN BSSR, Minsk

Shaikhatarov, K.; Lapshin, A.; Stolyarov, A. & lapshina, T. (1986). Disperse media photometer, Pat. SU 1435955 A1 G 01J 1/44

Shifrin, K. (1971). The study of substance properties from single scattering (in Russian), In: Theoretical and Applied Problems of Light Scattering, B.I. Stepanov, A.P. Ivanov, (Ed.), pp. 228–244, Nauka I Tekhnika, Minsk

Titov, S. & Muravlev, E. (2008). The use of a Videosprint/C/G4 digital camera in the studies of dynamic processes (in Russian), *All-Russian Conference "Perspectives of Designing and Applying High-energy Condensed Materials"*, pp. 173–179, ISBN 978-5-9257-0134-8, Biysk, Russia, 25-26 Sep 2008

Van de Hulst, G. (1961). Light scattering by fine particles, Moscow

Vorozhtsov, B.; Potapov, M.; Pavlenko, A.; Lushev, V.; Galenko, Yu. & Khrustalev, Yu. (1997). A multifrequency laser measuring complex of monitoring atmospheric and industrial aerosols (in Russian), *Atmospheric and Oceanic Optics*, Vol. 10, No. 7, pp. 928-832, ISSN 0869–5695

Zuyev, E.; Kaul, B.; Samokhvalov, I.; Kirkov, K. & Tsanev, V. (1986). *The laser probing of industrial aerosols* (in Russian), Nauka, Novosibirsk

16

Atmospheric Clock Transfer Based on Femtosecond Frequency Combs

Lingze Duan and Ravi P. Gollapalli
The University of Alabama in Huntsville
USA

1. Introduction

Precise timing and frequency synchronization has become ubiquitously needed as the world enters the age of global connectivity. Today's communication and computer networks are running under the regulation of synchronized time bases to ensure efficient data routing and information transfer. As the carrier frequencies of these networks continue to rise to accommodate the ever-increasing data traffic, the need for high-fidelity clock distribution becomes imperative (Prucnal et al., 1986). Within the scope of basic sciences, ultra-stable timing dissemination also becomes increasingly important as more and more highly sophisticated instrument and research facilities, such as space-borne atomic clocks (Chan, 2006), long-baseline radio telescope arrays (Cliché & Shillue, 2006), and particle accelerator-based X-ray pulse sources (Altarelli, 2007), require unprecedented level of synchronization in order to explore the uncharted physical parameter regimes.

Recently, there has been a growing interest in the research of high-fidelity, remote transfer of reference clock signals using femtosecond frequency combs (Foreman et al., 2007; Kim et al., 2008). The rationale behind the idea is that frequency combs can conveniently acquire their timing stability from optical atomic clocks, the next generation timing standard, and transfer the stability of an atomic clock at a fixed frequency onto a number of spectral lines (as many as ~10^5-10^6) over a broad spectrum in *both* optical and microwave frequency regions. These features give frequency combs enormous advantage as a carrier of timing and frequency references over conventional single-frequency lasers, which can only transfer a single channel of clock signal in *either* the optical or the microwave frequency region.

Over the last two years, we have proposed and experimentally demonstrated atmospheric remote clock transfer based on a femtosecond frequency comb (Alatawi et al. 2009; Gollapalli & Duan, 2010). It is a complementary scheme to the previously studied fiber-optic remote delivery of frequency combs (Holman et al., 2004). Compared to fiber-optic delivery, transferring frequency combs through the atmosphere does not rely on the existence of a wired system and hence reduces the cost while offering a much better flexibility. It is particularly desirable for timing synchronization between moving transmitters and receivers, such as motor vehicles, ships, airplanes and satellites, and it is the preferred scheme when a close-range, short-term transfer link is needed. Moreover, future space-terrestrial network will need high-fidelity optical up-and-down links to synchronize space-borne atomic clocks with ground-based atomic clocks (Chan, 2006). Pulse laser-based timing synchronization offers a viable solution to the establishment of such links.

Photodetection plays a key role in laser-based synchronization systems because it converts the timing signals stored in the photons into electronic form so the information can be read by the instruments under synchronization. Photodetectors are essential components in the study of laser remote clock transfer. Their unique impacts on the characterization of atmospheric frequency comb transmission are two-fold. First, the "square-law" nature of photodetectors leads to two completely different system configurations for the transfer of timing signals in the radio/microwave frequency (RF) range and in the optical frequency range, i.e. RF heterodyne and optical heterodyne. Second, atmospheric optical effects such as beam wander and speckle induce strong amplitude noise in the heterodyne signal, which enters the phase noise characterization as a limiting noise source due to the power-to-phase coupling effect in photodetectors.

In this chapter, we will give an overview of the recent progress in our study of atmospheric clock transfer based on femtosecond frequency combs. We shall first review the concept and the basic characteristics of femtosecond frequency combs. Then we will discuss the technical specifics associated with the atmospheric clock transfer experiment, especially the roles of photodetectors in different heterodyne schemes and the impact of power-to-phase coupling in the phase noise measurement. The two sections following that will be devoted to the experimental results, first with the clock signal in the RF range (~10^9 Hz) and then with the clock in the optical frequency range (~10^{14} Hz). We shall compare our results with similar experiments based on cw (continuous-wave) lasers and summarize at the end.

2. Femtosecond frequency combs

2.1 The concept of frequency combs

The concept of optical frequency combs was initially introduced as a way to bring frequency metrology into the optical region (Udem et al., 2002). The advance in precision spectroscopy and optical sensing requires that optical frequencies, which fall in the range of 100–1000 terahertz (THz) for the IR-Visible-UV band, be measured with accuracies comparable to microwaves. Such a level of accuracy can only be achieved with a metrological link that can directly reference optical frequencies to Cs atomic clocks, which operate at about 9.2 GHz (Hall & Ye, 2003). The emergence of solid-state femtosecond lasers in the late 1990s allowed this concept to be experimentally realized (Jones et al., 2000). In essence, a femtosecond frequency comb is simply a frequency-stabilized ultrafast laser (Holzwarth et al., 2000). Such lasers are able to produce optical pulses as short as a few femtoseconds (10^{-15} s) at very fast pulse repetition rates (~ 0.1–10 GHz). The result is a pulse train with a constant pulse interval, as shown in Fig. 1 (a). The spectrum of such a pulse train consists of a series of equally spaced spectral lines (often referred to as comb lines), spanning across an extremely broad spectral range owing to the ultra-short temporal profiles of the femtosecond pulses.

A conceptual illustration of such a supercontinuum spectrum is depicted in Fig. 1 (b). With proper electronic control, the pulse repetition rate (i.e. the spacing of the comb lines) and the carrier-envelope offset (CEO) frequency (i.e. the offset frequency of the comb when it is extended to zero), both in the microwave frequency range, can be stabilized against an atomic reference, such as a Cs clock (Udem et al., 2001). This establishes a grid of well-defined spectral markers through the simple relation of $f_m = f_{CEO} + mf_R$, where f_R and

f_{CEO} are the repetition rate and the CEO frequency, respectively, and m is the index number of the comb line in concern. Unknown spectral features within the span of the supercontinuum can be compared with these comb lines through interference and their exact frequencies can be determined with kilohertz-level resolutions or even better (Fortier et al., 2006).

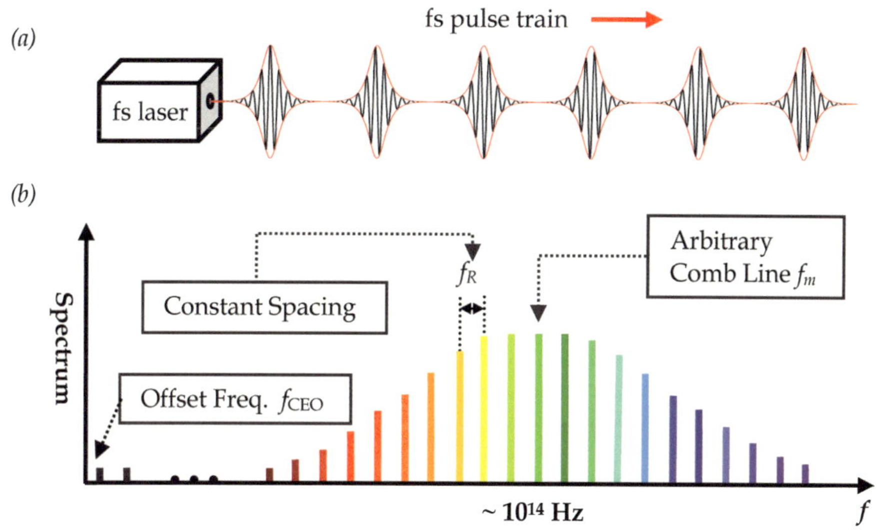

Fig. 1. Concept of femtosecond frequency combs. (a) time-domain picture – a femtosecond pulse train with a fixed pulse interval; (b) frequency-domain picture – a supercontinuum that consists of a series of equally-spaced spectral lines. The frequency of an arbitrary comb line has two degrees of freedom defined by $f_m = f_{CEO} + mf_R$. Locking f_R and f_{CEO}, or any two independent combinations of them to atomic clocks will transfer the frequency accuracy of the clocks onto the entire spectrum.

2.2 Key characteristics of frequency combs

Once stabilized, a femtosecond frequency comb offers some unique features unattainable with other types of laser systems. For example, it combines a supercontinuum spectrum, often spanning across more than an octave, with ultrahigh spectral precision for individual comb lines, which in the past could only be achieved with cw lasers operating at a single frequency. Sub-hertz linewidth across the entire comb spectrum (e.g. 580–1080 nm) has been reported (Bartels et al., 2004). This feature allows a frequency comb to combine the accuracy of a single-frequency laser and the flexibility of a supercontinuum source, and hence leads to a wide range of applications such as trace-molecule detection (Thorpe et al., 2006), precise ranging (Ye, 2004) and extrasolar planet search (Li et al., 2008).

Another interesting feature for a femtosecond frequency comb is that, besides the precisely defined grid of *optical* frequencies, the comb also carries a grid of highly stable *microwave* frequencies stemming from the constant pulse repetition rate. These frequency references can be easily accessed via photodetection. A photodetector essentially converts the ultrastable timing of the pulses into a series of discrete spectral lines consisting of the harmonics of the repetition rate. A femtosecond frequency comb hence can be treated as a carrier of timing/frequency references, or "clocks". But unlike conventional clock carriers, such as cw lasers or RF communication channels, which can only carry one single frequency at a time, a frequency comb simultaneously carries multiple (~ 10^5–10^6 in some cases) clock signals in both microwave and optical frequency ranges. Such a capability offers femtosecond frequency combs unprecedented versatility in remote clock transfer (Foreman et al., 2007), which we shall focus on in the rest of the chapter.

2.3 Frequency comb-based clock transfer

One important recent development in the application of femtosecond frequency combs is the demonstration of comb-based high-fidelity remote clock transfer through optical fibres (Holman et al., 2004; Holman et al., 2005; Foreman et al., 2007; Kim et al., 2008). In the microwave frequency range, a fractional frequency instability of $<9\times10^{-15}$ for a 1-s averaging time has been achieved through a 6.9-km installed fibre link with the help of active noise cancellation (Holman et al., 2005). In the optical frequency range, two lasers connected by a 300-m fibre link have been synchronized for 12 hours with a relative timing instability of 9×10^{-21} (Kim et al., 2008). These results are comparable to or exceeding the performance of the conventional techniques, which use a modulated cw laser to transfer a single microwave clock and rely on an ultra-stable single-frequency laser to deliver an optical frequency reference (Foreman et al., 2007). Thus, femtosecond frequency combs are not only efficient and versatile, but effective as well in terms of remote clock delivery via fibre networks.

The success of comb-based clock transfer in fibre logically leads to the question about free-space clock distribution using femtosecond lasers. This will not be a trivial extension from the work done in fibre because the transmission medium now becomes air, which is much more dynamic than glass. New phenomena are expected to dominate and new strategies are needed. In fact, the research on optical pulse propagation in the air dates back to the 1970s. In the context of optical communications, various models of pulse propagation in turbulent random media have been studied (Su & Plonus, 1971; Liu et al., 1974; Hong et al., 1977; Young et al., 1998). Today, femtosecond transmission in the atmosphere continues to attract much attention because of its potential applications in remote sensing and Light Detection and Ranging (LIDAR). Topics in this area that attract major research interest include nonlinear effects, ionization, atmosphere turbulence, and their impact on pulse propagation (Akozbek et al., 2000; Mlejnek et al., 1999; Sprangle et al., 2002; Rodriguez et al., 2004). However, most of the previous work on atmospheric propagation of femtosecond pulses focuses on the interactions between air and individual pulses. Little attention has been given to the propagation of a pulse train as a whole in the atmosphere, especially in terms of the precision degradation of its Fourier frequencies. The work described in the following aims to address this important aspect through experimental characterizations.

3. Atmospheric remote transfer of frequency combs

Before we begin the discussion about specific experiments, it is necessary to review several general technical aspects associated with these experiments so that readers can gain a better understanding of the physical background behind some of the techniques involved.

3.1 Noise sources in atmospheric transmission

The key question we seek to answer through the current research is: How much extra noise will be added to the clock signals as the femtosecond pulse train propagates through the air? To answer the question, we must first take a look at the potential noise sources in the atmosphere. Two major mechanisms can cause the degradation of clock precision in the air - temporal fluctuation of the refractive index and scattering (Ishimaru, 1978). Air turbulence creates random density fluctuations, resulting in a time-dependent average refractive index over the path of the laser beam, which leads to variation of pulse arrival time. On the other hand, density inhomogeneities in the air, along with scattering particles such as aerosols, may cause pulses being randomly scattered, as shown in Fig. 2. This may lead to pulse distortion and dephasing, which degrade the timing precision of the original pulse train and broaden the comb lines.

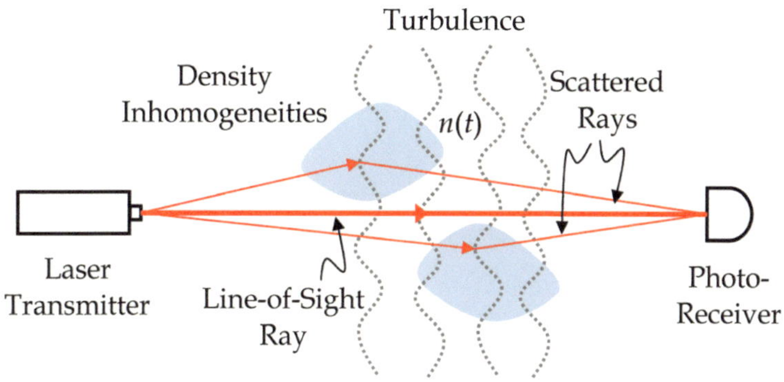

Fig. 2. The effects of index fluctuation and scattering on atmospheric optical transmission.

For ground-level short-distance (e.g. < 1 km) transmission links, one can assume clear air and weak turbulence (Andrews & Phillips, 2005). Under such conditions, it has been shown that scattering plays a far less important role in affecting the pulse arrival time compared to pulse wandering due to index fluctuation (Liu & Yeh, 1980). The transmission distance in our experiments is about 60 m, and we usually avoid taking measurement under extreme weather conditions such as rainy, foggy or highly windy days. Therefore, it is justified to ignore scattering in our research and focus only on the index fluctuation in line-of-sight beam propagation. In addition, it should be pointed out here that turbulence and scattering also cause beam pointing fluctuation and speckle, which affect the received pulse amplitude as well. Special care must be taken in the design of the optical system to minimize the impact of these effects. Specific discussions about some of these measures will be made in Section 3.4.

3.2 Excess phase noise and frequency instability

For an infinite periodical pulse train, the temporal variation of the pulse arrival time due to atmospheric transmission broadens the spectral lines in the Fourier domain, introducing excess phase noise to the initially highly precise grid of Fourier frequencies. The phase noise spectrum can be determined via a heterodyne process, which compares the transmitted pulse train with its copy from a reference path that induces negligible phase noise. By spectrally analyzing the heterodyne signal, we can measure the power-spectral density (PSD) of the phase-noise generated sidebands, $S\varphi(f)$, representing the mean-squared phase fluctuation at a Fourier frequency f away from the centre frequency with a 1-Hz bandwidth of measurement (Foreman et al., 2007),

$$S_{\varphi}(f) = [\delta\varphi(f)]^2 \quad (\text{rad}^2/\text{Hz}) \,. \tag{1}$$

Sometimes it is more convenient to characterize phase noise in terms of pulse timing jitter, which is defined as

$$\delta T(f) = \frac{\delta\varphi(f)}{2\pi\nu_0} \quad (\text{s}/\sqrt{\text{Hz}}) \,, \tag{2}$$

where ν_0 is the centre frequency. Integrating the spectral density of the timing jitter over the measurement bandwidth, from f_l to f_h, leads to the total rms timing jitter of the pulses

$$T_{rms} = \sqrt{\int_{f_l}^{f_h} [\delta T(f)]^2 df} \quad (\text{s}) \,. \tag{3}$$

The spectrum of the excess phase noise or timing jitter essentially characterizes the short-term (usually < 1 s) stability of the clock transmission link.

The second aspect in the characterization of clock stability is how much the time-averaged frequency varies over different lengths of the averaging period. It is usually used to describe the stability of a frequency signal over relatively longer term, e.g. > 1 s. Experimentally, this is done by measuring the Allan deviation of the frequency signal over an averaging time τ, which is defined as

$$\sigma_y(\tau) = \left\langle \frac{1}{2} [\overline{y}(t+\tau) - \overline{y}(t)]^2 \right\rangle^{1/2} \,, \tag{4}$$

where $\overline{y}(t)$ is the time average of the instantaneous fractional frequency deviation from the centre frequency (Allan, 1966).

The phase noise/timing jitter spectrum and the Allan deviation characterize clock instability from different perspectives. They are the main quantities we seek to determine through our experiments.

3.3 RF heterodyne vs. optical heterodyne

The heterodyne technique has long been used in RF technologies to shift information signals from one frequency channel to another (Nahin, 2001). In the current research, in order to study the phase noise sidebands due to pulse propagating in the atmosphere, we need to shift the noise sidebands to lower frequencies where a fast Fourier transform (FFT) analyzer can effectively resolve the noise spectrum. Conceptually, this can be easily done by beating the noisy signal with a noise-free local reference, which has a frequency identical or close to the centre frequency of the noisy signal. At the output of the heterodyne mixer, the noise sidebands are shifted to both the sum and the difference frequencies. With a low-pass filter, we can select the downshifted bands at the difference frequency. In the case the difference frequency is zero, i.e., the noisy signal and the reference have the same centre frequency, the noise sidebands are shifted to the base band, and the lower sideband is folded into the positive side by the FFT. Fig. 3 (a) & (b) give a conceptual sketch of this heterodyne process. The reference signal can be obtained by splitting a portion of the original clock at the output of the source. This also greatly reduces the impact of the source noise to the measurement as the source noise is common-mode and hence is cancelled at the difference frequency.

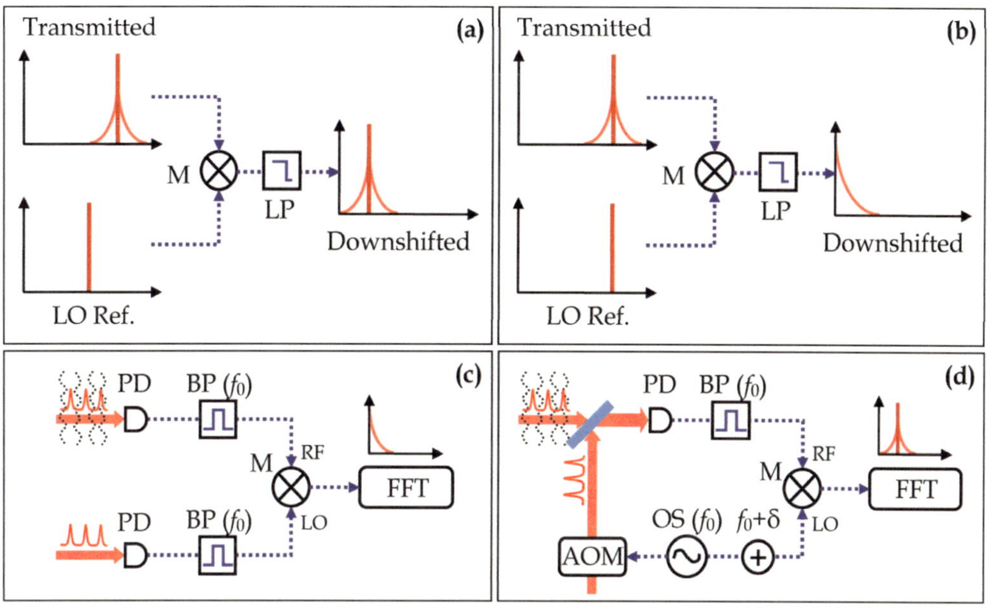

Fig. 3. Concept of heterodyne noise measurement with (a) different centre frequencies and (b) same centre frequency. Configurations of (c) RF heterodyne and (d) optical heterodyne. BP, bandpass filter; LP, lowpass filter; M, mixer; OS, oscillator; PD, photodiode.

In practice, two different heterodyne configurations are used for the two clock frequency ranges, namely RF heterodyne and optical heterodyne. The necessity originates from the use of square-law detectors. Photodetectors, such as photodiodes, are known to be square-law detectors. They cannot directly respond to the optical field due to the short optical cycles (~ fs). Instead, they respond to the quantized excitations of the optical field, i.e. photons. This is equivalent to finding the time-averaged intensity, which is proportional to the "square" of the optical field (Davis, 1996). As a result, the optical cycle is averaged out and the detector can only probe the slow variation of the field envelope. However, a square-law detector can respond to the beat frequency of two optical fields with a small frequency difference when they are incident on the detector simultaneously. To understand this effect, we shall assume optical waves A and B with a slight frequency offset $\Delta\omega$ superpose on a photodetector. The total field can be written as

$$\vec{E} = \vec{A}\cos\omega t + \vec{B}\cos(\omega + \Delta\omega)t \,. \quad (5)$$

The detector responds to the time-averaged intensity I, which can be expressed as

$$I \equiv \varepsilon_0 c < \vec{E}\cdot\vec{E} > \; = \varepsilon_0 c\left(\frac{A^2}{2}+\frac{B^2}{2}+\vec{A}\cdot\vec{B}\cos\Delta\omega t\right), \tag{6}$$

where <> indicates time-average, and ε_0 and c are free-space permittivity and speed of light, respectively. The last term on the right hand side is the interference term, which oscillates at the beat frequency. It can be detected by the photodetector as long as it falls within the detector's bandwidth. From this point of view, a photodetector can be used as a heterodyne mixer to generate an electric oscillation at the difference frequency of the two optical fields. This feature of photodetectors becomes the basis of optical heterodyne.

In Fig. 3 (c) and (d), we use our experiments as examples to illustrate the concepts of RF heterodyne and optical heterodyne. In the RF heterodyne, two identical photodetectors convert the femtosecond pulse trains from both the transmitted arm and the reference arm into a series of electric oscillations at the harmonics of the pulse repetition rate. One of these frequency components is selected by bandpass filers in both arms as the clock under study. The transmitted clock with excess phase noise is then mixed with the reference clock at a double-balanced mixer to generate the heterodyne beat signal, which is spectrally analyzed by an FFT analyzer to give the noise spectrum. In the optical heterodyne, the pulse trains from the two arms are first mixed on a photodetector. The reference beam gains a slight frequency shift f_0 at an acousto-optical modulator (AOM). As a result, the photodetector produces a beat signal at f_0. This beat signal is further downshifted via RF heterodyne to make it accessible by the FFT analyzer. As we shall discuss in Section 5, for optical clock transfer, it is more convenient to keep the final beat signal away from the base band so that the full line shape can be seen.

One unique feature of optical heterodyne with femtosecond frequency combs is the process of multiheterodyne. When two frequency combs are mixed on a photodetector, each pair of corresponding comb lines from the two arms produces a beat signal. The total heterodyne signal is a coherent superposition of all these beat notes from individual comb lines. The analysis of multiheterodyne of frequency combs is out of the scope of this chapter. In-depth discussions can be found elsewhere (Gollapalli & Duan, 2011).

3.4 Power-to-phase coupling in photodiodes

When a photodiode is directly used to extract microwave signals from a modulated optical signal, additional phase noise is generated in the spectra of the microwave signals due to power-to-phase conversion in photodiodes (Tulchinsky & Williams, 2005). When the optical signal is an ultrafast pulse train, this excess phase noise can be found at the harmonics of the pulse repetition rate, causing extra fluctuations to the microwave clock signals extracted from the pulse train (Ivanov et al., 2003). The exact physical mechanisms of the power-to-phase conversion in photodiodes are not completely clear. One likely cause is the saturation of photodetectors (Liu et al., 1999). When the average power of the optical signal reaches certain level, space-charge build-up in the depletion region of the photodiode reduces the electric field in the depletion region, leading to a reduction of the velocity of photogenerated carriers and hence a phase delay in the photocurrent. Clear correlations between the arrival times of the photocurrent pulses and the incident optical power have been observed experimentally (Bartels et al., 2005).

The presence of power-to-phase coupling in photodetectors results in a detector-induced excess phase noise, which, in the case of microwave clock transfer, could mask the excess phase noise due to transmission. Two unique factors affecting the quality of atmospheric laser communication are beam wander and speckle (Andrews & Phillips, 2005). Both effects have been visually observed in our experiment. One of their consequences is optical power fluctuation on the photodetector. The excess phase noise due to power-to-phase coupling can make a significant contribution to the total measured phase noise. In order to evaluate the impact of this detector-induced phase noise, we use a microwave power detector to monitor the power of the extracted clock signal and measure the correlation between the detected phase noise and the clock power (see Section 4.2 and Fig. 5 for description of the experimental setup). Under linear operation, the power of the recovered clock scales as the square of the optical power received by the photodiode. However, the fluctuation of the clock power can be treated as linearly proportional to the fluctuation of the optical power near the average power level. Therefore, the correlation between the phase noise and the fluctuation of the clock power actually reflects the correlation between the phase noise and the optical power fluctuation. Correlation is determined by recording the coherence function between the measured phase noise and clock power during each data run. Here "coherence" is defined as

$$C_{yx}(f) = \frac{\Phi_{yx}(f)}{\sqrt{\Phi_{yy}(f) \cdot \Phi_{xx}(f)}}, \tag{7}$$

where $\Phi_{xx}(f)$ and $\Phi_{yy}(f)$ are the power spectral densities of time series x and y (as functions of Fourier frequency f), and $\Phi_{yx}(f)$ is the cross-power spectral density between the two series. By definition, the coherence function ranges between 0 and 1.

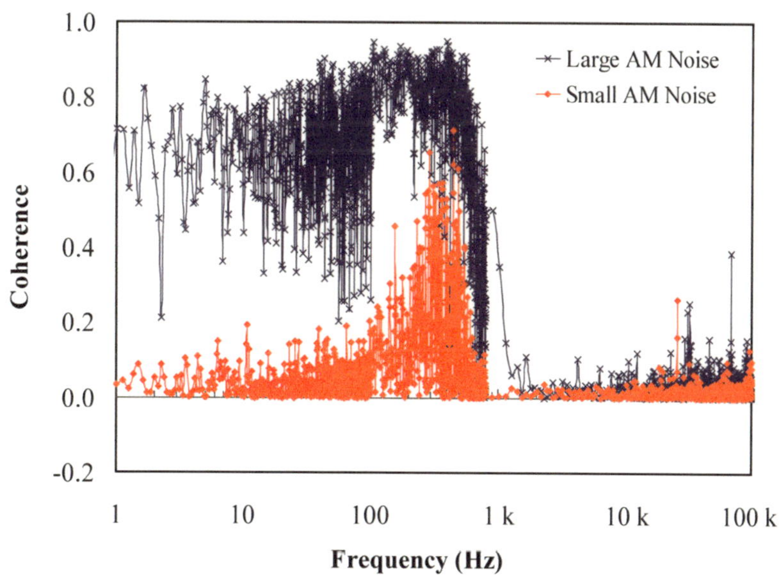

Fig. 4. The correlation (in terms of coherence function) between the measured phase and clock power indicates the impact of power-to-phase coupling in the photodiode.

Fig. 4 shows the measured phase-power coherence under two experimental configurations. The blue cross trace is obtained with a fibre-coupled photodiode and a fibre collimator receiving the transmitted laser beam. Since the coupling efficiency of the collimator critically depends on beam pointing and transverse intensity distribution of the incident beam, the power coupled to the detector is very sensitive to beam wander and speckle. As a result, a significant amount of phase noise is generated by the photodiode due to power-to-phase coupling. This is indicated by the coherence values close to 1 below 1 kHz. Apparently, such a system configuration would not allow us to effectively evaluate the transmission-induced phase noise because of the overwhelming contribution from the detector. Therefore, in the second configuration, we use large-diameter optics (e.g. 2 inch) in the receiving system and focus the beam directly onto the photodiode with the size of the focus much smaller than the active area of the detector. Such a configuration keeps the receiving system insensitive to the beam pointing drift and the transverse beam-profile variation. As the red diamond trace shows, the coherence function is close to zero over a wide frequency range, indicating little phase-power correlation at these frequencies. A moderate peak of coherence (~ 0.5) at a few hundred hertz can be attributed to temporal fluctuation of the irradiance due to scintillation of the laser beam (Andrews & Phillips, 2005). But its impact to the phase noise measurement is very limited because the noise spectrum rolls off quickly at such high frequencies.

4. Atmospheric transfer of microwave timing signals

As pointed out in Section 2, the ultrastable pulsing rate of a femtosecond frequency comb can be used as a clock signal. The repetition rates of most common femtosecond lasers range from tens of MHz to 1 GHz or so. Their harmonics, however, can be much higher. A femtosecond pulse train inherently carries all these harmonics simultaneously, and their retrieval as timing references is only limited by the bandwidths of the photodetectors used.

4.1 A rooftop transmission link

In order to study frequency comb-based atmospheric clock transfer, we established an outdoor laser transmission link. The link is located on the roof of our laboratory building on

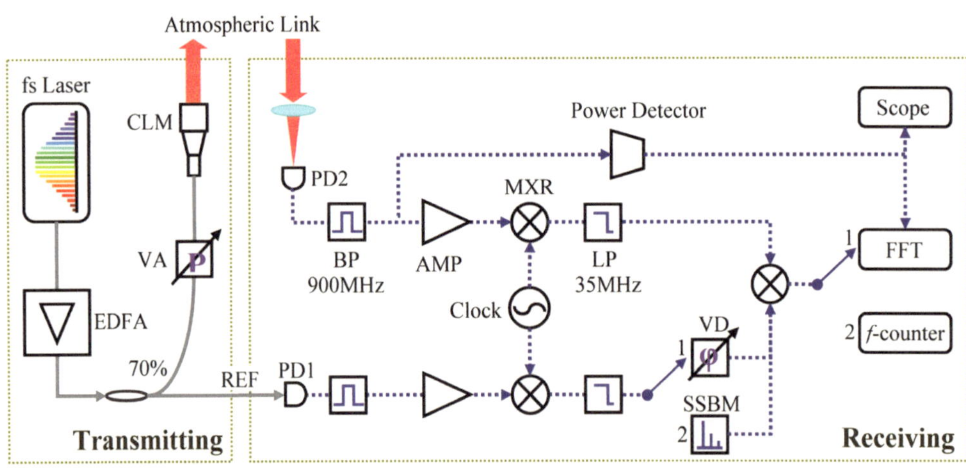

Fig. 5. Schematic layout of the test system for atmospheric microwave timing transfer. AMP, RF amplifiers; BP, band-pass filters; LP, low-pass filters; MXR, double-balanced mixers, PD, photodiodes; VA, variable attenuator; and VD, variable delay.

the campus of the University of Alabama in Huntsville. The four-floor building has an observation platform about 20 meters above the ground. There is no high-rise building or other tall structure nearby to create local turbulence. A sturdy steel tripod is anchored on the platform to house a metal beam reflector (a 2-inch gold mirror). The laser beam is launched from a nearby astronomical observatory into the open air and the reflector sends the beam back to the observatory, where all the signal processing and measurement takes place. The round-trip propagation distance is about 60 m.

4.2 Experimental setup

Fig. 5 shows the schematic of the experimental setup for microwave timing transfer. The system is divided into transmitting and receiving subsystems. In the transmitting part, a femtosecond fiber laser (Precision Photonics FFL-1560) generates a train of 150-fs pulses centered at 1560 nm, with a 90-MHz repetition rate and a 4-mW average power. An erbium doped fiber amplifier (EDFA) boosts the average power to 100 mW. A 70:30 broadband fiber coupler directs the majority of this power into a fiber collimator, which launches a 7 mm-diameter beam into the atmospheric transmission link. A fiber-coupled variable attenuator is inserted before the collimator as a power regulator to provide precise control over the total optical power reaching the receiving photodiode.

In the receiving subsystem, the transmitted beam is tightly focused onto a high-speed photodiode, which recovers the repetition frequency of the femtosecond laser as well as its harmonics. The 10th harmonic at 900 MHz is chosen as the microwave clock under test and is selected by a bandpass filter. To further reject the side modes, a local clock is used to beat the 900 MHz signal down to 35 MHz, where sharp low-pass filters can effectively remove all the remaining harmonics. Meanwhile, a reference clock signal is obtained by coupling a small portion of the EDFA output directly into the receiving subsystem via optical fiber (the REF path in Fig. 1) and then using a microwave circuit similar to the transmitted clock. The resulted frequency signal serves two sets of measurement. In the phase noise measurement, the reference clock passes through an adjustable delay line to gain a proper phase before it beats the transmitted clock in quadrature at a double-balanced mixer. The generated phase signal is frequency analyzed by a Fast Fourier Transform (FFT) analyzer (SRS SR785) and timing jitter spectral density can be calculated from the phase noise spectrum. In frequency stability measurement, the reference clock is frequency-shifted by 500 kHz through a single-sideband modulator (SSBM) and then mixed with the transmitted clock. This leads to a 500-kHz beat note, which is then measured with a frequency counter (SRS SR620) to determine its stability. It should be noted here that, although the repetition rate of the femtosecond laser is not stabilized in the test, the noise from the laser and the EDFA does not affect the measurement because it is common mode in the above heterodyne scheme. This ensures correct characterization of the excess clock instability due to the atmospheric propagation.

4.3 Laboratory test

In order to verify that the sensitivity of our experimental system is sufficient to measure the excess phase noise due to small index fluctuation in the air, we have first conducted a proof-of-principle test inside the lab, where the airflow can be better controlled. The main source of index fluctuation in this case is the air-conditioning (AC) airflow, predominantly moving downward across the laser beam. To verify the coincidence between the AC airflow and the phase noise, we fix the total length of the transmission link (~10 m) but change the effective beam path exposed in the AC airflow by inserting rigid plastic tubes (2.5 cm inner diameter) in the beam path. We find that the tubes can effectively control the measured phase noise. In addition, we have also made measurement with the AC system off and found the noise to be much lower. These facts confirm the effectiveness of our measurement system (Alatawi et al. 2009). Fig. 6 (a) shows the spectra of the transmission-induced timing jitter at four effective beam path lengths over a Fourier frequency range of 10^{-2} - 10^{5} Hz, along with the system noise (measured with a negligible beam path in the air). The system noise, mainly attributed to the RF amplifiers and the mixers, has a $1/f$ power frequency dependence below 1 kHz,

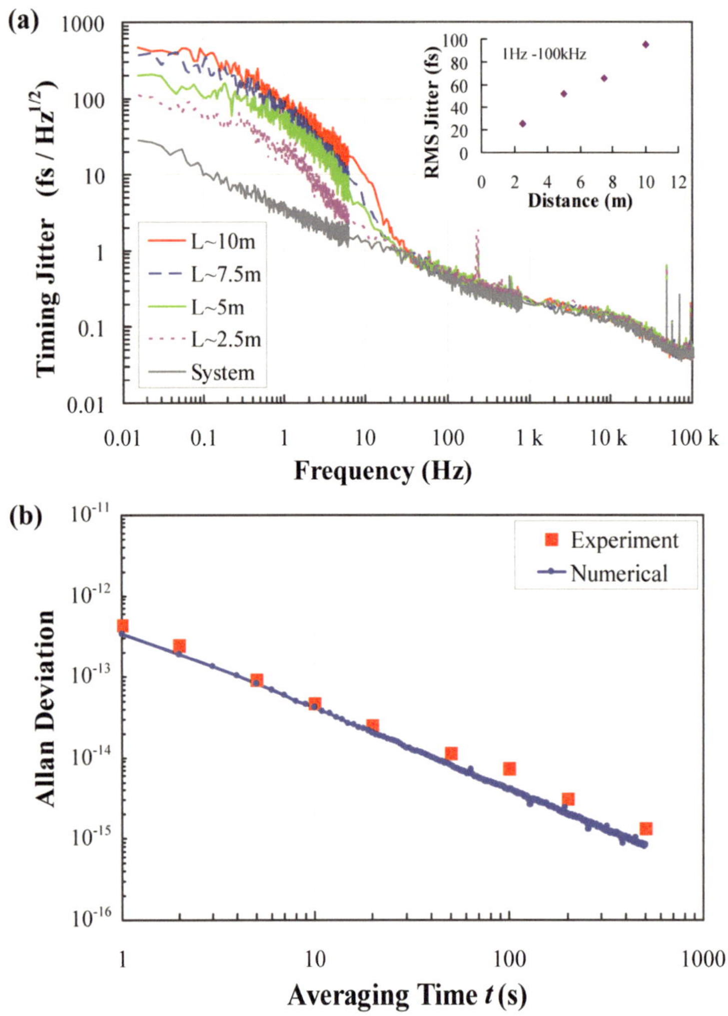

Fig. 6. (a) Spectral density of the excess timing jitter for four effective transmission distances, along with the system noise floor; inset: the total rms jitter integrated from 1 Hz to 100 kHz vs. effective beam path length in AC airflow. (b) The measured Allan deviation with a 10-m effective beam path and the numerical estimation based on the phase noise measurement.

indicating a Flicker phase noise that commonly exists in RF amplifiers within this frequency range (Halford et al., 1968). As the frequency increases, the noise floor gradually shifts toward white noise. The transmission-induced phase noise only becomes appreciably above the system noise below a few tens of hertz. As the Fourier frequency goes down, the power law of the noise gradually changes and eventually shows a trend toward f^0 below 0.1 Hz. Such behaviour is markedly different from what has been observed in fibre (Holman et al., 2005). To further verify the dependence of the excess phase noise on the AC airflow, we have calculated the rms timing jitter at different effective beam path lengths by integrating the timing jitter spectra in Fig. 6 (a) from 1 Hz to 100 kHz. For effective path lengths of 2.5, 5, 7.5, and 10 m, the total rms jitters are 37.5, 58.9, 71.2, and 99.0 fs, respectively, and the total rms jitter caused by the system noise is 27.5 fs. Fig. 6 (a) inset shows a near linear relation between the total *actual* rms timing jitter (system noise excluded) and the beam path length exposed in the airflow, confirming a direct correlation between airflow and phase noise.

The long-term transfer instability for 10-m effective beam path is evaluated by measuring the Allan deviation σ_y of the beat note frequency at a set of different averaging times. Fig. 6 (b) shows a typical set of experimental data, along with a set of numerical data based on the timing jitter measurement shown in Fig. 6 (a). The calculation uses the relation between the spectral density of phase noise and the Allan deviation (Rutman & Walls, 1991)

$$\sigma_y^2(\tau) = 2\int_0^{f_h} S_y(f)\frac{\sin^4(\pi\tau f)}{(\pi\tau f)^2}df , \tag{8}$$

where f_h is the bandwidth of the system low-pass filter. The excellent agreement between the two sets of data demonstrates the effectiveness of our measurement in assessing the excess clock fluctuation caused by atmospheric transmission.

4.4 Outdoor test

Once the measurement scheme passes the lab test, we apply it to the rooftop transmission link. Excess phase noise has been measured at different times of a day and under various weather conditions (except rainy days). The corresponding timing jitter is then calculated from the phase noise spectrum using (2). Fig. 7 (a) shows several typical traces of the jitter spectral density, along with the system noise baseline. The excess timing jitter is above the baseline only at frequencies below several hundred hertz. The magnitude and frequency dependence of the jitter spectra are strongly affected by the weather conditions, especially the wind speed, leading to a group of different spectral traces under otherwise similar conditions. The system baseline is mainly attributed to the RF amplifiers in the receiving system, and the noise spikes around 10-200 Hz are believed to be due to electric interference caused by the utility circuitry in the observatory. Fig. 7 (b) shows the rms jitter integrated from 1 Hz to the frequency in concern for all five traces in (a). Clearly, most contributions to the rms jitter come from noise below 100 Hz, indicating the dominance of slow phase modulations. The total rms jitters integrated from 1 Hz to 100 kHz range from several hundred femtoseconds to about two picoseconds. The system noise proves to have a negligible effect in the measurement as evident from its sub-100-fs effective rms jitter.

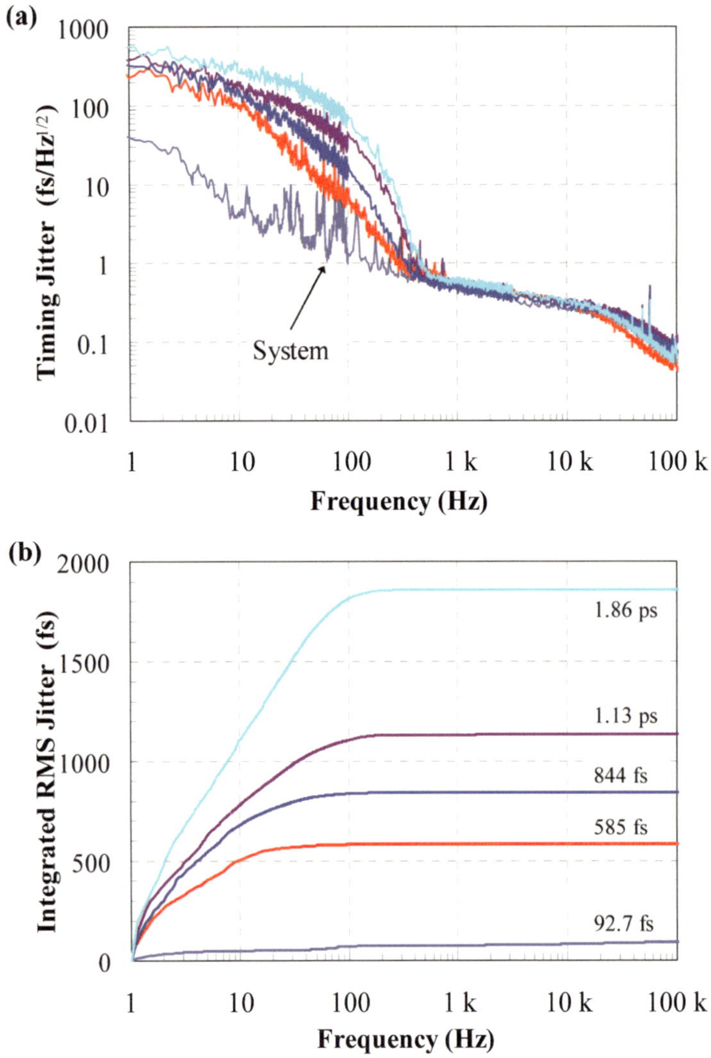

Fig. 7. Transmission-induced pulse timing jitter after a femtosecond frequency comb propagates 60 m through open atmosphere under various weather conditions. (a) Typical timing jitter spectra along with the system noise. (b) The integrated rms jitters for the five spectra in (a). The numbers are the RMS jitters integrated from 1 Hz to 100 kHz.

The result of the Allan deviation measurement is shown in Fig. 8, which also includes the trace from the 10-m indoor test for the purpose of reference. The fractional frequency stability after 60-m atmospheric transmission is in the order of a few parts per trillion at a 1-s averaging time τ. This is comparable to the frequency stability of most commercial atomic timing references, such as Cs or Rb clocks. At longer averaging times, the Allan deviation falls at a rate close to τ^{-1}, indicating white phase fluctuations (IEEE, 1983). This is different from fiber optic transmission, where the Allan deviation falls at $\tau^{-1/2}$ (Holman et al., 2004). It should also be noted that most atomic timing references also have a $\tau^{-1/2}$ behaviour. As a result, transferring a master microwave timing reference through free space becomes more advantageous than maintaining a less-precise local clock when the clock signal is averaged over longer time.

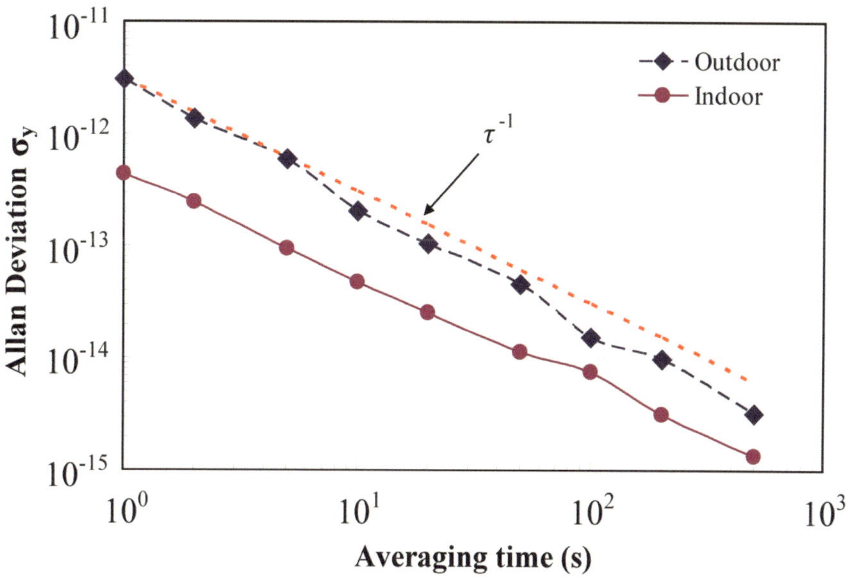

Fig. 8. A typical set of measured Allan deviation (diamond) of the 900 MHz clock signal after 60-m atmospheric propagation shows an approximate τ^{-1} dependence over the averaging time τ. The baseline data (circle) are obtained under indoor conditions with a much lower wind speed.

4.5 Discussion

The propagation of an optical pulse train through an atmospheric communication channel is susceptible to the refractive-index fluctuation caused by clear-air turbulence (Su & Plonus, 1971). From the rms timing jitter, we can derive the rms fluctuation of the group index n_g by using the relation $\Delta n_g = (c / L)\Delta T$, where c is the speed of light in vacuum, L is the total propagation distance, and ΔT represents the rms timing jitter. By using $\Delta T = 2$ ps and $L = 60$ m, we find the value of Δn_g to be 1×10^{-5}. Meanwhile, it has been shown that $\Delta n_g \approx a \cdot \Delta n$, where Δn is the fluctuation of the phase index and the proportional constant a is approximately equal to 3 in the visible and near infrared wavelength range (Ciddor, 1996; Ciddor & Hill, 1999). This leads to an estimated rms phase index fluctuation of several parts per million, which agrees with the well-known scale of such fluctuation due to clear-air turbulence (Shaik, 1988).

The measured timing transfer stability is compared with a similar rooftop experiment over a 100-m open link using the conventional carrier-modulation scheme (Sprenger et al., 2009). The fractional instability at 1 s measured in our test is several times smaller than the result presented in the reference. Moreover, their Allan deviations appear to have an averaging-time dependence close to $\tau^{-1/2}$ below 100 s, while our result is close to τ^{-1}. Such a difference in Allan deviation behaviour indicates possible difference in the underlying mechanism of instability. In fact, as pointed out by the authors, the earlier experiment is likely limited by the stability of the frequency synthesizers, and therefore only offers an upper bound of the propagation-induced instability. This seems to be supported by the fact that the electronic timing instrument usually shows a $\tau^{-1/2}$ characteristic. Such a system-noise limitation is partly because of the low clock frequency (80 MHz) used in the earlier work. In comparison, our experiment is free from such a restriction, as shown by the baseline data, because the wide bandwidth of the femtosecond pulses allows the use of a higher harmonic of the repetition rate as the clock signal.

5. Atmospheric transfer of optical frequencies

While clock signals in the microwave frequency range can be delivered by a femtosecond frequency comb and conveniently extracted with a photodetector, it is the ability to carry many optical frequency references that makes frequency combs very attractive as a clock transfer means. This is not only because the much higher frequencies of optical clocks lead to better fractional stability, but because optical frequency references can directly benefit from the next-generation atomic clocks, which will work in the optical frequency range.

5.1 Experimental setup

In order to demonstrate the capability of a femtosecond frequency comb to transfer optical frequency references through free space, we convert our rooftop experimental setup shown in Fig. 5 into an optical heterodyne configuration. Fig. 9 shows the new schematic layout. The system is again divided into transmitting and receiving subsystems. The transmitting subsystem is similar to the one used in the microwave clock transfer tests. The receiving

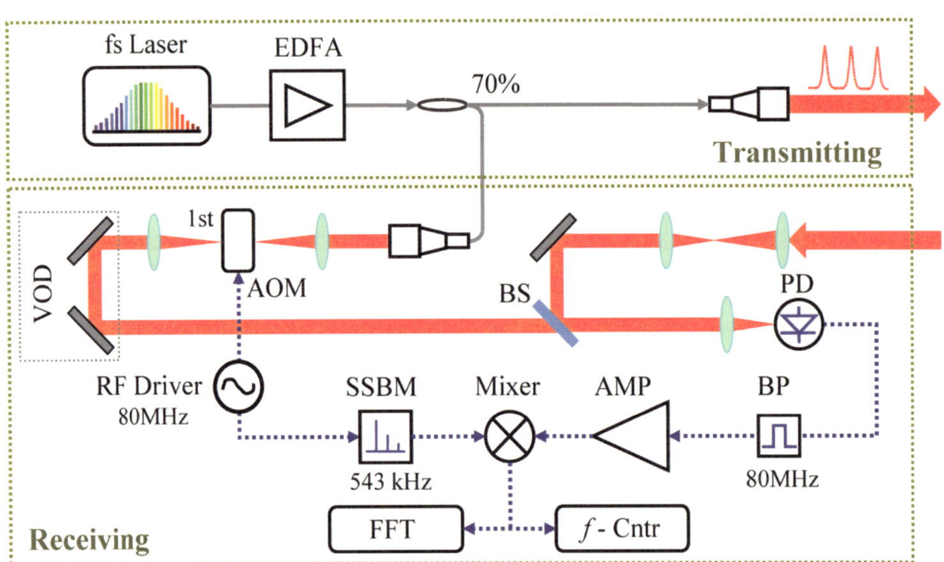

Fig. 9. Schematic layout of the experimental system for atmospheric transfer of optical frequency references. AMP, microwave amplifiers; BS, beam splitter; PD, photodiode; SSBM, single-side band modulator; and VOD, variable optical delay.

subsystem features an optical heterodyne configuration as discussed in Section 3.3. The reference beam passes through an AOM, which is driven by a RF driver at 80 MHz, and the first-order deflection is selected. Meanwhile, the beam transmitted through the atmospheric link is coupled into the receiving subsystem by a lens pair, which converts the expanded beam into a comparable size to the reference beam. The two beams are then combined into collinear propagation at a broadband beam splitter and focused onto a photodiode. A variable optical delay line in the reference arm ensures temporal overlapping of the transmitted pulses and the reference pulses. The photodiode produces an 80-MHz beat note from the two beams. The beat note is then phase-compared with a portion of the original 80-MHz driving signal at a double-balanced mixer. In order to measure the linewidth of the beat note, a small frequency shift (about 543 kHz) is added to the original 80-MHz signal through a single-sideband modulator (SSBM). The resulted 543-kHz beat signal is analyzed by an FFT analyzer for direct spectral width measurement and a frequency counter for frequency stability measurement. It should be noted here that the design of the receiving optical system has been made with the consideration of minimizing the impact of speckle and beam wander as discussed in Section 3.4.

5.2 Experimental results

The frequency transfer test was performed over the course of three months in the summer of 2010. Data have been taken under low-wind conditions with the wind speed below 2 m/s. Strong phase fluctuations in the multiheterodyne signal have been observed as pictured in Fig. 10 (a) inset. Time-domain phase traces like Fig. 10 (a) inset are taken by mixing the 80-MHz beat note from the photodiode directly with the original 80-MHz AOM driving signal without the SSBM. The fast oscillation of the beat note phase indicates a phase modulation index much greater than 2π. In order to examine the Fourier spectrum of the beat note, we add a small frequency offset of 543 kHz using the SSBM onto the 80-MHz reference from the AOM driver so that the beat signals from the mixer are shifted out of the base band. The resulted beat note spectrum is shown in Fig. 10 (a), plotted against the offset frequency from 543 kHz. Also plotted is the same beat signal without atmospheric transmission, realized by diverting the pulse train in the transmission arm through a fiber link of about 40 m long. The fiber is cut in such a way that it produces the same amount of time delay as the 60-m atmospheric link. Such an arrangement allows us to assess the contribution of the laser and the amplifier in our phase noise measurement.

Due to the imbalance of the interferometer, amplitude and phase noise from the laser and the EDFA could enter the heterodyne signal. One particular concern is that, because of the free-running laser, the drift of the laser repetition rate and carrier-envelope offset frequency may also broaden the beat note. With the fiber link, transmission-induced excess phase noise is much smaller (Holman et al., 2004), allowing us to assess the upper limit of the noise contribution from the laser and the amplifier. In Fig. 10 (a), both traces have a resolution bandwidth of 16 Hz, and are averaged over 1 s. The resolution limited sharp peak in the spectrum from fiber transmission indicates that the femtosecond laser and the EDFA has a negligible effect in the current excess phase noise characterization. In comparison, the atmospheric transmission causes significant spectral broadening. The broadened beat note spectrum has a full width of approximately 1 kHz.

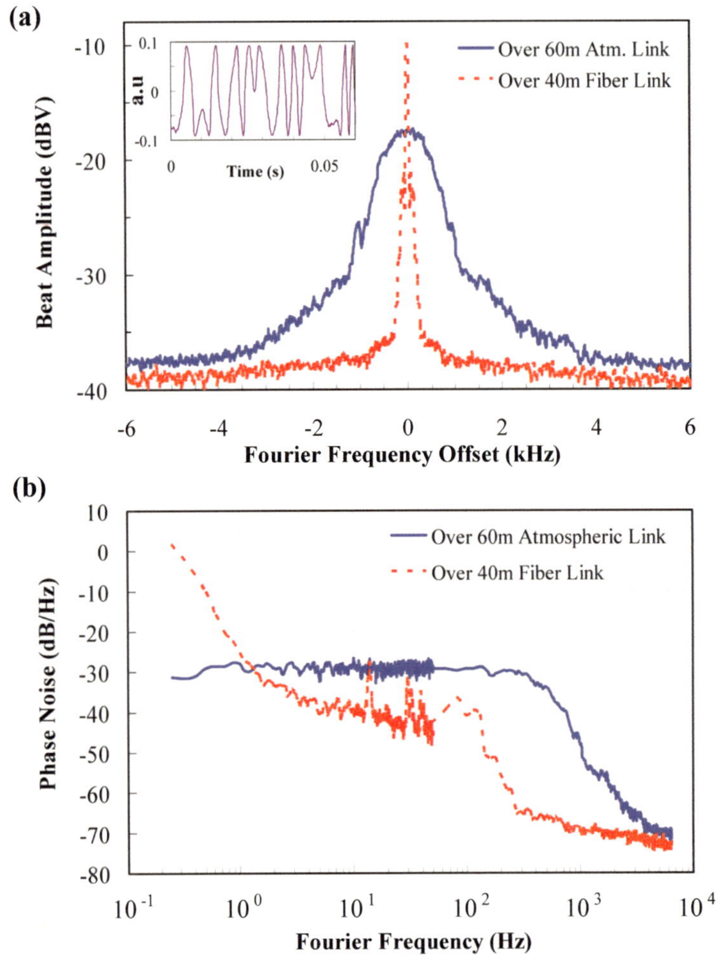

Fig. 10. (a) Fourier spectra of the 543-kHz heterodyne beat note for 60-m atmospheric transfer (solid) and 40-m fiber optic transfer (dashed). Significant linewidth broadening (~ kHz) is caused by the atmospheric propagation in comparison with fiber transmission. Inset: A time-domain trace of the homodyne beat note for atmospheric transfer demonstrates the large phase fluctuation. (b) Phase noise spectra of the 543-kHz beat note with the above two transfer cases.

A close look at the noise characteristics of the two transmission cases near the nominal frequency can be gained through their phase noise spectra, which are assembled from data over two frequency spans and are shown in Fig. 10 (b). It has been shown that such heterodyne phase noise can be understood as the zeroth-order excess phase noise of all the comb lines (Gollapalli & Duan, 2011). Since the beat note from the atmospheric transmission has no coherent spike as the carrier, the two curves in Fig. 10 (b) are normalized to their total signal powers (i.e., the integration of the beat note spectra) instead of the carrier power. A frequency-independent noise spectrum is seen below 300 Hz for the beat note from the atmospheric transmission. It is followed by a quick roll-off above 500 Hz. Such characteristics seem to agree with the spectral features for a slow phase modulation with a very large modulation index. In comparison, transmission in fibre with the same time delay results in a much smaller phase noise. The spikes between 10–100 Hz are believed to be the result of acoustic noise and electronic interference from the power circuits inside the lab. The above comparison shows that atmospheric fluctuations add a significant amount of phase noise to the optical frequency references transmitted through the air. As a result, atmospheric transfer of optical frequencies in general suffers much larger short-term fluctuations when compared with fibre-optic transfer.

For many applications, however, a more relevant parameter is the long-term stability of the frequency transfer scheme, which can be evaluated by measuring the Allan deviation of the 543-kHz beat note. Shown in Fig. 11 (a) are several sets of measured data of the Allan deviation versus the averaging time through the 60-m transmission link under various

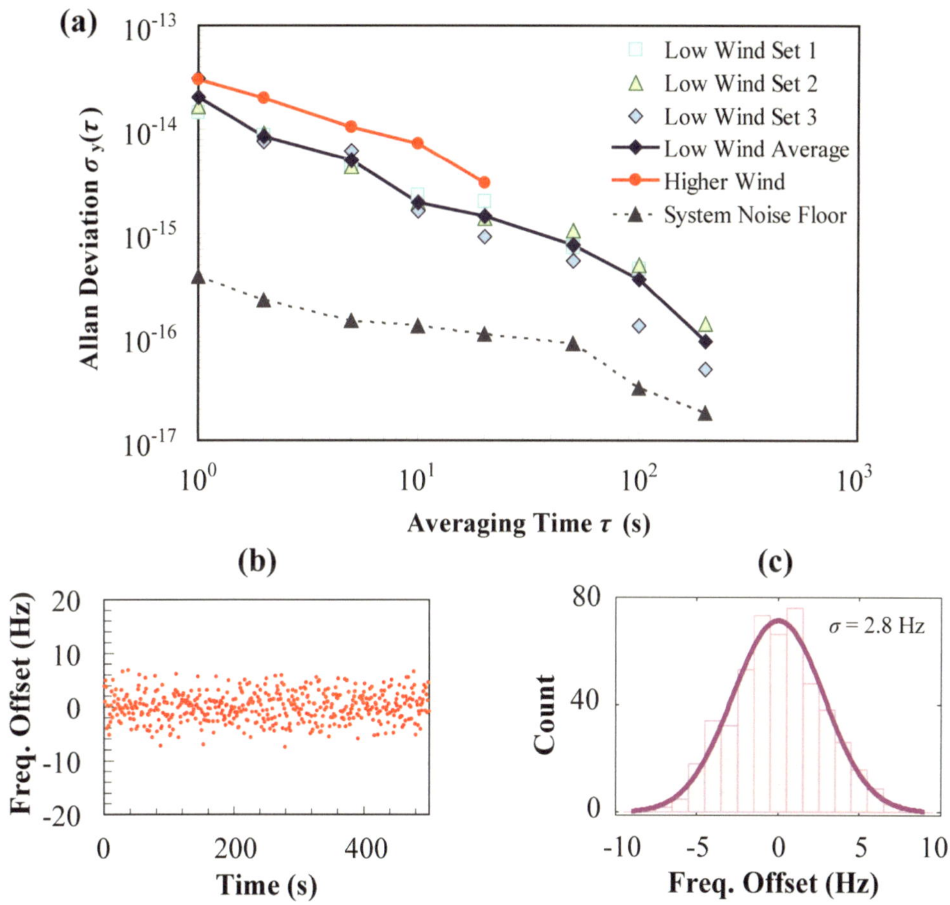

Fig. 11. (a) Allan deviations of the 543-kHz heterodyne beat note for 60-m atmospheric transfer under various weather conditions and the system noise floor. (b) A data set of 500 consecutive frequency measurements at 1-s gate time, plotted relative to the nominal frequency. (c) The histogram of (b), overlaid with a fitting to the normal distribution, leads to a standard deviation of 2.8 Hz for the frequency references transferred across the atmospheric link.

weather conditions. Among them, three sets were taken under low-wind conditions, with the wind speed generally staying below 5 mph (miles-per-hour), or 2.2 m/s, throughout the data collection periods (except for Set #1 at 200 s, which is not included here). The average of these data shows an overall behaviour of the Allan deviation close to τ^{-1} within the measurement range. The fractional frequency stability at 1 s is about 2×10^{-14}. When averaged over 100 s, the stability improves to 10^{-16} level. Over even longer time, however, the beat note often develops large power fluctuation, which prevents a reliable measurement of the Allan deviation with an averaging time longer than 200 s. The fluctuation of the beat note power is believed to be caused by the transmitted and the reference pulses drifting away from their optimum overlapping position due to the change of the effective path length of the transmission link. To demonstrate the effect of wind, we also include here a set of data taken under stronger wind, with wind speed ranging between 7–14 mph over the data collection period. Below 20 s, the higher-wind data stay roughly 3 dB above the average low-wind data and display a similar τ^{-1} dependence. No Allan deviation was successfully measured above 20 s due to the power fluctuation of the beat note. In addition, a set of baseline data is also included in Fig. 11 (a) to mark the system-limited measurement resolution. It is obtained by reflecting the transmission beam immediately into the receiving subsystem so the length of atmospheric propagation is negligible.

To further verify the scale of the frequency fluctuation, we take a close look at the measured beat note frequencies. Fig. 11 (b) shows consecutive readings of the frequency counter over several minutes with a 1-s gate time, taken under the low-wind condition. The frequencies have been offset to the nominal value of 543 kHz. A histogram analysis shows that the frequencies have a normal distribution around the nominal frequency, as shown in Fig. 11 (c), and a Gaussian fitting results in a standard deviation of 2.8 Hz. This frequency fluctuation level is in agreement with the Allan deviation measurement for low wind.

5.3 Discussion

To put our work into context, it is interesting to compare the above results with some of the previous reports on optical frequency transfer in the atmosphere. Sprenger et al. made a similar rooftop demonstration using a cw diode laser over a 100-m atmospheric link at the ground level (Sprenger et al., 2009). The group achieved an Allan deviation of 2×10^{-13} at 1 s. In addition, a statistics of the beat note frequency measured with a 1-s gate time shows a 70.5-Hz full width at half maximum. Compared with this work, the current study reports a frequency instability one order of magnitude lower at 1 s. Apart from the apparent shorter transmission distance, the smaller frequency fluctuation in the current work is likely due to the low-wind condition (< 2 m/s), which means lower index fluctuation due to air turbulence. Such a fact is evident in Fig. 11 (b), where continuous frequency measurement over several minutes shows a consistently low frequency fluctuation. More recently, Djerroud et al. have reported a coherent optical link across 5 km of turbulent atmosphere based on a cw Nd:YAG laser (Djerroud et al., 2010). A remarkably low Allan deviation of 1.3×10^{-14} is achieved at 1 s, and the stability further improves to 2×10^{-15} at 100 s. The authors, however, did point out that the experiment took place at an observatory 1323 m above the sea level, which could account, at least in part, for the high degree of frequency stability over a long distance. Based on the standard Kolmogorov model of atmospheric turbulence, the power spectral density of index fluctuation, Φ_n, is given by $\Phi_n(\kappa) = 0.033C_n^2\kappa^{-11/3}$, where C_n is the index structure constant and κ is the spatial wave number. Experimentally, it has been shown that C_n strongly depends on altitude (Ishimaru, 1978). For example, the altitude of Huntsville is 200 m, which corresponds to a C_n of approximately 10^{-7} $m^{-1/3}$ for sunny days. At 1323 m, however, C_n is about 10^{-8} $m^{-1/3}$ under similar conditions, which indicates a refractive index fluctuation roughly two orders of magnitude smaller than it at 200 m altitude.

6. Summary

In summary, we have experimentally demonstrated the transfer of clock signals in both microwave and optical frequency ranges over an open atmospheric transmission link using a femtosecond frequency comb. The excess phase noise due to the atmospheric propagation is successfully characterized in both cases, and the both results show large, slow-frequency phase modulations due to the index fluctuation of air. The fractional frequency stability for a 60-m transmission under typical calm weather conditions is in the order of a few parts per 10^{12} for microwave clock transfer and a few parts per 10^{14} for optical clock transfer, both with a 1-s averaging time. The much better fractional stability for optical clock transfer is mainly attributed to the much higher clock frequencies, which demonstrates the advantage of shifting reference clocks toward higher frequencies. Our measurement also shows an approximate τ^{-1} dependence of the Allan deviation on the averaging time in both frequency ranges up to 200 s. This is an encouraging finding for atmospheric clock transfer because it means that, over longer time, transferring a highly stable master clock via free space can be more advantageous than maintaining a less-precise local clock, which normally has a $\tau^{-1/2}$ dependence. Further study under more diverse weather conditions and over longer transmission distances is needed to confirm this results.

Two key properties of photodetectors play critical roles in the current research. The square-law detector allows us to perform optical heterodyne to characterize the excess phase noise in optical frequency transfer. Meanwhile, power-to-phase coupling in photodetectors puts a requirement on the receiving optics to keep the impact of beam wander and speckle low.

Further improvement of the clock transfer stability would require the use of active noise cancellation. Such a scheme has been demonstrated to be able to lower the transfer-induced instability by a factor of 10 in fiber-optic systems (Holman et al., 2005). To apply a similar technique in free space, retroreflection is needed to allow the returning beam to travel exactly the same path back in order to avoid differences in phase noise caused by different paths. In addition, the locking system likely needs to have a much wider bandwidth compared to the fiber-optic scheme even for a moderate transmission distance (e.g. 100 m) because of the relatively fast turbulence-induced index fluctuation in the air.

7. Acknowledgment

We would like to acknowledge Mr. Ted Rogers and Dr. Robert Lindquist of the Center for Applied Optics, the University of Alabama in Huntsville, for their assistance in the setup of the atmospherics transmission link. Ms Ayshah Alatawi made key contributions to the lab test of the microwave clock transfer. The work was supported in part by grants from UAH.

8. References

Akozbek, N.; Bowden, C. M.; Talebpour, A. & Chin, S. L. (2000). Femtosecond pulse propagation in air: Variational analysis, *Physics Review E*, Vol.61, pp. 4540-4549

Alatawi, A.; Gollapalli, R. P. & Duan, L. (2009). Radio frequency clock delivery via free-space frequency comb transmission, *Opt. Lett.*, Vol.34, pp. 3346-3348

Allan, D. W. (1966). Statistics of atomic frequency standards, *Proc. IEEE*, Vol.54, pp. 221-230

Altarelli, M (eds) et al. (2007). XFEL: The European X-Ray Free Electron Laser, *Technical Design Report DESY* 2006-097 (DESY, Hamburg, 2007), at http//xfel.desy.de

Andrews, L. C. & Phillips, R. L. (2005). *Laser beam propagation through random media*, SPIE Press, 2nd Ed.

Bartels, A.; Oates, C. W.; Hollberg, L. & Diddams, S. A. (2004). Stabilization of femtosecond laser frequency combs with subhertz residual linewidths, *Optics Letters*, Vol.29, pp. 1081-1083

Bartels, A.; Diddams, S. A.; Oates, C. W.; Wilpers, G.; Bergquist, J. C.; Oskay, W. H. & Hollberg, L. (2005). Femtosecond-laser-based synthesis of ultrastable microwave signals from optical frequency references, *Optics Letters*, Vol.30, pp. 667-669

Brookner, E. (1970). Atmosphere propagation and communication channel model for laser wavelengths, *IEEE Trans. Commun. Technol.*, COM-18, pp. 396-416

Chan, V. W. S. (2006). Free-space optical communications, *J. Lightwave. Technol.* Vol.24, pp. 4750-4762

Ciddor, P. E. (1996). Refractive index of air: new equations for the visible and near infrared, *Appl. Opt.*, Vol. 35, pp. 1566-1573

Ciddor P. E. and Hill, R. J. (1999). Refractive index of air. 2. group index, *Appl. Opt.*, Vol.38, pp. 1663-1667

Cliche, J. F. & Shillue, B. (2006). Precision timing control for radioastronomy, *IEEE Control Sys. Mag.* Vol.26, pp. 19-26

Davis, C. C. 1996. *Lasers and Electro-Optics Fundamentals and Engineering*, Cambridge University Press

Djerroud, K.; Acef, O.; Clairon, A.; Lemonde, P.; Man, C. N.; Samain, E. & Wolf, P. (2010). Coherent optical link through the turbulent atmosphere, *Opt. Lett.*, Vol.35, pp. 1479-1481

Foreman, S. M.; Holman, K. W.; Hudson, D. D.; Jones, D. J. & Ye, J. (2007). Remote transfer of ultrastable frequency references via fibre networks, *Review of Scientific Instruments*, Vol.78, pp. 021101

Fortier, T. M.; Le Coq, Y.; Stalnaker, J. E.; Ortega, D.; Diddams, S. A.; Oates, C. W. & Hollberg, L. (2006). Kilohertz-resolution spectroscopy of cold atoms with an optical frequency comb, *Physical Review Letters*, Vol.97, pp. 163905

Gollapalli, R. P. & Duan, L. (2010). Atmospheric timing transfer using a femtosecond frequency comb, *IEEE Photon. Journal*, Vol.2, pp. 904-910

Gollapalli, R. P. & Duan, L. (2011). Multiheterodyne Characterization of Excess Phase Noise in Atmospheric Transfer of a Femtosecond-Laser Frequency Comb, *Journal of Lightwave technology*, Vol.29, pp. 3401-3407.

Halford, D.; Wainwright, A. E. &Barnes, J. A. (1968). Flicker noise of phase in RF amplifiers and frequency multipliers: characterization, cause, and cure, *22nd Annual Symposium on Frequency Control,* pp. 340.

Hall, J. L. & Ye, J. (2003). Optical frequency standards and measurement, *IEEE Trans. Instrum. Meas.,* Vol.52, pp. 227-230

Holman, K. W.; Jones, D. J.; Hudson, D. D. & Ye, J. (2004). Precise frequency transfer through a fiber network by use of 1.5μm mode-locked sources, *Optics Letters,* Vol.29, pp. 1554-1556

Holman, K. W.; Hudson, D. D., Ye, J. & Jones, D. J. (2005). Remote transfer of a high-stability and ultralow-jitter timing signal, *Optics Letters,* Vol.30, pp. 1225-1227

Holzwarth, R.; Udem, T.; Hansch, T. W.; Knight, J. C.; Wadsworth, W. J. & Russell, P. St. J. (2000). Optical frequency synthesizer for precision spectroscopy, *Physical Review Letters,* Vol.85, pp. 2264-2267

Hong, S. T.; Sreenivasiah, I. & Ishimaru, A. (1977). Plane wave pulse propagation through random media, *IEEE Transactions on Antennas and Propagation,* AP-25, pp. 822-828

IEEE Std. 1139-1988, IEEE standard definitions of physical quantities for fundamental frequency and time metrology (IEEE, 1983).

Ishimaru, A. 1978. *Wave Propagation and Scattering in Random Media,* Academic, New York

Ivanov, E. N.; Diddams, S. A. & Hollberg, L. (2003). Analysis of noise mechanisms limiting the frequency stability of microwave signals generated with a femtosecond laser, *IEEE J. Select. Topics Quantum Electron,* Vol.9, pp. 1059-1065

Jones, D. J.; Diddams, S. A.; Ranka, J. K.; Stentz, A.; Windeler, R. S.; Hall, J. L. & Cundiff, S. T. (2000). Carrier-envelope phase control of femtosecond mode-locked lasers and direct optical frequency synthesis, *Science,* Vol.288, pp. 635-639

Kim, J.; Cox, J. A.; Chen, J. & Kaertner, F. X. (2008). Drift-free femtosecond timing synchronization of remote optical and microwave sources, *Nature Photonics,* Vol.2, pp. 733-736

Li, C. H. et al. (2008). A laser frequency comb that enables radial velocity measurements with a precision of 1 cm/s^{-1}, *Nature,* Vol.452, pp. 610-612,

Liu, C. H.; Wernik, A. W. & Yeh, K. C. (1974). Propagation of pulse trains through a random medium, *IEEE Transactions on Antennas and Propagation,* Vol.22, pp. 624-627

Liu, P. L.; Williams, K. J.; Frankel, M. Y. & Esman, R. D. (1999). Saturation characteristics of fast photodetectors, *IEEE Trans. Microwave Theory Tech,* Vol.47, pp. 1297-1303

Mlejnek, M.; Kolesik, M.; Moloney, J. V. & Wright, E. M. (1999). Optically turbulent femtosecond light guide in air, *Physics Review Letters,* Vol.83, pp. 2938-2941

Nahin, P. J. 2001. *The Science of Radio,* AIP Press, New York: Springer-Verlag

Prucnal, P.; Santoro, M. & Sehgal, S. (1986). Ultrafast All-Optical Synchronous Multiple Access Fiber Networks, *Selected Areas in Communications,* Vol.4, pp. 1484-1493

Ricklin, J. C. & Davidson, F. M. (2003). Atmospheric optical communication with a Gaussian Shell beam. *J. Opt. Soc. Am. A,* Vol.20, pp. 856-866

Rodriguez, M.; Bourayou, R.; Mejean, G.; Kasparian, J.; Yu, J.; Salmon, E.; Scholz, A.; Stecklum, B.; Eisloffel, J.; Laux, U.; Hatzes, A. P.; Sauerbrey, R.; Woste, L. & Wolf, J. (2004). Kilometer-range nonlinear propagation of femtosecond laser pulses, *Physics Review E,* Vol.69, pp. 036607

Rutman, J. & Walls, F. (1991). Characterization of frequency stability in precision frequency sources, *Proc. IEEE,* Vol.79, pp. 952-960

Shaik, K. S. (1988). Atmospheric propagation effects relevant to optical communications, *TDA Progress Report*, pp. 42–94

Sprangle, P.; Penano, J. R. & Hafizi, B. (2002). Propagation of intense short laser pulses in the atmosphere, *Physics Review E*, Vol.66, pp. 046418

Sprenger, B. ; Zhang, J.; Lu, Z. H. & Wang, L. J. (2009). Atmospheric transfer of optical and radio frequency clock signals, *Opt. Lett.*, Vol.34, pp. 965-967

Su, H. H. & Plonus, M. A. (1971). Optical-pulse propagation in a turbulent medium, *J. Opt. Soc. Am.* Vol.61, pp. 256-260

Thorpe, M. J.; Moll, K. D.; Jones, R. J.; Safdi, B. & Ye, J. (2006). Broadband cavity ringdown spectroscopy for sensitive and rapid molecular detection, *Science*, Vol.311, pp. 1595-1599

Tulchinsky, D. A. & Williams, K. J. (2005). Excess amplitude and excess phase noise of RF photodiodes operated in compression, *IEEE Photonics Technology Letters*, Vol.17, pp. 654-656

Udem, Th.; Diddams, S. A.; Vogel, K. R.; Oates, C.W.; Curtis, E. A.; Lee, W. D.; Itano, W. M.;Drullinger, R. E.; Bergquist, J.C. & Hollberg, L. (2001). Absolute frequency measurements of the Hg and Ca optical clock transitions with a femtosecond laser, *Physical Review Letters*, Vol.86, pp. 4996-4999

Udem, T.; Holzwarth, R. & Hansch, T. W. (2002). Optical frequency metrology, *Nature*, Vol.416, pp. 233-237

Ye, J. (2004). Absolute measurement of a long, arbitrary distance to less than an optical fringe, *Optics Letters*, Vol.29, pp. 1153-1155

Young, C. Y.; Andrews, L. C. & Ishimaru, A. (1998). Time-of-arrival fluctuations of a space-time Gaussian pulse in weak optical turbulence: An analytic solution, *Applied Optics*, Vol.37, pp. 7655-7660

17

Measurement of the Polarization State of a Weak Signal Field by Homodyne Detection

Sun-Hyun Youn
Department of Physics, Chonnam National University, Gwangju
Korea

1. Introduction

Information carried by an optical beam of light can usually be conveyed in the form of a temporal modulation of the intensity, phase, frequency or the polarization of the constituent mode(s). In that regard, homodyne detection is one of the most popular and standard method to measure the quantum mechanical properties of light Yuen & Shapiro (1978); Yuen et al. (1979); Yuen & Shapiro (1980); Yurke (1985). The quantum theory of homodyne detection originated principally from the works of Yuen and Shapiro et al. Yuen & Shapiro (1978); Yuen et al. (1979). In this method, a weak signal field is combined with a strong local oscillator field at a beam splitter, and the resulting signal is measured as a photocurrent. With the homodyne detection method, the quantum state of the signal field, such as the quadrature amplitude of the squeezed state, is easily measured Slusher et al. (1985); Wu et al. (1986); Polzik et al. (1992), and the quasiprobability distribution function can be measured by using the so-called optical homodyne tomography Smithey et al. (1993); Banaszek & W'odkiewicz (1996); Wallentowitz & Vogel (1996); Youn et al. (2001). Furthermore, optical homodyne detection is used to eavesdrop on the quantum key in quantum cryptography Hirano et al. (2000).

In quantum information science, the photon is a useful source for manipulating quantum information. To now, it has not been easy to make a consistent single-photon source, so the signal photon state is usually a weak coherent state. Therefore, it is very important to obtain the quantum state of an unknown signal field. The quantum mechanical properties of an unknown signal field, such as the amplitude squeezed state and the quadrature squeezed state, can be characterized by using the quasiprobability distribution (Wigner distribution) or the density matrix. The quasi probability distribution defines the statistical characteristics of the signal field . The well-known method to obtain quasi probability distributions (Wigner distributions) or density matrices is optical homodyne tomography Leonhardt (1997); Schiller et al. (1996).

Any kind of state reconstruction technique, however, in optical homodyne tomography requires repeated measurements of an ensemble of equally prepared signals. Therefore, this method is not adequate for finding the polarization state of a signal field that is changing pulse by pulse. Our novel scheme of polarization-modulated homodyne detection can obtain the polarization state of the signal field in a single-shot scheme. Even if the quasi probability of the signal field is not known in detail for a given single pulse, the varying polarization state of the signal field can be obtained pulse by pulse. In this respect, our novel scheme has an advantage: It can determine the varying polarization state of the signal against the usual optical homodyne tomography.

In our earlier work Youn & Noh (2003), we proposed a polarization-controllable homodyne detection scheme for a local oscillator field whose polarization state and global phase delay were changed by using an electro-optic modulator and a piezo- electric device, respectively. In that scheme, the polarization angle and the global phase of the local oscillator field must be scanned to obtain information on the polarization state of the signal field. Therefore, the scan time is non-zero, which limits the amount of polarization information that can be obtained from the pulsed signal field. The ordinary polarization modulated homodyne detection method, which requires no scan time, can obtain polarization information on a one-time pulsed signal field Youn (2005). Polarization-modulated ordinary homodyne detection does have an advantage in that it needs only one charged coupled device, but it also has a disadvantage in that it is not free from the noise of the local oscillator field.

In section 2, we propose a polarization-modulated balanced homodyne detection method. By inserting a set of wedged wave plates in the local oscillator port, we can modulate the relative phase of the two orthogonal components as well as the overall phase retardation of the local oscillator electric field Youn & Bae (2006). Using this spatially modulated local oscillator field in the homodyne detection scheme, we can obtain the amplitudes and the relative phase of the two orthogonal components of the signal electric field. We propose a practical method to measure the polarization state of the weak signal in a single-shot scheme without any scanning time. Our proposal will be one of the essential techniques in quantum information science for reading the polarization state of an unknown signal field in a single-shot scheme.

In section 3, a polarization-modulated homodyne detection scheme using photodetectors that measure four (temporally) simultaneous photocurrent signals corresponding to four (spatially) different quadrant-shaped combinations of wave plates is proposed Youn & Jain (2009). As an extension of our previous work, we essentially do a spatial phase modulation of the two orthogonal polarization components of the local oscillator (LO) electric field by inserting a system of waveplates in the path. Information about the Jones vector associated with the signal electric field can then be obtained, and this characterization can be performed in real-time, i.e., on a single-shot basis. In particular, we also articulate an analysis to discriminate between some typical polarization states, which has implications for the security offered by standard quantum cryptographic systems.

In our current work on polarization-modulated balanced homodyne detection, we are able to determine the polarization state of the signal field on a single-shot basis. This method could be described as a hybrid of ordinary homodyne detection and quantum polarization tomography, such as the one in Ref. James et al. (2001), employing an ensemble measurement of Stokes parameters for characterizing single qubits. However, while our scheme doesn't disseminate information about the quasi-probability distribution of the signal field for a given single pulse, it's usefulness comes into play in the determination of the varying polarization state of an arbitrary and unknown signal field in real-time.

In section 4, we discuss the applicability of the two methods for measuring the polarization state of a weak signal field.

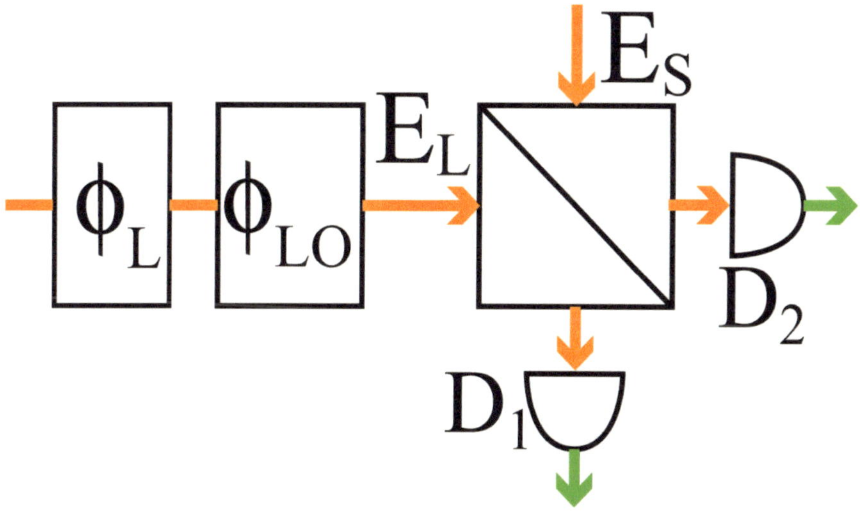

Fig. 1. Schematic diagram of polarization-controllable balanced homodyne detection. ϕ_L represents the polarization state of the local oscillator. ϕ_{LO} is the phase difference between the signal field and the local oscillator field (BS: Beam splitter, D: detector, E: Electric field).

2. Novel scheme of polarization-modulated balanced homodyne detection for measuring the polarization state of a weak field

2.1 Polarization-modulated Homodyne detection

In general, the polarizations of the signal field and the local oscillator are assumed to be linear and equal to each other. We propose a scheme for polarization-controlled balanced homodyne detection Youn & Noh (2003). When the polarizations of the signal and the local oscillator are different from each other, even the ordinary homodyne detection scheme can obtain various information about the polarization states of the signal field by varying the polarization of the local oscillator. In this paper, we propose an experimental scheme that can determine the polarization state of the input signal field by using polarization-modulated balanced homodyne detection.

The scheme of polarization-modulated homodyne detection is shown in Fig. 1. The signal and the local oscillator fields, both with arbitrary polarizations, are combined at a beam splitter (BS). The electric field in the local oscillator is given as Yariv (1989)

$$\vec{\mathcal{E}}_{LO}(\vec{r},t) = \left(\frac{2\hbar\omega}{\epsilon}\right)^{1/2} \left[\vec{E}_{LO}(\vec{r})\cos(\omega t + \phi_{LO}) + \frac{1}{2}\Delta\vec{E}_{LO}(\vec{r},t)\right] , \tag{1}$$

where ϵ is the dielectric constant and ω is the frequency of the local oscillator field. In Eq. (1), $\Delta\vec{E}_{LO}(\vec{r},t)$ is a fluctuation term, with the time average of $\Delta\vec{E}_{LO}(\vec{r},t)$ vanishing. The monochromatic term can be decomposed into two frequency parts as follows:

$$\begin{aligned} \vec{E}_{LO}(\vec{r},t) &\equiv \left(\frac{2\hbar\omega}{\epsilon}\right)^{1/2} \vec{E}_{LO}(\vec{r})\cos(\omega t + \phi_{LO}) \\ &= \vec{E}^{(+)}_{LO}(\vec{r},t) + \vec{E}^{(-)}_{LO}(\vec{r},t) , \end{aligned} \tag{2}$$

where ϕ_{LO} is the overall phase of the local oscillator field relative to the signal field, and the positive ($\vec{E}_{LO}^{(+)}$) and the negative ($\vec{E}_{LO}^{(-)}$) frequency component are given by

$$\vec{E}_{LO}^{(+)}(\vec{r},t) = \left(\vec{E}_{LO}^{(-)}(\vec{r},t)\right)^* = \left(\frac{\hbar\omega}{2\epsilon}\right)^{1/2} \vec{E}_{LO}(\vec{r})e^{-i\omega t}e^{-i\phi_{LO}} . \tag{3}$$

In addition, the fluctuating term can be decomposed into positive and negative frequency parts as

$$\Delta\vec{E}_{LO}(\vec{r},t) = \Delta\vec{E}_{LO}^{(+)}(\vec{r},t) + \Delta\vec{E}_{LO}^{(-)}(\vec{r},t) . \tag{4}$$

We also put the signal field in the signal port as

$$\vec{\mathcal{E}}_s(\vec{r},t) = \vec{\mathcal{E}}_s^{(+)}(\vec{r},t) + \vec{\mathcal{E}}_s^{(-)}(\vec{r},t) , \tag{5}$$

where

$$\vec{\mathcal{E}}_s^{(+)}(\vec{r},t) = \left(\vec{\mathcal{E}}_s^{(-)}(\vec{r},t)\right)^* = \vec{E}_s^{(+)}(\vec{r},t) ,$$

$$\vec{E}_s^{(+)}(\vec{r},t) = \left(\frac{\hbar\omega}{2\epsilon}\right)^{1/2} \vec{E}_s(\vec{r})e^{-i\omega t} . \tag{6}$$

The electric fields of the local oscillator and the signal are combined at a 50-50 BS, and the resultant electric fields at ports 1 and 2 are given by

$$\begin{aligned}\vec{E}_1(t) &= \frac{1}{\sqrt{2}}\left[\vec{\mathcal{E}}_{LO}(\vec{r},t) - \vec{\mathcal{E}}_s(\vec{r},t)\right] \\ &= \vec{E}_1^{(+)}(t) + \vec{E}_1^{(-)}(t) , \\ \vec{E}_2(t) &= \frac{1}{\sqrt{2}}\left[\vec{\mathcal{E}}_{LO}(\vec{r},t) + \vec{\mathcal{E}}_s(\vec{r},t)\right] \\ &= \vec{E}_2^{(+)}(t) + \vec{E}_2^{(-)}(t) ,\end{aligned} \tag{7}$$

respectively, where $\vec{E}_{1,2}^{(\pm)}(t)$ can be written as

$$\begin{aligned}\vec{E}_1^{(+)}(t) &= \left(\vec{E}_1^{(-)}(t)\right)^* \\ &= \frac{1}{\sqrt{2}}\left[\vec{E}_{LO}^{(+)}(\vec{r},t) + \left(\frac{\hbar\omega}{2\epsilon}\right)^{1/2} \Delta\vec{E}_{LO}^{(+)}(\vec{r},t) - \vec{E}_s^{(+)}(\vec{r},t)\right] , \\ \vec{E}_2^{(+)}(t) &= \left(\vec{E}_2^{(-)}(t)\right)^* \\ &= \frac{1}{\sqrt{2}}\left[\vec{E}_{LO}^{(+)}(\vec{r},t) + \left(\frac{\hbar\omega}{2\epsilon}\right)^{1/2} \Delta\vec{E}_{LO}^{(+)}(\vec{r},t) + \vec{E}_s^{(+)}(\vec{r},t)\right] .\end{aligned} \tag{8}$$

Since the incident photon flux operator is proportional to the product of electric field operators which are normally ordered and since one photoelectron is generated from an incident photon, based on the assumption of ideal photodetectors, the current measured from port 1 after integration over the detection area becomes

$$\begin{aligned} I_1(t) &= \frac{2e\sigma_{det}}{\hbar\omega}\sqrt{\frac{\epsilon}{\mu}}\vec{E}_1^{(-)}(t)\cdot\vec{E}_1^{(+)}(t) \\ &= \frac{ec\sigma_{det}}{2}\left\{\vec{E}^*_{LO}(\vec{r})e^{i\phi_{LO}+i\omega t}+\left[\Delta\vec{E}^*_{LO}(\vec{r})e^{i\omega t}-\vec{E}^*_s(\vec{r})e^{i\omega t}\right]\right\} \\ &\cdot\left\{\vec{E}_{LO}(\vec{r})e^{-i\phi_{LO}-i\omega t}+\left[\Delta\vec{E}_{LO}(\vec{r})e^{-i\omega t}-\vec{E}_s(\vec{r})e^{-i\omega t}\right]\right\}, \end{aligned} \tag{9}$$

where e is the electron charge, c is the speed of light in vacuum, and σ_{det} is the area of the detector. The current at port 2, $(I_2(t))$, is obtained similarly as above. Under the assumption that the fluctuation of the local oscillator field and the intensity of the signal field is much smaller than the mean intensity of the local oscillator field, the current difference between I_1 and I_2 can be given by

$$\begin{aligned} &I_1(t)-I_2(t)= \\ &-ec\sigma_{det}\left\{\vec{E}_{LO}(\vec{r})\cdot\vec{E}_s^{(+)}(\vec{r},t)e^{-i\phi_{LO}-i\omega t}+\vec{E}^*_{LO}(\vec{r})\cdot\vec{E}_s^{(-)}(\vec{r},t)e^{+i\phi_{LO}+i\omega t}\right\}. \end{aligned} \tag{10}$$

If we set $\vec{E}_{LO}(\vec{r})=A\hat{e}_L/\sqrt{V}$, where $\hat{e}_L$ is the Jones polarization vector, V is the mode volume, and A is a constant related to the intensity of the local oscillator field, the current becomes

$$\begin{aligned} &I(t)\equiv I_1(t)-I_2(t)= \\ &-\frac{ecA\sigma_{det}}{\sqrt{V}}\left[\vec{E}_s^{(-)}(\vec{r},t)\cdot\hat{e}_L e^{-i(\phi_{LO}+\omega t)}+\vec{E}_s^{(+)}(\vec{r},t)\cdot\hat{e}^*_L e^{+i(\phi_{LO}+\omega t)}\right]. \end{aligned} \tag{11}$$

In the homodyne detection scheme, the frequency of the signal field is the same as that of the local oscillator field, as in Eq. (6). Therefore, the current can be expressed by

$$I(t)=-\frac{ecA\sigma_{det}}{\sqrt{V}}\left[E^*_s(\vec{r})e^{-i\phi_{LO}}\hat{e}^*_s\cdot\hat{e}_L+E_s(\vec{r})e^{+i\phi_{LO}}\hat{e}_s\cdot\hat{e}^*_L\right], \tag{12}$$

where $\hat{e}_s$ is the Jones vector associated with the polarization of the signal field. Equation (12) is the final result for the measured current difference. In the general case, when both electric fields are arbitrarily polarized,

$$\hat{e}_s = a_1e^{i\delta_1}\hat{i}+a_2e^{i\delta_2}\hat{j}, \tag{13}$$

$$\hat{e}_L = \cos\theta_L\hat{i}+\sin\theta_L e^{i\phi_L}\hat{j}, \tag{14}$$

we have the current as

$$I=-\frac{2ecA\sigma_{det}}{\sqrt{V}}|E_s|\left[a_1\cos\theta_L\cos(\delta_1+\phi_{LO})+a_2\sin\theta_L\cos(\delta_2+\phi_{LO}-\phi_L)\right], \tag{15}$$

where the amplitudes a_1 and a_2 are non-negative real numbers which satisfy the normalization condition $a_1^2 + a_2^2 = 1$, and the phase factor δ_1 and δ_2 are real numbers.

2.2 Method to find the polarization state of the signal field

Information on the input polarization is obtained as follows: For a given signal field defined by $(a_1, a_2, \delta_1 - \delta_2)$, the intensity distribution I in Eq. (15) depends on three controllable parameters, θ_L, ϕ_L, and ϕ_{LO}, that are related to the local oscillator field. If we scan over (ϕ_L, ϕ_{LO}), we can obtain full information on the polarization state of the signal field Youn & Noh (2003). Usually, the parameters ϕ_L, and ϕ_{LO} are controlled by using an electro-optic modulator or a piezo electric material. An electro-optic modulator changes the polarization state of the local oscillator field, and a piezo electric material changes the path length associated with the phase delay of the local oscillator field relative to the signal field. However, it is difficult to get a full scan for a single-pulse signal because the scan needs non-zero time. In this work, we propose a new scheme which does not require any scan time. We made a spatially modulated local oscillator field to perform the scan for the ($\phi_L(t), \phi_{LO}(t)$) space in real spatial coordinates ($\phi_L(x,y), \phi_{LO}(x,y)$). To get spatially dependent photocurrents for a single shot-scheme, we only have to insert wave plates in the optical path of the local oscillator field, as shown in Fig. 2. Fig. 2 shows a device consisting of three wave plates and gives ϕ_L and ϕ_{LO} at once. The first two wave plates consist of two isotropic wedges whose refractive indices are n_L and n_R, respectively. The first and the second wedges are sliced by the plane $z = x \tan\alpha$, and the length from the first wedge and to the second surface at $y = 0$ is d_1, as in Fig. 2.

On the other hand, the third wave plate is a uniaxial crystal, such as calcite, whose refractive indices can be n_o and n_e. The adjacent surfaces of the second and the third wedges are sliced by the plane $z = y \tan\beta$, and at that position, the length from the second surface at $y = 0$ to the output surface of the third wedge is d_2, as in Fig. 2. The refractive indices of the third wedge are $n_x = n_o$ and $n_y = n_e$.

A ray passing horizontally to the right, $+z$, through the device at some arbitrary point (x, y) will traverse a thickness of $x \tan\alpha$ in the first wedge, $d_1 + y \tan\beta - x \tan\alpha$ in the second one, and $d_2 - y \tan\beta$ in the third one . The beam path delay imparted to the wave by the first wedge is $2\pi n_L x \tan\alpha/\lambda$, and that by the second wedge is $2\pi n_R(d + y \tan\beta - x \tan\alpha)/\lambda$. On the other hand, the refractive index of the third wedge depends on the polarization axis, so we have to calculate the wave retardation for two polarization axes. Let Γ^o be the wave retardation due to the three wave plates for the $x - axis$ linearly polarized light; then Γ^o becomes

$$\Gamma^o(x,y) = \frac{2\pi}{\lambda}\{n_L x \tan\alpha + n_R(d_1 + y\tan\beta - x\tan\alpha) + n_o(d_2 - y\tan\beta)\} . \tag{16}$$

and Γ^e, the wave retardation caused by the three wave plates for $y - axis$ linearly polarized light becomes

$$\Gamma^e(x,y) = \frac{2\pi}{\lambda}\{n_L x \tan\alpha + n_R(d_1 + y\tan\beta - x\tan\alpha) + n_e(d_2 - y\tan\beta)\} . \tag{17}$$

After passing through the three wedges, the incident ray falls on (x, y) as the initial Jones vector $\frac{1}{\sqrt{2}}(\hat{x} + \hat{y})$ changes into

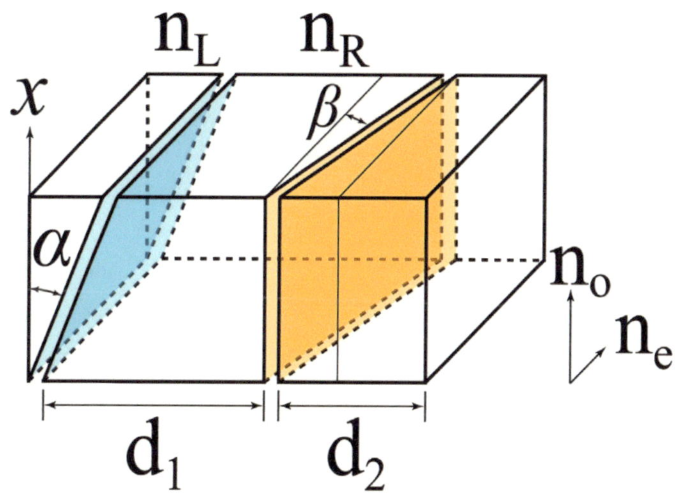

Fig. 2. Schematic diagram of two sets of wave plates with two isotropic wedges and one uniaxial wedge. The first and the second adjacent surfaces are sliced by the plane $z = x\tan\alpha$, and the refractive indices of the first and the second wedges are n_L and n_R, respectively. The second and the third adjacent surfaces are sliced by the plane $z = y\tan\beta$, and the refractive indices of the third wedge are $n_x = n_o$ and $n_y = n_e$.

$$\begin{pmatrix} \hat{e}_L^x(x,y) \\ \hat{e}_L^y(x,y) \end{pmatrix} = \begin{pmatrix} e^{i\Gamma^o(x,y)} & 0 \\ 0 & e^{i\Gamma^e(x,y)} \end{pmatrix} \begin{pmatrix} \frac{1}{\sqrt{2}} \\ \frac{1}{\sqrt{2}} \end{pmatrix}$$
$$\equiv e^{i\Phi_{LO}(x,y)} \begin{pmatrix} 1 & 0 \\ 0 & e^{i\Phi_L(x,y)} \end{pmatrix} \begin{pmatrix} \frac{1}{\sqrt{2}} \\ \frac{1}{\sqrt{2}} \end{pmatrix}, \tag{18}$$

where the phase factors are defined as

$$\Phi_{LO}(x,y) = \Gamma^o(x,y)$$
$$= \frac{2\pi}{\lambda}\{x(n_L - n_R)\tan\alpha + y(n_R - n_o)\tan\beta + n_R d_1 + n_o d_2\}, \tag{19}$$

$$\Phi_L(y) = \Gamma^e(x,y) - \Gamma^o(x,y)$$
$$= \frac{2\pi}{\lambda}\{y(n_o - n_e)\tan\beta + (n_e - n_o)n_o\}\,. \tag{20}$$

The overall phase delay between the signal field and the local oscillator field, and the relative phase difference between the two polarization directions $(\hat{x}, \hat{y})$ are spatially modulated by one set of wave plates. In other words, by inserting one set of wave plates, we can obtain the intensity distribution of the photocurrent difference over the entire range of the two phases

ϕ_L and ϕ_{LO}. Comparing Eq. (18) and Eq. (14), the spatially dependent phtocurrent becomes

$$I(x,y) = -\frac{\sqrt{2}ecA\sigma_{det}}{\sqrt{V}}|E_s|\left[a_1\cos(\delta_1+\phi_{LO}(x,y))+a_2\cos(\delta_2+\phi_{LO}(x,y)-\phi_L(x,y))\right]$$

$$= -\frac{\sqrt{2}ecA\sigma_{det}}{\sqrt{V}}|E_s|[a_1\cos(\delta_1+\frac{2\pi}{\lambda}\{(n_L-n_R)x\tan\alpha+(n_R-n_o)y\tan\beta+n_Rd_1+n_od_2\})$$
$$+a_2\cos(\delta_2+\frac{2\pi}{\lambda}\{(n_L-n_R)x\tan\alpha$$
$$+(n_R-2n_o+n_e)y\tan\beta+n_Rd_1-(2n_o+n_e)d_2\})], \quad (21)$$

where we put $\theta_L = \pi/4$ because the initial Jones vector of the local oscillator field is $\frac{1}{\sqrt{2}}(\hat{x}+\hat{y})$. Although we can find three parameters $(a_1, a_2, \delta_1-\delta_2)$ from the modulated intensity distribution in Eq. (21), it is better to match the refractive index of the second wave plate (n_R) with the refractive index of the third wave plate, $n_R = n_o$; then, the intensity distribution in Eq. (21) becomes

$$I(x,y) = -\frac{\sqrt{2}ecA\sigma_{det}}{\sqrt{V}}|E_s|[a_1\cos(\delta_1+\frac{2\pi}{\lambda}\{(n_L-n_o)x\tan\alpha\}+\Delta_1)$$
$$+a_2\cos(\delta_2+\frac{2\pi}{\lambda}\{(n_L-n_o)x\tan\alpha+(n_e-n_o)y\tan\beta\}+\Delta_2)], \quad (22)$$

where,

$$\Delta_1 = \frac{2\pi n_o(d_1+d_2)}{\lambda}, \quad (23)$$

$$\Delta_2 = \frac{2\pi(n_od_1-(2n_o+n_e)d_2)}{\lambda}. \quad (24)$$

Since the intensity distribution $I(x,y)$ depends on x and y independently , we can find the maximum value of the current I_{max} in Eq. (22) and let the values x_{max} and y_{max} be the x and y values that will yield the maximum current. When the measured values of x_{max} and y_{max} are used, the difference in the phase factor of the input polarization can be expressed as

$$\delta_2-\delta_1 = \frac{2\pi(n_e-n_o)y_{max}\tan\beta}{\lambda}+\Delta_2-\Delta_1+2m\pi, \quad (25)$$

where m is an integer. We can also obtain the magnitude of the polarization component of the input beam as

$$a_1 = \frac{I_0+I_\pi}{\sqrt{2\left(I_0^2+I_\pi^2\right)}},$$
$$a_2 = \frac{I_0-I_\pi}{\sqrt{2\left(I_0^2+I_\pi^2\right)}}, \quad (26)$$

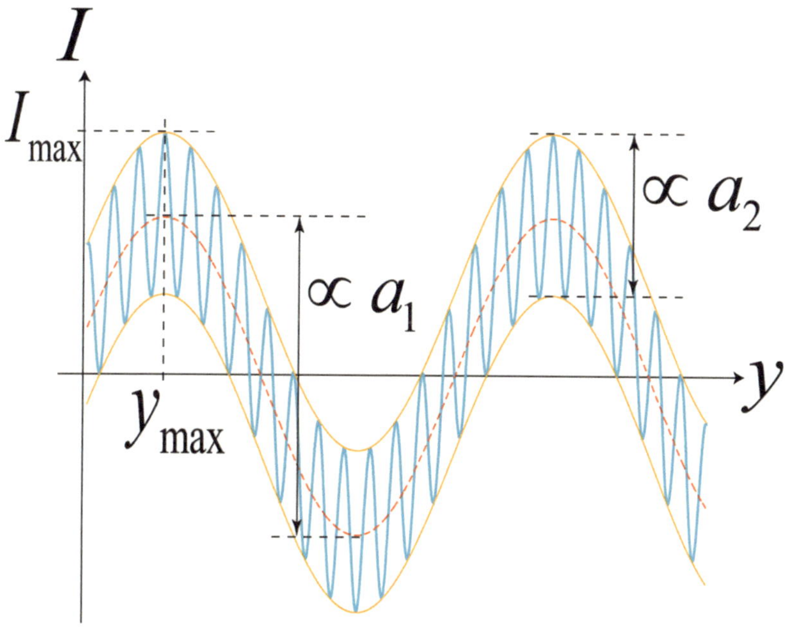

Fig. 3. Intensity modulation dependence on y. There are long- and short-term modulations whose modulation amplitudes depend on a_1 and a_2, respectively.

where

$$I_\eta \equiv I(x_{max}, y_{max} + \frac{\lambda}{2\pi(n_e - n_o)\tan\beta}\eta). \tag{27}$$

Thus, we can obtain full information on the polarization state of the input signal field, as shown in Eq. (25) and Eq. (26). Furthermore, since our results are obtained after scanning the entire space of the phase angles ϕ_L and ϕ_{LO}, we do not have to fix the relative phase angle ϕ_{LO} between the signal field and the local oscillator field.

Our three-wave-plate system may be simplified if we let the refractive index of the first wedge (n_L) be the same as that of the second wedge, $n_R = n_L$. Then, the modulated intensity distribution becomes

$$I(y) = -\frac{\sqrt{2}ecA\sigma_{det}}{\sqrt{V}}|E_s|[a_1\cos(\delta_1 + \frac{2\pi}{\lambda}\{(n_R - n_o)y\tan\beta)\} + \Delta_1)$$
$$+a_2\cos(\delta_2 + \frac{2\pi}{\lambda}\{(n_R - 2n_o + n_e)y\tan\beta\} + \Delta_2)]. \tag{28}$$

When $|n_R - n_o| << |n_R - 2n_o + n_e|$, there are fast and slow modulation frequencies in Eq. (28), as shown in Fig. 3. From the modulated data, it becomes simple to find the values of the unknown parameters (a_1, a_2, and $\delta_1 - \delta_2$) by using a fast Fourier transform or least square fitting method W. H. Press (1993).

Besides the complicated algorithm, a simple method is used to find the values of four parameters. From the intensity modulation data, we can roughly find the parameter a_2,

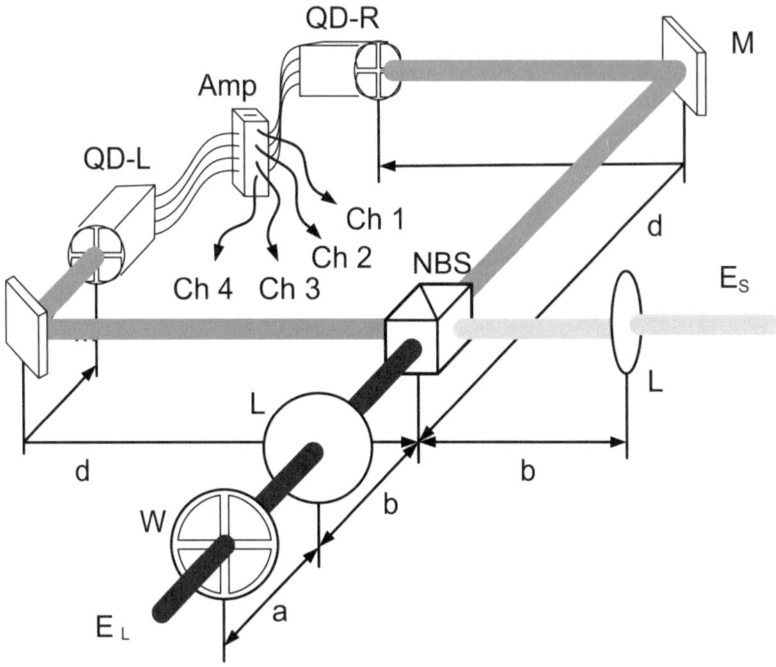

Fig. 4. Polarization-modulated homodyne detection. The organization of the waveplate quadrants for modulating the LO is explained in Fig. 5.

which is related to the short-term modulation amplitude. Furthermore, the total modulation amplitude is proportional to the sum of the two modulation amplitudes $(a_1 + a_2)$ as in Fig. 3. On the other hand, the phase factor δ_1 is also approximately determined at the interpolated intensity modulation peak I_{max} at y_{max} in Fig. 3. Then, the phase factor δ_1 becomes

$$\delta_1 = \frac{2\pi}{\lambda} \tan\beta(n_o - n_R)y_{max} - \Delta_1 + 2q\pi, \tag{29}$$

where q is an integer.

With the three parameters a_1, a_2, and δ_1, we can decide the value of the final phase factor δ_2 from the intensity modulation equation, Eq. (28). This experimental setup is very practical and easy for determining the polarization of a weak signal field.

3. Polarization-modulated quadrant homodyne detector for single-shot measurement of the polarization state of a weak signal field

3.1 Spatial modulation of the local oscillator field

Our scheme of polarization-modulated balanced homodyne detection (BHD) is shown in Fig. 4. A signal having an unknown polarization state and a local oscillator with its polarization modulated deterministically, as per the scheme explained in the next few paragraphs, impinge on a 50:50 non-polarizing beam splitter. The spatial modulation of the LO polarization is carried out by using a four-quadrant double wave plate system.

Fig. 5. Quadrant wave plate assembly: Quadrants cut from different HWPs (or QWPs) with their optical axis pre-rotated as per the requirement are glued again together to form a composite HWP (or QWP).

A de-constructed and simplified view of one of such waveplates is shown in Fig. 5. Basically, birefringent half and quarter wave plates (HWP and QWP - and at most four of each) are all cut into quadrants first. Prior to this operation, they may have to be rotated as per the polarization change that is to be induced onto the incoming field by the participating quadrant. A composite HWP (or QWP) is, thus, prepared with the four quadrants glued back together - this is similar to the idea used to make a custom-made phase modulator in the experiment by Bachor Bachor (2006). Fig. 5 shows four quadrants A-D assembled from three different (say Half) wave plates. The optical axis of the quadrants are a priori rotated by $\pi/4$, $\pi/2$, $-\pi/4$ and $\pi/2$ with respect to the horizontal. Of course, these rotation angles could really be arbitrary, but we refer to these specific values because they are employed in our computational analysis, presented in the next section. Also, it is easily observed that B and D can be cutout from one single HWP with the optical axis rotated by $\pi/2$, thus the need for three waveplates, in all.

With the transfer function $\Gamma[\psi]$ of a waveplate, which induces a phase shift of 2ψ between the two orthogonal polarization components, being given by

$$\mathbf{\Gamma}[\psi] = \begin{pmatrix} e^{-i\psi} & 0 \\ 0 & e^{i\psi} \end{pmatrix}$$

and a general rotation matrix given by

$$\mathbf{R}[\theta] = \begin{pmatrix} \cos\theta & -\sin\theta \\ \sin\theta & \cos\theta \end{pmatrix}$$

the output polarization state of the LO field (with input being denoted by $\hat{e}_L$), after having passed through a combination of a HWP and a QWP rotated at angles α and β, respectively, is given by

$$\hat{\mathbf{e}}_{\mathrm{L}}^{\mathrm{out}} = \mathbf{R}[-\alpha]\mathbf{\Gamma}[\pi/2]\mathbf{R}[\alpha]\mathbf{R}[-\beta]\mathbf{\Gamma}[\pi/4]\mathbf{R}[\beta]\hat{\mathbf{e}}_{\mathrm{L}}. \tag{30}$$

3.2 Homodyne detection & polarization state measurement

As can be seen from Fig. 4, the lenses positioned in the Signal and the LO arm have a focal length f such that

$$\frac{1}{a} + \frac{1}{(b+d)} = \frac{1}{f}. \tag{31}$$

Careful alignment of these lenses serves to put one image plane each at the quadrant-based detectors QD-L and QD-R, as shown in the Fig. 4. The interference between the four spatial polarization-modulated modes of LO and the signal field (assumed to be uniform spatially) at the beam splitter is thus captured just like in an ordinary homodyne detection experiment, and taking into the account the inversion (A $\leftrightarrow$ C and B $\leftrightarrow$ D), the corresponding difference photocurrent signals are amplified and obtained in channels 1-4. With the usual assumptions for a balanced homodyne detection analysis, i.e., the intensity of the signal field being much smaller than the mean intensity of the LO field, and the photodetector pair being ideal (i.e. $\eta_L = \eta_R = 1$), a generic expression for the difference photocurrent $I(t) \propto I_L(t)$ - $I_R(t)$ is Youn & Bae (2006)

$$I(t) = -\frac{ecK\sigma_{det}}{\sqrt{V}}\left[E_S(\vec{r})e^{i\phi_{LO}}\hat{e}_L^* \cdot \hat{e}_S + E_S^*(\vec{r})e^{-i\phi_{LO}}\hat{e}_L \cdot \hat{e}_S^*\right], \tag{32}$$

where
e is the electronic charge,
c is the speed of light in vacuum,
K is a constant dependent on the local oscillator's field intensity,
σ_{det} is the area of the detector,
V is the mode volume,
$E_S(\vec{r})$ is the complex amplitude of the signal electric field,
$\hat{e}_S$ and $\hat{e}_L$ are the Jones vectors associated with the polarization of the Signal and the LO field, respectively, and ϕ_{LO} is overall phase of the LO field (relative to signal).
Again, in the general case, both electric fields are arbitrarily polarized, i.e.,

$$\hat{e}_S = a_1 e^{i\delta_1}\hat{x} + a_2 e^{i\delta_2}\hat{y}, \tag{33}$$

$$\hat{e}_L = b_1 e^{i\phi_{Lx}}\hat{x} + b_2 e^{i\phi_{Ly}}\hat{y}, \tag{34}$$

where a_i, b_i are real numbers and the coefficients satisfy $a_1^2 + a_2^2 = 1, b_1^2 + b_2^2 = 1$. In addition, if we also consider the vacuum field fluctuations that are quantum mechanically independent from the signal field, then its Jones polarization vector would be orthogonal to $\hat{e}_S$, and would thus be of the form

$$\hat{e}_V = -a_2 e^{i\delta_1}\hat{x} + a_1 e^{i\delta_2}\hat{y}. \tag{35}$$

Now, if a symmetric input polarization is employed in the local oscillator beam, i.e., $\hat{e}_L = 1/\sqrt{2}\ \hat{x} + 1/\sqrt{2}\ \hat{y}$, then evaluating Eq. (30) yields the components for the output polarization, $\hat{e}_L^{out}$, in quadrants A, B, C and D, with pre-chosen values of α (same distribution as portrayed in Fig. 5 and β. Table 1 lists these polarization states:

Quadrant	α	β	$(\hat{e}_{Lx}^{out}, \hat{e}_{Ly}^{out})$
A	$\frac{\pi}{4}$	$-\frac{\pi}{4}$	$(\frac{1}{\sqrt{2}}, \frac{1}{\sqrt{2}})$
B	$\frac{\pi}{2}$	0	$(\frac{1}{\sqrt{2}}, \frac{i}{\sqrt{2}})$
C	$-\frac{\pi}{4}$	$-\frac{\pi}{4}$	$(-\frac{i}{\sqrt{2}}, -\frac{i}{\sqrt{2}})$
D	$\frac{\pi}{2}$	$-\frac{\pi}{4}$	$(-\frac{i}{\sqrt{2}}, \frac{1}{\sqrt{2}})$

Table 1. Output polarization with HWP & QWP rotated by α & β, respectively.

Further, considering a suitable form of Eq. (32) that accounts for the effect of vacuum, as well as substituting $\hat{e}_L^{out}$ for $\hat{e}_L$, in the same, yields (on simplification)

$$I(t) = -\frac{2ecK\sigma_{det}}{\sqrt{V}}\Big(b_1(a_1E_S - a_2E_V)\cos(\delta_1 - \phi_{Lx} + \phi_{LO}) + b_2(a_2E_S + a_1E_V)\cos(\delta_2 - \phi_{Ly} + \phi_{LO})\Big). \tag{36}$$

Here, the $\vec{r}$ dependence in E_S and E_V has been dropped for convenience sake, and ϕ_{Lx} and ϕ_{Ly} refer to the phases of the polarization components of the LO *after* the modulation, as dictated by the wave-plate assembly. Also, for non-classical fields such as the squeezed vacuum state, the difference photocurrent that constitutes terms arising from LO quantum noise and the quadrature amplitude of the signal (enhanced by the power in LO) has a statistical average of zero, i.e., $\langle \mathrm{I(t)} \rangle = 0$. For a weak coherent field however, $\langle \mathrm{I(t)} \rangle$ is finite and this gives us a ground to make measurements merely on I(t), instead of the usual $\langle \mathrm{I}^2\mathrm{(t)} \rangle$. Finally, in Eq. (36), setting $\delta_1 + \phi_{LO} \to \delta_1$ and $\delta_2 + \phi_{LO} \to \delta_2$, or equivalently absorbing the effect of ϕ_{LO} in δ_1 and δ_2 by reducing it to zero and computing ϕ_{Lx} and ϕ_{Ly} from the last column of table 1, the expressions for the difference photocurrents are produced in channels 1-4 corresponding to quadrants A-D:

Quadrant	ϕ_{Lx}^{out}	ϕ_{Ly}^{out}	I_{out}^v	I_{out}
A	0	0	$(a_1E_s - a_2E_v)\cos\delta_1 + (a_2E_s + a_1E_v)\cos\delta_2$	$a_1\cos\delta_1 + a_2\cos\delta_2$
B	0	$\frac{\pi}{2}$	$-(a_1E_s - a_2E_v)\sin\delta_1 - (a_2E_s + a_1E_v)\sin\delta_2$	$-a_1\sin\delta_1 - a_2\sin\delta_2$
C	$-\frac{\pi}{2}$	$-\frac{\pi}{2}$	$(a_1E_s - a_2E_v)\cos\delta_1 + (a_2E_s + a_1E_v)\sin\delta_2$	$a_1\cos\delta_1 + a_2\sin\delta_2$
D	$-\frac{\pi}{2}$	0	$-(a_1E_s - a_2E_v)\sin\delta_1 + (a_2E_s + a_1E_v)\cos\delta_2$	$-a_1\sin\delta_1 + a_2\cos\delta_2$

Table 2. Photocurrent expressions for the four different quadrants. The last column corresponds to the case when vacuum fluctuations are totally neglected, i.e., when the signal field is dominant.

In general, an arbitrary (but fixed) value of the LO phase should be considered in Eq. (36). Then, replacing ϕ_{LO} by ψ and an appropriate phase factor (so as to preserve the orthogonality between the *representation* of the signal polarization), the four channels' difference photocurrents for various (assumed) polarization states for the signal field can be found, as displayed in table 3.

$SNo.$	$\hat{e}_s$	A	B	C	D
1	(1,0)	$\cos\psi$	$-\sin\psi$	$\cos\psi$	$-\sin\psi$
2	(0,1)	$\cos(\psi-\frac{\pi}{4})$	$-\sin(\psi-\frac{\pi}{4})$	$\sin(\psi-\frac{\pi}{4})$	$\cos(\psi-\frac{\pi}{4})$
3	$\frac{1}{\sqrt{2}}(1,1)$	$\sqrt{2}\cos\psi$	$-\sqrt{2}\sin\psi$	$\sin(\psi+\frac{\pi}{4})$	$\cos(\psi+\frac{\pi}{4})$
4	$\frac{1}{\sqrt{2}}(1,-1)$	0	0	$\cos(\psi+\frac{\pi}{4})$	$-\sin(\psi+\frac{\pi}{4})$
5	$\frac{1}{\sqrt{2}}(1,i)$	$\cos(\psi+\frac{\pi}{4})$	$-\sin(\psi+\frac{\pi}{4})$	$\sqrt{2}\cos\psi$	$-\sqrt{2}\sin\psi$
6	$\frac{1}{\sqrt{2}}(1,-i)$	$\sin(\psi+\frac{\pi}{4})$	$\cos(\psi+\frac{\pi}{4})$	0	0

Table 3. Measured photocurrent values for various $\hat{e}_s$ in the four quadrants. A phase factor of $\frac{\pi}{4}$ is present so that the representation of $\hat{e}_s$ = (1,0) is orthogonal to $\hat{e}_s$ = (0,1).

Thus, the comparative knowledge from this should make it possible to derive information about the six different signal polarization states, which also happen to be the most fundamental ($|H\rangle$, $|V\rangle$, $|+\rangle$, $|-\rangle$, $|R\rangle$, and $|L\rangle$) in quantum information science. For example, for an experimental run, if no modulation is observed in channels A & B simultaneously, then $\hat{e}_s = \frac{1}{\sqrt{2}}(1,-1)$. A note of caution: since a knowledge of the absolute phase is not possible, the above expressions have been derived considering a relative phase of $\delta = \delta_2 - \delta_1$ (refer to Eq. (33)); hence, the information is true up to a global phase. Further, using Table 3, we can now also compute error functions, $er_i(\psi)$, that indicate the deviation of the theoretical values from the *experimentally* measured values, i.e.

$$er_i(\psi) = \sum_{j=1}^{4}(n_{ij} - o_j)^2, \tag{37}$$

where o_j is the actual value measured in the jth channel, and n_{ij} is an element of the 6x4 matrix that is contained in Table 3, with rows indicated by the signal polarization and columns by the four quadrants: e.g., $n_{32} = -\sqrt{2}\sin\psi$. Thus, for a given observation, Eq. (37) yields six different plots as functions of ψ and the unknown signal field polarization is indicated by the curve with the absolute minimum amongst the six.

4. Conclusion and discussion

For an ensemble of equally prepared signals, the well-established optical homodyne tomography method give us the quasi probability distribution of the signal field, and the quasi probability distribution defines the statistical characteristics of the signal field, such as the amplitude squeezed state and quadrature squeezed state Leonhardt (1997). This method, however, is not adequate for obtaining information on a signal field that is changing pulse by pulse. Our polarization-modulated homodyne detection scheme is able to obtain the polarization state of a signal field in a single-shot scheme. Although the quasi probability of the signal field for a given single pulse cannot be known, we can determine the varying polarization state of the signal field pulse by pulse. Our novel scheme can determine the varying polarization state of the signal and can be used in quantum information science.

We propose a novel homodyne detection method to measure a polarization state of a weak field. At first we introduced a novel scheme of polarization modulated balanced homodyne detection method. By inserting a set of wedged wave plate in the local oscillator port, we can modulated the relative phase of the two orthogonal components as well as the overall phase retardation of the local oscillator electric field. Using this spatially modulated local oscillator field in the homodyne detection scheme, we can obtain the amplitudes and the relative phase of the two orthogonal components of the signal electric field. It's a practical method to measure the polarization state of the weak signal in a single-shot scheme without any scanning time.

Note that, of course, other schemes may be able to measure the state of the polarization of a pulsed weak signal at once: for example, dividing the signal into many beams, measuring them with polarization-measurement setups, and finally deriving the polarization state of the input signal electric field. However, it is not practical, especially, for very weak beams because a divided beam is too weak for measurement, so the loss is not negligible. In our scheme, in some sense, we also divided the signal field spatially, but the strong local oscillator field plays a role in measuring the weak beam by a kind of amplification as in the usual homodyne detection scheme.

The second polarization-modulated homodyne detection scheme uses photodetectors that measure four simultaneous photocurrent signals corresponding to four different quadrant shaped combination of wave plates. We make a spatial phase modulation of the two orthogonal polarization components of the local oscillator electric field by inserting a system of wave plates in the path. Information about the Jones vector associated with the signal field can then be obtained, and characterization can be performed in real time, i.e., on a single shot basis.

The subtle aspect in our scheme is representing the polarization of photons, i.e., a qubit system, in a higher (four) dimensional space, thus allowing for a better discrimination between any two polarization states. To elaborate, if the quadrant-measured values are taken as components of a four-dim vector, then an orthogonality between two different signal polarization states is preserved in this new representation. Further, it includes the power of homodyne detection, which allows a measurement of a very weak or highly attenuated field, by amplifying it sufficiently. This new scheme might have application in bio-physics, where we have to measure the polarization change of the very weak beam from a single molecule.

5. Acknowledgements

I would like to thank Samyong Bae and Nitin Jain for their valuable work for this article.

6. References

H. P. Yuen and J. H. Shapiro, IEEE Trans. Inf. Theory 24, 657 (1978).
J. H. Shapiro, H. P. Yuen, and J. A. Machado-Matta, IEEE Trans. Inf. Theory 25, 179 (1979).
H. P. Yuen and J. H. Shapiro, IEEE Trans. Inf. Theory 26, 78 (1980).
B. Yurke, Phys. Rev. A 32, 311 (1985).
R. E. Slusher, L. W. Hollberg, B. Yurke, J. C. Mertz, and J. F. Valley, Phys. Rev. Lett. 55, 2409 (1985).
L-A. Wu, H. J. Kimble, J. H. Hall, and H. Wu, Phys. Rev. Lett. 57, 2520 (1986).
E. S. Polzik, J. Carri, and H. J. Kimble, Phys. Rev. Lett. 68, 3020 (1992).
D. T. Smithey, M. Beck, M. G. Raymer, and A. Faridani, Phys. Rev. Lett. 70, 1244 (1993).

K. Banaszek and K. W'odkiewicz, Phys. Rev. Lett. 76, 4344 (1996).
S. Wallentowitz and W. Vogel, Phys. Rev. A 53, 4528 (1996).
S. H. Youn, Y. T. Chough and K. An, J. Korean Phys. Soc. 39, 255, (2001).
T. Hirano, T. Konishi, and R. Namiki, quant-ph/0008037.
A. Yariv, *Quantum Electronics* (Wiley, New York, 1989).
U. Leonhardt, *Measuring the Quantum State of Light*, (Cambridge University Press, Cambridge, 1997).
S. Schiller, G. Breitenbach, S. F. Pereira, T. Muller, and J. Mlynek, Phys. Rev. Lett. 77, 2933 (1996).
D. S. Krahmer and U. Leonhardt, Phys. Rev. A 55, 3275 (1997).
H. R. Noh and S.H. Youn, J. Korean Phys. Soc. 43, 1029 (2003).
E. Hecht, *Optics* (Addison Wesley, New York, 2002).
S. H. Youn, J. Korean Phys. Soc. 47, 803 (2005)
W. H. Press *et al.*, *Numerical Recipes* second edition, (Cambridge University Press, Cambridge, England, 1993).
M. Munroe, D. Boggavarapu, M. E. Anderson and M. G. Raymer, Phys. Rev. A 52, R924 (1995).
C.H. Bennet and G. Brassard, Proceedings of IEEE Intl. Conf. on Computers Systems and Signal Processing, Bangalore India, December 175-179 (1984).
S.H. Youn and Nitin Jain, J. Korean Phys. Soc. 54, 29 (2009).
S.H. Youn and Samyong Bae, J. Korean Phys. Soc. 48, 397 (2006).
Daniel F. V. James, Paul G. Kwiat, William J. Munro, and A. G. White, Phys. Rev. A 64, 052312 (2001)
H-A. Bachor, J. Mod. Opt. 53, 5-6, 597 (2006).

18

Spin Photodetector: Conversion of Light Polarization Information into Electric Voltage Using Inverse Spin Hall Effect

Kazuya Ando and Eiji Saitoh
Institute for Materials Research, Tohoku University
Japan

1. Introduction

Recent developments in optical and material science have led to remarkable industrial applications, such as optical data recording and optical communication. The scope of the conventional optical technology can be extended by exploring simple and effective methods for detecting light circular polarization; light circular polarization carries single-photon information, making it essential in future optical technology, including quantum cryptography and quantum communication.

Light circular polarization is coupled with electron spins in semiconductors (Meier, 1984). When circularly polarized light is absorbed in a semiconductor crystal, the angular momentum of the light is transferred to the semiconductor, inducing spin-polarized carriers though the optical selection rules for interband transitions [see Fig. 1(a)]. This process allows conversion of light circular polarization into electron-spin polarization, enabling the integration of light-polarization information into spintronic technologies.

If one can convert electron spin information into an electric signal, light circular polarization information can be measured through the above process. Recently, in the field of spintronics, a powerful technique for detecting electron spin information has been established, which utilizes the inverse spin Hall effect (ISHE) (Saitoh, 2006; Valenzuela, 2006; Kimura, 2007). The ISHE converts a spin current, a flow of electron spins in a solid, into an electric field through the spin-orbit interaction, enabling the transcription of electron-spin information into an electric voltage. This suggests that light-polarization information can be converted into an electric signal by combining the optical selection rules and the ISHE.

This chapter describes the conversion of light circular polarization information into an electric voltage in a Pt/GaAs structure though the optical generation of spin-polarized carriers and the ISHE: the photoinduced ISHE (Ando, 2010).

2. Optical excitation of spin-polarized carriers in semiconductors

When circularly polarized light is absorbed in a semiconductor, the angular momentum of the light is transferred to the material, which polarizes carrier spins in the semiconductor through the spin-orbit interaction (Meier, 1984). This optical generation of spin-polarized carriers has been a powerful technique for exploring spin physics in direct band gap semiconductors, such as GaAs. In GaAs, the valence band maximum and the conduction band minimum are at Γ with an energy gap $E_g = 1.43$ eV at room temperature. The valence band (*p* symmetry) splits into fourfold degenerate $P_{3/2}$ and twofold degenerate $P_{1/2}$ states, which lie $\Delta = 0.34$ eV below $P_{3/2}$ at Γ, whereas the conduction band (*s* symmetry) is twofold degenerate $S_{1/2}$ as schematically shown in Fig. 1(a). In the fourfold degenerate $P_{3/2}$ state, holes can occupy states with values of angular momentum $m_j = \pm 1/2, \pm 3/2$, corresponding to light hole (LH) and heavy hole (HH) sates, respectively (see Fig. 1(a)). Let $|J, m_j\rangle$ be the Bloch states according to the total angular momentum J and its projection onto the positive *z* axis m_j. The band wave functions can be expressed as listed in Table 1, where $|S\rangle$, $|X\rangle$, $|Y\rangle$, and $|Z\rangle$ are the wave functions with the symmetry of *s*, p_x, p_y, and p_x orbitals. The interband transitions satisfy the selection rule $\Delta m_j = \pm 1$, reflecting absorption of the photon's original angular momentum. The probability of a transition involving a LH or HH state is weighted by the square of the corresponding matrix element connecting it to the appropriate electron state, so that the relative intensity of the optical transition between the heavy and the light hole subbands and the conduction band induced by circularly polarized light illumination is 3. Thus absorption of photons with angular momentum +1 produces three spin-down ($m_j = -1/2$) electrons for every one spin-up ($m_j = +1/2$) electron, resulting in an electron population with a spin polarization of 50% in a bulk material, where the HH and LH states are degenerate. The relative transition rates are summarized in Fig. 1(b). Therefore, because of the difference in the relative intensity, a spin-polarized carriers can be generated by the illumination of circularly polarized light. Note that the resulting electron spin is oriented parallel or anti-parallel to the propagation direction of the incident photon.

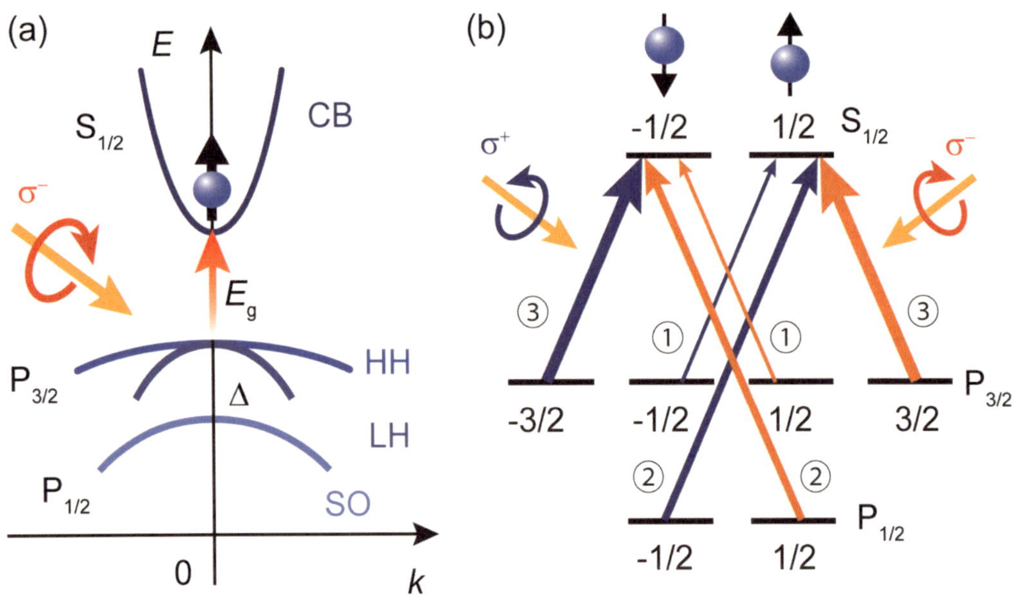

Fig. 1. (a) Optical generation of spin-polarized carriers in semiconductors. (b) Interband transitions for right and left circularly polarized light illumination.

	$\vert J, m_j\rangle$	wave function
CB	$\vert 1/2, 1/2\rangle$	$\vert S\uparrow\rangle$
	$\vert 1/2, -1/2\rangle$	$\vert S\downarrow\rangle$
HH	$\vert 3/2, 3/2\rangle$	$\vert (1/2)^{(1/2)}(X+iY)\uparrow\rangle$
	$\vert 3/2, -3/2\rangle$	$\vert (1/2)^{(1/2)}(X-iY)\downarrow\rangle$
LH	$\vert 3/2, 1/2\rangle$	$\vert (1/6)^{(1/2)}[(X+iY)\downarrow +2Z\uparrow]\rangle$
	$\vert 3/2, -1/2\rangle$	$\vert -(1/6)^{(1/2)}[(X-iY)\uparrow -2Z\downarrow]\rangle$
SO	$\vert 1/2, 1/2\rangle$	$\vert -(1/3)^{(1/2)}[(X+iY)\downarrow -Z\uparrow]\rangle$
	$\vert 1/2, -1/2\rangle$	$\vert (1/3)^{(1/2)}[(X-iY)\uparrow +Z\downarrow]\rangle$

Table 1. Wave functions for the conduction band (CB), heavy hole (HH), light hole (LH), and spin-orbit split-off band (SO).

3. Spin current and inverse spin Hall effect

A spin current is a flow of electron spins in a solid. One of the driving forces for a spin current is a gradient of the difference in the spin-dependent electrochemical potential $\nabla\mu_\sigma$ for spin up ($\sigma = \uparrow$) and spin down ($\sigma = \downarrow$). Here, $\mu_\sigma = \mu_\sigma^c - e\phi$, where μ_σ^c is the chemical potential. A current density for spin channel σ is expressed as

$$\mathbf{j}_\sigma = \frac{\sigma_\sigma}{e}\nabla\mu_\sigma, \tag{1}$$

where σ_σ is the electrical conductivity for spin up ($\sigma = \uparrow$) and spin down ($\sigma = \downarrow$) channel. Here, a charge current, a flow of electron charge, is the sum of the current for $\sigma = \uparrow$ and $\downarrow$ as $\mathbf{j}_c = \mathbf{j}_\uparrow + \mathbf{j}_\downarrow$:

$$\mathbf{j}_c = \frac{1}{e}\nabla\left(\sigma_\uparrow\mu_\uparrow + \sigma_\downarrow\mu_\downarrow\right). \tag{2}$$

This flow is schematically illustrated in Fig. 2(a). This flow carries electron charge while the flow of spins is cancelled. In contrast, the opposite flow of $\mathbf{j}_\uparrow$ and $\mathbf{j}_\downarrow$, $\mathbf{j}_s = \mathbf{j}_\uparrow - \mathbf{j}_\downarrow$, or

$$\mathbf{j}_s = \frac{1}{e}\nabla\left(\sigma_\uparrow\mu_\uparrow - \sigma_\downarrow\mu_\downarrow\right), \tag{3}$$

carries electron spins without a charge current. This is a spin current. In nonmagnetic materials, a spin current is expressed as $\mathbf{j}_s = (\sigma_N/2e)\nabla(\mu_\uparrow - \mu_\downarrow)$, since the electrical conductivity is spin-independent: $\sigma_\uparrow = \sigma_\downarrow = \sigma_N/2$.

Since charge ρ is a conserved quantity, the continuity equation of charge is described as

$$\frac{d}{dt}\rho = -\nabla\cdot\mathbf{j}_c. \tag{4}$$

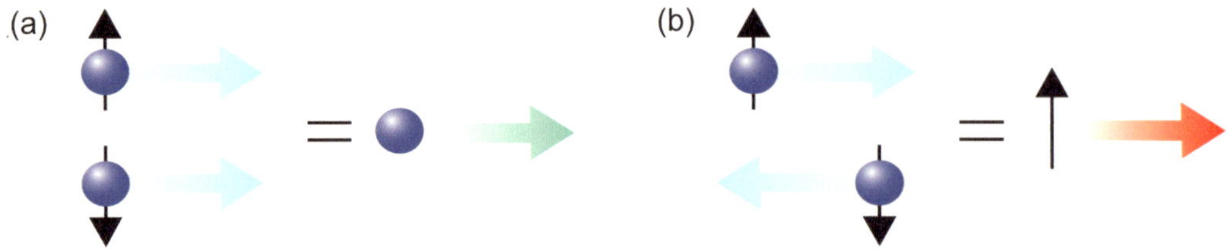

Fig. 2. (a) A schematic illustration of a charge current. (b) A schematic illustration of a spin current.

In contrast, spins are not conserved; a spin current decays typically in a length scale of nm to μm. Therefore, the continuity equation of spins are written as

$$\frac{d}{dt}M_z = -\nabla \cdot \mathbf{j}_s + \mathcal{T}_z, \tag{5}$$

where M_z is the z component of magnetization. z is defined as the quantization axis. Here, $\mathcal{T}_z = e(n_\uparrow - \overline{n}_\uparrow)/\tau_{\uparrow\downarrow} - e(n_\downarrow - \overline{n}_\downarrow)/\tau_{\downarrow\uparrow}$ represents spin relaxation. $\overline{n}_\sigma$ is the equilibrium carrier density with spin σ and $\tau_{\sigma\sigma'}$ is the scattering time of an electron from spin state from σ to σ'. Note that the detailed balance principle imposes that $N_\uparrow/\tau_{\uparrow\downarrow} = N_\downarrow/\tau_{\downarrow\uparrow}$, so that in equilibrium no net spin scattering takes place, where N_σ denotes the spin dependent density of states at the Fermi energy. This indicates that, in general, in a ferromagnet, $\tau_{\uparrow\downarrow}$ and $\tau_{\downarrow\uparrow}$ are not the same. In the equilibrium condition, $d\rho/dt = dM_z/dt = 0$, using the continuity equations, one finds the spin-diffusion equations:

$$\nabla^2(\sigma_\uparrow \mu_\uparrow + \sigma_\downarrow \mu_\downarrow) = 0, \tag{6}$$

$$\nabla^2(\mu_\uparrow - \mu_\downarrow) = \frac{1}{\lambda^2}(\mu_\uparrow - \mu_\downarrow), \tag{7}$$

where $\lambda = \sqrt{D\tau_{sf}}$ is the spin diffusion length. $D = D_\uparrow D_\downarrow(N_\uparrow + N_\downarrow)/(N_\uparrow D_\uparrow + N_\downarrow D_\downarrow)$ is the diffusion constant. The spin relaxation time τ_{sf} is given by $1/\tau_{sf} = 1/\tau_{\uparrow\downarrow} + 1/\tau_{\downarrow\uparrow}$. By solving the diffusion equations, one can obtain the spatial variation of spin currents generated by $\mu_\uparrow - \mu_\downarrow$. A spin current generated by $\mu_\uparrow - \mu_\downarrow$ decays as $e^{-\lambda/x}$. Thus a spin current play a key role only in a system with the scale of λ.

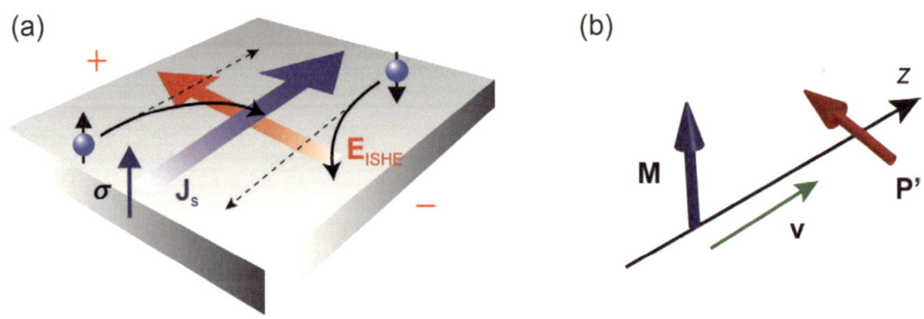

Fig. 3. (a) A schematic illustration of the inverse spin Hall effect. (b) Conversion of magnetic moment **M** into electric polarization **P**'.

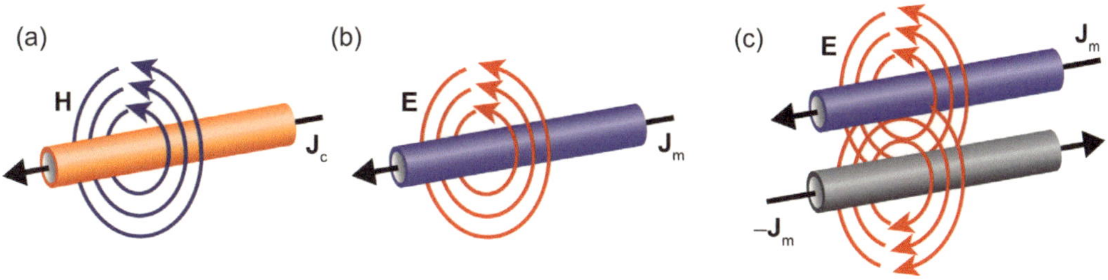

Fig. 4. (a) A schematic illustration of the magnetic-field **H** generation from a charge current $\mathbf{j}_c$ according to Ampere's law. (b) A schematic illustration of the electric-field **E** generation from a hypothetical magnetic-monopole current $\mathbf{j}_m$ according to the electromagnetic duality and Ampere's law. (c) A schematic illustration of the electric-field **E** generation from a pair of hypothetical magnetic-monopole currents, $\mathbf{j}_m$ and $-\mathbf{j}_m$, or a spin current.

A spin current can be detected electrically using the inverse spin Hall effect (ISHE), conversion of a spin current into an electric field [see Fig. 3(a)]. The ISHE has the same symmetry as that of the relativistic transformation of magnetic moment into electric polarization, which is derived from the Lorentz transformation, as follows. Consider a magnet with the magnetic moment **M** moving at a constant velocity **v** along the z axis with respect to an observer [see Fig. 3(b)]. This motion of the magnet is a flow of angular momentum, meaning an existence of a "spin current". In the observer's coordinate system, the Lorentz transformation converts a part of this magnetic moment **M** into an electric dipole moment **P'** as

$$\mathbf{P}' = -\frac{1}{\sqrt{1-(v/c)^2}}(\epsilon_0 \mathbf{v} \times \mathbf{M}), \tag{8}$$

where c and ε_0 are the light velocity and the electric constant, respectively. This indicates that electric polarization perpendicular to the direction of the magnetic-moment velocity is induced.

This electric-polarization generation can also be regarded as the spin-current version of Ampere's law as follows. As shown in Fig. 4(a), when a charge current $\mathbf{j}_c$ flows, a circular magnetic field **H** is induced around the charge current, according to Ampere's law: rot$\mathbf{H} = \mathbf{j}_c$. If a hypothetical magnetic monopole flows, a circular electric field **E** is expected to be induced around the monopole current $\mathbf{j}_m$ according to rot$\mathbf{E} = \mathbf{j}_m$ [see Fig. 4(b)], from the electromagnetic duality. Although this monopole has never been observed in reality, a spin current can be regarded as a pair of the hypothetical monopole currents flowing in the opposite directions along the spin current spatial direction. Therefore, a spin current may generate an electric field and this field is the superposition of the two electric fields induced by this pair of the monopole current, as shown in Fig. 4(c). This spin-current-induced electric field is identical to the field induced by the dipole moment described by Eq. (8).

In this way, electromagnetism and relativity predict that a spin current generates an electric field. According to Eq. (8), however, this electric field is too weak in a vacuum to be detected in reality. In a solid with strong spin-orbit interaction, in contrast, a similar but strong conversion between spin currents and electric fields appears, which is the ISHE.

In a solid, existence of a spin current can be modelled as that two electrons with opposite spins travel in opposite directions along the spin-current spatial direction $\mathbf{j}_s$, as shown in Fig. 3(a). Here, σ denotes the spin polarization vector of the spin current. The spin-orbit interaction bends these two electrons in the same direction and induces an electromotive force $\mathbf{E}_{ISHE}$ transverse to $\mathbf{j}_s$ and σ, which is the ISHE. The relation among $\mathbf{j}_s$, $\mathbf{E}_{ISHE}$, and σ is therefore given by (Saitoh, 2006)

$$\mathbf{E}_{\mathrm{ISHE}} = D_{\mathrm{ISHE}} \mathbf{J}_s \times \sigma, \tag{9}$$

where D_{ISHE} is the ISHE efficiency. This equation is similar to Eq. (8) but this effect may be enhanced by the strong spin-orbit interaction in solids.

The ISHE was recently observed using a spin-pumping method operated by ferromagnetic resonance (FMR) and by a non-local method in metallic nanostructures (Saitoh, 2006; Valenzuela, 2006; Kimura, 2007). Since the ISHE enables the electric detection of a spin current, it will be useful for exploring spin currents in condensed matter.

4. Photoinduced inverse spin Hall effect: Experiment

The combination of the optical generation of spin-polarized carriers and the ISHE enables direct conversion of light-polarization information into electric voltage in a Pt/GaAs interface (Ando, 2010). Figure 5(a) shows a schematic illustration of the Pt/GaAs sample. Here, the thickness of the Pt layer is 5 nm. The Pt layer was sputtered on a Si-doped GaAs substrate with a doping concentration of $N_D = 4.7 \times 10^{18}$ cm^{-3}. The surface of the GaAs layer was cleaned by chemical etching immediately before the sputtering. Two electrodes are attached to the ends of the Pt layer as shown in Fig. 5(a). During the measurement, circularly polarized light with a wavelength of $\lambda = 670$ nm and a power of $I_i = 10$ mW was illuminated to the Pt/GaAs sample as shown in Fig. 5(a). In the GaAs layer, electrons with a spin polarization σ along the light propagation direction are excited to the conduction band by the circularly polarized light due to the optical selection rule. Here, note that hole spin polarization plays a minor role in this setup, since it relaxes in ~ 100 fs, which is much faster than the relaxation time of ~ 35 ps for electron spin polarization (Hilton, 2002; Kimel, 2001). This spin polarization of electrons then travels into the Pt layer across the interface as a pure spin current. The injected spin current is converted into an electric voltage by the ISHE in the Pt layer due to the strong spin-orbit interaction in Pt (Ando, 2008). Here, note that the angle of the light illumination to the normal axis of the film plane is set at $\theta_0 = 65°$ to obtain the photoinduced ISHE signal, since the electric voltage due to the photoinduced ISHE is proportional to $j_s \sin\theta_0$ because of the relation $\mathbf{E}_{ISHE} \propto \mathbf{j}_s \times \sigma$, where the spin polarization σ is directed along the light propagation direction. The difference in the generated voltage between illumination with right circularly polarized (RCP) and left circularly polarized (LCP) light, $V^R - V^L$, was measured by a polarization-lock-in technique using a photoelastic modulator operated at 50 kHz. The difference in the intensities between RCP and LCP light incident on the sample was confirmed to be vanishingly small. All the measurements were performed at room temperature at zero applied bias across the junction.

In-plane light illumination angle θ dependence of $V^R - V^L$ for the Pt/GaAs sample is shown in Fig. 5(b), where the in-plane angle θ is defined in Fig. 5(a). Figure 5(b) shows that $V^R - V^L$ varies systematically by changing the illumination angle θ. Notable is that this variation is well reproduced using a function proportional to $\cos\theta$, as expected for the photoinduced ISHE. The relation of the ISHE, $\mathbf{E}_{ISHE} \propto \mathbf{j}_s \times \sigma$, indicates that the electric voltage due to the photoinduced ISHE is proportional to $|\mathbf{j}_s \times \sigma|_x \propto \cos\theta$, since σ and $\mathbf{j}_s$ are directed along the light propagation direction and the z axis, respectively. Here, $|\mathbf{j}_s \times \sigma|_x$ denotes the x component of $\mathbf{j}_s \times \sigma$ [see Fig. 5(a)]. This electromotive force was found to be disappeared in a Cu/GaAs system, where the Pt layer is replaced by Cu with very weak ISHE, supporting that ISHE is responsible for the observed electric voltage.

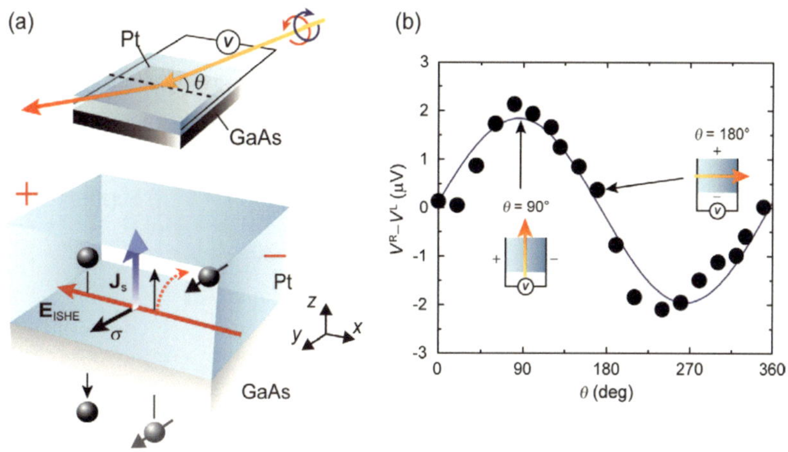

Fig. 5. (a) A schematic illustration of the Pt/GaAs hybrid structure and the photoinduced ISHE in the Pt/GaAs system. (b) In-plane illumination angle θ dependence of $V_R - V_L$ measured for the Pt/GaAs hybrid structure.

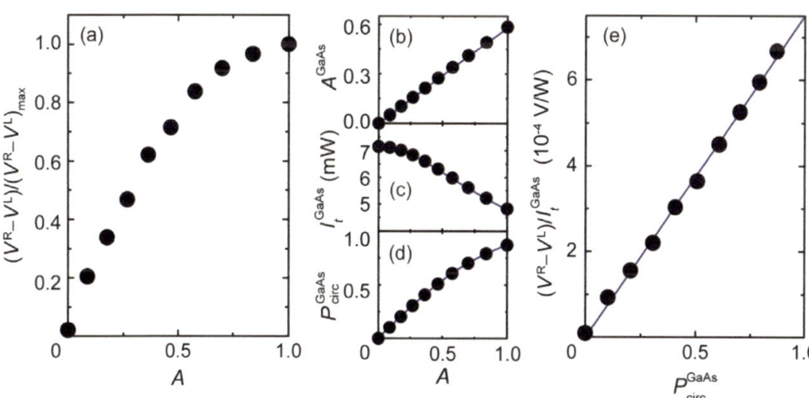

Fig. 6. (a) Ellipticity A of the illuminated light dependence of $V_R - V_L$. (b) A dependence of the ellipticity A^{GaAs} of the light injected into the GaAs layer. (c) A dependence of the intensity I_t^{GaAs} of the light injected into the GaAs layer. (d) A dependence of the degree of circular polarization P_{circ}^{GaAs} of the light injected into the GaAs layer. (e) P_{circ}^{GaAs} dependence of $(V^R - V^L)/I_t^{GaAs}$.

The observed electric voltage signal depends strongly on the ellipticity of the illuminated light polarization. Here, the ellipticity A is defined as the ratio of the minor to major radiuses of the elliptically polarized light. Figure 6(a) shows the illuminated-light ellipticity A dependence of $V^R - V^L$. As shown in Fig. 6(a), the $V^R - V^L$ signal increases with the ellipticity A of the illuminated light. This supports that this signal is induced by the photoinduced ISHE, since the angular momentum component of a photon along the light propagation direction is zero (maximized) when $A = 0$ (1).

5. Photoinduced inverse spin Hall effect: Theory

The A dependence of $V^R - V^L$ shown in Fig. 6(a) demonstrates that the electric voltage observed in the Pt/GaAs junction is induced by the circularly polarized light illumination. However, the variation of the electric voltage with respect to A is not straightforward to understand; the $V^R - V^L$ signal is not linear to A. In the following, we discuss in detail on the experimental result by calculating the polarization of the light injected into the GaAs layer.

The propagation of light in a multilayer film is characterized by the optical admittance $Y^{s(p)} = C^{s(p)}/B^{s(p)}$, where s(p) denotes s(p) polarized light. $B^{s(p)}$ and $C^{s(p)}$ are expressed as

$$\begin{pmatrix} B^{s(p)} \\ C^{s(p)} \end{pmatrix} = \begin{pmatrix} \cos\delta & (i\sin\delta)/\eta_1^{s(p)} \\ i\eta_1^{s(p)}\sin\delta & \cos\delta \end{pmatrix} \begin{pmatrix} 1 \\ \eta_2^{s(p)} \end{pmatrix}, \tag{10}$$

where $\delta = 2\pi n_1 d_1 \cos\theta_1/\lambda$, $\eta_r^{p} = n_r(\varepsilon_0/\mu_0)^{1/2}/\cos\theta_r$, and $\eta_r^{s} = n_r(\varepsilon_0/\mu_0)^{1/2}\cos\theta_r$ (r = 0, 1, 2). Here, n_0, n_1, and n_2 are the complex refractive indices for air, Pt, and GaAs, respectively. d_1 is the thickness of the Pt layer and θ_r is the incident angle of the light defined as shown in Fig. 7. Using $B^{s(p)}$ and $C^{s(p)}$, the transmittance $T^{s(p)} \equiv I_t^{s(p)}/I_i^{s(p)}$ and the transmission coefficient $\tau^{s(p)} \equiv E_t^{s(p)}/E_i^{s(p)}$ are obtained as

$$T^{s(p)} = \frac{4\eta_0^{s(p)}\Re[\eta_2^{s(p)}]}{(\eta_0^{s(p)}B^{s(p)} + C^{s(p)})(\eta_0^{s(p)}B^{s(p)} + C^{s(p)})^*}, \tag{11}$$

$$\tau^{s} = \frac{2\eta_0^{s}}{\eta_0^{s}B^{s} + C^{s}}, \quad \tau^{p} = \frac{2\eta_0^{p}}{\eta_0^{p}B^{p} + C^{p}}\frac{\cos\theta_0}{\cos\theta_2}, \tag{12}$$

where $I_{i(t)}^{s(p)}$ and $E_{i(t)}^{s(p)}$ are the illuminated (transmitted) light intensity and the amplitude of the electric field of s(p) polarized light [see Fig. 7], respectively. Here, $\Re[\eta_2^{s(p)}]$ is the real part of $\eta_2^{s(p)}$. Using Eqs. (11) and (12) with the parameters shown in Table 2, the transmittance $T^{s(p)}$ and the transmission coefficient $\tau^{s(p)}$ for the Pt/GaAs system are obtained as shown in Table 3. The calculated transmission coefficients $\tau^{s(p)}$ show that the transmission of the s and p polarized light is different. This indicates that the ellipticity of the illuminated to the sample is changed during the propagation of the film. The relation between the ellipticity A^{GaAs} of the light injected into the GaAs layer and the ellipticity A of the illuminated light is shown in Fig. 6(b). Here, A^{GaAs} is obtained using

$A^{\mathrm{GaAs}} \simeq (\Re[\tau^{\mathrm{s}}]/\Re[\tau^{\mathrm{p}}])A$. From the value of the ellipticity A, the degree of circular polarization P_{circ}, the difference in the numbers between RCP and LCP photons, can be written as,

$$P_{\mathrm{circ}} \equiv \frac{I^{+} - I^{-}}{I^{+} + I^{-}} = \frac{2A}{1 + A^2}, \tag{13}$$

where I^+ and I^- are the intensities of the RCP and LCP light, respectively. The degree of circular polarization $P_{\mathrm{circ}}^{\mathrm{GaAs}}$ of the light injected into the GaAs layer is shown in Fig. 6(d), which is obtained from the ellipticity shown in Fig. 6(b) using Eq. (13). Here, notable is that the degree of circular polarization $P_{\mathrm{circ}}^{\mathrm{GaAs}}$ of the light injected into the GaAs layer is proportional to the electron spin polarization generated by the circularly polarized light. The propagation of the circularly polarized light also changes the intensity of the light as $I_{\mathrm{t}}^{\mathrm{GaAs}} = T^{\mathrm{s}} I_{\mathrm{i}}^{\mathrm{s}} + T^{\mathrm{p}} I_{\mathrm{i}}^{\mathrm{p}}$. Figure 6(c) shows the light ellipticity A dependence of the intensity I_t^{GaAs} of the light injected into the GaAs layer obtained from

$$I_{\mathrm{t}}^{\mathrm{GaAs}} = \left(T^{\mathrm{s}} \frac{A^2}{1 + A^2} + T^{\mathrm{p}} \frac{1}{1 + A^2} \right) I_{\mathrm{i}}. \tag{14}$$

Here, $I_{\mathrm{i}} = I_{\mathrm{i}}^{\mathrm{s}} + I_{\mathrm{i}}^{\mathrm{p}}$ is the illuminated light intensity. Since the electric voltage due to the photoinduced ISHE is expected to be proportional to the intensity of the absorbed light, or the number of spin-polarized carriers generated by the circularly polarized light, one should calculate $(V^{\mathrm{R}} - V^{\mathrm{L}}) / I_{\mathrm{t}}^{\mathrm{GaAs}}$ to compare the electric voltage induced by the circularly polarized light for different A. The $P_{\mathrm{circ}}^{\mathrm{GaAs}}$ dependence of $(V^{\mathrm{R}} - V^{\mathrm{L}}) / I_{\mathrm{t}}^{\mathrm{GaAs}}$ is shown in Fig. 6(e). As shown in Fig. 6(e), $(V^{\mathrm{R}} - V^{\mathrm{L}}) / I_{\mathrm{t}}^{\mathrm{GaAs}}$ is proportional to $P_{\mathrm{circ}}^{\mathrm{GaAs}}$, or the electron spin polarization. This is consistent with the prediction of the photoinduced ISHE. Thus both the light illumination angle and light ellipticity dependence of the electric voltage support that the electric voltage is induced by the ISHE driven by photoinduced spin-polarized carriers.

n_0	n_1	n_2	θ_0 (deg)	d_1 (nm)	λ (nm)
1.00	$2.12 - 4.00i$	$3.79 - 0.157i$	65	5	670

Table 2. The parameters used in the calculation. n_0, n_1, and n_2 are the complex refractive indices for air, Pt, and GaAs, respectively (Adachi, 1993; Ordal, 1983). θ_0 is the incident angle of the illumination to the normal axis of the film plane. d_1 is the thickness of the Pt layer and λ is the wavelength of the light.

T^s	τ^s	T^p	τ^p
0.249	$0.168 + 0.0200i$	0.715	$0.287 + 0.00589i$

Table 3. The transmittance $T^{s(p)}$ and the transmission coefficient $\tau^{s(p)}$ for the Pt/GaAs hybrid structure.

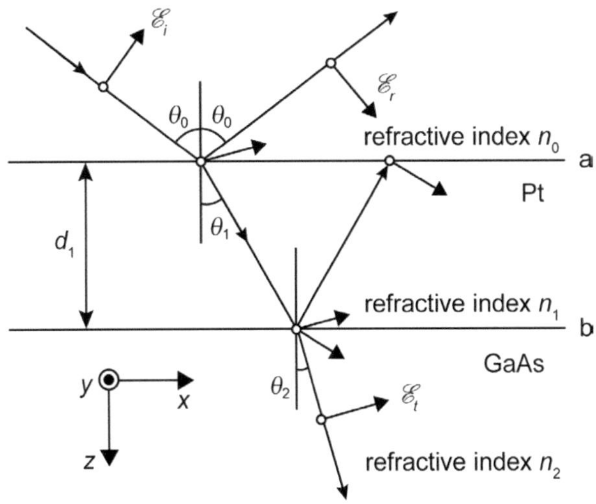

Fig. 7. The definition of θ_0, θ_1, and θ_2.

The photoinduced ISHE allows direct conversion of the circular-polarization information P_{circ} of the illuminated light into an electric voltage. The relation between $V^R - V^L$ and the circular-polarization information P_{circ} of the illuminated light can be argued from the linear dependence of $(V^R - V^L)/I_t^{GaAs}$ on P_{circ}^{GaAs} shown in Fig. 6(e). For simplicity, we assume that the imaginary parts of n_2 and $\tau^{s(p)}$ are negligibly small: $n_2 \equiv \Re[n_2]$ and $\tau^{s(p)} \equiv \Re[\tau^{s(p)}]$ [see Tables 2 and 3]. From Eqs. (11), (12), (13), and (14), one obtains

$$I_t^{GaAs} = \left((\tau^s)^2 \frac{A^2}{1+A^2} + (\tau^p)^2 \frac{1}{1+A^2} \right) \frac{n_2 \cos\theta_2}{n_0 \cos\theta_0} I_i, \quad (15)$$

$$P_{circ}^{GaAs} = \frac{2\tau^s \tau^p A}{(\tau^p)^2 + (\tau^s)^2 A^2}, \quad (16)$$

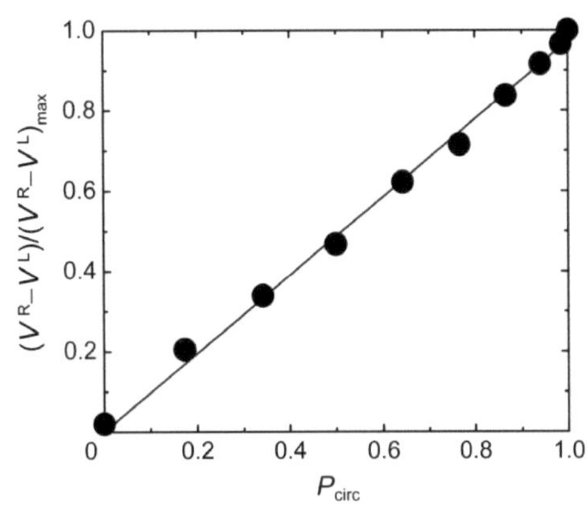

Fig. 8. The degree of circular polarization of the illuminated light ellipticity A of the illuminated light P_{circ} dependence of $V_R - V_L$.

and thus

$$V^{R} - V^{L} = \left(Q \frac{\tau^{s} \tau^{p} n_2 \cos\theta_2}{n_0 \cos\theta_0} I_i \right) P_{circ}. \tag{17}$$

Here, $Q \equiv (V^{R} - V^{L}) / (I_t^{GaAs} P_{circ}^{GaAs})$ is the proportionality constant as seen from Fig. 6(e) and we used $A^{GaAs} \simeq (\tau^{s} / \tau^{p}) A$. Since the proportionality constant Q is proportional to $\sin\theta_2$ because of Eq. (9), we obtain

$$V^{R} - V^{L} = (Q' \tau^{s} \tau^{p} \cos\theta_2 \tan\theta_0 I_i) P_{circ}, \tag{18}$$

where $Q \equiv Q' \sin\theta_2 = Q'(n_0 / n_2) \sin\theta_0$. Equation (18) shows that, in spite of the inequalities of $I_t^{GaAs} \neq I_i$ and $P_{circ}^{GaAs} \neq P_{circ}$ due to the presence of the top Pt layer and oblique illumination, the output signal $V^R - V^L$ is proportional to the degree of circular polarization P_{circ} of the illuminated light outside the sample. This indicates that the photoinduced ISHE can be used as a spin photodetector: the direct conversion of circular polarization information into electric voltage. This function is demonstrated experimentally in Fig. 8, in which $V^R - V^L$ is proportional to the degree of circular polarization of the illuminated light outside the sample.

6. Conclusion

The photoinduced inverse spin Hall effect provides a simple way for detecting light circular polarization through a spin current. This phenomenon enables the direct conversion of light-polarization information into electric voltage in a Pt/GaAs junction. This technique will be useful both in spintronics and photonics, promising significant advances in optical technology.

7. Acknowledgment

The authors thank to M. Morikawa, T. Trypiniotis, Y. Fujikawa, C. H. W. Barnes, and H. Kurebayashi for valuable discussions.

8. References

Adachi, S. (1993). *Properties of Aluminium Gallium Arsenide*, Inspec.

Ando, K., Takahashi, S., Harii, K., Sasage, K., Ieda, J., Maekawa, S. & Saitoh, E. (2008). Electric manipulation of spin relaxation using the spin Hall effect, *Physical Review Letters* Vol. 101: 036601.

Ando, K., Morikawa, M., Trypiniotis, T., Fujikawa, Y., Barnes, C. H. W. & Saitoh, E. (2010). Photoinduced inverse spin Hall effect Conversion of light-polarization information into electric voltage, *Applied Physics Letters* Vol. 96: 082502.

Hilton, D. J. & Tang, C. L. (2002). Optical orientation and femtosecond relaxation of spin-polarized holes in GaAs, *Physical Review Letters* Vol. 89: 146601.

Kimel, A. V., Bentivegna, F., Gridnev, V. N., Pavlov, V. V., Pisarev, R. V. & Rasing, T. (2001). Room-temperature ultrafast carrier and spin dynamics in GaAs probed by the photoinduced magneto-optical Kerr effect, *Physical Review* B Vol. 63:235201.

Kimura, T., Otani, Y., Sato, T., Takahashi, S. & Maekawa, S. (2007). Room-temperature reversible spin Hall effect, *Physical Review Letters* Vol. 98: 156601.

Meier F. & Zakharchenya, B. P. (1984). *Optical orientation*, North-Holland.

Ordal, M. A., Long, L. L., Bell, R. J., Bell, S. E., Bell, R. R., Alexander, J. R. W. & Ward, C. A. (1983). Optical properties of the metals Al, Co, Cu, Au, Fe, Pb, Ni, Pd, Pt, Ag, Ti, and W in the infrared and far infrared, *Applied Optics* Vol. 22: 1099.

Saitoh, E., Ueda, M., Miyajima, H. & Tatara, G. (2006). Conversion of spin current into charge current at room temperature: Inverse spin Hall effect, *Applied Physics Letters* Vol. 88: 182509.

Valenzuela, S. O. & Tinkham, M. (2006). Direct electronic measurement of the spin Hall effect, *Nature* Vol. 442: 176.

19

Shape of the Coherent Population Trapping Resonances Registered in Fluorescence

Sanka Gateva and Georgi Todorov
Institute of Electronics, Bulgarian Academy of Sciences
Bulgaria

1. Introduction

There has been permanent interest in investigations of new magnetic sensors and their various applications (Edelstein, 2007). Last years there is a rapid progress in development of magneto-optical sensors because of their sensitivity and potential for miniaturization. Magnetometers, based on magneto-optical sensors have high sensitivity - comparable to, or even surpassing this of the SQUIDs (Superconducting Quantum Interference Devices) (Kominis et al., 2003; Dang et al., 2010; Savukov, 2010; Knappe, 2010). Microfabrication of components using the techniques of Micro-Electro-Mechanical Systems (MEMS) developed for atomic clocks (Knappe, 2004) gives the opportunity for building small, low consuming, low cost and non-cryogenic (as SQUIDs) sensors (Griffith et al., 2010). Coherent optical effects can be applied for magnetic field detection and offer perspectives for development of high-precision optical magnetometers (Cox et al., 2011; Kitching et al., 2011). These magnetometers are appropriate for geomagnetic, space, nuclear and biological magnetic field measurements (cardio and brain magnetic field imaging), environmental monitoring, magnetic microscopy, investigations of fundamental physics, etc. Coherent magneto-optical resonances have many applications not only in magnetometry, but in high-resolution spectroscopy, lasing without inversion, laser cooling, ultraslow group velocity propagation of light, etc. (Gao, 2009).

Magneto-optical resonances can be prepared and registered in different ways (Budker&Romalis, 2007 and references therein). Most frequently Coherent–Population-Trapping (CPT) is observed when two hyperfine levels of the ground state of alkali atoms are coupled by two laser fields to a common excited level. When the frequency difference between the laser fields equals the frequency difference between the two ground states, the atoms are prepared in a non-absorbing state, which can be registered as a fluorescence quenching and transparency enhancement in spectral interval narrower than the natural width of the observed optical transition (Arimondo, 1996).

In degenerate two-level systems coherent states can be created by means of Hanle effect configuration (Alzetta et al., 1976). In this case the coherent non-absorbing state is prepared on two Zeeman sublevels of one hyperfine level by monochromatic laser field (the so called single frequency CPT). Hanle configuration is important for performing significantly simplified experiments and to build practical devices as well.

For a lot of applications in medicine, quantum information experiments, materials characterization, read out of stored memory domains etc., magnetic field imaging is essential (Bison et al., 2009; Johnson et al., 2010; Kimble, 2008; Lukin, 2003; Mikhailov et al., 2009; Romalis, 2011; Xia et al., 2006). The most common techniques for monitoring the flow phenomena involve fluorescence detection (Xu et al., 2008) because it is sensitive, non-invasive, and allows in-line analysis of multiple step processes.

For all applications, where narrow signals and high signal-to-noise ratios are important, ensuring reliable operation requires good knowledge of the resonance shape and the internal and external factors influencing it.

In this chapter, the influence of different factors on the shape of the CPT resonances obtained by means of the Hanle effect configuration and registered in fluorescence is analysed in uncoated, room temperature vacuum cells from point of view of cell diagnostics and building high-sensitive magneto-optical sensors.

2. Shape of the single frequency CPT resonances

2.1 Experimental shapes

2.1.1 Experimental set-up

The experimental set-up geometry is shown in Fig. 1. The resonances were measured in uncoated vacuum cells containing a natural mixture of Rb isotopes at room temperature (22°C). A single-frequency linearly polarized in **z** direction diode laser beam (2 mm in diameter, 22 mW in power) was propagating along the cell's axis **x**. Its frequency and emission spectrum were controlled by observing fluorescence from a second Rb vapor cell and a Fabry-Perot spectrum analyzer. A magnetic field B_{scan}, created by a solenoid, was applied collinearly to the laser beam. As the shape of the CPT resonance is very sensitive to stray magnetic fields (Huss et al., 2006), the gas cell and the solenoid were placed in a 3 layer μ-metal magnetic shield. The fluorescence was detected perpendicular to the laser beam direction and to the light polarization vector (in **y** direction) by a photodiode. In all experiments it was placed close to the front window of the cell because the CPT resonances registered in fluorescence are sensitive to the position of the photodiode along the cell (Godone et al., 2002). The signals from the photodiode were amplified and stored in a PC, which also controlled the magnetic field scan. The main sources of errors were the accuracy of the laser power measurements and the poor signal-to-noise ratio at low laser powers.

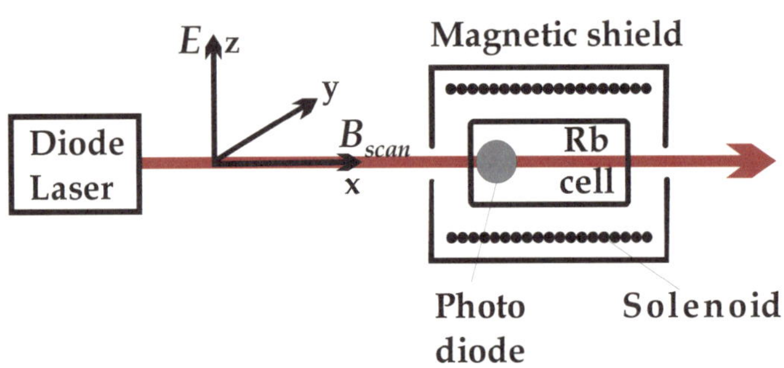

Fig. 1. Experimental setup geometry.

The resonances were examined in 4 Rb vacuum cells with internal dimensions: cell **A** (length l_A=4.5 cm, diameter d_A=2.9 cm); cell **B** (length l_B=2.0 cm, diameter d_B=2.0 cm); cell **C** (length l_C=4.4 cm, diameter d_C=1.0 cm) and cell **D** (length l_D=4.3 cm, diameter d_D=2.1 cm).

All investigations were performed on the degenerate two-level system of the (F_g=2 – F_e=1) transition of the ^{87}Rb D_1 line because the D_1 line consists of hyperfine transitions at which only dark resonances can be observed, the fluorescence of the F_g=2 – F_e=1 transition is practically not overlapping with another and it has the highest contrast (55%) (Dancheva et al., 2000).

As the spectra in the Hanle effect configuration were measured by tuning the magnetic field, all experimental spectra are plotted as function of the magnetic field. The conversion of magnetic field to frequency is straightforward on the basis of the Zeeman splitting between adjacent magnetic sublevels of the ground level ~0.7 MHz/G (Steck, 2009).

2.1.2 Experimental shapes

The measurement of the CPT resonances in four uncoated Rb vacuum cells with different dimensions (the dimensions of the cells **A**, **B**, **C** and **D** are given in Section 2.1.1) has shown that they are different not only in width but also in shape. In Fig. 2 the normalized CPT signals measured in cell **A** and cell **D** are compared in 1.25 G and 40 mG tuning ranges. In cell **A** the resonance width (FWHM) is about 50 mG, while in cell **D** it is 250 mG (Fig. 2a). In cell **D** the resonance has triangular shape (Fig. 2b scattered circles curve), while in cell **A** the resonance has a narrow Lorentzian structure of the order of few mG centred at zero magnetic field superimposed on a broad pedestal of the order of a few hundred mG (Fig. 2 dashed line curve).

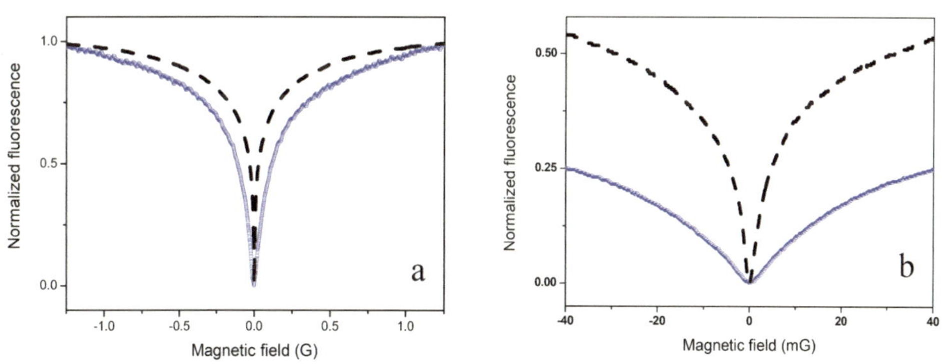

Fig. 2. Normalized CPT resonance shapes in cell **A** (dash line curve) and cell **D** (scattered circles curve) in 1.25 G (a) and 40 mG (b) tuning range. Power density 75 mW/cm^2.

In order to obtain information about the processes leading to the narrow structure formation, measurement of the resonance shapes in all cells in dependence on the laser beam diameter, laser power density, additional constant magnetic fields, and the laser frequency position relative to the centre of the Doppler broadened line profile were performed (Alipieva et al., 2003). The analysis of the shapes, measured at the same geometry of excitation and registration, has shown that the origin of the narrow structure is connected with processes in the vacuum cell. The width of the resonances is different in different cells and it is about 2 orders smaller than the transit-time broadening, which is of the order of 100 kHz (Thomas & Quivers, 1980). The width of the narrow Lorentzian structure of the CPT resonance is in agreement with the assumption that the broadening is mainly affected by the relaxation processes due to atomic collisions with the walls of the cell.

In Fig. 3 the measured narrow resonance widths of the four cells are plotted as function of the mean distance between two collisions with the cell walls L. The red line is the calculated resonance width (FWHM) $\Delta_L = 1/2\pi\tau$ determined by the mean time between two collisions with the cell walls $\tau = L/v = 4V/Sv$ (Corney, 1977) where v is the mean thermal velocity, V is the volume, S is the surface area of the cell and L is the mean distance between 2 collisions with the cell walls. The time of atom-wall interaction during the collision is small enough (Bouchiat & Brossel, 1966) and is neglected. The measured values of the resonance width are in good agreement with the evaluated for three of the cells (cell **A**, **B** and **C**). Only in cell **D** there is no narrow structure and in Fig. 3 this is designated by a dashed line at the corresponding to cell **D** mean distance between collisions with the cell walls. The different dimensions of the cell can not be the reason for the existence of the narrow resonance. The estimated mean path between two collisions with the cell walls L is 2.1 cm in cell **A** and 1.6 cm in cell **D**. As the measurements are made at the same temperature, the mean velocity of the atoms is the same. Then, the evaluated difference between the widths of the eventual narrow resonances is of the order of 25-30% and is detectable.

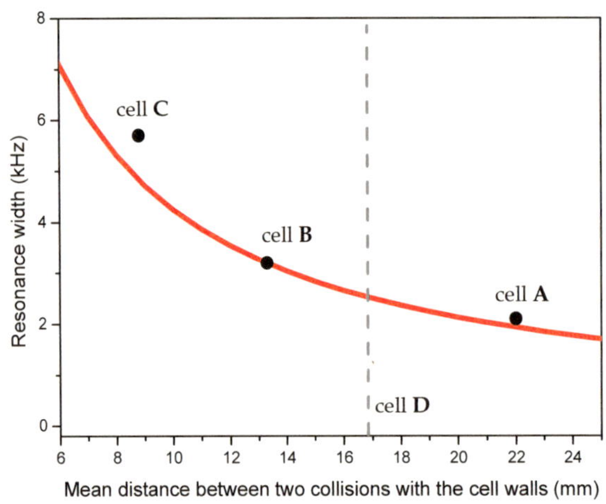

Fig. 3. FWHM of the narrow resonances (black circles) as function of the dimensions of the cell (mean distance between two collisions with the cell walls) and the calculated dependence of the width of the resonance, result of relaxation on the cell walls (red line). Laser beam diameter 2 mm, P_{las}= 75 mW/cm^2.

In Fig. 4 the power dependences of the amplitudes and widths of the narrow and the wide structures of the CPT resonances measured in cell **A** are shown. The amplitudes of both signals increase with power. The amplitude of the wide structure increases fast with power and at intensities larger than 150 mW/cm^2 saturates. The amplitude of the narrow structure is lower and does not saturate up to a power density of 1 W/cm^2.

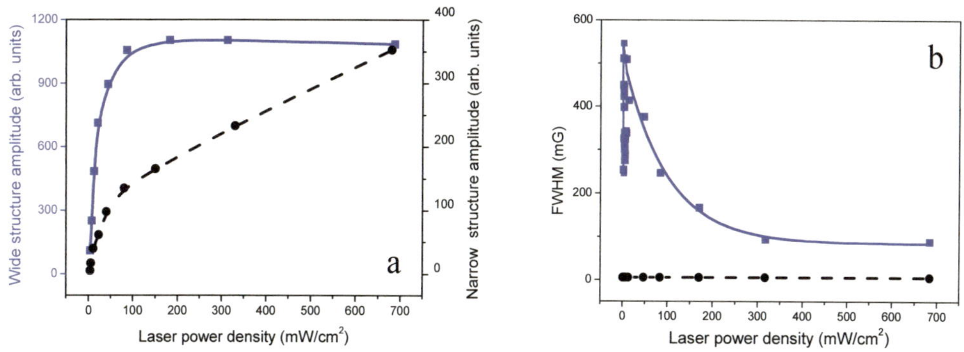

Fig. 4. Amplitudes (a) and widths (b) of the narrow (black circles) and wide (blue squares) structure of the CPT resonance in dependence on the laser power density. (The lines are only to guide the eyes.)

The width of the narrow structure does not depend on the laser power, while the wide structure width does - at low powers it increases, and at laser power densities higher than 1 mW/cm^2, there is no power broadening, but a resonance-line narrowing with the power - an exponential decrease to 80 mG. A possible explanation of this narrowing in our experiment is the high pumping rate. The CPT signal width is proportional to the pumping rate Ω^2/γ^* (γ^* is the total decay rate of the excited state defined not only by the spontaneous decay but by all relaxation processes) (Levi et al., 2000). When the laser power density is increased, the saturation of the absorption cannot be neglected, the decay rate of the excited state γ^* increases due to the stimulated emission and a resonance width narrowing is observed.

The comparison of the narrow structure registered in fluorescence and transmission (Gateva et al., 2005) shows that the narrow structure width is the same in fluorescence and transmission. In fluorescence, the narrow resonances can be observed in the whole range of laser powers, while in transmission, in a very small range of powers because of the absorption saturation.

The measured dual structure resonance shapes and the power dependences of the two components make reasonable the assumption that the signal in these cells is formed by different subensembles of atoms interacting with different light fields - the laser beam and a weak light field, for example, of scattered light.

This assumption is confirmed by the series of experiments with cell **A** with an expanded beam filling the whole volume of cell **A**. The dependence of the shape of the CPT resonances on the laser beam diameter is given in Fig. 5a. (The diameter of the laser beam with laser power density 0.3 mW/cm^2 was changed by different diaphragms in front of the cell.) The resonances broaden with decreasing the diameter, but there is a narrow range (about 3 mG in width) around the zero magnetic field, where their shapes are Lorentzian and equal in width. In the limits of the accuracy of the measurement this width is equal to the narrow component width and to the non power broadened CPT resonance of the expanded to 30 mm laser beam (Fig. 5b). In Fig. 5b the power dependence of the resonance width with

30 mm laser beam is shown. The initial intensity of the laser beam was reduced by a neutral filter from maximum to 1/128. In this case, the resonance width is determined by the relaxation processes due to atomic collisions with the walls of the cell and the power broadening.

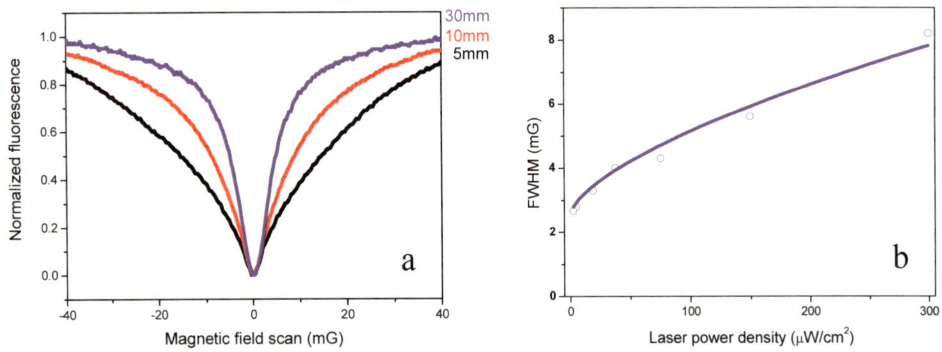

Fig. 5. CPT resonances with expanded laser beam:
(a) shape of the resonances at different diameters of the laser beam and reduced laser power density 0.3 mW/cm^2;
(b) power dependence of the resonance width with 30 mm laser beam diameter.

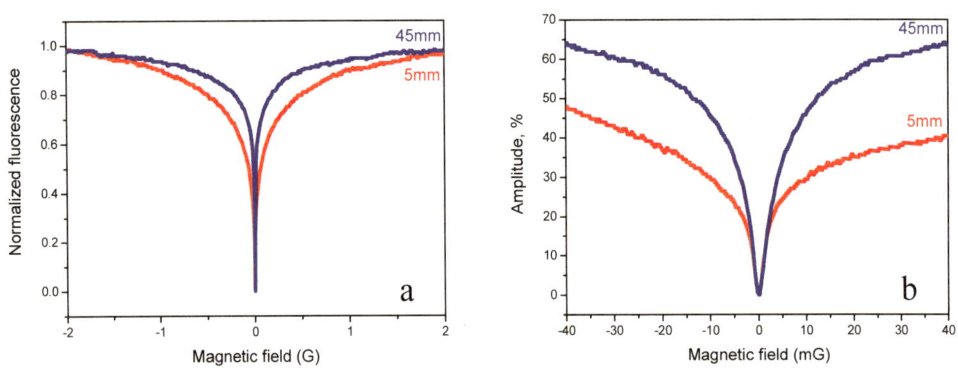

Fig. 6. CPT resonances registered at 5 mm and 45 mm along the cell in two different tuning ranges: a) 2 G and b) 40 mG. Beam diameter 2 mm.

In Fig. 6 the resonance shapes registered with laser beam diameter 2 mm at 5 mm and 45 mm distance from the input window are given. The comparison of the shapes of the resonances in a 2 G tuning range (Fig.6a) shows, that the resonance at the end of the cell is narrower. This result coincides with the theoretical and experimental results of Godone et al. (Godone et al., 2002) and is due to the lower laser power at the end of the cell and the smaller power broadening.

The comparison of the signals in a 10 mG range shows that there is a narrow range (about 3 mG in width) around the zero magnetic field, where their shapes coincide.

2.2 Theoretical description

The shape and width of the CPT resonances and their dependence on the input power has been studied experimentally and theoretically in many papers (Auzinsh, 2009a,b and references therein). Most of the experimental investigations were performed at low laser power, the shape of the resonances was Lorentzian and the dependence of the resonance width Γ on the laser power density was linear up to a few mW/cm^2. The slope of this dependence was smaller than the predicted by the theory (Arimondo, 1996):

$$\Gamma = \gamma_\varphi + \Omega^2/\gamma_f \tag{1}$$

where γ_φ is the ground state coherence relaxation rate; Ω, the Rabi frequency; γ_f , the population decay rate from the excited state into the ground states.

There are different decoherence processes influencing the resonance shape - transit time broadening, population exchange, atom-atom and atom-wall collisions, transverse laser intensity distribution etc. The investigations of the resonance shapes at different conditions have shown that not always the power dependence of the resonance width is linear (Figueroa et al., 2006; Javan et al., 2002; Ye&Zibrov, 2002).

An analytical expression for the FWHM of the EIT resonances Γ_{EIT} in the case of a Doppler broadened medium in the linear with respect to the probe field approximation, when the population exchange between the ground states is the main source of decoherence, was obtained by Javan et al. (Javan et al., 2002):

$$\Gamma_{EIT}^2 = \frac{\gamma_\varphi}{\gamma_f}\Omega^2(1+x)\left\{1+\left[1+\frac{4x}{(1+x)^2}\right]^{1/2}\right\} \tag{2}$$

under the condition $\left(\gamma_f / w_D\right)^2 << x << \gamma_f / \gamma_\varphi$. The parameter $x = \Omega^2 / \Omega_{in\,\mathrm{hom}}^2 = \Omega^2\gamma_f / 2\gamma_\varphi w_D^2$ represents the degree of optical pumping within the inhomogeneous linewidth w_D. The dependence of Γ_{EIT} on the laser power density is not linear at low intensities -when $x<<1$, $\Gamma_{EIT} \approx \sqrt{\frac{2\gamma_\varphi}{\gamma_f}}.\Omega$, and linear at high intensities - when x>>1, $\Gamma_{EIT} \approx \frac{\Omega^2}{\mathrm{w_D}}$.

The investigation of the influence of the light beam transverse intensity distribution on the line shape has shown that due to the Gaussian shape of the beam, at high pumping rates the shape of the resonances is no more Lorentzian (Pfleghaar et al., 1993). The obtained by Levi et al. (Levi et al., 2000) analytical lineshape is:

$$\Pi(\delta) = \frac{1}{2}\ln\frac{\Omega^2 / \gamma_f\gamma_\varphi}{1+(\delta / \gamma_\varphi)^2} \tag{3}$$

where δ is the two-photon detuning, while by Taichenachev et al. (Taichenachev et al., 2004) it is:

$$R_G = \pi r_0^2(S_0 - \delta \arctan[S_0\delta / (1+S_0+\delta^2)] + \frac{1}{2}\ln\{(1+\delta^2) / [(1+S_0)^2+\delta^2]\}) \tag{4}$$

where $S_0 = \dfrac{\Omega^2}{\gamma_f \cdot \gamma_\varphi}$.

At high power densities ($S_0 >> 1$) it is

$$R_G \propto \{1-(\delta/S_0)\arctan(S_0/\delta)\} \tag{5}$$

These equations describe well the standard (bichromatical) CPT resonance registered in absorption.

To understand the reason for the observed peculiarities in the shapes and widths of the Hanle - CPT resonances in different cells we have performed a numerical modelling of the influence of the experimental conditions. The theoretical description is based on the standard semiclassical approach: the atomic system is described by the statistical operator $\hat{\rho}$ in density matrix representation (Landau & Lifshitz, 1965) and laser and magnetic fields are considered classically.

The Hamiltonian of the system $\hat{H}$ is sum of the operator of the free atom $\hat{H}_0$, the operator of magneto-dipole interaction $\hat{H}'$ and the operator of interaction with the laser radiation $\hat{V}$. Terms describing the atomic relaxation and excitation transfer are added to the Liuville/Neuman equation:

$$\dot{\hat{\rho}} = -\frac{i}{h}\left[\hat{H},\hat{\rho}\right] + \left(\dot{\hat{\rho}}\right)_{relax} + \left(\dot{\hat{\rho}}\right)_{tr} + \hat{N} \tag{6}$$

where $\hat{H} = \hat{H}_0 + \hat{H}' + \hat{V}$.

At weak magnetic fields the magneto-dipole interaction is smaller than the hyperfine splitting and the hyperfine splitting can be included in the main Hamiltonian $\hat{H}_0$. In this case the magneto-dipole operator can be written as: $\hat{H}' = \mu_B g_{F_S}\left[\vec{F}_s, \vec{H}\right]$, where μ_B is the Bohr magneton, g_{F_s} is the Lande factor, F_s is the total moment of the atom in the s state, ($s = f, \varphi$; f- upper, φ - lower state), and $\vec{H}$ is the vector of the magnetic field. The complete magnetic field includes the scanning magnetic field and a residual magnetic field H_{str} .

The operator of interaction with the laser radiation $\hat{V}$ is described with the scalar product of the vector operator of the dipole moment $\vec{d}$ and the electric vector $\vec{E}$ of the resonant to the atomic transition $\varphi \to f$ light:

$$\hat{V} = -\left(\vec{d}.\vec{E}\right), \qquad \vec{E} = E\vec{e}_Q \exp\{-i(\omega_{las}t - kz)\} + c.c. \tag{7}$$

where: $\vec{e}_Q$ $(Q = 0, \pm 1)$ are the circular components of the laser field, ω_{las} is the frequency of the laser light and $\vec{k}$ is the wave vector.

Applying the irreducible tensor operator (ITO) formalism, the density matrix $\rho_{nn'}$ in (n,n′) representation can be decomposed into polarization moment (PM) representation with tensor components ρ_q^k (Dyakonov & Perel, 1966)

$$\rho_q^k = (2F_{\rho'}+1)^{1/2} \sum_{n,n'} (-1)^{F_\rho - n} \begin{pmatrix} F_\rho & k & F_{\rho'} \\ -n & q & n' \end{pmatrix} \rho_{nn'} \tag{8}$$

Here $\rho = f,\varphi,\xi$ denote the components (f_q^k) and (φ_q^k) of the upper (f, F_f) and lower (φ, F_φ) levels respectively and the optical components (ξ_q^k).

The parentheses denote the 3j-symbol. Since q =n - n′, the ρ_q^k components with q = 0 are linear combination of the sublevels population. For example, for k=0, $\rho_0^0 = \sum_n \rho_{nn}$ is the total population and for k=2, q=0, $\rho_0^2 = \sum_{n=-F_\rho}^{F_\rho} f_{nn} \frac{3n^2 - F_\rho(F_\rho+1)}{[(2F_\rho+3)(F_\rho+1)(2F_\rho-1)]^{1/2}}$ is the longitudinal alignment, etc.

The relaxation of the atomic system is described phenomenologically by sum of two terms and in the common case includes radiation and collisional processes. Besides the clear physical meaning of the tensor components in this (k,q) representation, the relaxation matrix is diagonal in the main cases and all relaxation parameters for different ρ, $\gamma_\rho(k,q) = \gamma_\rho(k)$ $(\rho = f,\varphi,\xi)$ for the k-th tensor component of the upper (f) and lower (φ) levels and the decay of the optical coherence (ξ) depend on the rank k of the components only. The decay of the lower level $\gamma_\varphi(0)$ includes the time-of-flight influence. The second term $(\dot{\rho})_{tr}$ describes the excitation transfer by spontaneous emission $\Gamma_{F_f F_\varphi}(\kappa)$ from the upper level to the lower one and includes the transfer of Zeeman coherence.

According to Ducloy & Dumont (Ducloy & Dumont, 1970) and taking into account Dyakonov & Perels's normalization of the irreducible tensor operators (Dyakonov & Perel, 1966)

$$\Gamma_{F_f F_\varphi}(k) = (-1)^{F_f+F_\varphi+k+1}[\gamma_f(0)(2F_f+1)(2F_\varphi+1)(2J_f+1)] \cdot \\ \cdot \begin{Bmatrix} F_f & F_\varphi & 1 \\ J_\varphi & J_f & I \end{Bmatrix}^2 \sqrt{(2F_f+1)(2F_\varphi+1)} \begin{Bmatrix} F_f & F_f & k \\ F_\varphi & F_\varphi & 1 \end{Bmatrix} \tag{9}$$

Here $\gamma_f(0)$ is the total probability of decay of the upper state to the lower states.

It is worth to note that the relaxation constant $\Gamma_{F_f F_\varphi}(k)$ describes the "losses" in the channel $F_f \to F_\varphi$. If the branching ratio $\Gamma_{F_f F_\varphi}(0)/\gamma_f(0)$ is close to 1, the atomic system is closed. In the particular case of ^{87}Rb D_1 line transition $F_f = 1 \to F_\varphi = 2$, the ratio $\Gamma_{F_f F_\varphi}(0)/\gamma_f(0)$ =5/6.

For the matrix element of the dipole transitions between hf states in ITO representation one can write (Alexandrov et al., 1991):

$$d_{F_f F_\varphi} = \left\| d_{f\varphi} \right\| (-1)^{2F_\varphi+J_f+I+1} \left[(2\Gamma_f+1)(2F_\varphi+1)\right]^{1/2} \begin{Bmatrix} I & J_f & F_f \\ 1 & F_\varphi & J_\varphi \end{Bmatrix} (-1)^{F_\varphi+m} \begin{Bmatrix} F_\varphi & 1 & F_f \\ -m & Q & \mu \end{Bmatrix} e_Q \tag{10}$$

The brackets denote 6j-Wigner symbols, $\| d_{f\varphi} \|$ is the reduced matrix element of the dipole $F_f \to F_\varphi$ transition.

The initial conditions are defined by the operator $\hat{N} = N_s(2F_s + 1)W(\mathrm{v})$, which describes the population of the resonant to the laser light levels. It is supposed that the atomic ensemble is in equilibrium and the velocity distribution of the atoms is described by a Maxwell function $W(v)$.

$$W(v) = (u\sqrt{\pi})^{-1} e^{(-v^2/u^2)} \tag{11}$$

Taking into account the stated above, the system of equations describing the ground state (φ), the exited state (f) and the optical coherency (ξ) for arbitrary angular moments is:

$$\begin{aligned} \dot{f}_q^k + \gamma_f(k) f_q^k = i\mu_B g_f h^{-1} \Big\{ qH_0 f_q^k + \left[\frac{1}{2}(k+q)(k-q+1)\right]^{1/2} H_1 f_{q-1}^k - \\ -\left[\frac{1}{2}(k-q)(k+q+1)\right]^{1/2} H_{-1} f_{q+1}^k \Big\} + L_q^k + \left(2F_f + 1\right) N_f W(v) \delta_{k0} \delta_{q0} \end{aligned} \tag{12a}$$

$$\begin{aligned} \dot{\varphi}_q^k + \gamma_\varphi(k) \varphi_q^k = i\mu_B g_\varphi h^{-1} \Big\{ qH_0 \varphi_q^k + \left[\frac{1}{2}(k+q)(k-q+1)\right]^{1/2} H_1 \varphi_{q-1}^k - \\ -\left[\frac{1}{2}(k-q)(k+q+1)\right]^{1/2} H_{-1} \varphi_{q+1}^k \Big\} + M_q^k + \left(2F_\varphi + 1\right) N_\varphi W(v) \delta_{k0} \delta_{q0} + \Gamma_{F_f F_\varphi}(k) f_q^k \end{aligned} \tag{12b}$$

$$\begin{aligned} \dot{\xi}_q^k + \left(\gamma_\xi(k) + i\omega_0\right) \xi_q^k = ih^{-1} \sum_{k'Q} (-1)^Q H_{-Q} (-1)^{F_f + F_\varphi + q} (2k'+1) \begin{pmatrix} k' & 1 & k \\ -q' & Q & q \end{pmatrix} \\ \left[\mu_B g_\varphi \left(\varphi \| j \| \varphi\right) \begin{Bmatrix} F_f & F_\varphi & k' \\ 1 & k & F_\varphi \end{Bmatrix} + (-1)^{k+k'} \mu_B g_f \left(f \| j \| f\right) \begin{Bmatrix} F_\varphi & F_f & k' \\ 1 & k & F_f \end{Bmatrix} \right] \xi_{q'}^{k'} + G_q^k \end{aligned} \tag{12c}$$

Where:

$$L_q^\kappa = ih^{-1} (2\mathrm{F}_\varphi + 1)^{-1/2} \sum_{\kappa' q' Q} E_{-Q} \mathrm{C}_{qq'Q}^{\kappa\kappa'} \left[\mathrm{d}\xi_{q'}^{\kappa'} + \mathrm{d}^* (\xi_{q'}^{\kappa'})^* (-1)^{\kappa + \kappa' + q'} \right]$$

with $\mathrm{C}_{qq'Q}^{\kappa\kappa'} = (-1)^{2\mathrm{F}_\varphi + q'} (2\mathrm{F}_f + 1)^{1/2} (2\kappa' + 1) \begin{Bmatrix} \kappa' & 1 & \kappa \\ F_f & F_f & F_\varphi \end{Bmatrix} \begin{pmatrix} k' & 1 & \kappa \\ -q' & Q & q \end{pmatrix};$

$$M_q^\kappa = (-1)^\kappa ih^{-1} (2\mathrm{F}_\varphi + 1)^{-1/2} \sum_{\kappa' q' Q} E_{-Q} (-1)^{\kappa'} \mathrm{B}_{qq'Q}^{\kappa\kappa'} \left[\mathrm{d}\xi_{q'}^{\kappa'} + \mathrm{d}^* (\xi_{q'}^{\kappa'})^* (-1)^{\kappa + \kappa' + q'} \right]$$

with $\mathrm{B}_{qq'Q}^{\kappa\kappa'} = (-1)^{2\mathrm{F}_f + q'} (2\mathrm{F}_\varphi + 1)^{1/2} (2\kappa' + 1) \begin{Bmatrix} \kappa' & 1 & \kappa \\ F_\varphi & F_\varphi & F_f \end{Bmatrix} \begin{pmatrix} \kappa' & 1 & \kappa \\ -q' & Q & q \end{pmatrix};$

$$G_q^\kappa = ih^{-1} (2\mathrm{F}_f + 1)^{-1/2} \mathrm{d}^* \sum_{\kappa' q' Q} E_{-Q} \left[\mathrm{S}_{qq'Q}^{\kappa\kappa'} f_{q'}^{\kappa'} + (-1)^{\kappa + \kappa'} \mathrm{R}_{qq'Q}^{\kappa\kappa'} \varphi_{q'}^{\kappa'} \right]$$

with $\mathrm{R}^{\kappa\kappa'}_{qq'Q} = (-1)^{2F_\varphi+q'}(2F_f+1)^{1/2}(2\kappa'+1)\begin{Bmatrix}\kappa' & 1 & \kappa \\ F_f & F_\varphi & F_\varphi\end{Bmatrix}\begin{pmatrix}\kappa' & 1 & \kappa \\ -q' & Q & q\end{pmatrix}$

and $\mathrm{S}^{\kappa\kappa'}_{qq'Q} = (-1)^{2F_f+q'}(2F_\varphi+1)^{1/2}(2\kappa'+1)\begin{Bmatrix}\kappa' & 1 & \kappa \\ F_\varphi & F_f & F_f\end{Bmatrix}\begin{pmatrix}\kappa' & 1 & \kappa \\ -q' & Q & q\end{pmatrix}$

Here ω_0 is the resonance frequency for the given transition and $d \equiv \| d_{f\varphi} \|$, $(\rho\|j\|\rho) = [(2F_\rho+1)(F_\rho+1)F_\rho]^{1/2}$.

This basic system of equations (Eqs. 12) is practically analogous to the well known equations (Decomps et al., 1976; Dyakonov&Perel, 1966).

The system of Eqs. 12 is specified for the atomic transition $F_f=1 \rightarrow F_\varphi=2$ and the experimental geometry. The quantization axis is chosen parallel to the electric vector $E(\equiv E_z=E_0)$ of the laser field (see the experimental setup in Fig. 1). The scanned magnetic field $\boldsymbol{H}_{scan}(\equiv H_x)$ is perpendicular to this axis. The influence of the laboratory magnetic field is taken into account including two orthogonal components H_z and H_y into the equations.

Using a rotating wave approximation (RWA) and assuming a single-frequency laser field (Eq. 7), the system of equations is reduced to an algebraic one. The obtained system of equations is solved numerically at different parameters: Rabi frequency ($\Omega = d \cdot E/h$), relaxation constant of the levels f and φ $((\gamma_f(k), \gamma_\varphi(k))$, spontaneous emission transfer coefficients $\Gamma_{f\varphi}(k)$ and stray field components (H_y, H_z). All these parameters and the magnetic fields are expressed in units $\gamma_f(0)$. The initial (unbroadened by the laser field) low level population relaxation rate constant $\gamma_\varphi(0)$ is determined by the time of flight of the atoms. To describe better the experimental conditions the program is additionally modified to take into account the Gaussian intensity distribution of the irradiating beam and the Maxwell velocity distribution of the atoms. The solution for a given variable (ρ_q^k) is taken after summarizing the partial solutions for the sub-ensemble of atoms with velocities in a given interval. The integration step is varied and the integration region is chosen to be > 40 $\gamma_f(0)$. The obtained solutions for a given power are summarized on the laser beam intensity distribution.

The intensity of the registered in our experiments fluorescence $I_{F_eF_g}(\vec{n}(\theta,\varphi))$ from the upper to the lower level in direction $\vec{n}(\theta,\varphi)$ is described with:

$$I_{F_fF_\varphi}(\vec{n}) = C_0(-1)^{F_e+F_g}(2F_f+1)^{-1/2}\left|d^2_{f\varphi}\right| \sum(2\kappa+1)\begin{Bmatrix}1 & 1 & \kappa \\ F_f & F_f & F_\varphi\end{Bmatrix} \sum_q (-1)^q f_q^\kappa \Phi^\kappa_{-q}(\vec{n}(\theta,\varphi)) \quad (13)$$

where f_q^k are the tensor components describing the upper level, $\Phi_q^\kappa(\vec{n}(\theta,\varphi))$ is the observation tensor (Alexandrov et al., 1991), θ is the angle between the laser light polarization and the direction of registration (the inclination angle) and φ is the azimuth angle (Alexandrov et al., 1991; Dyakonov&Perel, 1966).

In a dipole approximation the fluorescence is defined only by the tensor components f_q^k with rank $k \leq 2$. The unpolarized fluorescence intensity $I_{f\varphi}^{unpol}$ for θ=0 can be written as:

$$I_{f\varphi}^{unpol} = C_0[\frac{f_0^0}{\sqrt{2F_f+1}} + (-1)^{F_f+F_\varphi+1}\sqrt{30}\begin{Bmatrix} 1 & 1 & 2 \\ F_f & F_f & F_\varphi \end{Bmatrix} f_0^2] \tag{14}$$

Only the two tensor components for the upper level f_0^0 and f_0^2 describe the signal observed in this case. In linear approximation, both components do not depend on the magnetic field. The magnetic field dependence is a result of the transfer of coherence created on the low level and it is a typical nonlinear effect.

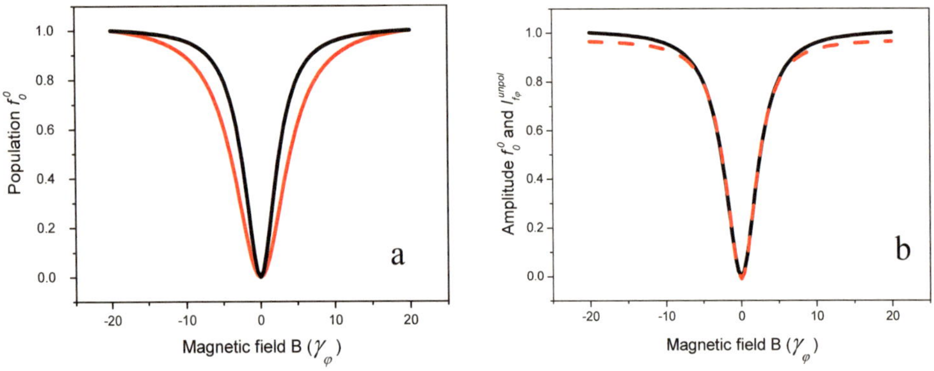

Fig. 7. Comparison of the theoretical upper level population f_0^0 : (a) calculated with (black line) and without (red line) integration over the Doppler velocity distribution and (b) with the unpolarized intensity $I_{f\varphi}^{unpol}$ (dashed red line).

It is well known that the Maxwell velocity distribution leads to narrowing of the CPT resonance (Firstenberg et al., 2007) and that the resonance shape remains quasi Lorentzian. The change in the upper level population shape and width after velocity integration over the Doppler distribution is illustrated in Fig. 7a (Petrov et al., 2007).

The main part of the unpolarized fluorescence intensity is determined by the upper level population f_0^0. The influence of the longitudinal alignment f_0^2 on the resonance shape is mainly on the wings of the resonance (Fig. 7b). It should be noted that the comparison of the numerical calculated shape of the population and intensity with a Lorentzian shape shows that there is a small specific structure in the vicinity of the zero magnetic field. This small structure is result of the High Rank Polarization Moments (HRPM) influence on the observables tensor components (see bellow).

The influence of the Gaussian distribution of the laser beam intensity on the shape of the CPT resonances is more essential. This has been discussed in a large number of papers (Knappe et al., 2001; Levi et al., 2000; Pfleghaar et al., 1993; Taichenaichev et al., 2004) and we will not illustrate it here.

The photon reabsorption is not included in this model, because in our experiments the medium is optically thin on the length scale of the laser light-atom interaction zone (Matsko et al., 2001).

The numerically calculated resonance shape which takes into account the Gaussian distribution of the laser beam intensity, the experimental geometry of the excitation and the velocity distribution of the atoms coincides with the shape of the fluorescence resonance in cell **D** (Fig. 8a) and the pedestal of the resonance in cell **A** (Fig. 8b). This shape is in very good coincidence with the shape calculated with Eq. 5 (Taichenaichev et al., 2004). The difference between the experimental and theoretical shapes in cell **A** (Fig. 8b) is practically a Lorentzian which width is of the order of $\Delta_L=1/(2\pi\tau)$, defined by the mean time τ between two atom collisions with the cell's walls. This is the narrow structure from Fig. 2.

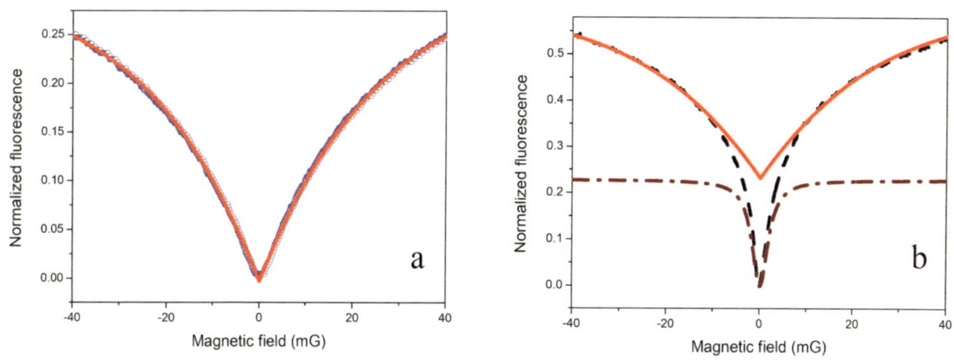

Fig. 8. Comparison of the experimental shapes with theoretical shapes:
a) experimental in cell **D** (scattered circles curve) and theoretical (solid curve);
b) experimental in cell **A** (dash line curve), theoretical (solid curve) and the difference between them (dash-dot line curve).

The comparison of the theoretical and experimental shapes in Fig. 8a shows that there is a small difference around zero magnetic field. Similar difference between the theoretical and experimental shapes was reported earlier (for example Pfleghaar et al., 1993; Taichenachev et al., 2004) and in all these cases the influence of the HRPM was not taken into account.

When the laser power density (the resonance excitation) is increased, together with the transfer of the quadrupole coherence from the ground to excited state, the influence of the multiphoton interactions will increase and high order (high rank) coherences will be created.

Our theoretical evaluations of the influence of the HRPM on the shape of the CPT resonances (Gateva et al., 2007, 2008a) proved that HRPM conversion (Okunevich, 2001) cause the CPT resonance shape peculiarities at the center of the resonance.

For a plane wave the difference from the Lorentzian shape is small and it is only around the center (at 0 magnetic field) of the resonance, with the shape and amplitude of this difference corresponding to multiphoton resonances. Although the moments of rank 4 (hexadecapole moments) (Decomps et al., 1976) do not influence directly the spontaneous emission, they are converted into the f_q^2 components of the upper level of rank 2 thus influencing the spontaneous emission.

In Fig. 9 the scheme of the connection of different rang polarization moments and the "conversion path" in population and longitudinal alignment for our scheme of excitation is illustrated (Polischuk et. al., 2011).

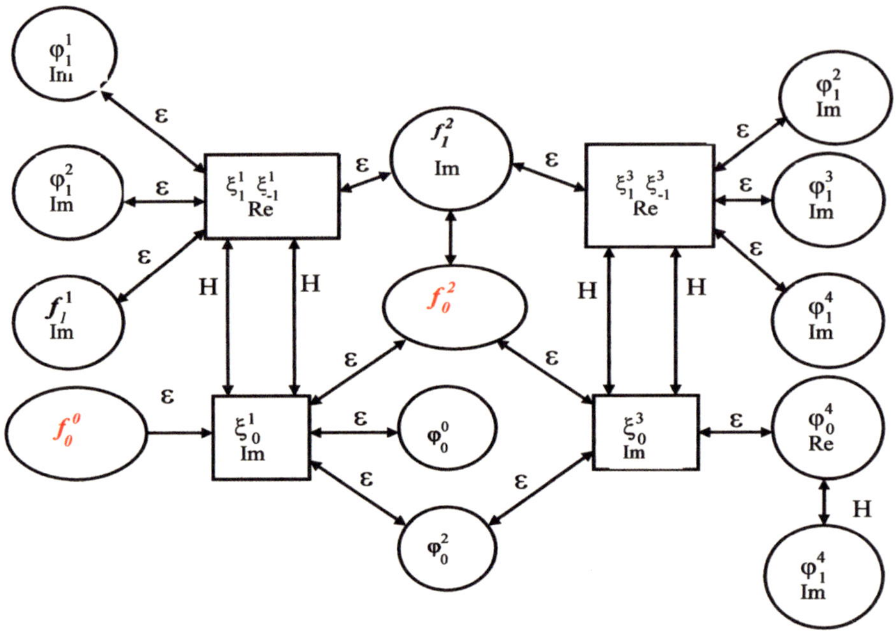

Fig. 9. Partial diagram of the mutual conversion of the tensor components forming the population (f_0^0) and the longitudinal alignment (f_0^2) (ε describes the interaction with the laser field, H - with the scanned magnetic field).

In Fig. 10 the calculated shape of the CPT resonances at different Rabi frequencies is given. At relatively low powers (Fig. 10, curve 1) the shape is close to Lorentzian; as the power is increased, the resonance flattens at the centre and an inverse narrow structure appears (Fig. 10, curves 2-4).

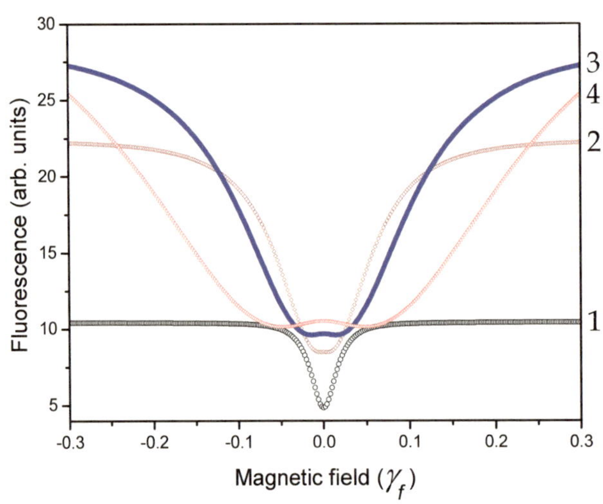

Fig. 10. Calculated shape of the CPT resonances at different Rabi frequencies $\Omega=\gamma_f(1)$; $\Omega=3\gamma_f$ (2); $\Omega=5\gamma_f(3)$; $\Omega=10\gamma_f(4)$.

At low power density of 1.7 mW/cm² (for a laser line width 50 MHz, corresponding to reduced Rabi frequency Ω=1 MHz), the measured CPT resonance shape can not be visually distinguished from a Lorentzian, but in the difference between the experimental shape and its Lorentzian fit there is a specific structure at the center of the shape (Fig. 11a). This structure is similar to the difference between the calculated resonance and its Lorentzian fit at Ω=1 MHz (Fig. 11b).

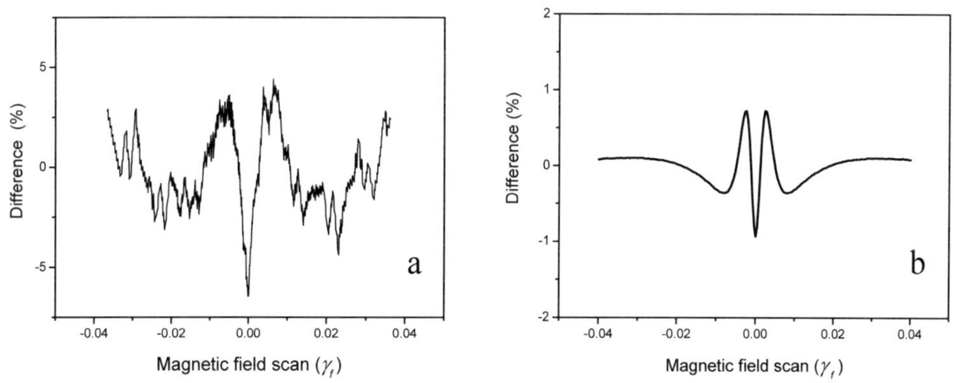

Fig. 11. Difference between:
(a) the experimental resonance shape and its Lorentzian fit
(b) the calculated resonance shape and its Lorentzian profile fit (b) at Ω=1 MHz. (The magnetic field scan is presented in γ_f units (γ_f=6 MHz is the radiation width of the upper level).

In the case of a strong laser field, at power density 1 W/cm² (corresponding to a reduced Rabi frequency Ω =25 MHz), when the measured resonance has a complex shape, the comparison of the experimental and theoretical shapes is more complicated because the amplitude of the narrow resonance increases with power (Section 2.1.2).

To observe the influence of the HRPM, the narrow structure must be eliminated. As the narrow structure is very sensitive to magnetic fields perpendicular to the polarization vector $\boldsymbol{E}(\equiv E_z)$ and $\boldsymbol{B}_{scan}(\equiv B_x)$ (Alipieva et al., 2003), magnetic fields B_y of the order of 10 mG destroy it, while the theoretical evaluations and experimental results show, that at this magnetic fields the changes in the broader structure due to the HRPM influence are practically not affected. The comparison of the shape of the resonances observed at B_y=0 (Fig. 12 curve 1) and the shape of the resonance at 32 mG (Fig. 12 curve 2) shows that at this power density the amplitude of the CPT resonance in a magnetic field is reduced twice and an additional inverted structure (about 5% in amplitude) is observed. At higher magnetic fields this inverted structure disappears (Fig. 12, curve 3).

The developed numerical model which takes into account the geometry of the excitation, the velocity distribution of the atoms and the Gaussian distribution of the laser beam intensity describes only the broad pedestal, not the narrow component from Section 2.1.2. The observed in fluorescence narrow structure of the resonance is with Lorentzian shape, it is not radiation broadened and its amplitude increases with the laser power. Its width (FWHM) Δ_L does not change with the laser beam diameter and corresponds to a relaxation time, equal to that due to atomic collisions with the cell walls.

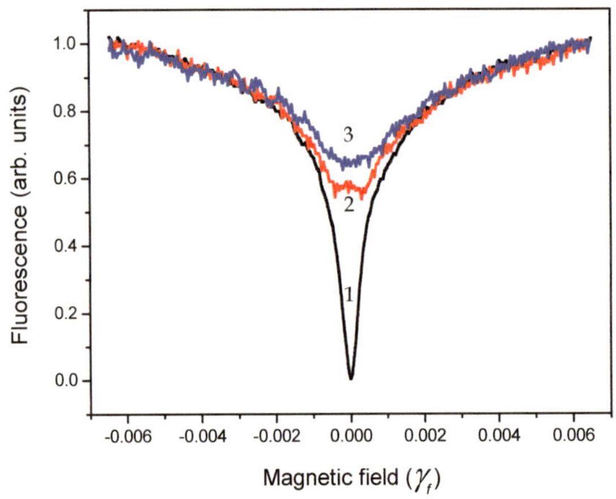

Fig. 12. CPT signals measured at different transverse to the polarization vector and propagation direction magnetic field B_y: 0 mG (curve 1); 32 mG (curve 2); 72 mG (curve 3).

Narrow structures in the CPT resonances are reported as result of diffusion-induced Ramsey narrowing (Xiao et al., 2006, 2008). In this case a sharp central peak on a broad pedestal was observed in the Electromagnetically Induced Transparency (EIT) resonance shapes. The broad pedestal is associated with the single pass interaction time and is power broadened. The sharp central peak is the central Ramsey fringe, which adds coherently for all Ramsey sequences. Its width changes with the laser beam diameter. At low laser power, small beam diameter and low buffer gas pressure the sharp central peak is not Lorentzian in shape and is insensitive to power broadening. At high laser intensity the central peak loses its contrast and is Lorentzian in shape and power broadened.

The described characteristics of the narrow structure in Fig. 2 and 8 show that it is not result of diffusion-induced Ramsey narrowing. The resonance shapes, measured at different geometries of excitation and registration (Gateva et al., 2008b, 2011), show that the narrow structure at the centre of the resonance can be considered as a result of a weak field - atom interaction, probably scattered light in the whole cell volume.

One of the possible scattering processes influencing the CPT resonance shape is Rayleigh scattering because in this case the scattered light maintains coherence with the incident beam and in this way a diffusion of the coherent light is created. For example, Rayleigh scattering has been used for studying the properties of cold atoms (Datsyuk et al., 2006) and optical lattices (Carminati et al., 2003 and references therein).

2.3 CPT resonances and Rayleigh scattering

For linear polarization of the incident light, the power of the Rayleigh-scattering light P_R registered at 90° to the laser beam direction is (Boyd, 1992; Measures, 1984)

$$P_R(\lambda,\theta) = P_0 \; \alpha \; N \; \sigma(\lambda) \sin^2 \theta \tag{15}$$

where P_0 is the incident light power, α is the registration efficiency coefficient, N is the scattering particles density, $\sigma(\lambda,\theta) = \sigma(\lambda) \sin^2 \theta$ is the differential cross-section of Rayleigh scattering of polarized light with wavelength λ in the plane perpendicular to the beam direction and θ is the angle between the laser light polarization and the direction of registration. According to Equation (9) the Rayleigh scattering light has a maximum at θ =90° and is zero at θ =0°.

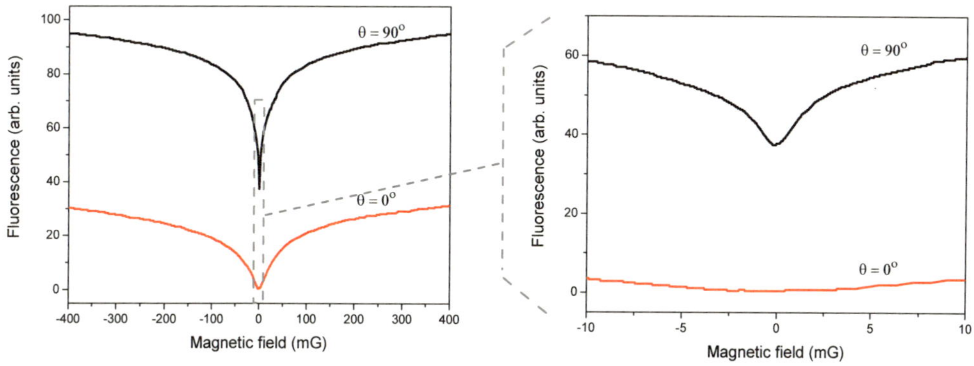

Fig. 13. Shape of the measured CPT resonances given in two different scales at angles θ =0° and 90°.

In Fig. 13 are given the shapes of the CPT resonances measured at angles θ =0° and 90° when the laser is tuned on the F_g=2 – F_e=1 transition of the ^{87}Rb D_1 line. The results are presented in two different scales. The comparison of the resonances on the broader scale shows that at the wings they are with the same shape and that this shape is in good coincidence with the calculated with the model from Section 2.2.

The comparison of the shapes in Fig. 13 on the narrower scale shows, that at θ =90° there is a narrow structure. This narrow structure is Lorentzian in shape with width of the order of the defined by the mean time between 2 collisions with the cell walls Δ_L. The amplitude of this narrow structure (Fig. 14a) has the typical $\sin^2 \theta$ dependence on θ and it is maximum at θ =90° and minimum at θ =0°.

In Rb vacuum cell at room temperature the Rb pressure is of the order of 4.10^{-5} Pa and at this pressure the Rayleigh scattered light from Rb atoms can not be registered in our experiment (Boyd, 1992; Measures, 1984) - this is the case of cell **D**. However, if the vacuum cell is not pumped very well, there will be some residual air, water and oil vapor, as well as rare but relatively strongly scattering submicron particles, which will scatter the light - this is the case of cells **A**, **B** and **C**. These cells are very old (more than 20 years), manufactured with an oil diffusion pump connected by a liquid nitrogen cold trap to the cell. Cell **D** is a new one, which was sealed off some months before the experiments. It was manufactured using an oil-free vacuum installation - a turbomolecular pump and ion pump. Checking the vacuum with a Tesla coil (high voltage transformer) shows that there is no light emission (glowing) in cell **D,** while the cell glass is glowing in cell **A**, cell **B** and cell **C**.

A demonstration of the existence of Rayleigh scattered light is the measured angular dependence of the scattered light power when the laser is tuned out of line. It has the typical $\sin^2\theta$ dependence from Equation 9 with maximum at θ =90° and minimum at θ =0° (Fig. 14b).

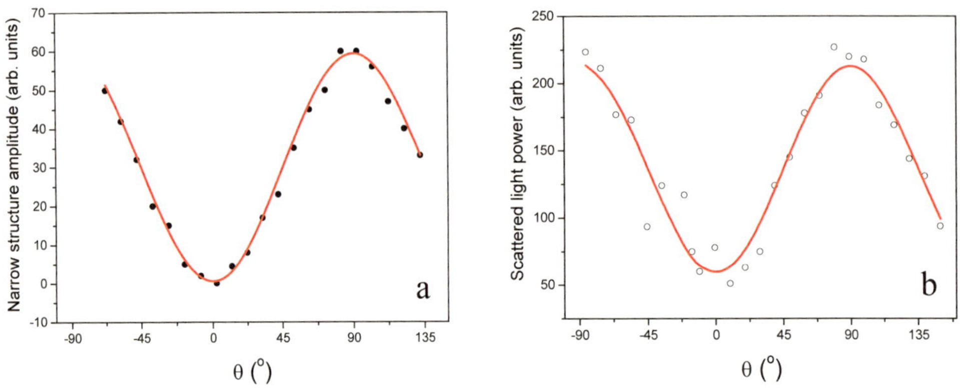

Fig. 14. Angular dependence on θ, the angle between the direction of observation and laser light polarization at 90° to the laser beam of:
(a) the narrow Lorentzian structure amplitude;
(b) the measured scattered light power when the laser is out of line.

Qualitatively, the observed narrow resonance can be described by including into the numerical model from Section 3.2 a weak, resonant, polarized light field in the cell volume. Such an approach corresponds to the supposition that the signal in the cell is formed by two subensembles of atoms interacting with different light fields. The general solution for the fluorescence intensity is sum of the numerical integrations of the density matrix equations for each of the subensembles.

For the subensemble interacting with the laser beam the evaluations following the model described in Section 2.2 have shown that this part of the signal practically does not depend on the observation direction. Although the components of the observation tensor $\Phi_q^2(\vec{n})$ depend on the direction of observation, the performed evaluations for the investigated transition have shown that the main contribution to the intensity of the unpolarized fluorescence has the population f_0^0 and $I_{F_f F_\varphi}$ practically does not depend on the observation direction.

For the subensemble interacting with the weak Rayleigh scattered light, the angular distribution of the Rayleigh scattered light has to be included in the excitation tensor. This angular dependence in the excitation reflects on all tensor components f_q^k and in this way on the fluorescence signal $I_{F_f F_\varphi}(\vec{n})$.

As the scattered light is very low in power, the resonances are with Lorentzian shape and not power broadened. When the Rabi frequency of the light field Ω is small enough, so that $\Omega^2 / \gamma_f < \gamma_\varphi$ (Eq. 1), there is no power broadening. So long as the mean free path of the atom is longer than the cell dimensions, γ_φ is defined by the mean time between two successive collisions with the cell walls. Since the weak field is due to Rayleigh scattering, the amplitude of the Lorentzian will depend on the density of the scattering particles and can be used as indicator of the level of the vacuum cleanness of the cell.

Another confirmation that the shape of the resonances is sum of the resonance shapes of different subensembles of atoms is done in experiment where the shape of the resonances is measured at two different geometries of observation (Gateva et al., 2008b, 2011). In the first case, the photodiode is on the cell wall and the observation field of view is considerably larger than the laser beam diameter (Fig. 15, solid curve), but the angle of view is sufficiently small to ensure satisfactory angular resolution of observation. In the second case a lens is used in front of the photodiode in order to restrict the observation field of view just to the laser beam volume (Fig. 15, scattered circles curve). Because of the limited dimensions of the μ-metal shield, these measurements were performed in a smaller cell, cell **B**. The comparison of the shape of the resonances in these two cases shows that the broad pedestal doesn't change, while the shape and the width of the narrow structure change (Fig. 15). In the first case (Fig. 15, solid curve), when the fluorescence mostly of atoms out of the laser beam is registered, the narrow structure is Lorentzian in shape and its width Δ_L is defined by the relaxation on the cell walls. In the second case (Fig. 15, scattered circles curve), when fluorescence mostly from the laser beam volume is registered, the measured narrow structure is narrower than Δ_L. For explanation of such narrowing the influence of the diffusion-induced Ramsey narrowing (Xiao et all., 2006), which is result of the impurities in the cell and/or on the cell walls, has to be taken into account. Diffusion induced Ramsey narrowing resonances have been measured in Na vacuum cell, too (Gozzini et al., 2011).

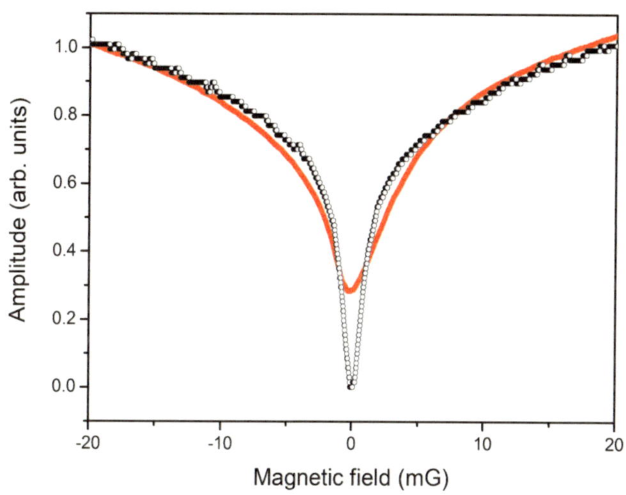

Fig. 15. Narrow structure of the CPT resonance at different geometries of registration: when the photodiode is on the cell wall (solid curve) and when the laser beam is projected by a lens on the photodiode (scattered circles curve).

This experiment can explain the difference in the narrow resonance shapes in the works of Alipieva et al. and Xiao et al. (Alipieva et al., 2003; Xiao et al., 2006). When the resonance is registered in transmission (EIT) only atoms from the laser beam volume contribute to the signal. In this case the diffusion-induced Ramsey narrowing is responsible for the narrow structure. If the signal is registered in fluorescence (Alipieva et al., 2003) the signal is mainly from atoms out of the laser beam volume, which interact with the Rayleigh scattered light. In this case the narrow component is Lorentzian and its width is defined by the relaxation time due to atomic collisions with the cell walls.

3. Conclusion

The investigation of the influence of different factors on the shape of the CPT resonances excited in Hanle effect configuration in uncoated, room temperature vacuum cell and registered in fluorescence has shown that the resonance has a complex shape. The performed numerical calculations, which take into account the Gaussian distribution of the laser beam intensity, the experimental geometry of the excitation, the velocity distribution of the atoms, the high rank polarization moment, the stray magnetic fields influence, and the Rayleigh scattering, describe very well the experimental shapes. The results show that the developed theoretical modelling can be applied for analysis of the influence of the different factors and analysis of the cell quality. The Rayleigh scattering can be used for analysis of the vacuum cleanness of the cell and the quality of the cell walls.

On the other hand these narrow resonances can be applied for building magneto-optical sensors. From the point of view of applications in magnetometry, where narrow signals and high signal-to-noise ratios are important, the narrow Rayleigh structure of the CPT resonance measured in fluorescence offer good possibilities: is not power broadened, its width do not depend on the position of the detector and its amplitude increases with the power. The complex structure of the resonance expands the range of the measured magnetic fields.

4. Acknowledgements

The authors are pleased to acknowledge the financial support of part of the investigations by the Bulgarian NCSR (Grant No. DO02-108/22.05.2009).

5. References

Alexandrov, E. B.; Chaika, M. P. & Hvostenko, G. I. (1991). *Interference of atomic states*, Springer-Verlag, ISBN-10: 354053752X, ISBN-13: 978-3540537526

Alipieva, E.; Gateva, S.; Taskova, E. & Cartaleva, S. (2003). Narrow structure in the coherent population trapping resonance in Rb, *Optics Letters*, Vol.28, No.19, pp. 1817-1819, ISSN 0146-9592

Alzetta, G.; Gozzini, A.; Moi, L. & Orriols, G. (1976). An experimental method for the observation of r.f. transitions and laser beat resonances in oriented Na vapour, *Il Nuovo Cimento B Series 11*, Vol.36, No.1, pp. 5-20, ISSN 0369-3546

Arimondo, E. (1996). Coherent population tapping in laser spectroscopy, *Progress in Optics*, Vol.35, pp. 257-354, ISSN 0079-6638

Auzinsh, M.; Budker, D. & Rochester, S.M. (2009a). Light-induced polarization effects in atoms with partially resolved hyperfine structure and applications to absorption, fluorescence, and nonlinear magneto-optical rotation, *Physical Review A - Atomic, Molecular, and Optical Physics*, Vol. 80, No. 5, art. no. 053406, pp. 1-22, ISSN 1050-2947

Auzinsh, M.; Ferber, R.; Gahbauer, F.; Jarmola, A. & Kalvans, L. (2009b). Nonlinear magneto-optical resonances at D1 excitation of Rb 85 and Rb87 for partially resolved hyperfine F levels, *Physical Review A - Atomic, Molecular, and Optical Physics*, Vol. 79 No. 5, art. no. 053404, pp. 1-9, ISSN 1050-2947.

Bison, G; Castanga, N.; Hofer, A.; Knowles, P.; Schenker, J.L.; P., Schenker, J.-L.; Kasprzak, M.; Saudan, H. & Weis, A. (2009). A room temperature 19-channel magnetic field mapping device for cardiac signals. *Applied Physics Letters*, Vol.95, No.17, art. no. 173701, pp. 1-3, ISSN 0003-6951

Bouchiat, M.A. & Brossel, J. (1966). Relaxation of optically pumped rb atoms on paraffin-coated walls. *Physical Review*, Vol.147, No., pp. 41-54, ISSN 0031-899X

Boyd, R. W. (1992). *Nonlinear Optics*, Academic Press, ISBN 0123694701

Budker, D. & Romalis, M. (2007). Optical magnetometry, *Nature Physics*, Vol.3, No.4, pp. 227-234, ISSN 17452473

Carminati, F.-R.; Sanchez-Palencia, L.; Schiavoni, M.; Renzoni, F. & Grynberg, G. (2003). Rayleigh Scattering and Atomic Dynamics in Dissipative Optical Lattices, *Physical Review Letters*, Vol.90, art.no. 043901, ISSN 0031-9007

Corney, A. (1977). *Atomic and Laser Spectroscopy*, Clarendon, Oxford, ISBN 9780198511380

Cox, K.; Yudin, V.I.; Taichnachev, A.V.; Novikova, I. & Mikhailov, E.E. (2011). Measurements of the magnetic field vector using multiple electromagnetically induced transparency resonances in Rb vapor, *Physical Review A - Atomic, Molecular, and Optical Physics*, Vol.83, No.1, art. no. 015801, pp. 1-4, ISSN 1050-2947.

Dancheva, Y.; Alzetta, G.; Cartaleva, S.; Taslakov, M. & Andreeva, Ch. (2000). Coherent effects on the Zeeman sublevels of hyperfine states in optical pumping of Rb by monomode diode laser, *Optics Communications*, Vol.178, No.1, pp. 103-110, ISSN 0030-4018

Dang, H.B.; Maloof, A.C. & Romalis, M.V. (2010). Ultrahigh sensitivity magnetic field and magnetization measurements with an atomic magnetometer, *Applied Physics Letters*, Vol.97, No.15, art. no. 151110, ISSN 0003-6951

Datsyuk, V. M.; Sokolov, I. M.; Kupriyanov, D. V. & Havey, M. D. (2006). Diffuse light scattering dynamics under conditions of electromagnetically induced transparency *Physical Review A - Atomic, Molecular, and Optical Physics*, Vol.74, art.no. 043812, ISSN 0031-9007

Decomps B., Dumont M. & Ducloy M. (1976). *Laser Spectroscopy of Atoms and Molecules*, Springer-Verlag, Berlin, ISBN 3-540-07324-8

D'yakonov, M. I. & Perel, V. I. (1966). On the theory of the gas laser in a magnetic field, *Optics and Spectroscopy*, Vol.20, No.3, pp. 472-480, ISSN 0030-400X

Ducloy, M. & Dumont, M. (1970). Etude du transfert d'excitation par emission spontanee, *Le journal de Physique*, Vol.31, pp. 419-427, ISSN 0302-0738

Edelstein, A. (2007). Advances in magnetometry. *Journal of Physics Condensed Matter*, Vol.19, No.16, art. no. 165217, ISSN 0953-8984.

Firstenberg, O.; Shuker, M.; Ben-Kish, A.; Fredkin, D.R.; Davidson, N. & Ron, A. (2007). Theory of Dicke narrowing in coherent population trapping. *Physical Review A - Atomic, Molecular, and Optical Physics*, Vol. 76, No. 1, art. no. 013818, pp.1-6, ISSN 1050-2947.

Figueroa, E.; Vewinger, F.; Appel, J. & Lvovsky, A.I. (2006) . Decoherence of electromagnetically induced transparency in atomic vapor. *Optics Letters*, Vol.31, No.17, pp. 2625-2627, ISSN 0146-9592

Gao, J.-Y.; Xiao, M. & Zhu, Y. (eds.) (2009). *Atomic coherence and Its Potential Application*, Bentham Science Publishers Ltd., ISBN: 978-1-60805-085-7

Gateva, S.; Alipieva, E. & Taskova, E. (2005). Power dependence of the coherent-population-trapping resonances registered in fluorescence and transmission: Resonance-width narrowing effects. *Physical Review A - Atomic, Molecular, and Optical Physics*, Vol. 72, No.1, art.no. 025805, pp.1-4, ISSN 1050-2947

Gateva, S.; Petrov, L.; Alipieva, E.; Todorov, G.; Domelunksen, V. & Polischuk, V. (2007). Shape of the coherent population-trapping resonances and high-rank polarization moments. *Physical Review A - Atomic, Molecular, and Optical Physics*, Vol. 76, No. 2, art. no. 025401, pp. 1-4, ISSN 1050-2947

Gateva, S.; Alipieva, E.; Petrov, L.; Taskova, E. & Todorov, G. (2008a). Single frequency coherent-population-trapping resonances for magnetic field measurement. *J. Optoelectronics and Advanced Materials*, Vol. 10, No.1, pp. 98-103, ISSN 1454-4164

Gateva, S.; Alipieva, E.; Domelunksen, V.; Polischuk, V.; Taskova, E.; Slavov, D.; Todorov, G. (2008b). Shape of the coherent-population-trapping resonances registered in fluorescence at different experimental geometries. *Proceedings of SPIE - The International Society for Optical Engineering*, Vol. 7027, art. no. 70270I, pp. 1-12, ISBN 9780819472410.

Gateva, S.; Gurdev, L.; Alipieva, E.; Taskova, E. & Todorov, G. (2011). Narrow structure in the coherent population trapping resonances in rubidium and Rayleigh scattering. *Journal of Physics B: Atomic, Molecular and Optical Physics*, Vol. 44, No. 3, art. no. 035401, pp. 1-6 , ISSN 0953-4075

Godone, A.; Levi, F.; Micalizio, S. & Vanier, J. (2002). Dark-line in optically-thick vapors: Inversion phenomena and line width narrowing, *The European Physical Journal D*, Vol.18, No.1, pp. 5-13, ISSN 1434-6060

Gozzini, S.; Marmugi, L.; Lucchesini, A.; Gateva, S.; Cartaleva, S. & Nasyrov, K. (2011). Narrow structure in the coherent population trapping resonance in sodium, *Physical Review A - Atomic, Molecular, and Optical Physics*, Vol. 84, No. 1, art. no. 013812, pp. 1-9, ISSN 1050-2947

Griffith, W.C.; Knappe, S. & Kitching, J. (2010). Femtotesla atomic magnetometry in a microfabricated vapor cell, *Optics Express*, Vol.18, No.26, pp. 27167-27172, ISSN 1094-4087.

Huss, A.; Lammegger, R.; Windholz, L.; Alipieva, E.; Gateva, S.; Petrov, L.; Taskova, E. & Todorov, G. (2006). Polarization-dependent sensitivity of level-crossing, coherent-population-trapping resonances to stray magnetic fields. *Journal of the Optical Society of America B: Optical Physics*, Vol.23, no.9, pp. 1729-1736, ISSN 0740-3224

Javan, A.; Kocharovskaya, O.; Lee, H. & Scully, M.O. (2002). Narrowing of electromagnetically induced transparency resonance in a Doppler-broadened medium. *Physical Review A - Atomic, Molecular, and Optical Physics*, Vol.66, No. 1, pp. 138051-138054, ISSN 1050-2947

Johnson, C.; Schwindt, P.D.D. & Weisend, M. (2010). Magnetoencephalography with a two-color pump-probe, fiber-coupled atomic magnetometer. *Applied Physics Letters*, Vol.97, No.24, art. no. 243703, ISSN 0003-6951

Kimble, H.J. (2008). The quantum internet. *Nature*, Vol.453, pp. 1023-1030, ISSN 0028-0836

Kitching, J.; Knappe, S. & Donley, E.A. (2011). Atomic sensors - A review. *IEEE Sensors Journal*, Vol. 11, No.9, art. no. 5778937, pp. 1749-1758. ISSN 1530-437X.

Knappe, S.; Wynands, R.; Kitching, J.; Robinson, H. G. & Hollberg, L. (2001). Characterization of coherent population-trapping resonances as atomic frequency references, *Journal of the Optical Society of America B: Optical Physics*, Vol.18, No.11, pp. 1545-1553, ISSN 0740-3224

Knappe, S.; Shah, V.; Schwindt, P.D.D.; Hollberg, L.; Kitching, J.; Liew, L. & Moreland, J. (2004). Microfabricated atomic clock, *Applied Physics Letters*, Vol. 85, No. 9, pp. 1460-1462, ISSN 0003-6951

Knappe, S.; Sander, T.M.; Kosch, O.; Wiekhorst, F.; Kitching, J. & Trahms, L. (2010). Cross-validation of microfabricated atomic magnetometers with superconducting quantum interference devices for biomagnetic applications, *Applied Physics Letters*, Vol. 97, art. no. 133703, pp. 1-3, ISSN 0003-6951

Kominis, I. K.; Kornack, T. W.; Allred, J. C. & Romalis, M. V. (2003). A subfemtotesla multichannel atomic magnetometer. *Nature*, Vol. 422, No.6932, pp. 596-599, ISSN 0028-0836

Landau, L.D. & Lifshitz, E.M. (1965). *Quantum Mechanics: Non-Relativistic Theory*, Pergamon Press, ISBN 978-0-080-20940-1

Levi, F.; Godone, A.; Vanier, J.; Micalizio, S. & Modugno, G. (2000). Line-shape of dark line and maser emission profile in CPT. *The European Physical Journal D*, Vol.12, No.1, pp. 53-59, ISSN 1434-6060

Lukin, M.D. (2003). Colloquium: Trapping and manipulating photon states in atomic ensembles. *Reviews of Modern Physics*, Vol.75, pp. 457-472, ISSN 0034-6861

Matsko, A. B.; Novikova, I.; Scully, M. O. & Welch, G. R. (2001). Radiation trapping in coherent media. *Physical Review Letters*, Vol.87, No.13, art. no. 133601, ISSN 1079-7114

Measures, R. M. (1984). *Laser Remote Sensing: Fundamentals and Applications*, Wiley ISBN 0471081930 (ISBN13: 9780471081937), New York

Mikhailov, E.; Novikova, I,; Havey, M.D. & Narducci, F.A. (2009). Magnetic field imaging with atomic Rb vapour. *Optics Letters*, Vol.34, No.22, pp. 3529-3531, ISSN 0146-9592

Okunevich, A. I., (2001). On the possibility of registration of hexadecapole transversal components of the atoms in fluorescence. *Optics and Spectroscopy*, Vol.91, No.2, pp. 193-200, ISSN 0030-400X

Petrov, L.; Slavov, D.; Arsov, V.; Domelunksen, V.; Polischuk, V. & Todorov, G. (2007) High rank polarization moments in a Doppler broadened 87Rb transition, *Proceedings of SPIE - The International Society for Optical Engineering*, Vol. 6604, pp. 66040H1-66040H5, ISBN 9780819467423

Pfleghaar, E.; Wurster, J.; Kanorsky, S.I. & Weis, A. (1993). Time of flight effects in nonlinear magneto-optical spectroscopy, *Optics Communications*, Vol. 99, No. 5-6, pp. 303-308. ISSN 0030-4018.

Polischuk, V.; Domelunksen, V.; Alipieva, E. & Todorov, G. (2011). MatLab based modelling of nonlinear interaction of Rb87 atoms with polarized radiation. *unpublished*

Romalis, M.V. & Dang, H.B. (2011). Atomic magnetometers for materials characterization. *Materials Today*, Vol.14, no.6, pp. 258-262, ISSN 1369-7021

Savukov, I. (2010). Ultra-Sensitive Optical Atomic Magnetometers and Their Applications, In: *Advances in Optical and Photonic Devices*, Ki Young Kim, (Ed.), 329-352, InTech, ISBN 978-953-7619-76-3, Croatia

Steck, D.A. (2009) *Rubidium 87 D Line Data* (available online http://steck.us/alkalidata)

Taichenachev, A.V.; Tumaikin, A. M.; Yudin, V. I.; Stähler, M.; Wynands, R.; Kitching, J. & Hollberg L. (2004). Nonlinear-resonance line shapes: Dependence on the transverse intensity distribution of a light beam. *Physical Review A - Atomic, Molecular, and Optical Physics,* Vol.69, art. no. 024501, ISSN 1050-294711

Thomas, J. E. & Quivers, W. W. (1980). Transit-time effects in optically pumped coupled three-level systems, *Physical Review A - Atomic, Molecular, and Optical Physics* 22(5), 2115-2121, ISSN 1050-294711

Xia, H.; Baranga, A.B.-A.; Hoffman, D. & Romalis, M.V. (2006). Magnetoencephalography with an atomic magnetometer. *Applied Physics Letters,* Vol.89, art. no. 211104, ISSN 0003-6951

Xiao, Y.; Novikova, I.; Phillips, D. F. & Walsworth, R. L. (2006) Diffusion-Induced Ramsey Narrowing. *Physical Review Letters,* Vol.96, art. no. 043601, ISSN 1079-7114

Xiao, Y.; Novikova, I.; Phillips, D. F. & Walsworth, R. L. (2008). Repeated interaction model for diffusion-induced Ramsey narrowing, *Optics Express,* Vol.16, pp. 14128-14141, ISSN 1094-4087

Xu, S.; Crawford, C.W.; Rochester, S.; Yashchuk, V.; Budker, D. & Pines, A. (2008). Submillimeter-resolution magnetic resonance imaging at the Earth's magnetic field with an atomic magnetometer. *Physical Review A - Atomic, Molecular, and Optical Physics,* Vol. 78 No.1, art. no. 013404, pp. 1-4, ISSN 1050-294711.

Ye, C.Y. & Zibrov, A. S. (2002). Width of the electromagnetically induced transparency resonance in atomic vapor. *Physical Review A - Atomic, Molecular, and Optical Physics,* Vol. 65, No. 2, art.no. 023806, pp. 1-5, ISSN 1050-2947

20

Photodetectors in Calorimeters for the Linear Collider

Jaroslav Cvach and CALICE Collaboration
Institute of Physics of the ASCR, v.v.i., Praha
Czech Republic

1. Introduction

The next high energy accelerator that will be built after the Large Hadron Collider currently operating at CERN is expected to be a Linear Collider (LC). It is envisaged to be an electron-positron colliding beams facility with the centre-of-mass energy of 500–3000 GeV (ILC, 2007, Vol. 3; CLIC, 2011). The detector at the LC will have to deal with a large dynamic range in particle energy, complexity of final states and small signal-to-background ratio. The track density in collimated jets can be as high as one per mm^2 at a radius of 1.5 cm, and the accelerator induced backgrounds produce typical hit densities of the order of $0.03/mm^2$ per bunch crossing at a radius of 1.5 cm, and $0.003/cm^2$ at a radius of 30 cm. The physics requirements demand detector performance parameters which must be substantially better than at previous experiments on accelerators at the large electron-positron collider in CERN and at the Stanford linear collider in SLAC. The goal is to achieve the energy resolution for jets $\sigma/E < 3\%$ in wide range of energies starting from the mass of the intermediate bosons W and Z, $E \sim m_W, m_Z \sim 80$ GeV to several hundred GeV. This condition is often translated into the formula used for the energy resolution of the sandwich calorimeter $\sigma/E \sim 30\%/\sqrt{E}$ (jet energy E is in GeV).

The experimental technique proposed to reach these goals was formulated ten years ago and is now known as the particle flow algorithm (PFA) (Brient & Videau, 2001; Cvach, 2002; Morgunov, 2002). The PFA combines the information from tracking and calorimetry to obtain an optimal estimate of the flow of particles from the interaction vertex and of the original parton four-momenta. The subdetectors must have excellent spatial granularity to enable a PFA algorithm which resolves energy depositions of almost overlapping particles, combines redundant measurements properly (e.g. of electrons in tracking and the electromagnetic calorimeter or of charged pions in tracking and calorimetry) and provides other corrections (e.g. calorimeter software compensation). The calorimeter for experiments at the LC must be realized as a dense and hermetic sampling calorimeter with a very high granularity, where one can efficiently separate the contributions of different particles in a jet and use the best suited detector to measure their four-momenta.

In this article we review the performance of modern photodetectors used in already built calorimeter prototypes with plastic scintillator as the active medium. Calorimeters are constructed as a sandwich with an absorber between the scintillator layers to decrease the calorimeter size.

2. Calorimeter prototypes optimized for the PFA

The CALICE collaboration (Calice, n.d.) has undertaken an intensive R&D program starting in 2001 to prove feasibility of the particle flow approach for the improvement of detector energy resolution by a factor of approximately two for jets produced in e^+e^- collisions at centre of mass energies 50-1000 GeV. The corresponding effort was done both on the subdetector side as well as on the software side - reconstruction and simulation programs. We shall concentrate in this article on calorimeters optimized for the PFA. For effective functioning of a calorimeter, the vertex tracker and tracker standing in front of calorimeters are important. We refer e.g. to (ILC, 2007, Vol. 4) where relevant solutions for trackers are discussed in detail.

The first completed prototype of the electromagnetic calorimeter ECAL was SiW ECAL (Repond et. al., 2008). The large silicon diode sensors of 1x1 cm^2 area made on high resistive Si wafers 0.5 mm thick define the size of the active calorimeter cell. A sensitive area of 18x18 cm^2 of silicon was inserted between absorber tungsten plates 1.2-4.8 mm thick. In total 30 layers of Si + W represent 24 radiation lengths X_0. The total number of cells was 9720. As this calorimeter does not use photodetectors, we refer for further details to the latest publication (Adloff et al., 2010b).

As an alternative to the SiW ECAL, a scintillator ECAL was built which used scintillator strips read by Multi-Pixel Photon Counter (MPPC) photodetectors. The absorber is also tungsten and the sensitive area was 18x18 cm^2. The scintillator strips are 1 cm wide and oriented alternatively in x and y directions to make effectively 1 cm^2 cells comparable to SiW ECAL. Thus good comparison between two ECAL realisations can be done. We refer in detail about the scintillator ECAL in section 4.

The first completed prototype of the hadron calorimeter was a 1-m^3 steel-scintillator calorimeter with SiPMs as photodetectors that provide analog amplitude. The optimisation studies for the PFA defined (Thomson, 2009) as the best tile size 3x3 cm^2. For the economical and practical reasons, the prototype was built from tiles of three different sizes made of 5 mm thick scintillator placed between steel absorber plates 16 mm thick. 38 sandwich layers have a depth of 1 m and represent 5.3 interaction lengths. The details are given in section 5.

Recently two versions of calorimeter with the scintillator replaced by the gas detector - Resistive Plate Chamber (RPC) - were completed and are currently tested in beam. As the RPC provides digital amplitude from a 1 cm^2 cell, we call this type of calorimeter digital. Further details are given in references (Bilki et al., 2009; Laktineh, 2011).

3. Silicon photomultiplier

The CALICE collaboration contributed to the first massive use of a novel multi-pixel silicon photodetector operated in the Geiger mode. There are several producers able to deliver this photodetector, e.g. Hamamatsu Photonics Japan (first commercial producer), MEPHI/PULSAR and Dubna/Micron in Russia, Photonique in Switzerland, SENSL in Ireland, ITC IRST Trento in Italy, Zecotek in Singapur, MPI Semiconductor Laboratory Munich in Germany, SensL Cork in Ireland, STmicroelectronics in Italy, Novel Device Laboratory Beijing in China, RMD Boston in U.S.A., and possibly others.

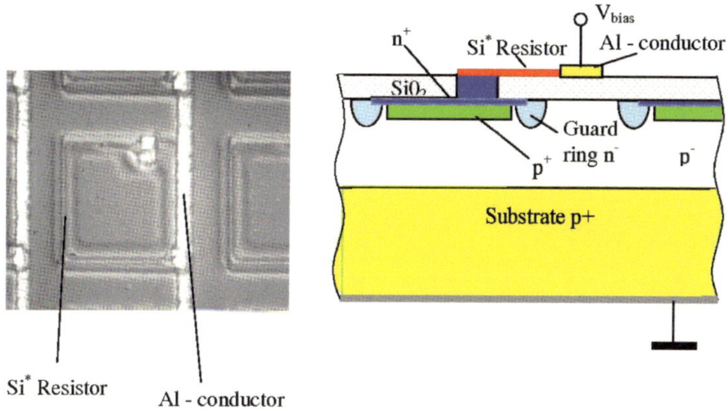

Fig. 1. (a) View of a pixel in an array of the silicon photomultiplier. (b) A cross section of a pixel of SiPM showing the layers. The Si* resistor serves as a quenching resistor, the SiO2 resistor reduces the inter pixel cross talk.

We shall here report about experimentation with the Silicon PhotoMultiplier (SiPM) from MEPHI/PULSAR which delivered more than 10000 pieces for the Analog Hadron CALorimeter (AHCAL) and the Tail Catcher and Muon Tracker (TCMT) and the Pixelated Photon Counter from Hamamatsu whose 2160 pieces of the "1600-pixel MPPC" were used in the Scintillator Electromagnetic CALorimeter (ScECAL).

The silicon photomultiplier is an array of small pixels connected on a common substrate (see Fig. 1). Each cell has its quenching resistor of the order of a MΩ. A common bias voltage is applied to all cells 10-20% above the breaking voltage. The cells fire independently when a carrier is liberated in the depletion layer or by a photon arriving on the surface or thermally. The output signal equals to the sum of fired cells. At small signals the detector works as an analog photodetector, at large signals the detector saturates due to the limited number of pixels (Bondarenko et al., 2000; Buzhan et al., 2001) The photon detection efficiency is $\varepsilon = QE\ \varepsilon_{geom}$, where quantum efficiency QE is 0.5-0.8 and ε_{geom} is geometrical factor giving the effective sensitive area of the pixel. The geometrical factor increased from originally several percents up to 80% nowadays (Musienko, 2011). Each pixel works as a digital device - several photons hitting the same pixel produce the same signal as a single photon. Optical cross-talk between pixels causes adjacent pixels to be fired. This increases gain fluctuation, increases noise and excess noise factors. Due to the Geiger discharge, the pulse is short, typically several ns long. The photodetector for the AHCAL prototype was developed in collaboration of Moscow Institutes MEPhI and ITEP with PULSAR factory near Moscow and with support from DESY Hamburg. They called it silicon photomultiplier because of its high gain. The photosensitive area of 1.1 mm x 1.1 mm holds 1156 pixels each having an area 32 µm x 32 µm (sensitive area 24 µm x 24 µm). The single pixel signal is determined by the total charge collected during the Geiger discharge $Q = C_{pixel}\Delta V$ where ΔV is the voltage above the breakdown voltage V_{break}. For the pixel capacitance C_{pixel} = 50 fF and ΔV = 3V the signal reaches Q = 150 fC ~ 10^6 electrons. This defines the gain of ~ 10^6. The working voltage equals to $V_{bias} = V_{break} + \Delta V$ ~ 50-60 V. The cross talk increases with ΔV, the typical value is

20%. The Geiger discharge time is ~ 500 ps. The pixel recovery time is defined by the product C_{pixel} R_{pixel} , where R_{pixel} is the pixel quenching resistor (400 kΩ - 10 MΩ). The quenching resistor interrupts the discharge in a pixel. For smaller resistor values the quenching time is ~ 20 ns.

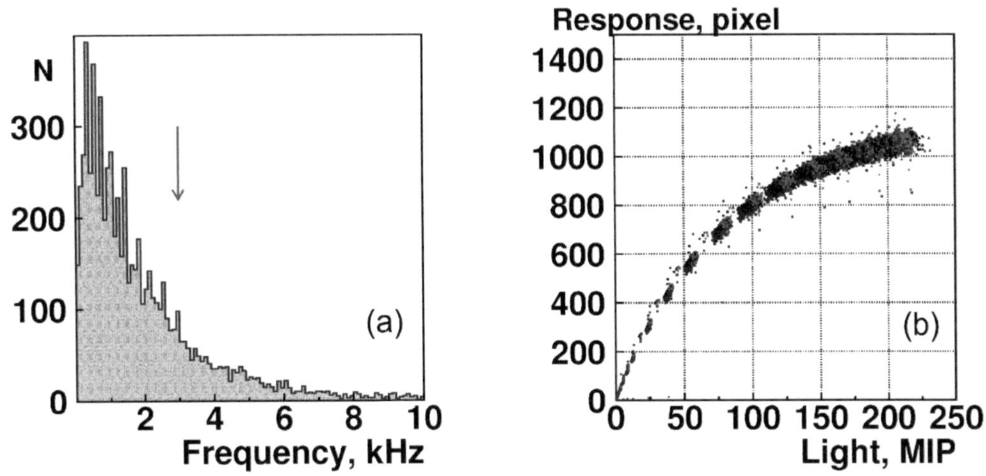

Fig. 2. (a) Distribution of noise above ½ MIP threshold for 10000 SiPMs produced in MEPHi/PULSAR. The arrow at 3 kHz shows selection cut. SiPMs with higher noise were discarded. (b) The response curve for the selected SiPMs. The curve shows the SiPM signal in number of fired pixels as a function of the LED light expressed in MIPs.

The aim was to keep it below 100 ns with the signal width of ~ 10 ns. For short recovery times comparable to the signal width, the SiPM can fire repeatedly causing undesirable inter pixel crosstalk. The pixels are connected in parallel to one readout channel therefore the output pulse is proportional to the number of detected photons (Danilov, 2007b).

More than 10000 SiPM have been produced by the MEPhI/PULSAR group and were tested at ITEP. The SiPMs were illuminated with calibrated light from a UV LED (Ultra-Violet Light Emitting Diode). The light was brought by a Kuraray Y11 WLS fibre. For each SiPM the working voltage was chosen individually to fire 15 pixels for LED light pulse corresponding to the signal which produces a minimum ionizing particle (MIP) - in practice a muon - in a scintillator tile. The gain, the dark rate, the inter-pixel crosstalk, the noise above a threshold of the ½ MIP and the non-linear response function were measured for all SiPMs. The SiPMs had to fulfil selection criteria on gain > 4.10^5, noise at ½ MIP threshold < 3 kHz, cross talk probability < 0.35 and dark current < 2.10^{-6} A. As an example in Fig. 2a the noise distribution is shown, the arrow gives the position for the selection cut.

SiPMs were tested for radiation hardness on the proton accelerator at ITEP Moscow at a beam energy of 200 MeV. The dark current of SiPMs increased with the proton flux Φ as expected. At doses $\Phi \sim 10^{10}$ protons/cm^2 the detector does not see individual photoelectron peaks and still resolves MIPs at $\Phi \sim 10^{11}$ protons/cm^2. This radiation hardness is sufficient for doses the AHCAL will obtain at the LC (Danilov, 2007a).

Photodetectors for the ScECAL were developed by cooperation between the Shinshu University and the Hamamatsu Photonics Company. The MPPC has a photon detection area of 1 mm x 1 mm with 1600 pixels. The delivery was realised in two batches. The smaller batch in 2006 comprised 532 pieces for test purposes of a small calorimeter in the DESY test beam. The main delivery of ~ 2000 pieces was carried out in 2008. Some of the MPPCs from the DESY test set up (448 pieces) were used also in the ScECAL. The characteristics of MPPCs were measured in the pulse mode in a test stand using LED flashes controlled by a pulser. The MPPC temperature was kept at 25°C in a thermostat. The distribution of gain and crosstalk for three MPPC groups is shown in Fig. 3 (Sakuma, 2010). The MPPC properties depend on the production batch but within the same batch they exhibit low spread. The different MPPC gain between two production batches was equalized in the calorimeter by adjusting the individual bias voltage setting for each MPPC. This is done by the front-end ASIC chip.

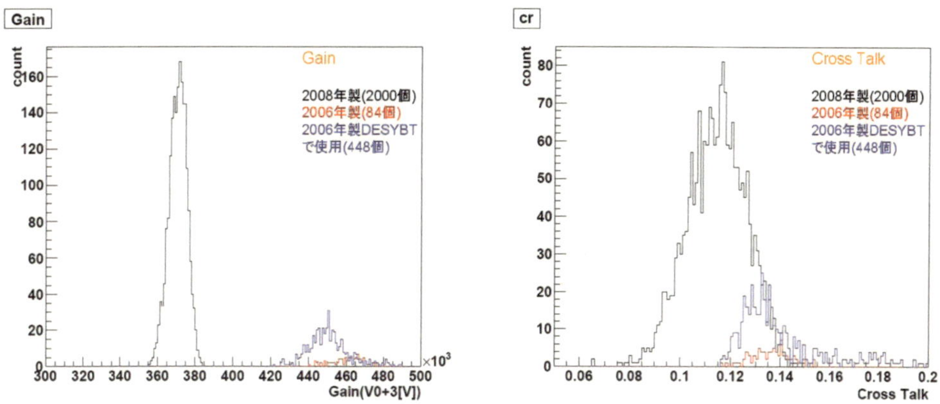

Fig. 3. Distribution of gain (left) and cross-talk (right) for MPPCs produced in Hamamatsu and used in the ScECAL. The production was done in years 2006 and 2008. 448 MPPCs from 2006 and all from 2008 were used in the ScECAL.

The SiPM becomes saturated at large photoelectron intensities N_{ph} due to the limited number of pixels N_{avail}. From the probability considerations the number of fired pixels N_{pix} increases exponentially with N_{ph} as $N_{pix} = N_{avail}\ (1 - \exp(-N_{ph} / N_{avail}))$. The measured dependence of the response curve N_{pix} (N_{ph}) for 10000 SiPMs is given in Fig. 2b. The inverse of this dependence is used as a correction function of the measured signal N_{pix} to get the effective number of photoelectrons N_{ph}. For more sophisticated treatment of the saturation correction see section 8.

4. Scintillator electromagnetic calorimeter

The most important role of the Electromagnetic Calorimeter (ECAL) is to identify photons in jets. The majority of photons are produced by two-photon decay of the neutral pions. The distance between two photons decreases with increasing pion energy and it is several centimetres at the distance 210 cm from the decay point for pion energy several tens of GeV. Therefore, the ECAL granularity around 1 cm will allow resolution of photons from pions up to energies of about 50 GeV.

The ScECAL was built with 1 cm x 4.5 cm scintillator strips 0.3 cm thick assembled from 18 pieces in 4 rows into a plane. Each strip is wrapped in reflecting foil to isolate it from the neighbour and to improve the light collection. Scintillation light is read by a wavelength shifting (WLS) fibre placed in the middle of the strip along its longer side to improve the homogeneity of the light collection. The calorimeter is made of 30 layers with strips alternately rotated by 90 degrees. In between two scintillator planes, tungsten absorber plane 0.35 cm thick, is inserted (see Fig. 4).

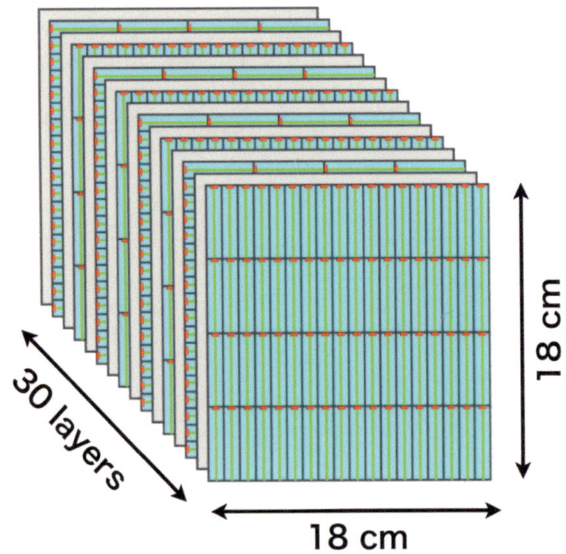

Fig. 4. Scintillator ECAL structure. The scintillator strips (green) are arranged in 4 rows by 18 pieces to form a plane. WLS fibers in yellow lead light to MPPC (red spot) at the strip end. The W absorber planes are shown in grey.

The WLS double clad fibre contains Y11 wavelength shifter and emits light at 550 nm. The fibre is read at one end by a MPPC. The MPPC has a photon detection area of 1 mm x 1 mm with 1600 pixels. The MPPCs were soldered on a flat cable and mounted at the end of each fibre. Test were also made where the MPPCs were attached directly to the tile as the loss in the light collection from the missing WLS can be partly regained by a better match in peak sensitivity of MPPC (at 400 nm) to the scintillation light. The number of photoelectrons dropped to one half but a significant simplification in construction can be achieved.

The readout of the calorimeter uses the same electronics as the AHCAL and it is described in the next section. The basic eighteen channel front-end ASIC chip reads out in this case one row of scintillators. The prototype of the ScECAL had 2160 MPPCs and it was built in collaboration of Shinshu, Kobe, Tsukuba, Tokyo, Niagata Universities in Japan and Kyungpook National University in Korea (Kotera, 2010).

The calorimeter has been tested in muon and electron beams at energies 1-32 GeV, pion beam 2-60 GeV energy (including neutral pions) in Fermilab in 2008-9. The detector was calibrated using muons to measure the energy response of a MIP. The light yield of 23 photoelectrons was measured for a minimum ionizing particle. The data were further corrected for the saturation (see section 8). The response of the whole calorimeter to several electron beam energies is shown in Fig. 5. The data show clear Gaussian behaviour for individual energy.

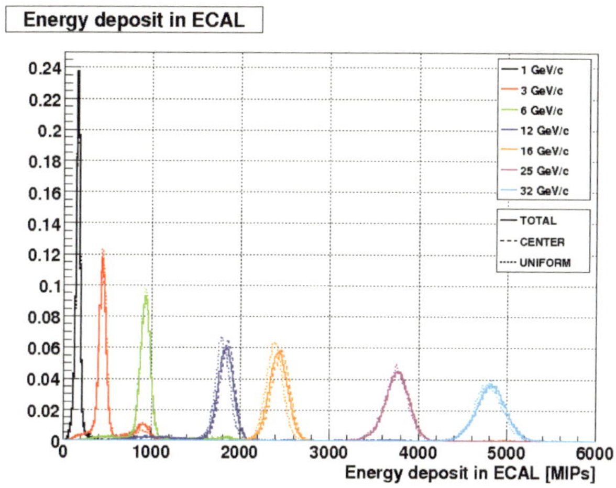

Fig. 5. The corrected energy spectra measured with ScECAL at energies 1–32 GeV.

The data were then purified for small contamination from other particles in the beam namely muons and electrons. The energy distribution was fitted for each energy with a Gaussian function including 90% of data around the Gaussian maximum to further suppress residual contamination (lowest energy point at 1 GeV required a more refined procedure). The deposited energy in the calorimeter is given as the mean of Gaussian function from the fit and the resolution by the standard deviation σ.

The mean of the Gaussian for each beam momentum is displayed in Fig. 6 (left). The points were fitted by a linear function forced to go through the origin. The points differ from the fit at most by 6% at the highest energies. The sources of the nonlinearity are temperature dependence of sensors, spread in saturation correction of sensors and spread in calibration

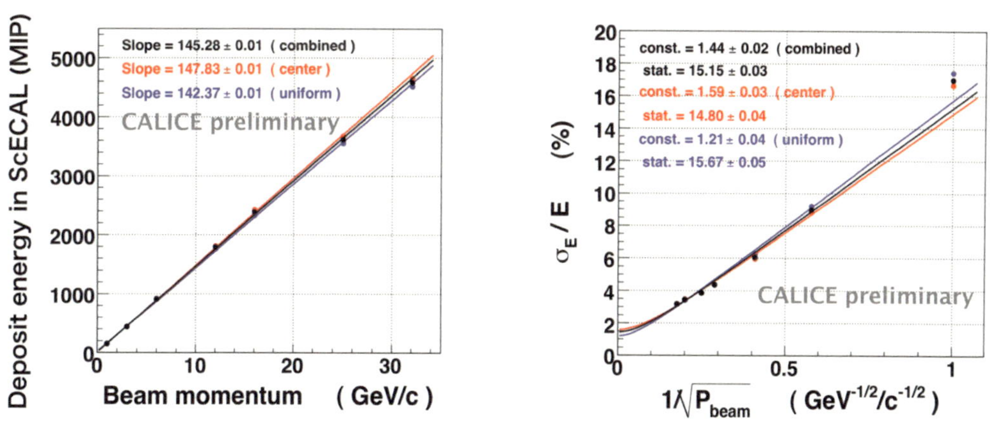

Fig. 6. (left) Linearity response of the ScECAL. Blue (red) data correspond to the beam shot uniformly on (at the center of) the front face of the calorimeter, black are data combined together. (right) The energy resolution of ScECAL as a function of the inverse of the square root of the beam momentum. The formula fitted to data points is quadratic and it is explained in the text.

factors for strips. The sources of the nonlinearity are being further investigated. The calorimeter energy resolution σ_E/E is plotted in Fig. 6 (right) as a function of $\sqrt{(1/P_{beam})}$. Data at different beam momenta were fitted to the quadratic formula $\sigma_E/E = \sigma_{const} \oplus \sigma_{stat}/\sqrt{(P_{beam})}$, where σ_{const} (σ_{stat}) represent constant and stochastic term in the calorimeter energy resolution and $A \oplus B = \sqrt{(A^2+B^2)}$. Non-zero σ_{const} describes the effect of non-uniformities in the calorimeter (mechanical or due to physics, i.e. different response to electrons and hadrons). The value of σ_{stat} depends on the ratio of scintillator and absorber thicknesses and it is called the 'stochastic coefficient'. It is due to the sampling fluctuations of particle showers in the active medium. The values obtained from tests are σ_{const} = (1.44±0.02)% and σ_{stat} = (15.15±0.03)%. The value σ_{const} is slightly worse than the value 1.07% obtained for a similar calorimeter using 1 cm x 1 cm Si diodes (Repond et al., 2008). The value σ_{stat} is better than the value 16.53% from the SiW calorimeter. The fact that the ScECAL uses 4 times fewer channels which are cheaper to produce and shows a similar performance makes this type of calorimeter a very attractive candidate for the electromagnetic calorimeter in the detector at the LC.

5. Analog hadron calorimeter

The most important task of the hadron calorimeter is to separate the contributions from deposits of charged and neutral hadrons in the hadron shower and to measure the energy of neutral particles. The choice of parts for the physics prototype was a compromise between performance, price and availability. The scintillator chosen comes from UNIPLAST Russia (polystyrene + 1.5% PTO + 0.01% POPOP), the WLS fibre of 1 mm diameter is Kuraray Y11(200) in a groove with ¼ circle for the smallest tiles and with 1 loop for the larger tiles. Before bending the fibre was heated to ~ 80° C.

One layer of the sampling structure of the tile AHCAL physics prototype calorimeter consists of a 16 mm thick absorber plate made from stainless steel and a steel cassette of 96 cm x 96 cm area with two 2 mm thick steel walls to house a plane made from 5 mm thick scintillator tiles. The calorimeter is built as a cube of approximately 1 m^3 volume. The total number of 38 layers corresponds to 5.3 interaction lengths λ_{int}. The view of one plane with scintillator tiles is shown in Fig. 7 left. In the centre where the test beam passes through, one hundred small 3 cm x 3 cm tiles is surrounded by three rows of tiles 6 cm x 6 cm and finally at the perimeter by one row of tiles 12 cm x 12 cm. The tiles at 4 corners were removed as the occupancy is expected to be low. Each tile is coupled via a WLS fibre inserted in a groove to a SiPM via an air gap (Fig. 7 right). The tiles are covered from top and bottom with a super-radiant foil VM2000 from 3M. The tile side edges were matted in order to provide better homogeneity of the light on the photodetector.

One detector plane comprises 216 scintillator tiles closely packed inside a frame made from aluminium bar 10 mm x 8 mm. The SiPMs are connected to the front-end electronics via 50 Ω micro-coax cables that carry both signal and bias voltage. Each tile is also connected to a clear fibre which brings calibration light from LED. Both coax cables and fibres lay on a support FR4 plate covered with a mylar foil for insulation. It means that SiPMs are not soldered on a PCB as this construction is too fragile to carry the mass of the scintillator. The coax cables are connected on one side of the plane to the front-end electronics, eighteen optical fibres are attached on the other side to one LED mounted on the calibration and monitoring board (Adloff et al., 2010a).

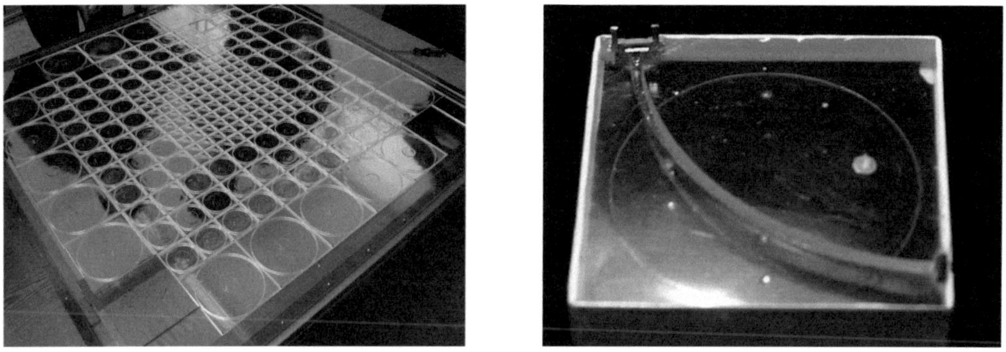

Fig. 7. (left) Scintillator tile layer of the AHCAL calorimeter prototype. The tile of three different sizes can be recognized as well as the WLS fibers of the circular shape inside the tiles. The photodetector SiPM is placed in the tile corner and powered and read-out via connections to micro-coax cable below the scintillators. (right) Scintillator tile 3 cm x 3 cm with a WLS fiber in a groove attached via an air gap to the photodetector SiPM placed in the upper left corner of the tile. Two SiPM contacts are used to bring the bias voltage and the readout of the signal. The other end of the WLS fiber is covered by a reflector to improve the overall light yield.

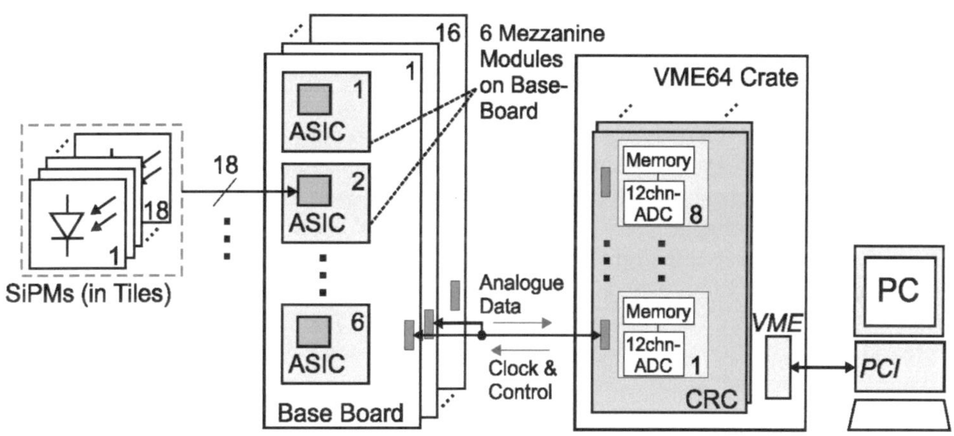

Fig. 8. Schematic view of the readout and data acquisition system of the AHCAL

A schematic view of the readout system is shown in Fig. 8. Eighteen SiPMs are connected to the same ASIC (de La Taille et al., 2005) which has for each SiPM preamplifier, shaper and sample-and-hold-circuit. It also allows individual bias voltage settings for each SiPM. It is made in the 0.8 μm Complementary Metal-Oxide-Semiconductor (CMOS) technology. The signals from twelve Application-Specific Integrated Circuits (ASICs) are fed into the CALICE Readout Card (CRC) and digitized by 16-bit Amplitude Digital Converters (ADCs). The data are stored via a PC, which controls also the data taking. The construction of the physics prototype and the readout and data acquisition system is a common effort of DESY, Hamburg University, Imperial College London, ITEP, Lebedev Physics Institute and MEPHI Moscow, LAL Orsay, and Institute of Physics Prague.

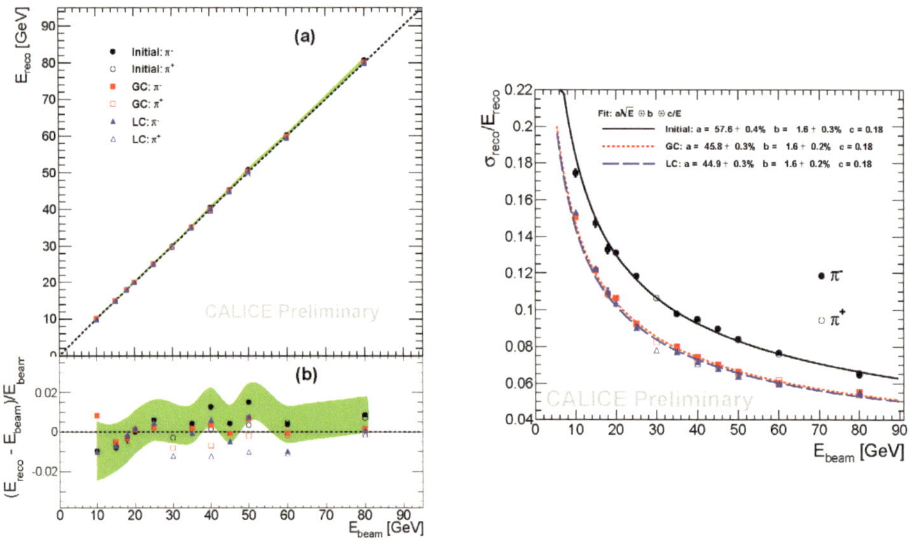

Fig. 9. (left a) Linearity of the AHCAL to pions. (left b) Relative residuals to beam energy versus beam energy. (right) Relative energy resolution of the AHCAL for pions versus beam energy. Black points and lines correspond to analysis without compensation, blue (red) points and curves show results of analysis after local (global) compensation.

The AHCAL was extensively exposed to beams of electrons and hadrons in years 2007-9. All detectors showed high reliability. The calibration and monitoring procedures were validated. At the beginning the calorimeter was calibrated to get the overall energy scale. With help of the wide muon beam each calorimeter cell was exposed to have at least 500 reconstructed muon tracks in the cell. These data were used to obtain the MIP conversion factor expressing the charge measured (in units of ADC counts) by the SiPM in units of MIPs. With low intensity LED light the individual photons (or pixels, see Fig. 10 in section 7) were recorded by the same data acquisition chain and also measured in ADC counts. The operation voltage of SiPMs was set at the value of ~ 13 pixels/MIP to equalize the SiPM response. The factor converting MIP to energy in MeV was obtained from the Monte Carlo simulation and allowed to express the charge measured by SiPM in energy scale in MeV. This energy was finally corrected for the SiPM saturation. The SiPMs proved their reliability and gain stability over the whole period and the whole HCAL showed its robustness during transport between laboratories at DESY, CERN and Fermilab. During the test beam periods at CERN the data were collected in the pion beam in the energy range 8-80 GeV and $e^{\pm}$ beams 6-45 GeV. The pion data were purified, reconstructed and calibrated using the standard CALICE chain (Adloff et al., 2011). To improve the linearity and the energy resolution of the AHCAL, two software correction methods were developed which compensate the AHCAL higher response to electrons e/π = 1.19. The algorithms use information on the shower substructure and reweight higher local energy deposits by suitable factors on event-by-event basis. The linearity and energy resolution of the complete CALICE setup (ECAL + AHCAL + TCMT) is shown in Fig. 9 for data (black) and data after software compensation correction by the local compensation (blue) and global compensation (red) methods. The response is linear within ± 1.5% with slight improvement due to compensation. The resolution formula fitted to data is $\sigma_E/E = \sigma_{const} \oplus \sigma_{stat}/\sqrt{(E_{beam})} \oplus 0.18/E_{beam}$, E_{beam} in GeV. The last term describes the noise width estimation of the test beam setup. In this case the effect of compensation is significant. The values of σ_{stat} = 57.6% (44.9%, 45.8%) for the data without (with local, global) the software compensation show that on average the compensation methods improve the energy resolution by 21%. The constant term σ_{const} = 1.6%. (Chadeeva, 2011)

6. Tail catcher and muon tracker

The TCMT is a sandwich calorimeter placed behind the AHCAL. It has also a volume of 1 m^3, but coarser structure than the AHCAL. Steel absorber plates have thickness 2 and 10 cm, the active layer consists of 100 cm long, 5 cm wide extruded scintillator strips 0.5 cm thick. The first 8 sections behind the AHCAL have a similar longitudinal segmentation as the AHCAL and will supplement the AHCAL measurement for the tail-end of the hadron showers. The last 8 coarser sections serve as a prototype of a possible muon detector for any design of the ILC detector. It represents 5.8 interaction lengths.

The scintillation light is read out by parallel WLS fibres of 1.2 mm diameter inserted in the co-extruded holes. The fibre is read at one end by a SiPM of the same type as in the AHCAL. The strips and SiPMs are inserted into cassettes made from 1 mm thick steel, the cassettes are inserted alternately in the x and y directions between absorber plates. The TCMT uses 320 SiPMs in total. Each cassette has inside a LED driver board with 20 UV LEDs, one per strip. The TCMT has been constructed by Fermilab, Northern Illinois University, and DESY (Chakraborty, 2005).

7. Monitoring and calibration

During the data taking with the calorimeter it is important to monitor the stability of the entire read-out system starting from the quality of light transfer between scintillator and photodetector to the read-out electronics. The photodetector gain G significantly changes with temperature as $dG/dT \sim$ -1.7%/K and with the bias voltage as $dG/dU \sim$ 2.5%/0.1V (Eigen, 2006), therefore the temperature of the photodetector must be periodically recorded.

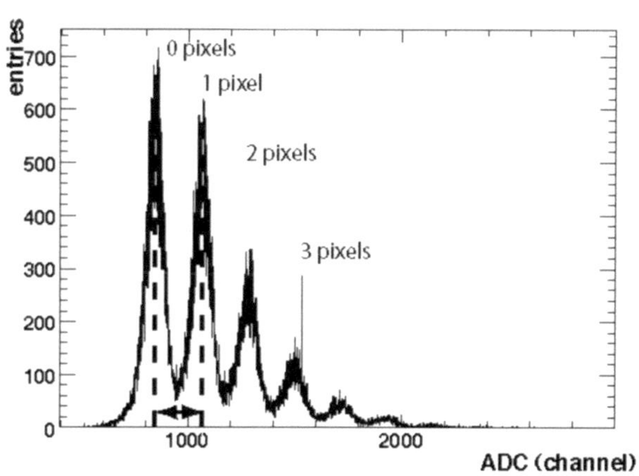

Fig. 10. SiPM response to low intensity light. The first peak corresponds to the pedestal, other peaks are signals from one, two and three photoelectrons. The peaks are equidistant, the distance indicated by ↔ is proportional to SiPM gain. The scale is in ADC counts corresponding to the charge of the SiPM signal (1 ADC count is 0.25 pC).

Also the stability of the SiPM response curve in the whole dynamic range must be known for offline corrections. These tasks are achieved with the calibration and monitoring system.

The photodetector gain and its saturation are monitored by UV LED flashes (λ = 400 nm) which are delivered to each tile via clear fibres. For the response curve measurement, the LED pulses must be tuneable in intensity from a few photoelectrons to thousands photoelectrons on the SiPM. The upper range value is given by the maximal expected rate of particles in a calorimeter cell which was obtained from shower simulations and is about 100 particles (MIPs). For the gain setting of about 13 photoelectrons/MIP, we arrive at the highest signals of ~ 1300 photoelectrons.

The LED light pulse is ~ 10 ns wide and of trapezoidal shape. Particle ionization in the scintillator tile produces typically one order of magnitude shorter light pulses. The LED light pulse intensity is proportional to the pulse length. To get intensity on the level of a thousand of photoelectrons it is easier to use longer pulses. The pulse length is limited by the after-pulsing property of the photodetector and by choosing the value of the quenching resistor, the afterpulsing can be almost avoided. The low intensity light is used for gain calibration and uses the unique sensitivity of SiPM to single photoelectrons (see Fig. 10).

The scheme of the electronic circuit for LED light pulses - LED driver - is displayed in Fig. 11. The design is particularly optimized for the generation of nearly rectangular fast pulses (rise and fall time ~1 ns) with variable amplitude in a large dynamical range. For these reasons we abandoned the classical approach of discharging a capacitor to the LED (Apuhn et. al., 1997). We adopted a scheme using the IXLD02 integrated circuit developed for semiconductor lasers. Based on the Tcalib signal from the calorimeter control, three signals are derived. The ENable, DRVpulse, and invDRVpulse control the output stage of LED driver. The ENable pulse wakes up the circuit. The LED is reversely polarized until the arrival of the DRVpulse which switches on the transistor Q1 for about 10 ns and the LED makes a flash. The role of the invDRVpulse is to improve the rise and fall time of the LED flash by switching complementary the transistor Q2 and since that, LED is quickly reversed till the end of ENable. The intensity of the LED light amplitude is controlled by the Vcalib amplitude in the control current sink CCS. The LED light emission characteristics do not change with the increase of the LED current in the range 10 µA to 10 mA (Polak, 2006).

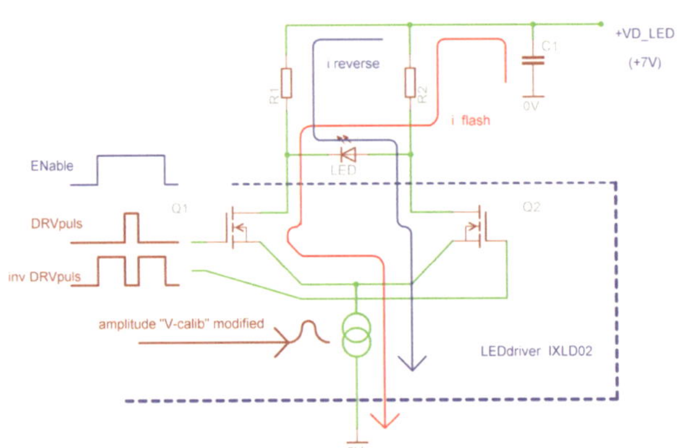

Fig. 11. The principle scheme of the LED driver. IXLD02 is the IC which produces rectangular pulses based on signals Ebable and DRV. The invDRVpulse improves the pulse shape. The pulse amplitude is controlled by the Vcalib value.

In order to reduce light fluctuations among LEDs, LEDs were sorted into groups with a similar light-intensity profile. One LED supplied 18 tiles and a monitoring PIN diode. Since PIN photodiodes have a gain of unity, an additional charge-sensitive preamplifier for the PIN photodiode readout was envisaged. The presence of a high-gain preamplifier directly on the board in the vicinity of the power signals for LEDs, however, has turned out to be a source of cross talk. One calorimeter plane of 216 tiles required 12 LEDs and 12 PIN diodes. The electronics providing LED pulses is laid on the Calibration and Monitoring Board (CMB) and located along one side of the calorimeter plane opposite to the read-out electronics and bias voltage supply. Another task of the CMB consists of reading out the temperature sensors via a 12 bit ADC. The temperature values are sent to the slow control system via a CANbus interface. One CMB operates seven temperature sensors, two sensors directly located on the readout board and five sensors distributed across the centre of the cassette. The sensors consisting of integrated circuits of type LM35DM produced by National Semiconductor are placed in a 1.5 mm high SMD socket. Their absolute accuracy is $< 0.6°C$. A microprocessor PIC 18F448 in association with a CAN controller interface (PCA82C250) provides the communication of the CMB with the slow control system. We have not observed any noise pickup in the CMB.

8. Saturation correction

As already mentioned in section 3, this photodetector is highly non-linear device due to finite number of pixels, the finite recovery time given by the value of internal resistor R_{pixel} and also due to the crosstalk. The number of fired pixels N_{pix} in the simplest case can be approximated by the formula $N_{pix} = N_{avail}\ (1 - \exp(-\varepsilon N_{ph} / N_{avail}))$ for number of photons N_{ph} impinging uniformly and simultaneously on the total number of available pixels N_{avail} of the photocathode, ε is the photon detection efficiency (defined in section 3). This simplest form has been used for the MPPCs of the ScECAL (Kotera, 2010). The number of pixels obtained from the fit to the measured response curve was $N_{avail} = 2424$ for photodetector with 1600 pixels. The large value of N_{avail} from the fit was attributed to the short recovery time of some pixels. The pixel then can fire for a second time on late photons radiated from the WLS fibre. In ref. (Adloff et al., 2011) where properties of SiPMs in the AHCAL are described, the sum of two exponentials was used which takes into account possible non homogeneity in the light transmission between the WLS fibre and the photodetector photocathode. The formula can in principle be extended to the sum of more exponentials to take into account other possible non-homogeneities. The response function for each SiPM was inverted and the correction function for one SiPM is shown in Fig. 12. The response of all SiPMs of the AHCAL clearly showed that on average only 80% of the photodetector pixels are used. This was explained by the different geometry cross-section of the WLS fibre with 1 mm diameter that covers effectively 78.5% of the square surface of the photocathode 1x1 mm^2.

The detailed study of the parameter N_{avail} was thereafter done for all SiPMs using only the response at the highest LED light intensities using the single exponential $N_{pix} = N_{avail}\ [1-\exp(-(X-C)*B)]$, X is the voltage on the LED. The constant C takes into account the LED starts to emit at voltage $X = C$. Fits to data allowed a value of N_{avail} to be obtained for 83% of SiPMs with a spread of 0.7% around the value 1156 pixels * 0.8 = 928 pixels (Zalesak, 2011). For the remaining SiPMs the nominal value of $N_{avail} = 928$ pixels can be used.

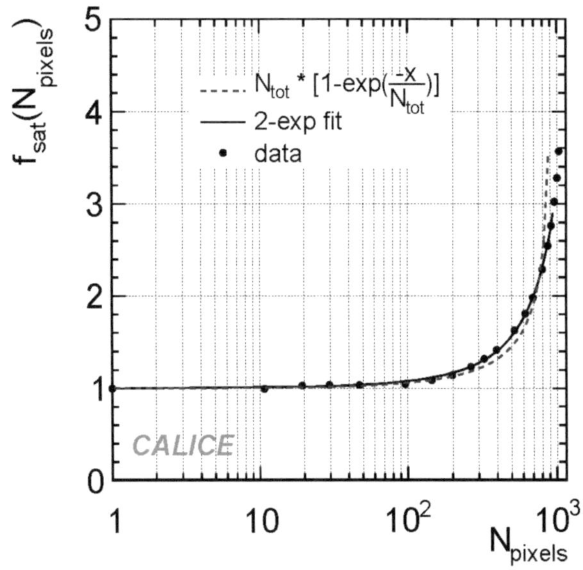

Fig. 12. The SiPM non-linearity correction function. The points are values for one SiPM. The solid line is double exponential fit, the dashed line is the single exponential fit, $N_{tot} = N_{avail}$ used in the text.

9. New developments

The experience with the physics prototypes will be used in the design of calorimeters for the full size detector at the LC. E.g. the hadron calorimeter will occupy volume of ~ 80 m^3 and with scintillator tiles 30 mm x 30 mm it will have ~ 3.2 10^6 channels (Abe et al., 2010). The readout electronics will be integrated in the calorimeter volume to minimize the dead space. The electronics has to have low power consumption to minimize the demands on cooling. The detector parts must be designed to allow the industrial production. As the next step technological prototypes are being designed and built and will meet most of these requirements.

9.1 Hadron calorimeter

The design of a half-octant of AHCAL is shown in Fig. 13. A single detection layer, one of 48 making the full depth of the calorimeter, consists typically of 16 HCAL Base Units (HBU). The HBU (Fig. 14) is the smallest component of the calorimeter with 144 scintillator tiles. To minimize the depth of the calorimeter, the tile thickness was decreased from 5 to 3 mm. Each tile will be equipped with a new type of SiPM (Sefkow, 2006).

The WLS fibre will be placed in a straight groove in the tile to avoid the time-consuming process of fibre bending (see Fig. 15). The thinner scintillator tile affects the SiPM design. We plan to use CPTA MRS APDs with the number of pixels reduced to ≤ 800 (Buanes et al., 2010). These SiPMs have smaller cross-talk and noise and lower sensitivity to temperature and voltage variations than SiPMs from PULSAR used in the AHCAL physics prototype. The latest tests in the positron beam showed a very good functionality with a signal of ~ 15 pixels/MIP.

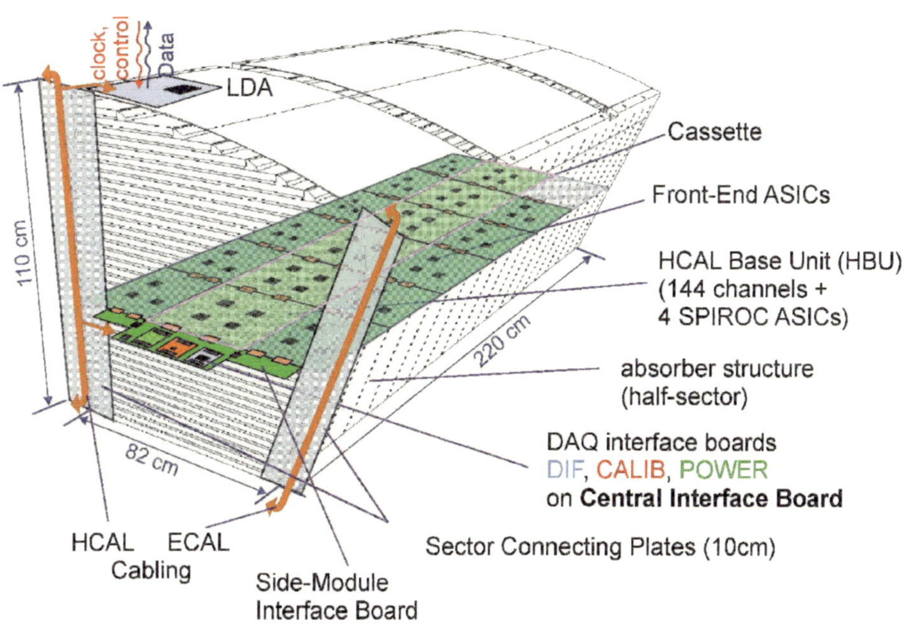

Fig. 13. View of the 1/16 of the AHCAL barrel wheel. One of 48 detection layers is shown in colour. It is made of HCAL base units HBUs. The service modules DAQ, DIF, CALIB and POWER are outside the detector volume.

Fig. 14. (left) Four scintillation tiles assembled on the bottom of the HBU. The top side of the HBU. (right) has the front-end electronics with 4 ASICs and calibration LEDs for 144 tiles. The HBU is connected from right to DIF (bottom), CALIB (middle) and POWER (top) boards.

The new front-end ASIC chip powers, reads and amplifies the signal from 36 photodetectors (twice more than in the physics prototype). A 12-bit ADC in the chip digitizes the photodetector amplitude and provides fully digitized output. The size of the chip is reduced to 30 mm^2 using the 0.35 µm technology. The chip can be operated in the power pulsing mode with significantly reduced power consumption 25 µW (40 µW) per channel (including

SiPM bias voltage). After packaging the ASIC size is 28x28x1.4 mm³ (Raux, 2008). The first experience with the ASIC performance was recently reported (Terwort, 2011).

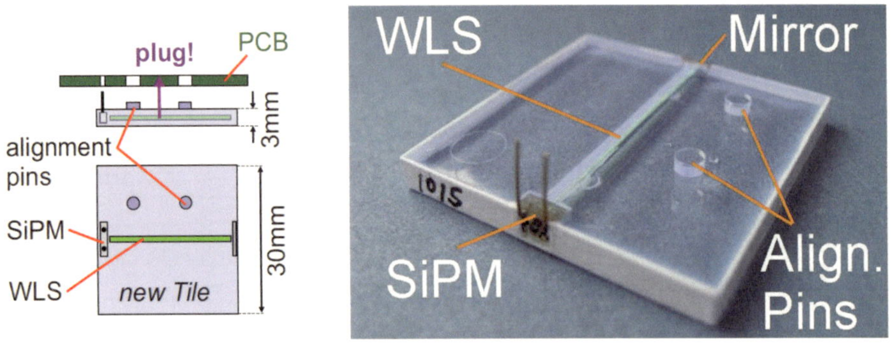

Fig. 15. The construction drawing (left) and the photo (right) of the scintillator tile with straight WLS fibre and SiPM attached to its end. The alignment pins in the tile fix the tile into the PCB.

The response of SiPMs is controlled by new calibration and monitoring system. Two concepts are currently under development:

- Each channel incorporates a SMD LED embedded on the HBU. LED illuminates the scintillator through a hole in the PCB. The control of LEDs is provided by the CALIB module on side of the HBU. This system is already implemented in the setup of Fig. 14. Each LED has its own driver - a low-component-count and effective circuit that discharges a capacitor through the LED. It was observed that pulse length depends on the LED type - blue LED generates long pulses ~40 ns and UV LEDs fast pulses ~8 ns long. This behaviour was explained by the different internal structure of LEDs which leads to different internal LED capacitance. At low intensity the driver generates single photon electron spectra for gain calibration purposes. The driver was finally tuned to deliver also high intensity light up to saturation mode of the SiPM with 796 pixels (Sauer et al., 2011).
- A luminous space LED with its driver placed on a special board. The light is distributed via notched fibres to rows of tiles (up to 72 tiles/fibre with the light output homogeneity of 20%). The fibre is positioned on the HBU on the top side. The notch flashes light through a hole in the PCB to the scintillator. The LED driver works as a heavily dumped quasi-resonant sine waves generator, where the first positive half-wave generates light and negative half-wave bias the LED negatively. Following sine waves are small and keep the LED reverse biased (non-shining mode) (Kvasnicka, 2011). It utilizes a toroidal inductor embedded in the PCB, and produces almost sinusoidal pulses which reduce the electromagnetic interference. For the toroid inductance 35 nH, the LED flash has fixed length of 3.5 ns. The LED amplitude is tuneable from low intensity light for the SiPM gain calibration to the high intensity signal equivalent to 200 MIPs in the tile. For the driver circuit scheme and the PCB with the circuit and toroidal inductor see Fig. 16.

As can be seen in Fig. 13 and in a greater detail in Fig. 14 (right), each detection plane is interfaced with the Data AcQuisition (DAQ) system via Detector InterFace (DIF) board. The calibration system is steered by the CALIB board and the power module (POWER) distributes necessary voltages and arranges the power pulsing regime. All modules are available and are currently under test to commission the communication between different systems. The detection plane or its part will undergo beam tests in the year 2012.

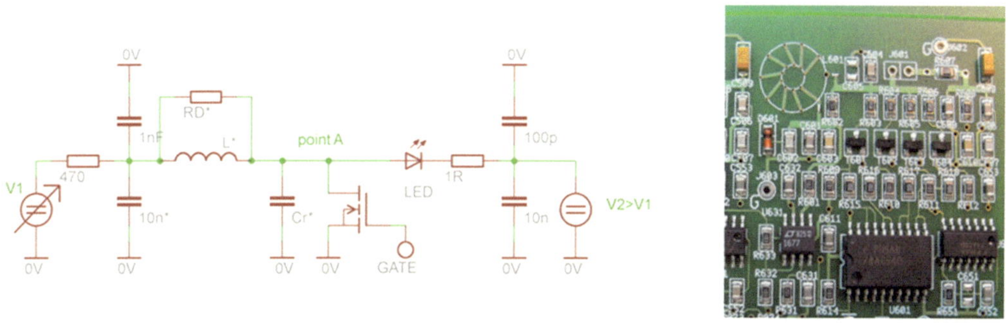

Fig. 16. (left) Driver circuit scheme. (right) LED driver on PCB with toroidal inductor at the top left.

9.2 Scintillator ECAL

The development of the next generation of the scintillator electromagnetic calorimeter will be done together with the SiW ECAL (Takeshita, 2011). To decrease the price of the calorimeter determined mainly by the price of the silicon sensors, it was decided to replace approximately half of the silicon sensor planes in the ECAL by scintillator planes with orthogonal strips. As the size of the silicon sensors was decreased from 10 mm x 10 mm to 5 mm x 5 mm, the width of scintillator strips will be correspondingly reduced from 10 mm to 5 mm. The scintillator thickness will decrease from 5 mm to 2 mm to better match the width of silicon layers. It is no longer possible to insert a WLS fibre inside such a thin scintillator. Therefore a new R&D program is envisaged aimed at the direct coupling of the MPPC to the tile side. This will further simplify production of large detector.

10. Conclusion

Prototypes of electromagnetic and hadron calorimeters with scintillator and embedded photodetectors for a detector at a future linear collider have been successfully built. The calorimeters with unprecedented granularity were successfully tested in electron, muon and hadron beams at accelerators in CERN and Fermilab. Commissioning and operation in the test beam demonstrated that calorimeters perform according to expectations. Test data were used to set up and tune the energy calibration of calorimeters. In case of the AHCAL, two new software correction methods were developed which compensate for the calorimeter's higher response to electrons. The application of both methods to the test beam data results in improvement in the energy resolution of the hadron calorimeter by 21% in the energy range 10–80 GeV.

New semiconductor photodetectors were developed for both calorimeters which provide gain of a classical photomultiplier but in a photodetector with millimetre dimensions and bias voltage of ~ 60 V. The new photodetector is embedded directly in the scintillator tile/bar and significantly reduces the calorimeter dead space and simplifies calorimeter construction. More than 10000 of these photodetectors in total were produced by

MEPHi/PULSAR (called silicon photomultipliers) and Hamamatsu Photonic (called pixelated photon counters). The successful operation of these detectors over a period of several years opened the door for their use in applications in material research and medicine.

The experience with the physics prototypes of calorimeters will be used in the design of calorimeters for the full size detector at the linear collider. Here the total number of channels will be of the order of millions. This brings new challenges especially on electronics which must be integrated in the calorimeter volume to minimize the dead space. The electronics must have low power consumption to minimize demands on cooling. The first versions of read-out ASICs were developed. The ASIC integrates functions performed previously by the data acquisition electronics. The next generation of calorimeter prototypes has started to be built and will be ready for beam tests in 2012.

11. Acknowledgment

I would like to thank to all my colleagues from the CALICE collaboration for the excellent work done during the construction of calorimeters, their operation and convincing results from the beam tests. Especially I want to thank to M. Danilov, G. Eigen, E. Garutti, K. Kotera, J. Kvasnička, V. Morgunov, I. Polák, F. Sefkow, D. Ward and J. Zálešák for discussions, comments, and providing me with the material for this contribution.

This work was supported by the Ministry of Education, Youth and Sports of the Czech Republic under the projects AV0 Z3407391, AV0 Z10100502, LC527 and LA09042.

12. References

Abe, T., et al. (2010). The International Large Detector: Letter of Intent, In: *FERMILAB-LOI-2010-03, FERMILAB-PUB-09-682-E, DESY-2009-87, KEK-REPORT-2009-6*, ISBN 978-3-935702-42-3

Adloff, C., et. al. (CALICE Collaboration). (2010a). Construction and commissioning of the CALICE analog calorimeter prototype. *Journal of Instrumentation*, Vol.2010, No. 5, (May 2010), p. P05004, ISSN 1748-0221, e-Print: arXiv:1003.2662 [physics.ins-det]

Adloff, C., et al. (CALICE Collaboration). (2010b). Study of the interactions of pions in the CALICE silicon-tungsten calorimeter prototype. *Journal of Instrumentation*, Vol.2010, No. 5, (May 2010), pp. P05007, ISSN 1748-0221, e-Print: arXiv:1004.4996 [physics.ins-det]

Adloff, C., et. al. (CALICE Collaboration). (2011). Electromagnetic response of a highly granular hadronic calorimeter. *Journal of Instrumentation*, Vol.2011, No. 6, (June 2011), p. P04003, ISSN 1748-0221, e-Print: arXiv:1012.4343 [physics.ins-det]

Apuhn R.-D., et. al. (1997). The H1 lead / scintillating fiber calorimeter, *Nuclear Instruments & Methods in Physics Research*, Vol.A386, (1997), pp.397-408, ISSN 0168-9002

Bilki, B., et al. (2009). Measurement of Positron Showers with a Digital Hadron Calorimeter. *Journal of Instrumentation*, Vol.2009, No.4, (April 2009), p. P04006, ISSN 1748-0221, e-Print: arXiv: 0902.1699 [physics.ins-det]

Bondarenko, G., Buzhan, P., Dolgoshein, B., Golovin, V., Gushin, E., Ilyin, A., Kaplin, V., Karakash, A., Klanner, R., Pokachalov, V., Popova, E. & Smirnov, S. (2000). Limited Geiger-mode microcell silicon photodiode: New results. *Nuclear Instruments & Methods in Physics Research*, Vol.A422, No.1-3, (March 2000), pp. 187-192, ISSN 0168-9002

Brient, J.-C. & Videau, H. (2001). Calorimetry at the future e+e- collider, *Proceedings of APS/DFB/DBP summer study on the future of particle physics,* Snowmass, Colorado, USA, June30-July21, 2001

Buanes, T., Danilov, M., Eigen, G., Göttlicher, P., Markin, O., Reinecke, M. & Tarkovski, E. (2010. The CALICE hadron scintillator tile calorimeter prototype, *Nuclear Instruments & Methods in Physics Research,* Vol.A623, No.1, (November 2010), pp. 342-4, ISSN 0168-9002

Buzhan, P., Dolgoshein, B., Ilyin, A., Kantserov V., Kaplin, V., Karakash, A., Pleshko, A., Popova, E., Smirnov, S., Volkov, Yu., Filatov, L., Klemin, S. & Kayumov, F. (2001). An advanced study of silicon photomultiplier. *ICFA Instrumentation Bulletin,* Vol.Fall2001, No.3, (2001), pp. 1-14, Available from: http://www.slac.stanford.edu/pubs/icfa/fall01/paper3/paper3b.html

Calice. (n.d.). CALICE Collaboration, Available from: https://twiki.cern.ch/twiki/bin/view/CALICE/CaliceCollaboration

Chadeeva, M. (2011). Software Compensation using the CALICE calorimeters, *Proceedings of the LCWS 2011,* Granada, Spain, September 26-30, 2011

Chakraborty, D. (2005). The Tail-Catcher/Muon Tracker for the CALICE Test Beam, *Proceedings of the 2005 International Linear Collider Workshop,* PSN 0919, Stanford, California, U.S.A, March 18 - 22, 2005. Available from: http://www.slac.stanford.edu/econf/C050318/proceedings.htm

CLIC. (2011). The Compact Linear Collider Study, Available from: http://clic-study.org

Cvach, J. (2002). Calorimetry at a future e^+e^- collider, *Proceedings of the 31th international conference on high energy physics (ICHEP),* pp. 922-926, ISBN 0-444-51343-4, Amsterdam, The Netherlands, July 23-29, 2002

Danilov, M. (2007a). Comparison of different multipixel Geiger photodiodes and tile-photodetector couplings, *Proceedings of the Linear Collider Workshop LCWS 2007 and ILC 2007,* pp. , ISBN 978-3-935702-27-0, Hamburg, Germany, May 30 - June 3, 2007. Available from: http://lcws07.desy.de/e14/index_eng.html

Danilov, M. (2007b). Scintillator tile hadron calorimeter with novel SiPM readout. *Nuclear Instruments & Methods in Physics Research,* Vol.A581, No.1-2, (October 2007), pp. 451-456, ISSN 0168-9002

Eigen, G. (2006). The CALICE scintillator HCAL test beam prototype, *Proceedings of the 12th International Conference on Calorimetry in High Energy Physics, AIP Conference Proceedings Series, High Energy Physics Subseries,* Vol. 867., 2006, XXIV, ISBN: 978-0-7354-0364-2, Chicago, Illinois, U.S.A., June 5-9, 2006

ILC. (2007). ILC Reference Design Report, August 2007, Available from: http://www.linearcollider.org/about/Publivations/Reference-Design_Report

Kotera, K. (2010). Study of the Granular Electromagnetic Calorimeter with PPDs and Scintillator Strips, *Proceedings of the 12th Vienna Conference on Instrumentation VCI2010,* Vienna, Austria, February 15 - 20, 2010. Available from http://indico.cern.ch/getFile.py/access?contribId=182&resId=0&materialId=paper&confId=51276

Kvasnicka, J. (2011). LED calibration systems for CALICE hadron calorimeter, *Proceedings of the 2nd conference on Technology and Instrumentation in Particle Physics TIPP 2011,* Chicago, Illinois, U.S.A., June 8-14, 2011

Laktineh, I. (2011). Construction of a technological semi-digital hadronic calorimeter using GRPC. *Journal of Physics: Conference Series,* Vol.293 (2011) p. 012077, ISSN 1742-6596

Morgunov, V.L. (2002). Calorimetry design with energy-flow concept, *Proceedings of the 10th conference on calorimetry (Calor02)*, pp. 70-84, ISBN 981-238-157-0, Pasadena, California, USA, March 25-29, 2002

Musienko, Yu. (2011). State of the art in SiPM's. *The Technology Transfer Network for Particle, Astroparticle and nuclear physics event*, CERN, Geneva, Switzerland, February 16 - 17, 2011. Available from: http://indico.cern.ch/getFile.py/access?contribId=11&sessionId=7&resId=0&materialId=slides&confId=117424

Polak, I. (2006). Development of Calibration system for AHCAL, *International Linear Collider (ILC) Workshop*, Valencia, Spain, November 6-10, 2006. Available from: http://www-hep2.fzu.cz/calice/files/ECFA_Valencia.Ivo_CMB_Devel_nov06.pdf]

Raux L. (2008). SPIROC Measurement: Silicon Photomultiplier Integrated Readout Chip for ILC, *Proceedings of the 2008 IEEE Nuclear Science Symposium (NSS08)*, Dresden Germany, 2008

Repond, J., et al. (CALICE Collaboration). (2008). Design and Electronics Commissioning of the Physics Prototype of a Si-W Electromagnetic Calorimeter for the International Linear Collider. *Journal of Instrumentation*, Vol.2008, No. 3, (March 2008), pp. P08001, ISSN 1748-0221, e-Print: arXiv:0805.4833 [physics.ins-det]

Sauer, J., Götze, M., Weber, S. & Zeitnitz, C. (2011). Concept and status of the LED calibration system, *CALICE Collaboration meeting*, CERN Geneva, Switzerland, May 19-21, 2011. Available from: http://indico.cern.ch/contributionDisplay.py?sessionId=10&contribId=7&confId=136864

Sakuma, T. (2010). *Performance Study of Prototype Fine-granular EM Calorimeter for ILC*. Shinshu University, Shinshu, Japan, Master thesis, 01.02.2010. Available from: http://hepl.shinshu-u.ac.jp/index.php?HE%20Lab%20Top%20Page%2FIntroduction%2FMaster%20Theses

Sefkow, F. (2006). MGPDs for calorimeter and muon systems: requirements and first experience in the CALICE test beam. *Proceedings of Science*, Vol.PD07 (2006), p. 003, ISSN 1824-8039

De La Taille, C., Martin-Chassard, G. & Raux, L. (2005). FLC−SIPM: Front-End Chip for SIPM Readout for ILC Analog HCAL, *Proceedings of the 2005 International Linear Collider Workshop*, PSN 0916, Stanford, California, U.S.A, March 18 - 22, 2005. Available from: http://www.slac.stanford.edu/econf/C050318/proceedings.htm

Thomson, M.A. (2009). Particle flow calorimetry and the PandoraPFA algorithm. *Nuclear Instruments & Methods in Physics Research*, Vol.A611, No.1, (November 2009), pp. 25-40, ISSN 0168-9002

Takeshita, T. (2011). ScECAL status report, *CALICE Collaboration meeting*, Heidelberg, Germany, September 14-16, 2011. Available from: http://ilcagenda.linearcollider.org/conferenceTimeTable.py?confId=5213#20110916

Terwort, M. (2011). Concept and status of the CALICE analog hadron calorimeter engineering prototype, *Proceedings of the 2nd conference on Technology and Instrumentation in Particle Physics TIPP 2011*, Chicago, Illinois, U.S.A., June 8-14, 2011

Zalesak, J. (2011). Calibration issues for the CALICE 1 m^3 AHCAL, *Proceedings of the LCWS 2011*, Granada, Spain, September 26-30, 2011